Decision Analysis For Petroleum Exploration

DECISION ANALYSIS FOR PETROLEUM EXPLORATION

Paul D. Newendorp

BOOKS

A Division of The Petroleum Publishing Company

Tulsa, Oklahoma

Library of Congress Catalog Card Number: 75-10936
International Standard Book Number 0-87814-064-6
Printed in U.S.A.

2 3 4 5 • 81 80 79 78 77

Preface

THIS BOOK is the outgrowth of about seven years of teaching adult education courses on petroleum exploration economics and risk analysis. It follows two earlier limited-edition books on the subject co-authored with business associate and former mentor, John Campbell. Topics covered in this book represent a composite of current attitudes and practices throughout the world for analyzing exploration prospects, and several new ideas and concepts not previously published.

The methods and logic of decision analysis are applicable to any type of decision choice involving risk and uncertainty. However the emphasis throughout this book is on the application of these methods to petroleum exploration decisions.

As a result the book will probably be of more interest to those in the petroleum industry who become involved in the analysis of drilling prospects. This will normally include petroleum geologists, engineers, geophysicists, and management personnel. The book stresses applications and de-emphasizes mathematics to the extent possible.

There are no complicated mathematical equations or calculus, and you do not have to be a mathematician or statistician to follow the discussions. I have added as many real-world examples as possible to assist in showing how the decision analysis ideas can be applied to petroleum exploration.

The first Chapter summarizes decision analysis and why it can be of use in analyzing drilling prospects. Chapters 2–5 cover various types of decision criteria for assessing and comparing the value of drilling prospects. Chapters 6–8 are directed to the difficult and complex subject of risk analysis – quantifying the degree of risk and uncertainty associated with the various exploration areas and prospects.

Over the years there has not been a great deal written on exploration risk analysis. Historically we've accounted for risk subjectively with adjectives such as "it's a sure thing," "it's a risky prospect," etc. Decision analysis is predicated on quantification of risk and uncertainty, and the techniques and methods to do this are discussed in these three important chapters.

The problems of implementing decision analysis into the corporate decision making process are discussed in Chapter 9. The final two Chapters discuss special types of analyses and certain open issues related to the application of decision analysis in the oil business.

Following Chapter 11 there are several short sections including a list of references for further reading. A glossary of terms and abbreviations is given, as well as a conversion table from metric units to the units commonly used in the U.S. and Canada (barrels, acres, cubic feet, etc.). The final portion of the book contains nine separate appendix sections. These appendices include various tables and special calculations useful in applying decision analysis to petroleum exploration. Each table is preceded by a brief discussion of how to use the table.

I would especially call your attention to a Case Study called the Blackduck Prospect. It is a three-part series describing a management committee's actions on an exploration prospect being presented by Henry Oilfinder. The series consists of the dialogue of conversation among the various managers as they consider whether to accept or reject the recommendation. These conversations should be helpful to highlight the key issues in prospect evaluation for which decision analysis can be useful.

The first part of the series appears at the end of Chapter 3, with succeeding parts at the end of Chapters 4 and 8. It is suggested that you read the Case Study in the order in which the parts appear. That is, read Chapter 4 before reading the second part, and read Chapters 6, 7, and 8 before reading the concluding part.

All reference to taxation (local, state, and national or federal) has been omitted from the book. Taxation is, of course, a fact of life and we clearly need to consider its effect when we analyze a drilling prospect. But all of the decision analysis ideas are equally valid on a 'before-tax' or 'after-tax' basis. So rather than muddying the water with discussions of the tax rules in every part of the world it seemed to be clearer to just omit taxation and focus on the methods themselves. Please remember, however, that you must certainly account for taxation when applying the ideas of this book to the real-world. Also, it goes without further saying that all of the methods discussed can be used with any of the monetary units or currencies of the world.

As we will see later some of the analysis methods can be done by computers much more conveniently than by hand calculations. I've assumed, however, that most people reading this will not be computer programmers themselves. Consequently I've generally omitted computer programming in the book. The one exception to this is a section in Chapter 8 describing the general approach to programming the risk analysis method called simulation.

Finally, I acknowledge with thanks the assistance given by Scientific Software Corporation of Denver in providing computer facilities to run a numerical example given in Chapter 8.

Paul D. Newendorp

Norman, Oklahoma
March, 1975

Contents

tributions, cumulative frequency distributions, single value parameters of distributions (mean, standard deviation, median, mode). Specific distributions: normal, lognormal, uniform, triangular, binomial, multinomial, hypergeometric. Bayes' Theorem: its use to revise risk estimates, the generalized solution to Bayes' Theorem, sequential sampling.

Decision Analysis for Petroleum Exploration

An introduction to the application of statistical decision theory concepts to the analysis of risk and uncertainty in petroleum exploration investment decisions.

1 Decision Analysis — Why Bother?

It is a truth very certain that when it is not in our power to determine what is true we ought to follow what is most probable.

— Descartes

VIRTUALLY all important business decisions are made under conditions of uncertainty. The decision maker must choose a specific course of action from among those available to him, even though the consequences of some, if not all, of the possible courses of action will depend on events that cannot be predicted with certainty. Decision analysis is a discipline consisting of various methods, techniques, and attitudes to help the decision maker choose wisely under these conditions of uncertainty.

Decision making under uncertainty implies that there are at least two possible outcomes that could occur if a particular course of action is chosen. For example, when the decision to test a seismic anomaly with a wildcat well is made it is not known with certainty what the outcome will be. Even if the well is successful in discovering a large new oilfield it is not entirely certain what the ultimate value of the reserves will be. In fact, petroleum exploration has frequently been given the dubious distinction of being the "classic" example of decision making under uncertainty.

Decision analysis methods provide new and much more comprehensive ways to evaluate and compare the degree of risk and uncertainty associated with each investment choice. The net result, hopefully, is that the decision maker is given a clearer, sharper insight of potential profitability and the likelihoods of achieving various levels of profitability than older, less formal methods of investment analysis.

The older methods of analyzing decision choices usually involved only cash flow considerations, such as computation of an average rate of return on invested capital. Thus, the new dimension that is added to the decision process with decision analysis is the quantitative consideration of risk and uncertainty and how these factors can be used in formulating investment strategies.

The fundamental concepts used in decision analysis were formulated over 300 years ago (1654). However, the application of these concepts in the general business sector did not really become apparent until the 1950's. And it has only been in the last 5–10 years that decision analysis has been seriously applied to petroleum exploration decisions.

Use of these methods and techniques in drilling decisions probably developed more by default than as the result of crusading efforts by a single person, group, or company. Reasons for this are quite evident in the industry — rising drilling costs, the need to search for petroleum in deeper horizons or in remote areas of the world, increasing government controls, etc.

These stresses and strains became so critical that most petroleum exploration decision makers were no longer satisfied to base their decisions on experience, intuition, rules of thumb, or similar "seat-of-the-pants" approaches. Instead, they recognized that better ways to evaluate and compare drilling investment strategies were needed.

Today it is the view of many decision makers that decision analysis offers these and other advantages over traditional ways of selecting drilling prospects. Hence the current interest in the application of decision analysis to petroleum exploration.

Decision analysis is a multi-disciplinary science that has various other synonyms: statistical decision theory, management science, operations research, and modern decision theory. It involves aspects of many traditional areas of learning — economics, business, finance, probability and statistics, computer science, engineering, and psychology, to name a few.

We will, of course, discuss all aspects of decision analysis in detail in later chapters, but for purposes of gaining an overview of the methodology we can summarize decision analysis as a series of steps:

   A. Define possible outcomes that could occur for each of the available decision choices, or alternatives.
   B. Evaluate profit or loss (or any other measure of value or worth) for each outcome.
   C. Determine or estimate the probability of occurrence of each possible outcome.
   D. Compute a weighted average profit (or measure of value) for each decision choice, where the weighting factors are the respec-

tive probabilities of occurrence of each outcome. This weighted average profit is called the expected value of the decision alternative, and is the comparative criterion used to accept or reject the alternative.

The new parts of this approach to analyzing drilling decisions are steps C and D. The analysis requires that the explorationist associate specific probabilities to the possible outcomes (dry hole, or various levels of reserves). The quantitative assessment of these probabilities is frequently called risk analysis. Anyone who has attempted to assess these probabilities in petroleum exploration is undoubtedly aware that risk analysis is no easy task.

Some have compared the "state-of-the-art" of risk analysis to the older, well documented profitability analysis by observing that we as a petroleum industry can compute a discounted rate of return to an accuracy of 50 decimal places; whereas we are fortunate if we can even estimate (guess?) the probabilities to the first significant digit! Risk analysis is, thus, a vital component in the broader discipline of decision analysis, and we will be devoting most of our attention in later chapters to discussions on how to quantify risk and uncertainty in the petroleum exploration context.

Persons exposed to decision analysis for the first time can often raise some rather penetrating questions. These include

- Decision analysis – why bother?
- Is it worth the effort and frustration of a formal analysis of risk and uncertainty?
- If we had used decision analysis would it have prevented the drilling of our last dry hole?
- If Dad Joiner had used decision analysis would he still have found the giant East Texas Field in 1930?
- Is decision analysis a passing fad? Can its use lead our corporation to some very bad errors (decisions)?

The answers to some of these questions should become obvious as we proceed in our discussions of decision analysis. Others must be resolved on a judgment basis by each individual explorationist and decision maker. The judgment should be based, in part, on the extent to which you feel decision analysis can provide a better evaluation of risk and uncertainty. This is a personal judgment which cannot be made by the author. Thus we may have to defer judgment about the value of decision analysis until we have had a chance to explore the subject in further detail.

Advantages of Decision Analysis

In considering the merits of decision analysis it is certainly important to note several distinct advantages that the new approach has over the less formal procedures used in the past.

1. Decision analysis forces a more explicit look at the possible outcomes that could occur if the decision maker accepts a given prospect.

Instead of saying the well will either find oil or be dry, the explorationist must also consider how much will be found, what are the different levels of reserves found, and what are the probabilities of each level of reserves occurring. Decision analysis forces the explorationist to look more at the pieces and how they relate to the whole, rather than just looking at the whole and trying to decide if the prospect is feasible.

2. Certain techniques of decision analysis provide excellent ways to evaluate the sensitivity of various factors relating to overall worth.

We will see that answers to "what if" questions can be obtained with relative ease using decision analysis. (What if the probability of finding gas is only 0.12 — can we justify the gamble of drilling? How many wells might have to be drilled to develop at least 500 billion cubic feet of reserves in _____ Basin? What if we encounter a blowout and the No. 1 Smith Well will cost an additional $400,000 to complete? etc.)

3. Decision analysis provides a means to compare the relative desirability of drilling prospects having varying degrees of risk and uncertainty.

The methods effectively place risk and uncertainty under a "common denominator" for comparative purposes. The decision maker is able to compare the relative merits of an offshore prospect with an onshore prospect using decision analysis. Or he can compare a structural prospect with a stratigraphic play, or a gas prospect with an oil prospect, etc.

4. Decision analysis is a convenient and unambiguous way to communicate judgments about risk and uncertainty.

Our later discussions will point out that there are only two ways to express judgments about the likelihoods of occurrence of various chance events — numerical probabilities or adjectives. The disadvantages of adjectives are obvious. Descriptions such as "it's a sure thing," "we have a good chance in the Triassic sands," or "it's a pretty risky prospect" are often ambiguous and imprecise. What is a "sure thing" in the judgment of one person may not necessarily be a "sure thing" to another. On the other hand, to say we have a 0.7 probability of a successful completion conveys a sharp, clear impression of risk to the decision maker. Decision analysis is based entirely on descriptions of risk and uncertainty in the form of probabilities.

5. Exceedingly complex investment options can be analyzed using decision analysis techniques.

This point will become obvious to the reader in succeeding chapters. Exploration decisions to move into a new offshore area, for example, involve many critical factors and subsequent (beyond time of initial decision) contingencies and managerial options. These and other very complex decisions can be evaluated using the systematic logic and techniques of decision analysis.

Misconceptions Regarding Decision Analysis

While it is important to note the advantages of decision analysis, you should also be aware of several misconceptions and misunderstandings relative to its use in petroleum exploration decisions.

One misconception is that decision analysis will eliminate risk in decision making. An advertising flier was received recently in which a publisher was promoting sale of a new book on decision making. It began:

Dear Sir: We have released to the business world an extraordinary book that puts profit-making on an entirely new basis, free of complicated statistics and free at last of uncertainty and risk. . . .

A bit later in the brochure it continued:

I simply want you to read it and use it. Try out this proven way to banish the risk and doubt that always accompanies uncharted planning and decision making. Then you'll be in a position to decide whether or not this book is the long-needed Guide to hazard-free profit planning all industry has been waiting for.

This book does not offer such answers. The truth of the matter is that risk and uncertainty cannot be eliminated from petroleum exploration decision making. There are no decision methods which can eliminate or even reduce risk. *The utility of these methods is not to reduce risk, but rather as tools to evaluate, quantify, and understand risk so that the explorationist and manager can devise a decision strategy which will minimize the firm's exposure to risk.*

This point can be emphasized using a somewhat facetious example. An intrepid decision maker standing on a sidewalk on one side of a street has as his objective getting across the street to the opposite sidewalk. In his pathway, however, is a huge, fire-breathing, hungry dragon pacing back and forth in the street. For our analogy the dragon represents risk and uncertainty. If decision analysis could reduce risk his first step would be to make some calculations, perhaps consult his trusty computer, and review his new management science book to find how to "tame dragons."

With sufficient computer printouts and the "decisions" in hand he would then be facing a docile animal which would pose no threat to his safe crossing of the street. But of course this is ridiculous—no matter how scientific or detailed his calculations are, the dragon will still be there! The way these methods can help our distressed decision maker is to provide a means for him to study and quantify risk and uncertainty (the dragon, that is) so that he can devise a strategy for reaching his objective that will minimize his exposure to risk and uncertainty.

For example, he could observe the habits of the dragon, its reactions, reflexes, etc. (this is analogous to analyzing past data with statistical concepts.) Perhaps he will conclude from such analyses that he could throw a piece of meat out in front of the dragon to divert its attention and then run across the street behind him and minimize his likelihood of being eaten. He would have thus devised a strategy to minimize his exposure to risk and uncertainty—one objective of the new decision analysis methods.

Another misconception is that these new methods will replace professional judgment. This, of course, is also a falsehood. It will be seen that these methods are intended to *supplement,* rather than replace the necessary judgments of geologists and engineers. This is not always an easy tenet to accept, however. The language of economics and risk analysis is numbers, and the explorationist must express his "feel" (hunch, judgment, experience, bias, etc.) for the prospect in terms of numbers.

This admittedly is no simple task. But when these methods are used properly converting "feel" to numbers should not dampen creativity or imagination any more than using written language to describe the prospect. Probably in the final analysis the explorationist has no alternative. If he doesn't quantify his judgment other people less qualified will provide the necessary numbers to get "answers" for management.

We must also remember that decision analysis is not an oil-finding tool. Decision analysis can not be judged on the basis of whether or not its use was instrumental in adding new reserves to the firm. Some persons tend to dismiss decision analysis in petroleum exploration once they realize it may not necessarily help them find oil. Remember again our earlier statement that the value of decision analysis is in giving the decision maker a better understanding of the risk and uncertainty associated with each of his investment choices so he can choose wisely when faced with the uncertainties of petroleum exploration.

A Final Comment

We have already briefly mentioned that the quantitative assessment of probabilities (risk analysis) is a very tough problem. We will unfortunately not be able to develop a single "handy-dandy" formula which will cure all the evaluation problems relating to capital investment decisions in petroleum exploration. In addition, it will become apparent in our later discussions that decision analysis methods have certain "pitfalls," or "traps" for the un-informed decision maker. This tends to create certain problems in implementing these ideas in the daily decision making process.

But despite these problems we must be careful to not dismiss the subject *a priori,* because there is very substantial evidence that decision analysis works in other types of business decisions and situations involving uncertain outcomes. Perhaps the most compelling argument for use of decision analysis in petroleum exploration is '*What is the alternative?*' When exploring for petroleum in very high risk and/or high cost areas the prudent decision maker must consider risk explicitly, together with subsequent contingencies and alternatives. And decision analysis is just about the only way to do this in many instances.

The ideas, concepts, methods, and attitudes of decision analysis are quite new to the petroleum industry. Because of this "newness" the understanding and effective application of decision analysis represents a challenge to the explorationist. You are encouraged to jump in and get your feet wet! Study and read about decision analysis. Find out how managers in your company are making decisions and what criteria they use. Think in terms of probabilities and statistical concepts. Form a position on decision analysis—for or against. At worst, these efforts should make you a better poker player! And, hopefully, you will also be able to better understand and cope with risk and uncertainty in capital investment decisions. The challenge is yours!

And by the way, be sure and watch out for the dragon. . . .

2 Measures of Profitability

. . . Management needs an objective means of measuring the economic worth of individual investment proposals in order to have a realistic basis for choosing among them and selecting those which will mean the most to the company's long-run prosperity.

—Joel Dean

THIS chapter considers the meaning and uses of measures of profitability, the parameters used by the decision maker to order, accept, reject, or compare investment proposals. These parameters are also frequently called profit indicators, decision criteria, economic yardsticks, or measures of investment worth.

Our objective in this chapter will be to discuss the principal "no-risk" measures of profitability, so called because they include no explicit statements of risk and uncertainty (probability numbers). These "no-risk" measures of value include payout, profit-to-investment ratios, rate of return, net present value, etc. The dimension of risk in measures of value will be introduced in Chapter 3.

Several comments are in order about the general structure of the treasury of a firm. A good understanding of cash flows in and out of the treasury is essential to the proper use and interpretation of various measures of value. Cash flows are here defined as movements of money into or out of the treasury. Expenditures for drilling costs, lease equipment, and revenues from the sale of crude and natural gas are examples of cash flows.

The basic treasury structure of a profit-oriented organization is shown in Figure 2.1. The pattern of cash flows applies equally well to large and small oil companies. The "bank" represents any source of funds other than those generated by previous investments. Examples include short and long term bank loans, debentures, and stock issues. The equity represents the assets of the firm.

The special project is the one under immediate consideration. The

9

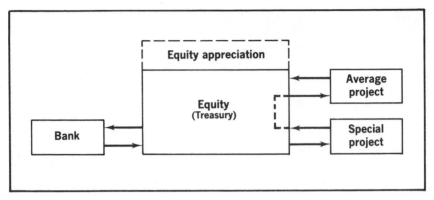

Figure 2.1 *Basic treasury structure of a profit-oriented firm or organization. Arrows indicate cash flows into and out of the treasury.*

average project represents the average alternative and/or future investment opportunities of the firm. It represents the result of policies and circumstances which will presumably be in effect during the projected future life of the special project.

Two observations about the organization of the firm's treasury are important: (*a*) *the objective of the firm is equity appreciation—growth of the total assets of the firm, and* (*b*) *money in the treasury loses all identity as to its source.* The future value of funds in the treasury is determined by the rate at which the funds can appreciate, or grow in value. This growth is obtained by continuous reinvestment of funds. The rate at which the composite, or total treasury, grows might be called the "appreciation rate," or "average opportunity rate." Hence, from a capital investment viewpoint the value of money received from any special project being considered is determined by the rate at which the money (revenues) can be reinvested to provide appreciation of the firm's total assets. Note that this rate may be different than the rate at which the specific project returns money to the treasury.

The second observation above is a logical corollary to the equity appreciation objective. A dollar received in the treasury is a dollar, whether it was received from gas sales of the Riley Gas Unit No. 1 or the Joe Smith Oil Well No. 6! In addition, a dollar received at some future point in time has a certain, single value to the firm today, regardless of the specific project which generated the dollar.

It should be noted at this point that these observations and, in fact, most of what follows in this chapter are based on the postulate of profit-motivated decisions. That is to say, we strive to invest our capital to make a given level of profit. Our motive for being in the petroleum exploration business is to make a financial profit. In recent times there has been an increasing number of people and groups who have questioned whether this should be the sole, or even principal reason for a firm's existence. Other motives such as the needs of society, protection of the world's environment, and providing energy at a minimum cost are sometimes cited by these persons as being of higher priority to justify a firm's existence than simply making a profit.

There are, of course, certain merits in these views. Just having raised the question has had its effect on corporate decision policy. Certainly today's decision maker is much more conscious of "non-profit" factors such as safeguards against pollution, creating jobs for unemployed, etc. than he was five or ten years ago. The problem in trying to resolve whether or how these factors should be used in the decision process is how to quantify them. How does one quantify the "needs of society" to determine if project A is better than project B? At this point in time it seems to be the consensus of most corporate decision makers that until better quantitative analysis methods become available which explicitly include these "non-profit" parameters the decision maker will have to consider them in a subjective (rather than quantitative) manner. Thus it seems that profitability considerations and motivations will continue to be the principal quantitative measure of value, at least for the immediate future.[1]

We must also qualify our following discussions of profitability criteria to recognize that we are evaluating the worth (profitability) of a single drilling investment at a time. We usually decide whether to accept or reject an exploration prospect on its own merits and profitability expectations, as opposed to comparing the given prospect to all other possible investment opportunities available at the moment.

This causes certain conflicts with the broader corporate accounting and financial dimensions of a firm. For example, most routine drilling prospects are analyzed and decided upon without explicit consideration of corporate debt/equity ratios, direct cost of capital, depreciation schedules and similar accounting and financial parameters.

The profitability computations normally include only the direct investments, gross revenues, direct taxes and royalties, and operating

1. The author is aware that these statements do not constitute defense of the "profit motive." Certainly the mentioned "non-profit" considerations are important and should be considered in investment decisions. The position to not consider them in this discussion resulted more from being unable to quantify the "needs of society" than by a rational justification of profit as the sole motive for a firm's existence.

costs. Thus our analysis of a given drilling prospect will consider the direct cash flows required and generated by the prospect, but usually will not involve some of the broader aspects of accounting and financial operations of the firm. This limitation is not critical for the day-to-day routine investment decisions, and to limit profitability computations to direct cash flows is an accepted practice in the business sector in general.

For investments which require huge sums of capital the limitation becomes restrictive, and it becomes important to consider sources and costs of capital, leveraged transactions, etc. These complex financial considerations are beyond the scope and intent of this book and will not be discussed herein. The interested reader is directed to the bibliography and several indicated references relating to leveraged transactions, etc.

CHARACTERISTICS OF MEASURES OF PROFITABILITY

There is probably no single measure of profitability that considers all of the factors or dimensions of investment projects that are pertinent to the decision maker. A company, therefore, must select the profitability parameter(s) which most nearly represents the financial aspects of its treasury. Listed below are some of the characteristics that a realistic measure of profitability should have.

Characteristics of a good measure of profitability

1. It must be suitable for comparing and ranking the profitability of investment opportunities.
2. The parameter should reflect the firm's "time-value" of capital. That is, it should realistically represent the fiscal policies of the firm, including future reinvestment opportunities.
3. The parameter should provide a means of telling whether profitability exceeds some minimum, such as the cost of capital and/or the firm's average earnings rate.
4. It should include quantitative statements of risk (probability numbers).
5. It would be desirable to have the parameter reflect other factors, such as corporate goals, decision maker's risk preferences, and the firm's asset position, if possible.

No attempt will be made to rank the desirability of the various profit indicators. Our purpose here is to describe each measure of value and summarize its advantages and limitations. Specific criteria covered include the following:

Payout
Profit-to-investment ratio
Rate of return
Net present value profit
Discounted profit-to-investment ratio
Appreciation of equity rate of return
Percentage gain on investment
"Risk-weighted" rates of return

The actual calculation of each of the measures of value will be illustrated using an example development well drilling prospect. The pertinent data and projected cash flows for the prospect are given in Figure 2.2. Federal taxation will be excluded from the analysis for reasons stated previously in the Preface.

Prospect Data

Prospect Name: "DECISION METHODS UNIT, WELL NO. 1"
Investment: $268,600 for completed well; $200,000 for dry hole
Estimated recoverable reserves: 234,000 Bbls; 234 MMCF gas
Estimated average producing rate during first two years: 150 BOPD
Future Expenditures: Pumping Unit, Year 3 – $10,000; Workover, Year 5 – $20,000
Working interest in proposed well: 100%
Average investment opportunity rate: 10%
Type of compounding used by the company: annual compounding, with cash flows assumed
    to occur at mid-year
Production and cash flow projection (Federal Income Tax not considered):

Year	Estimated Oil Production, Bbls	Annual Net Revenue*	Future Expenditures	Net Cash Flow
1	54,750	$132,900		$132,900
2	54,750	132,900		132,900
3	44,600	107,600	10,000	97,600
4	29,200	69,200		69,200
5	18,900	43,500	20,000	23,500
6	12,900	28,600		28,600
7	7,800	15,900		15,900
8	5,200	9,400		9,400
9	3,700	5,600		5,600
10	2,200	1,900		1,900
	234,000	$547,500	30,000	$517,500

* Annual Net Revenue = (Annual Gross Revenue – Royalty – Taxes – Operating expenses).

Figure 2.2 Data and projected cash flows for example drilling prospect.

PAYOUT

The payout of a capital investment project is defined as the length of time required to receive accumulated net revenues equal to the investment. Stated differently, it is the length of time it takes to get the invested capital back. As such, payout time is an approximate measure of the rate at which cash flows are generated early in the project. Payout time is a relatively simple number to calculate, and can be expressed in terms of "before tax" or "after tax" revenues. It tells the decision maker nothing about the rate of earnings after payout time and does not consider the total profitability of the investment opportunity. Consequently, it is not a sufficient criterion in itself to judge the worth of an investment.

If we were to set up a special account for a given investment project, we could keep track of the cumulative "project account balance" as a function of time. Such a "project account balance," when plotted graphically is called a cash position curve, as in Figure 2.3. Notice that the account is negative at time zero by an amount equal to the initial investment. As revenue is received from the project it is credited to the account. The length of time which elapses until the account balance is exactly zero is, by definition, payout time. All revenues received after payout represent new capital generated from the project. It is probably obvious that, all other factors equal, the decision maker would like to invest in projects having the shortest possible payout time.

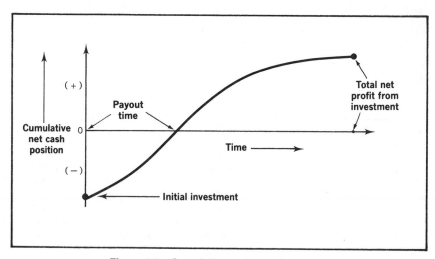

Figure 2.3 *Cumulative cash position curve.*

Payout calculation for example prospect

The net revenue in the first year is estimated to be $132,900 (from Column 5 of Fig. 2.2). Thus the unrecovered portion of the initial investment of $268,-600 at the end of Year 1 is $268,600 − $132,900 = $135,700. The unrecovered portion of the investment at the end of Year 2 is $135,700 minus the net revenue received in Year 2, or $135,700 − $132,900 = $2,800. The portion of Year 3 required to recover this remaining balance is $2,800/$97,600 = 0.029. Thus the payout time is *2.029 years.*

Payout time has been in wide use for many years as an integral part of the economic analysis of drilling prospects. It is a useful parameter to compare the relative rates of receipt of revenues early in the projects. But, as stated previously, it is not a parameter that reflects or measures all the dimensions of profitability which are relevant in capital expenditure decisions.

PROFIT-TO-INVESTMENT RATIO

One weakness of payout time as a measure of value is that it does not consider the *total profit* from the investment project. The profit-to-investment ratio is a measure that does reflect total profitability. It is defined as the ratio of total undiscounted net profit to investment. It is a dimensionless number relating the amount of new money generated from an investment project per dollar invested. It is sometimes called the return-on-investment, or simply ROI.

Undiscounted Profit-to-Investment Ratio (ROI) calculation for example prospect

$$\text{Undiscounted Net Profit} = \text{Total Net Cash Revenues} - \text{Investment}$$
$$= \$517,500 - \$268,600$$
$$= \$248,900$$

$$\begin{array}{l}\text{Profit-to-Investment} \\ \text{Ratio (ROI)}\end{array} = \frac{\$248,900}{\$268,600} = 0.927$$

Some companies use the ratio of net revenue-to-investment in lieu of the above ratio. This ratio is sometimes called "leverage." For our example drilling prospect the net revenue-to-investment ratio would be

$$\begin{array}{l}\text{Net Revenue-to-Investment} \\ \text{Ratio (leverage)}\end{array} = \frac{\$517,500}{\$268,600} = 1.927$$

The denominator of the ratio is usually the drilling cost of a well for a single well prospect. If expenditures extend over a period of time before any revenues are received, the ratio is sometimes computed using the maximum amount of cash invested in, but not yet recovered from the project as the term in the denominator. This investment term is called the "maximum out-of-pocket cash" and represents the lowest negative value on a cumulative cash position curve.

The ratio is simple to calculate, and can be expressed in "before tax" or "after tax" values. The parameter is a measure of total profit from the investment, and does not require a detailed cash flow. It is often used with payout time to give two relative profit indicators on routine drilling economics. The major weakness of the ratio is that it does not reflect the time-rate patterns of income from the project. Two investment opportunities could have the same payout time and profit-to-investment ratio yet have quite different cash-flowback patterns.

For example, consider Prospects A and B in Table 2.1. Each have the same payout times and the same profit-to-investment ratios. Yet most decision makers, if given a choice would prefer A over B because prospect

Table 2.1
Comparison of Two Hypothetical Drilling Prospects

	Prospect "A"	Prospect "B"
Estimated producing rate during early life of well	150 BOPD	150 BOPD
Operating Costs: $/month	$575	$575
Investment	$150,000	$150,000
Payout: months	14.3 ← same →	14.3
Recoverable Reserves	200,000 Bbls	243,000 Bbls
Producing Life	15 yrs.	30 yrs.
Total Net Revenues ($2.92/Bbl, 1/8 Royalty, 5% state production tax)	$485,500	$589,000
Total Operating Costs $\text{Total life} \times \dfrac{12 \text{ months}}{\text{year}} \times \dfrac{\$575}{\text{month}}$	−103,500	−207,000
	$382,000	$382,000
Workover, year 7	−20,000	−20,000
Initial Investment	−150,000	−150,000
NET PROFIT	$212,000	$212,000
PROFIT-TO-INVESTMENT RATIO ($212,000/$150,000)	1.41 ← same →	1.41

A returns total income in one-half the time. Thus it should be obvious that a missing dimension in the profit-to-investment ratio and payout parameters is the time-rate patterns of cash flows.

To stress this point in a different way, suppose you invest a dollar today to receive $3 in three years or $4 in 10 years. Which would you prefer? Most people say $3 in three years, even though the ratio is higher for the other option.

TIME-VALUE OF MONEY CONSIDERATIONS

The usual method of relating the time-rate patterns of future cash flows to some measure of profit is by use of "time-value of money" concepts – compounding and discounting. We will digress momentarily at this point to summarize the basic aspects of the time-value of money. All of the measures of value to be discussed subsequently (rate of return, net present value, etc.) will include the element of compounding or discounting.[1]

First, let us define a few terms.

C = value of principal sum, as of a specified time – usually the starting time reference of the evaluation, time zero

S = value of the principal sum plus interest at a future point in time, "n" years away

j = nominal annual interest rate, fraction

i = effective annual interest rate, fraction

n = number of years separating C and S

m = number of interest periods per year

i_m = effective interest rate per interest period: $i_m = j/m$

All of the time-value of money considerations are generally based on the following relation:

$$C(1 + i)^n = S \qquad (2.1)$$

Equation 2.1 is called the compound interest equation and relates the future value, S, to a principal amount of money today, C. The term $(1 + i)^n$ is called a compound interest factor. By dividing both sides of

1. Those readers inclined to study the history of compound interest concepts should read "A Brief History of Interest Calculations," by G. W. Smith, Journal of Industrial Engineering, Vol. 18, No. 10, October 1967, p. 569–574. This fascinating article traces the concept back to 2300 B.C., and traces important contributions leading to development of interest tables during the sixteenth century A.D.

equation 2.1 by $(1 + i)^n$ we get a modified form of the equation, called the present value equation:

$$C = \frac{S}{(1 + i)^n} = S\left[\frac{1}{(1 + i)^n}\right] \qquad (2.2)$$

The interest rate, i, is usually called the discount rate when used in the context of the above equation.

It is important to recognize that C and S are equivalent in value, even though separated in time by n years.

EXAMPLE: If $100 is invested for 4 years at 10% compounded annually, what will it appreciate in value to after 4 years?

$$C = \$100, \quad i = 0.10, \quad n = 4 \text{ years}, \quad S = ?$$

From equation 2.1:

$$\$100(1 + 0.10)^4 = S$$
$$\$100(1.4641) = S \qquad\qquad S = \underline{\$146.41}$$

Thus, $100 invested at 10% compound interest will appreciate in value to $146.41 at the end of 4 years. If 10% is the representative time-value of money, receiving $100 today has no greater, or lesser, value than receiving $146.41 in 4 years. To say that $100 today will become $146.41 is the same as saying that receiving $146.61 in 4 years has a *present value* or *present worth* of $100.

Equation 2.1 or 2.2 provides a means of comparing values of sums of money received at different times. In most petroleum evaluations the common point in time for comparing values of monetary sums (cash flows) is usually the present time, or time zero. Consequently, equation 2.2 will be used more frequently. The term in brackets in equation 2.2 is called a discount factor. These factors are available in tables.

Some general comments about compounding and discounting:

1. We normally speak of compound interest or discount rates in terms of the nominal interest rate per year. If the investment earns interest once a year (annual compounding) the nominal and effective interest rates (j and i) are the same. If interest is credited to the investment at periods less than a year, such as quarterly, the effective interest rate is slightly greater than the nominal. The following relation can be used to calculate the effective interest rate if the nominal rate and the frequency of compounding is known.

$$i = \left(1 + \frac{j}{m}\right)^m - 1 \qquad (2.3)$$

Of course the reason the effective annual rate is greater than the nominal rate if the period of compounding is more frequent than a year is that the interest credited for a partial part of the year earns interest itself for the remainder of the year.

EXAMPLE: A bank advertisement:

"Earn 4.93% each year when our 4.85% is compounded quarterly and maintained for a year." (Banks sometimes call the 4.93% the yield.) How did they calculate this?

j = nominal annual interest rate = 0.0485
m = 4, the number of interest periods per year
i = effective annual interest rate = ?

From equation 2.3:

$$i = \left(1 + \frac{0.0485}{4}\right)^4 - 1 = \underline{\underline{0.0493}}$$

Earning 4.85% compounded quarterly is equivalent to earning 4.93% compounded annually.

Another way to define the effective interest rate "i" is as follows: 'If $1 accumulates in one year to $(1 + i)$, then the effective annual interest is "i," or $100i\%$ if expressed as a percentage.'

2. If money is received (or returned) in a series of equal payments at the end of each succeeding interest period, the following equation can be used to calculate the present worth of the entire series of cash flows:

$$C = I\left[\frac{(1 + i_m)^{n'} - 1}{i_m(1 + i_m)^{n'}}\right] \qquad (2.4)$$

where I is the equal payment paid or received at the end of each interest period, and n' is the total number of interest periods. Equation 2.4 is the summation of equation 2.2 for a series of equal cash flows.

3. The time at which a cash flow is received (or disbursed) may or may not coincide with the timing of the interest periods. The concept of compounding and discounting is not dependent on the timing of when the cash flows are received. Thus we can talk about annual compounding and yet deposit money each day, or daily compounding but deposit money once every five years, etc.

4. It is sometimes convenient to use continuous compounding or discounting. This is the limiting case where the interest period approaches zero (and m, the number of interest periods per year, becomes infinite). Continuous compounding has certain computational advantages, and equations 2.1 and 2.2 take on a slightly different form having exponential terms. The subject is discussed in further detail in Appendix C.

5. The type of compounding (annual, monthly, daily, continuous, etc.) and the pattern of receipts (year-end, mid-year, monthly, uniform, etc.) used in cash flow analyses should be representative of the fiscal practices of the firm's treasury and the project being considered. Tables of discount factors for the following cases are included in the appendix sections of this reference:

	Appendix
Annual compounding, mid-year cash flows	A
Annual compounding, year-end cash flows	B
Continuous compounding, lump-sum cash flows at any point in time during year	C
Continuous compounding, uniform (continuous) annual cash flows	D

Each appendix section has an introduction describing how the tables are used and interpreted. Subsequent examples in this chapter utilize discount factors from Appendix A. Generally speaking, there is little difference in the discount factors for the various types of compounding at lower discount rates (less than about 25%). At higher discount rates the difference can become large, and it becomes necessary to use the type of compounding and cash flow pattern which most closely match the way money enters and leaves the treasury of the firm.

Having briefly considered the important aspects of the time-value of money, we now return to our discussion of measures of profitability. Recall that of the first two criteria discussed, payout and profit-to-investment ratio, a common weakness was that they did not measure or reflect the time-rate patterns of cash flows and the time-value of money. Rate of return is one profit indicator which considers these factors.

RATE OF RETURN

One of the more widely used profit indicators in recent years has been rate of return. It has been given many different names, including

discounted rate of return, internal yield, internal rate of return, profitability index (PI), DCF rate of return, and marginal efficiency of capital.
We will use just the term rate of return in this reference.

Definition of rate of return:

The interest rate which equates the value of all cash inflows to the cash outlays when these cash flows are discounted or compounded to a common
point in time. Stated differently, it is the interest rate which makes the present
value of net receipts equal to the present value of the investments.

The rate of return calculation is made after the series of anticipated
future cash flows to be received from the investment has been defined.
The calculation is a trial-and-error process which begins by selecting
an interest rate and discounting all the cash flows back to time zero—
i.e., finding the present value of all the future cash revenues generated
by the initial investment.

If the sum of the present values of future cash flows exceeds the
investment the discount rate selected was too low. If the sum of the
present values is exactly equal to the initial investment the discounting

Table 2.2
Rate of Return Calculation
"DECISION METHODS UNIT, WELL NO. 1" Prospect

Year	Net Cash Flow*	Discount Factor, 40%	40% Discounted Cash Flow
0	−$268,600	1.000	−$268,600
1	+ 132,900	0.845	+ 112,300
2	+ 132,900	0.604	+ 80,300
3	+ 97,600	0.431	+ 42,100
4	+ 69,200	0.308	+ 21,300
5	+ 23,500	0.220	+ 5,200
6	+ 28,600	0.157	+ 4,500
7	+ 15,900	0.112	+ 1,800
8	+ 9,400	0.080	+ 700
9	+ 5,600	0.057	+ 300
10	+ 1,900	0.041	+ 100
			0

RATE OF RETURN = 40%

* From Figure 2.2.
    The discount factors are from Appendix A (annual compounding,
mid-year cash flows). This table is the final trial of a series of trial-and-
error computations using varying values of the discounting rate.

rate selected is, by definition, the rate of return. If the sum is less than the initial investment the discount rate is too high, and a lower discount rate should be used for the second trial. By following such a trial-and-error procedure the rate of return for the example drilling prospect (Figure 2.2) was computed to be 40%, as shown in Table 2.2 on page 21.

The tabular calculation of Table 2.2 is a convenient way to compute rate of return. Notice that when the algebraic sum of the discounted cash flow column is zero the discount rate used is, by definition, the rate of return of the investment. In making the trial-and-error computations the rule to follow in selecting the rate for the next trial is as follows: if the sum is positive use a higher discount rate for the next trial. If it is negative, use a lower discount rate for the next trial.

A negative cash flow is considered as an expenditure or cash outflow. If any expenditures are anticipated after time zero they are simply included in determining the net cash flow for the appropriate year. The net cash flow terms after time zero do not have to all be positive. The tabular calculation of Table 2.2 really consists of solving equation 2.2 eleven separate times.

Mathematically, Table 2.2 is equivalent to solving the following rate of return equation:

$$C_0 + S_1\left[\frac{1}{(1 + i_r)^{1-0.5}}\right] + S_2\left[\frac{1}{(1 + i_r)^{2-0.5}}\right] + \cdots + S_{n'}\left[\frac{1}{(1 + i_r)^{n'-0.5}}\right] = 0$$

(2.5)

where C_0 = initial investment at time zero = −$268,600
    S_1 = net cash flow received at midpoint of year 1, = +$132,900
    S_2 = net cash flow received at midpoint of year 2, = +$132,900
        .
        .
        .
    n' = total number of years in cash flow = 10
    $S_{n'}$ = net cash flow received at midpoint of year n' = +$1,900
    i_r = rate of return, decimal fraction

Equation 2.5 is the rate of return equation for mid-year cash flows. If the cash flows were assumed to be received at a different point in time each year (such as year-end) the exponents of the discount factor terms would have to be modified. The discount factors (the terms in brackets) were obtained from Appendix A in lieu of solving the terms numerically. Of course, if the future projection of cash flows extends beyond year 10 equation 2.5 would simply have additional terms on the left-hand side of the equation. The common point in time used for comparison in the above example was time zero. This is usually how most rate of return calculations are made. The time reference, however, does not have to be time zero—it could be any point in time during the future cash flow.

So much for the mechanics of computing the rate of return—what does it mean? It means that the expenditure of $268,600 to purchase the future series of cash revenues has a profitability equivalent to investing $268,600 in a savings account (or any other type of investment, for that matter) which pays 40% interest compounded annually. The expenditure has an earnings rate equivalent to 40% per year compounded annually. Note that we are defining rate of return as the earnings rate of the *initial investment.*

Alternative definitions such as the rate of return being the earnings rate of the unrecovered balance, or the earnings rate of the undepreciated balance have been used in the past. However, these definitions require specification of capital recovery schedules, depreciation schedules and other accounting procedures not usually considered in capital investment decision making. Thus we stress that the rate of return is an indicator of the worth of the initial investment only, and is not dependent on any accounting procedures. This definition is now accepted almost universally for capital investment decision making.

Specific characteristics of the rate of return concept include the following:

1. Computation of rate of return requires a series of trial-and-error computations.

    The reason for this is that the mathematical equation of rate of return (such as equation 2.5 above) cannot be solved explicitly for i_r. This characteristic is of no real consequence if the rate of return is being calculated on a computer. However the repetitive calculations can become quite tedious if done by hand, particularly if cash flows extend over a long period of time.

    Several "short-cut" aids exist to minimize the trial-and-error computations. A useful rule of thumb for estimating the first discounting rate to try is to divide *100%* by the number of years to payout.[1] The result is sometimes a good "first guess" discount rate for oil and gas depletion type revenue schedules. For our

1. This rule of thumb was given by Megill, *An Introduction to Exploration Economics,* The Petroleum Publishing Company, Tulsa, 1971. The rule was based on his experience in dealing with depletion type patterns of revenue schedules and is only of use for these particular types of rate of return computations.

example drilling prospect the payout time was computed to be 2.029 years (page 15). Thus, by this rule of thumb the first discount rate to try in the trial-and-error sequence would be $100\%/2.029$ years $\cong 50\%$.

Another useful aid is to make two trials using different rates. Then prepare a graph (on coordinate graph paper) of the algebraic sum of the discounted cash flows (Column 4 of Table 2.2) versus the discount rates used. An approximate discount rate for the third trial would be the rate which corresponds to a discounted cash flow sum of zero when the two plotted points are connected by a straight line.

For example, in the trial-and-error sequence to compute the rate of return for our drilling prospect suppose we had, in fact, used 50% for the first trial. Making a 4-column computation like Table 2.2 the algebraic sum of the "50% discounted cash flows" (Column 4) works out to be −$26,930. As stated previously, if this sum is negative the next rate to use should be lower.

Suppose we tried 30% next. The algebraic sum of the "30% discounted cash flows" works out to be +$34,850. A positive sum indicated the rate used was too low. From these two trials we have bracketed the desired rate of return—we know it is less than 50%, but greater than 30%. To get a good approximate value of the desired rate of return we can plot these values on coordinate graph paper as in Figure 2.4.

Next we connect the two data points with a straight line and observe where the line (or its extrapolation if both points are above or below the x-axis) crosses the "zero discounted cash flow sum" point. For our numerical example this occurs at a discount rate of *41%*, a good value to use for the third trial.

You are reminded, however, that this graphical aid will be only an approximate value for the rate of return of the series of cash flows. The actual correlation between the discount rate and the sum of the discounted cash flows is non-linear, however over relatively limited ranges of discount rate the linear correlation will provide a good estimate for the third trial. In our example the 41% read from Figure 2.4 is, in fact, quite close to the actual rate of return of 40%, as shown in Table 2.2.

2. The rate of return concept introduces the "time-value" of money into the profitability criterion.

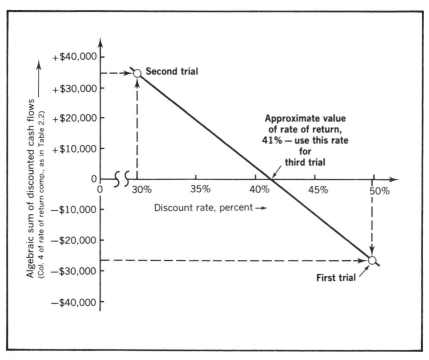

Figure 2.4 *Discounted cash flow sum vs discount rate. The discount rate to use on the third trial is located at the point where the linear extrapolation crosses the x-axis. This is only an approximation—but a useful one to minimize the number of repetitive trials required.*

To be able to compute rate of return the analyst must predict the cash flow time—rate schedule over the entire life of the investment project. Investment capital has a certain intrinsic "time-value" to the firm, and being able to incorporate this value into the profitability criteria used in decision making is an important step in achieving a more realistic estimate of project profitability.

3. Rate of return is a profit indicator that is independent of the magnitude of the cash flows.

In our numerical example of Table 2.2 if each of the cash flows of Column 2 had been millions of dollars (that is, $268.6 million, $132.9 million, etc.) the rate of return would still have been 40%. Some managers consider this an advantage of rate of return in that it provides a basis of comparison not dependent on the magnitudes of the cash flows. That is, they can use rate of return to compare the desirability of investing in a very expensive prospect with a low-cost drilling prospect. Others feel this characteristic is a disadvantage because the magnitudes of investment capital *are* important to consider in the decision process.

4. There are certain types of cash flows in which there is more than one discount rate which satisfies the definition of rate of return.

Examples of cash flow schedules which sometimes lead to multiple rates of return include some rate acceleration projects, and projects requiring a large expenditure at a later point in the life of the project. A necessary condition to have multiple rates of return is a second sign reversal in the cumulative cash position function, as shown in Figure 2.5.

A second sign reversal, however, is not a sufficient condition for having multiple rates of return. These two sentences mean that the cash flow must have a second sign reversal, but just hav-

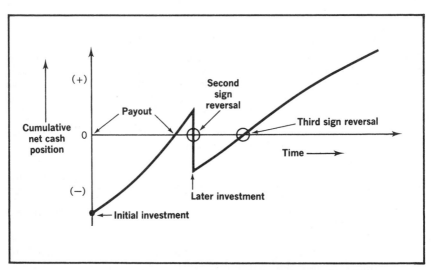

Figure 2.5 *A cumulative cash position curve which has a second and third sign reversal of the cumulative cash position function. Having at least a second sign reversal is a necessary condition (but not a sufficient condition) for multiple rates of return. The first sign reversal occurred at payout, when the function changed from negative to positive.*

ing the reversal is not sufficient to force multiple rates of return. It also depends on when the second reversal occurs and the magnitude of the negative cash flows causing the reversal. In cash flows having multiple rates of return there is no way to establish which (if any) of the multiple rates is the correct, or true rate of return.

In situations such as this the economic analysis should be made using other profitability criteria such as net present value at a specified discount rate (to be discussed in the next section of this chapter). In summary, if the negative cash flows (investments) are all at or near time zero, there will be only one discount rate which satisfies the rate of return definition. If the project involves later expenditures the analyst should be alert to the possibility of multiple rates. And also, in the analysis of rate acceleration projects using incremental cash flows the analyst should be especially watchful for multiple rates of return.

A very easy test to determine if multiple rates are possible is to compute a column of cumulative cash flows over the life of the project and observe the number of sign reversals. For our numerical example of this chapter we would use the cash flows of Column 2 in Table 2.2 as follows:

Year	Net Cash Flow (Col. 2, Table 2.2)	Cumulative Net Cash Flows	
0	−$268,600	−$268,600	
1	+ 132,900	− 135,700	
2	+ 132,900	− 2,800	First sign
3	+ 97,600	+ 94,800	reversal
4	+ 69,200	+ 164,000	
5	+ 23,500	+ 187,500	
6	+ 28,600	+ 216,100	No other sign reversals.
7	+ 15,900	+ 232,000	Thus, this cash flow
8	+ 9,400	+ 241,400	will only have one Rate
9	+ 5,600	+ 247,000	of Return (40%)
10	+ 1,900	+ 248,900	

5. A rate of return cannot be calculated for the following situations:
   a. Cash flows are all negative (For example: a dry hole).
   b. Cash flows are all positive (Investment is paid out of future revenues).
   c. Total undiscounted revenues are less than the investment.

(For example: a marginal producing well or field which is depleted before reaching payout.)

For these situations rate of return is mathematically undefined and cannot be computed. Negative rates of return have no meaning. A "zero rate of return" corresponds to a cash flow schedule in which the undiscounted revenues exactly equal the undiscounted investment. This would occur in the situation where the project generated no additional revenues after reaching payout time.

6. Cash flows received early in the project are weighted more heavily than later cash flows. This becomes particularly pronounced as the interest rate increases. In the rate of return example of this chapter the present value of one dollar received in year five was only $0.22! (Table 2.2). Cash flows received or disbursed late in the life of the project (after 15–20 years) have very little effect on the computed rate of return.

7. The computed rate of return is very sensitive to errors in estimating initial investment and the early cash revenues. If you are uncertain about drilling costs, for instance, it is highly recommended that several rates of return be computed for several possible values of initial investment. A small variation (on a percentage basis) in initial investment can sometimes cause a much larger (percentage-wise) variation in the resultant rate of return.

   (*Question for the reader: In our example of Table 2.2, if the drilling costs had been underestimated by 10%—i.e. actual cost was $268,600 + $26,860 = $295,460—would the rate of return be reduced by an equivalent 10%, from 40% to 36%?*)

8. Rate of return is a convenient measure of profitability to compare with a "minimum," such as a cost of capital or a corporate objective of a ___% annual growth rate. Management can easily relate a rate of return to interest and loan rates, etc., and this is one of the probable reasons for its wide popularity as an investment criterion.

9. Rate of return includes the implicit assumption that all cash flows will be reinvested at the computed rate of return when received. If they are not reinvested at that rate the initial expenditure will not have the earning power of the rate of return given. This is an

extremely important characteristic of rate of return which is often misunderstood or ignored by those who assume the criterion to be a realistic measure of profitability.

Recall our previous statement that investing $268,600 to buy or receive the series of future cash revenues was equivalent (in terms of profitability) to investing the same amount of money in a savings account or other investment opportunity paying the 40% rate of return interest rate compounded annually.

For this equivalence to hold, the revenues from the example drilling prospect must be reinvested at 40% when they are received. (You should prove that this is true. One way to prove the reinvestment assumption is to compute the future worth of $268,600 invested at 40% for 10 years. Then compute the future worth as of year 10 of all the anticipated cash flows from the drilling prospect reinvested at 40%—and compare the future values of each alternative.)

If the firm does not have unlimited reinvestment opportunities at the rate of return of 40% for the next 10 years, rate of return will not be a realistic measure of true profitability for the example drilling prospect.

Suppose our example project was accepted on the presumption that the $268,600 expenditure to drill the well had an overall profitability of 40% but when the ten annual revenues were received they went into the treasury and then out again in other "average" projects having a lower earnings rate, say 10%–15%. Then the earning power of the initial $268,600 would, in fact, be much lower than the 40%. The computed rate of return of 40% would thus have been an unrealistically high measure of earning power.

The reinvestment at higher rates (say 60 or 70%) would have resulted in the computed rate of 40% being an unrealistically *low* measure of earning power. The point here is that the computed rate of return of an investment is a realistic and accurate measure of profitability if, and only if, the future cash flows are reinvested at the computed rate the instant they are received.

If a firm anticipates that its future reinvestment opportunities are in the range of 8%–15% it becomes important to realize that high rates of return (say above 25%–30%) may provide a *relative* basis to accept or reject a project, but the high rates of earning power should not be considered as measures of *actual* earning power of the capital expenditures.

10. Rate of return is not a completely realistic parameter to rank desirability of competing investments because different "time-values-of money" are presumed for each project.

EXAMPLE: Suppose we had the following investment opportunities available:

Project	Computed Rate of Return	Discount Factor Used for Year 10 (Appendix A)
A	30%	0.083
B	50%	0.021
C	10%	0.404
D	40%	0.041
E	20%	0.177

Assuming each project had a future life exceeding 10 years we could look at the discount factors used to discount a dollar received in year 10 from each of the investment opportunities. Notice a dollar received in year 10 from project C has more value than a dollar received at the same time from project A. In reality a firm has a certain time-value of money that is the same for all investment opportunities. A dollar received in year 10 has a certain present value to the firm—regardless of its source. To make an accept-reject decision on these five projects would imply five different "time-values-of-money," an obviously unrealistic situation.

11. Risk (probability numbers) cannot be incorporated mathematically into the rate of return equation. Later in this chapter we will show how the element of risk can be treated empirically using rate of return. It will be seen, however, that it is not a completely adequate approach to including risk in our profit criterion. There is no mathematical relationship between payout time, profit-to-investment ratio and rate of return.

12. Rate of return can be computed on a "before tax" or "after tax" basis.

In summary, rate of return is certainly a more realistic measure of value than payout time and/or profit-to-investment ratio, primarily because it includes the time-value of money concept. It is a useful measure of the *relative* profitability of investments having approximately the same total life and cash-flow patterns. Its primary weaknesses as a measure of true profitability are the frequent problems of not satisfying the reinvestment assumption and the fact that it cannot be used to explicitly consider the dimension of risk and uncertainty.

NET PRESENT VALUE (NPV)

The profit criterion of net present value, NPV, is similar to rate of return except that a single, previously specified discount rate is used for all economic analyses. The single discount rate is usually called the average opportunity rate, and presumably represents the average earnings rate at which future revenues can be reinvested. Thus, one of the ad-

Table 2.3

Net Present Value (NPV) Calculation for Example Prospect
"DECISION METHODS UNIT, WELL NO. 1" Prospect

Year	Net Cash Flow	Discount Factor 10%	10% Discounted Cash Flow
0	−$268,600	1.000	−$268,600
1	+ 132,900	0.953	+ 126,700
2	+ 132,900	0.867	+ 115,200
3	+ 97,600	0.788	+ 76,900
4	+ 69,200	0.716	+ 49,500
5	+ 23,500	0.651	+ 15,300
6	+ 28,600	0.592	+ 16,900
7	+ 15,900	0.538	+ 8,600
8	+ 9,400	0.489	+ 4,600
9	+ 5,600	0.445	+ 2,500
10	+ 1,900	0.404	+ 800
			+$148,400 = NPV @ 10%

Average investment opportunity rate: 10%.

Net Cash Flows are from Figure 2.2.

Net Present Value Calculation, $i_0 = 0.10$.

The discount factors are from Appendix A (annual compounding, mid-year cash flows).

vantages of net present value over rate of return is that it is computed using a more realistic appraisal of future investment opportunities.

Notice that the NPV calculation of Table 2.3 is exactly like the calculation of rate of return except that the discount factors will always be based on the firm's average opportunity rate. The algebraic sum of the fourth column (discounted cash flow) is, by definition, the net present value discounted at the average opportunity rate (sometimes designated i_0).

If NPV is positive it means that the investment will earn a rate of return equal to i_0 *plus* an additional amount of cash money equal to the NPV as of time zero. If the average opportunity rate used is realistic of the firm's ability to invest capital, then it follows that investment opportunities which have a negative NPV should be rejected.

Characteristics of net present value as a measure of profitability include the following:

1. Computation is no longer a trial-and-error solution, and no multiple rates of return are possible. Can be computed on an "after-tax" or "before-tax" basis.

2. Net present value (NPV) has all features of rate of return regarding time-value of money concepts, plus the added fact that the reinvestment assumption is satisfied because the average opportunity rate presumably reflects future investment opportunities. NPV is suitable for use with probability numbers to consider risk in a quantitative and explicit manner (see Chapter 3).

3. If NPV $= 0$, then the investment is yielding a rate of return equal to discount rate used, i_0. If NPV is negative it means that the investment will yield a rate of return less than i_0. If positive, the sum represents present value cash worth *in excess* of making a rate of return equal to i_0. Or stated differently, a positive NPV is the amount of additional money which could be invested in the project and still realize a rate of return equal to i_0.

4. As with rate of return, the NPV is independent of absolute size of cash flows. To illustrate this point consider the following investment projects:

	Project "A"	Project "B"
NPV of Revenues	$2,500,000	$300,000
(less) Initial Invest.	− 2,400,000	− 200,000
NPV *Profit*	$ 100,000	$100,000

Both projects have the same NPV profit, and on the basis of NPV alone would be considered as equally attractive investments. Of course if the decision maker had to operate within a limited capital constraint he would probably consider Project "B" as more desirable since he can receive the same NPV profit with a much smaller initial investment. Therefore, the objective of maximizing NPV profit is not a completely adequate profitability criterion if there are limitations on the availability of capital.

5. The same discount rate does not have to be used for the entire period of cash flows. If fluctuations in the money market and/or future corporate investment opportunities are anticipated, a varying discount rate can be used. The calculations are not quite as straightforward, however. The example of Table 2.4 illustrates the correct way to calculate NPV using varying discount rates.

Table 2.4
Calculation of NPV Using Two Discount Rates

Find the NPV of the following cash flow using 12% rate of discounting for the first 10 years and 10% for the remaining 15 years. Assume annual compounding and mid-year cash flows.

ANTICIPATED FUTURE CASH FLOW

Year	Net Cash Flow ($M)	Year	Net Cash Flow ($M)
1	−$200	11	+$90
2	+ 150	12	+ 80
3	+ 150	13	+ 70
4	+ 140	14	+ 65
5	+ 140	15	+ 60
6	+ 130	16	+ 55
7	+ 130	17	+ 50
8	+ 120	18	+ 45
9	+ 110	19	+ 40
10	+ 100	20	+ 35
		21	+ 30
		22	+ 25
		23	+ 20
		24	+ 15
		25	+ 10

Step 1. Find discounted worth of Year 11–25 cash flows as of Year 10, T_{10}, @ 10%:

Actual Year	No. of Years After T_{10}	Net Cash Flow ($M)	10% Discount Factor (Appendix A)	Discounted Cash Flow @ 10% ($M)
11	1	+$90	0.953	$85.8
12	2	80	.867	69.4
13	3	70	.788	55.2
14	4	65	.716	46.6
15	5	60	.651	39.1
16	6	55	.592	32.6
17	7	50	.538	26.9
18	8	45	.489	22.0
19	9	40	.445	17.8
20	10	35	.404	14.1
21	11	30	.368	11.0
22	12	25	.334	8.4
23	13	20	.304	6.1
24	14	15	.276	4.1
25	15	10	.251	2.5

$441.6M Discounted value as of T_{10}

Step 2. Find present worth at T_0 of $441,600 received at year 10, discounted at 12%.

$$NPV_{T_0} = \$441,600 \left[\frac{1}{(1 + 0.12)^{10}} \right] = \$142,000$$

Step 3. Find present worth of Year 1–10 cash flows as of T_0 @ 12%.

Year	Net Cash Flow ($M)	12% Discount Factors (Appendix A)	Discounted Cash Flow @ 12% ($M)
1	−$200	.945	−$189.0
2	+ 150	.844	+ 126.5
3	150	.753	113.0
4	140	.673	94.3
5	140	.601	84.2
6	130	.536	69.8
7	130	.479	62.3
8	120	.427	51.3
9	110	.382	42.0
10	100	.341	34.1

+$488.5M Present value of cash flows of years 1–10, as of T_0, discounted at 12%.

Step 4. Add results of Steps 2 and 3 to find total present value as of T_0.

$$\text{NPV} = \$142,000 + \$488,500 = \underline{\underline{\$630,500}}$$

6. The net present value concept can be used to evaluate purchase versus lease-option alternatives in which all of the cash flow terms are negative. In this instance the preferred decision option will have the least negative present value.

7. Specifying the single discount rate to use for NPV computations is sometimes not an easy procedure. The usual first reaction is to check the overall corporate annual earnings rate in past years. This, of course, may not be realistic because the desired rate is what *future* invested capital will earn, not past results. Predictions of the rate at which future revenues can be reinvested involve an obvious element of uncertainty. The discount rate for NPV computations is usually set by top management after consideration of at least some of the following factors:

   a. If the firm is operating on borrowed capital the rate should at least exceed the interest rate being paid on the loan.

   b. If the capital comes from several sources (internally generated funds, short and long term debt, and equity sources) the determination of an average cost of capital is sometimes used as a basis for a minimum value of i_0.

   c. Corporate growth objectives — the specified rate at which management has set for growth of the firm's total assets.

   d. The risk of oil exploration as compared to less risky investments (?) such as chemical processing, refining, retail marketing. Note, however, that if risk in oil exploration can be adequately quantified and included in the NPV calculation it does not necessarily follow that a higher i_0 should be used for oil exploration investments.

   e. Future investment opportunities — are they limited or unlimited? What is the anticipated earnings rate of future investment capital?

Although the rate of return concept became very popular and widely used in the 1950's there is a noticeable trend at this time away from rate of return in favor of NPV as a measure of profitability. The principal reason cited by most decision makers is that NPV is a more realistic measure of profitability than rate of return.

This is because the reinvestment assumption is completely satisfied and all decisions are based on a single "time-value of money" system. NPV also has the distinct advantages of being meaningful for *all* types of cash flows (including those having all negative terms and those which do not reach payout) and completely compatible with risk factors (probabilities). Most firms now using NPV as a measure of profitability appear to be using discounting rates in the range of 9% to 15% for petroleum exploration investments.

Remember that our discussions thus far have been on a "no-risk" basis. If the analyst uses probabilities explicitly with NPV risk would be accounted for and the 9–15% discount rate would be used. However, some companies do not yet quantify risk explicitly, but rather, they account for risk by raising this discounting rate to some higher level, such as 25% or 30%. We will discuss this rather arbitrary approach to risk analysis at a later point. The range of 9%–15% should be interpreted as a "no-risk" future earnings rate and/or the discounting rate to use with an explicit analysis of risk and the expected value concept described in Chapter 3.

DISCOUNTED PROFIT-TO-INVESTMENT RATIO (DPR)

To sidestep the weakness of NPV being independent of the absolute size of the cash flows (item 4 in above section on net present value), it is advantageous to use the criterion called Discounted Profit-to-Investment Ratio, or DPR. This is simply the dimensionless ratio obtained by dividing NPV by the present value of the investment. The ratio is interpreted as the amount of discounted net profit generated in excess of the average opportunity rate per dollar invested. It is useful to select investment opportunities under the constraint of limited investment capital, when it is essential to gain the most profit per dollar invested.

Discounted Profit-to-Investment Ratio (DPR) calculation for example prospect

Discounted net profit = $148,400   (NPV computed previously in Table 2.3)
   (at 10%)
Investment = $268,600
Discounted profit-to-investment ratio (DPR) = $\dfrac{\$148,400}{\$268,600} = 0.553$

Characteristics of discounted profit-to-investment ratio include:

1. It has all of the advantages of NPV (such as realistic reinvestment rate, no multiple rates, not a trial-and-error solution, etc.) plus

providing a measure of profitability *per dollar invested.* This is a particularly important consideration for selecting projects from a list which contains more opportunities than the available funds can cover.

2. The DPR is interpreted as the amount of discounted NPV profit generated in excess of the average opportunity rate per dollar invested.

3. It is a suitable measure of value for ranking and comparing investment opportunities. Some writers on profitability analysis conclude that DPR is the most representative measure of true earning potential of an investment (of those we've considered thus far). It is sometimes called the "present-value index," or PVI.

4. DPR is not defined for NPV profits which are negative (on the assumption that if NPV was negative it would be considered a sub-optimal investment and rejected outright). Thus, the DPR ratio will always be positive or zero, but never negative. (This also assumes that the denominator, when defined as "investment," does not carry a negative sign.)

On final balance it should be reasonably obvious at this point that use of a single discount rate gives a more realistic measure of true profitability than does a rate of return. This implies the superiority of NPV. And since we rarely have an unlimited supply of money we must consider the importance of trying to choose capital investments that will give the maximum gain in profitability per unit of money invested. This inevitably leads to the investment strategy of maximizing the discounted profit-to-investment ratio. This sequence of logic approximately parallels the chronological development of capital budgeting criteria in the business sector over the past several decades.

OTHER MEASURES OF VALUE

There are a number of other measures of profitability that are occasionally used in petroleum investments. All of the profit indicators listed here are merely adaptations and/or modifications of the basic concepts of rate of return and net present value.

Appreciation of Equity Rate of Return

This is a modified rate of return which reflects the overall net earning power of an investment, if in fact, the cash flows are reinvested at some lower rate (such as the average opportunity rate) rather than at the true rate of return. Recall that in Table 2.2 we computed that the true rate of return of the $268,600 investment was 40%, and we stated that the cash flows had to be reinvested at 40% when received for this to be true.

Suppose, however, that the project was accepted on the basis of a 40% rate of return, but when the actual cash flows were received the firm only had reinvestment opportunities of 10%. It should be obvious that the overall profitability of the initial investment will be less – the question is what is the reduced equivalent rate of return of the initial investment assuming the cash flows were reinvested at only 10%.

The answer is the appreciation of equity rate of return, as computed in Table 2.5. This calculation is sometimes called the Baldwin Method and has also been given the term "Growth Rate of Return" in some references. The appreciation of equity rate of return will always have a numerical value between the true rate of return and the specified rate at which cash flows are actually reinvested.

It is, of course, perfectly logical to ask the following question:

> "If we anticipate a future reinvestment opportunity rate of i_0, why not express profitability in terms of net present value discounted at i_0 rate, rather than make the less meaningful calculation of rate of return or appreciation of equity?"

The probable appeal of the appreciation of equity rate of return is that it acknowledges the future reinvestment earnings rate and has the advantage of being expressed as an earnings rate which is analogous to interest rates paid by banks – a concept familiar to everyone. It can be shown that, in terms of an accept/reject criterion and/or a ranking criterion that the appreciation of equity rate of return will give the same ranking as discounted profit-to-investment ratio (DPR) using i_0 as the reinvestment discount rate.

On the other hand, some analysts and decision makers view the appreciation of equity rates as a second, possibly confusing rate as compared to the future reinvestment rate. These persons probably prefer to keep everything in terms of a single discount rate i_0 and the corresponding NPV. Undoubtedly this is one reason why the criterion is not more widely used.

It can also be shown that the computed i_{ae} is sensitive to the size and total duration of cash flows. For example if, for some strange reason, the

Table 2.5

Appreciation of Equity Rate of Return Calculation for Example Prospect

Year	Net Cash Flow	Number of Years Reinvested*	Compound Interest Factor, 10%	Appreciated Value of Net Cash Flow as of End of Project (Year 10)
1	$132,900	9.5	2.475	$328,900
2	132,900	8.5	2.247	298,600
3	97,600	7.5	2.045	199,600
4	69,200	6.5	1.859	128,600
5	23,500	5.5	1.689	39,700
6	28,600	4.5	1.536	43,900
7	15,900	3.5	1.397	22,200
8	9,400	2.5	1.269	11,900
9	5,600	1.5	1.153	6,500
10	1,900	0.5	1.049	2,000
				$1,081,900

Average future investment opportunity rate: 10%.

Net cash flows are from Figure 2.2.

* Recall that we assumed that annual cash flows were received at the midpoint of the year.

The compound interest factor, Column 4, is computed from the relation $(1 + i)^n$, where $i = 0.10$ and n is the number of years given in Column 3. For example, the compound interest factor for "year 4" is computed as $(1 + 0.1)^{6.5} = 1.859$.

The sum of Column 5, $1,081,900, is the total value of the cash flows as of the end of year 10 if each had been reinvested at 10% when received. (To state it differently, if each of the ten cash flows had been deposited in a savings account which paid 10% interest compounded annually the balance as of Year 10 would be $1,081,900.)

The appreciation of equity rate of return is the interest rate that the initial expenditure would have had to be invested at to be worth the same amount at Year 10 as the cash flows reinvested at i_0 rate. Thus, our final step is to solve the following equation:

$$\$268,600 \ (1 + i_{ae})^{10} = \$1,081,900$$

In words, this equation says that $268,600 (the original expenditure) invested at a rate equal to i_{ae} for 10 years with annual compounding will appreciate to $1,081,900. The interest rate i_{ae} is called the appreciation of equity rate of return. Solving the above equation gives a value of i_{ae} as

$$\text{Appreciation of equity rate of return} = i_{ae} = \underline{.1495}$$

Hence, the initial expenditure made the equivalent rate of return of about 15% if, in fact, all of the future cash flows were reinvested, when received, at only 10%. Compare this to the true rate of return calculated in Table 2.2 of 40%.

cash flows of Table 2.5 were extended an additional five years with receipt of $1/year in years 11–15, the magnitude of i_{ae} would be reduced, when in fact the true profitability (NPV) would be very slightly higher. To adjust for this fact some analysts use a previously determined future time horizon (such as year 10, or year 15) to compute future worth, regardless of the specific length of the projects being evaluated. Readers wishing to pursue this point in further detail should consult References 2.4 and 2.22.

    Another recent modification proposed that the rate to be used in decision making should be the appreciation of equity rate of return multiplied by two risk coefficients. Both coefficients range from 0 to 1 and are dimensionless. One coefficient accounts for possible (or probable) variation in the magnitudes of cash flows and the other coefficient accounts for uncertainties that might occur before payout is achieved. This approach is, in essence, accounting for certain risks and uncertainties via the multipler coefficients. (If there were no risks the coefficients would be 1.0 and the decision criterion would exactly equal i_{ae}. As risk increases regarding variations in cash flows and uncertainties relating to payout the coefficients would get smaller, resulting in a decision criterion earnings rate less than i_{ae}.) While there is an obvious need to consider risk and uncertainty in the decision process we will later discuss methods to incorporate risk and uncertainty into the decision criterion in a much more explicit way than the rather arbitrary, empirical nature of this modification. Reference 2.23 discusses the risk coefficients in further detail, for those interested in pursuing this approach.

Percentage Gain on Investment

    This profit indicator was proposed by Kaitz in 1967 (Reference 2.5). It is a measure of the gain an investment is expected to realize in excess of investing the same money in the average project, or at the average opportunity rate. The gain is expressed as a fixed number of dollars that can be withdrawn from each future cash revenue, with the remainder of the cash flows being sufficient to recover the initial investment plus interest at the average opportunity rate. The gain per year is computed using the relationship $(C_0)(i_{PGI})$, where C_0 is the initial investment (absolute value, not a negative term), and i_{PGI} is computed using equation 2.6.

$$i_{PGI} = \frac{1}{C_0} \left[\frac{(1+i_0)^{n'}(i_0)}{(1+i_0)^{n'} - 1} \right] (NPV) \qquad (2.6)$$

where C_0 = initial investment (absolute value, not a negative term)
   i_0 = average reinvestment opportunity rate, decimal fraction
   n' = total duration of cash flows, years
   NPV = Net present value at i_0

For our example drilling prospect

   C_0 = $268,600, the completed well cost
   i_0 = 0.10, the anticipated future reinvestment opportunity rate
   n' = 10 years, the total duration of cash flows
   NPV = $148,400, the net present value of the project discounted at i_0, Table 2.3

Thus,

$$i_{PGI} = \frac{1}{\$268,600} \left[\frac{(1+0.10)^{10}(0.10)}{(1+0.10)^{10} - 1} \right] (\$148,400)$$

$$i_{PGI} = 0.09$$

Finally, the gain per year in excess of making an earnings rate of i_0 is computed as

$$\text{Gain per year} = (C_0)(i_{PGI})$$
$$= \$268,600 \times 0.09 = \underline{\$24,174}$$

The percentage gain on investment concept is based on the premise that the gain (computed as $(C_0)(i_{PGI})$) can be subtracted from each annual cash flow and termed a profit in excess of the i_0 earnings rate. The balance of each flow, when reinvested at i_0 will return the initial investment plus interest at i_0 per cent. In summary, PGI is a measure of the level of profit in excess of rate of return equal to i_0, and is expressed as the amount of money which can be subtracted from each annual cash flow as "excess profit." The magnitude of this incremental cash gain would be used by the decision maker to choose from among competing investments, his criterion being to maximize this cash gain. The "excess profit" term is a constant amount for each year of the project.

The PGI concept is really just a modification of net present value, because it can be shown that the present value of receiving $24,174 each year for 10 years, discounted at 10% is identically the NPV calculated previously of $148,400. So we could define PGI in the reverse manner by saying it is the value computed by dividing the NPV into a series of equal annual payments to be received over the life of the project. (Note· the reader should prove to himself that the above statement of the relation between NPV and PGI is correct, as well as the fact that the remainder of each annual cash flow will result in a rate of return of exactly 10% for the example prospect.)

A final comment about PGI. We said it was the amount of money we could withdraw from each cash flow such that the balance, when reinvested at i_0 will return the investment plus interest. There is a separate calculation of the same type called Hoskold's Method which defines the amount which could be withdrawn each year such that the balance, when reinvested at i_0 would recover only the investment (no interest). This calculation is seldom used in exploration prospect analysis because it obviously overlooks the average project opportunity rate of earnings available to the firm.

"RISK-WEIGHTED" RATES OF RETURN

Initially we stated that all the measures of value discussed in this chapter were "no-risk" profitability criteria. That is, it was assumed that all of the projected cash flows and expenditures were guaranteed to occur at the exact times specified. We also stated under the characteristics of rate of return that there is no way to explicitly include statements of risk (probability numbers) in the rate of return equation. There is, however, a method to get an approximation of the effects of risk in a rate of return number by increasing the initial investment to include monetary losses for the situation when no cash revenues are received from the project.

This method can only be used for the special case where there are only two possible outcomes: (a) the investment results in generation of the entire and exact cash flows as originally projected with probability p or, (b) a cash expenditure results in no revenue with probability $(1 - p)$ of occurrence. The real-world situation of interest here would be the drilling of an exploratory or development well in which the projected series of cash flows is obtained or else the well is dry and abandoned with no future revenue.

To illustrate this concept, let us consider again our example drilling prospect. We have previously computed a rate of return of 40% for the completed well cost of $268,600. Now let us add to the analysis the risk assessment that *"we have a probability of 0.5 of the well being successful*

Table 2.6
"Risk-Weighted" Rate of Return Computation
DECISION METHODS UNIT, WELL NO. 1 Prospect
(Final Two Trials in the Trial-and-Error Computation)

Year	Net Cash Flows*	Discount Factors for 5%	5% Discounted Cash Flow	Discount Factor for 4%	4% Discounted Cash Flow
0	−$468,600	1.000	−$468,600	1.000	−$468,600
1	+ 132,900	.976	+ 129,700	.981	+ 130,400
2	+ 132,900	.929	+ 123,500	.943	+ 125,300
3	+ 97,600	.885	+ 86,400	.907	+ 88,500
4	+ 69,200	.843	+ 58,300	.872	+ 60,300
5	+ 23,500	.803	+ 18,900	.838	+ 19,700
6	+ 28,600	.765	+ 21,900	.806	+ 23,100
7	+ 15,900	.728	+ 11,600	.775	+ 12,300
8	+ 9,400	.694	+ 6,500	.745	+ 7,000
9	+ 5,600	.661	+ 3,700	.717	+ 4,000
10	+ 1,900	.629	+ 1,200	.689	+ 1,300
			− $6,900		+ $3,300

The "risk-weighted" rate of return is between 4% and 5%.

* From Figure 2.2, except that the time-zero investment is based on probability 0.5 of successful completion costing $268,600 and a probability $(1 - 0.5)$ of a dry hole costing $200,000.

The discount factors are from Appendix A (annual compounding, mid-year cash flows).

and generating the 10 year series of future cash revenues, and a probability of $(1 - 0.5) = 0.5$ of a dry hole costing $200,000."[1]

Speaking in very general terms this statement implies that over a series of such wells roughly one half of them would be successful completions and one half of them would be dry holes. Or, stated differently, it is reasonable to expect that over a series of such wells we would anticipate drilling one dry hole for each of the successful wells drilled. Thus, in effect, our total expected investment to secure the cash revenues will be $268,600 plus $200,000 (the dry hole cost), or $468,600. This total is then entered as the time-zero investment in a standard rate of return computation. The final trials for this example are given in Table 2.6.

1. For the moment the reader should not be concerned with the important issue of how does the analyst make such an assessment of risk, or specific definitions or interpretations of probability numbers. These topics will be discussed in much detail in subsequent chapters.

By adding the dry hole cost to the investment we see from Table 2.6 that the rate of return for our example prospect has dropped from 40% to slightly over 4%! Stated differently, if the 10 year schedule of revenues must stand an additional $200,000 of expenditures at time zero the earning power of the total investment is equivalent to only 4% per year compounded annually.

We now formalize the above discussion and numerical example as they relate to the "risk-weighted" rate of return concept. Let p = probability that the future cash flow will occur (decimal fraction), C_0 = investment at time zero that generates the future cash flow, $(1 - p)$ = probability of no future cash flow (decimal fraction), and C_D = investment loss if no future cash flow is generated. Then the "pseudo" investment as of time zero for the "risk-weighted" rate of return computation is given by equation 2.7.

$$\begin{array}{c}\text{Pseudo investment at time zero for} \\ \text{"risk-weighted" rate of return}\end{array} = C_0 + \left[\frac{1-p}{p}\right]C_D \qquad (2.7)$$

This equation is applicable for any value of p. In the context of drilling decisions C_0 is usually taken as the producing well cost, C_D as the dry hole cost, and p as the probability well will produce oil or gas.

This approach for computing rates of return for drilling prospects seems to have considerable appeal to the decision makers in some oil companies. The concept enables the decision maker to see the effects of what could be loosely called "the dry hole risks" while still retaining the appeal of expressing profitability as an annual earnings rate. The reader should be aware, however, of several important weaknesses of the concept:

A. The approach only considers two possibilities—the predicted future cash flow or zero cash return. Additional possible outcomes (such as a future cash flow reduced by one-half, or increased by 28%, etc.) cannot be considered. It's an "either-or" case and we know the petroleum exploration business just isn't that simple.

B. If the "pseudo investment" computed by equation 2.7 exceeds the undiscounted sum of the future net revenues the "risk-weighted" rate of return cannot be computed. (See item 5-c under characteristics of rate of return, page 27). For example, in our example prospect, the total undiscounted revenue from a successful completion was given as the sum of Column 6 in

Figure 2.2, $517,500. If the pseudo investment exceeds this value we would not have been able to compute a "risk-weighted" rate of return.

In our example, this circumstance would have occurred for all values of p less than about 0.45, (assuming, of course, that $C_0 = \$268,600$ and $C_D = \$200,000$). If the analyst had felt there was a probability of hitting oil of only 0.25 we would have been unable to use this criterion. (The reader should try the computation for p = .25 if he is still not convinced. Remember, there is no such thing as a negative rate of return!)

The obvious weakness here is that with the "risk-weighted" rate of return the decision maker is using a value system that can be computed and defined in some cases but is undefined (and hence a meaningless value criterion) in other instances.

C. It is *not* correct to say that, since the analyst estimated that only one-half of the time would the well be successful, we could compute the "risk-weighted" rate of return as $\frac{1}{2} \times 40\% = 20\%$, or $0.5 \times 40\% = 20\%$. *Such a computation is absolutely incorrect!* In general it is not permissable to multiply a rate of return by probability numbers.

The approach is certainly a step in the right direction because the analyst and decision maker are now at least considering the element of risk. But, for the reasons listed above it is, at best, an approximation and at worst, a criterion which cannot be computed. We will show in later chapters that there are, in fact, much more realistic ways to incorporate the element of risk into our decision parameters.

GENERAL COMMENTS ABOUT MEASURES OF PROFITABILITY

Our discussions so far regarding measures of profitability have been relatively brief, but hopefully you will now have an idea of the principal aspects of profitability criteria in common use today for prospect analysis. Before closing this discussion we list here several miscellaneous and general comments relating to profitability criteria.

1. We have discussed the most common measures of profitability in use today for petroleum exploration investments. There is probably no single method or calculation that completely describes all of the dimensions of profitability. There seems to

be a growing awareness, however, of the importance of the reinvestment assumption, resulting in a trend towards use of Net
Present Value discounted at i_0 as the most realistic profit indicator. Or, if the decision maker has a limited capital constraint
there seems to be increasing use of the Discounted Profit-to-
Investment Ratio (DPR) as the most realistic measure of project
profitability.

2. A useful method to display profitabilities of a set of decision
   options at various discount rates is a graph called a "Net Present
   Value Profile Curve."

   When interpreting Figure 2.6 remember that present value
   at a zero discount rate is simply total undiscounted net profit.
   Thus in this example, Project A will generate the most total
   profit. The discount rate at which the net present value is
   zero is by definition, the rate of return. Thus, while Project A
   has the highest total profit, it has the lowest rate of return, 24%,
   and Project C has the highest rate of return. If the firm's time
   value of money is 10%, Project A would be preferred (NPV =

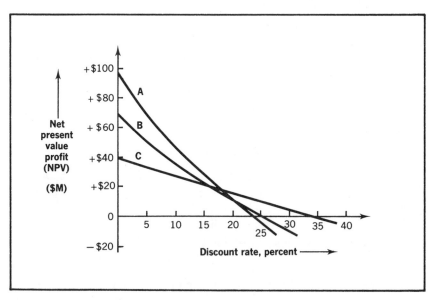

Figure 2.6 *Net present value profile curve.*

$48,000, versus $35,000 for Project B and $24,000 for Project C). The likely reason that Project C has the highest rate of return but the lowest total undiscounted profit is that it probably recovered the future cash flows at a very rapid rate.

This is a good example to show that projects are not necessarily ranked in the same order of preference using NPV as the ranking on the basis of rate of return. A Net Present Value Profile Curve does not resolve the issue of determining the "best" profit criterion (i.e. NPV at i_0 vs. rate of return), but it is frequently a useful pictorial representation of the effects of discount rate on the cash flow streams of competing investment projects.

3. All of the profitability computations described thus far are on a "no-risk" basis — with the anticipated cash flows occurring with certainty. The single exception to this was the "risk-weighted" rate of return computation. All of the criteria described can be based on "before-tax" cash flows or "after-tax" cash flows.

4. The measures of profitability based on some form of discounting give little or no value to revenues received after about 20 years. However, the petroleum industry seems to be moving in the direction of very long, extended types of cash flows (order of magnitude of 20–40 years). There is a very basic question as to whether discounted measures of value even adequately portray profitability of these investments.

To point out the problem, what is the value to you today of receiving $1 thirty years from now? Or to state it differently, how much would you be willing to pay today for the right to receive $1 thirty years from now? Tough question isn't it! Trying to find measures that realistically reflect profitability for extended cash flows is one of the open questions in decision making. We will discuss this point in further detail in Chapter 11.

5. In recent years the words "energy shortage" have nearly become household phrases to the general public. Of course the petroleum exploration decision maker has long been aware that he is dealing in a finite, depletable supply of oil fields. And as more and more oil fields are found, he is cognizant that the remaining number of undrilled fields is declining. On the other hand he is

also aware of the constantly expanding demand for oil and gas as a source of energy throughout the world.

These constraints have led some exploration decision makers to redefine their value system for decision making to one of trying to minimize their "replacement costs per barrel." This is in lieu of the traditional profitability criteria of trying to maximize rate of return, NPV profit, or discounted profit-to-investment ratio.

Thus, in deciding whether to drill prospect A or B he would choose the prospect in which the replacement costs (finding and development costs) are lowest. He is in effect saying that during the next x years our company expects to produce y barrels of oil, and in deciding where to explore for oil we should select areas where we can replace these y barrels at a minimum per-barrel cost.

The specific calculations for replacement costs involve discounting, cash flow projections, etc. as described in this chapter—the only difference being that the decision maker's objective is stated in a different manner. In the coming years the criterion of trying to minimize replacement costs per barrel will probably gain wider use in worldwide exploration activities.

6. With any type of discounted measure of value (rate of return, net present value, etc.) it is implicit that unlimited reinvestment opportunities will exist at all future times during the projected cash flows. If there are only *limited* investment opportunities most of the measures discussed become unrealistic. (Perhaps this is part of the dilemma of the petroleum industry today. See Chapter 11 for a further discussion of this point.)

7. Some analysts prefer using discounting formulas that are based on two separate discount rates—one for discounting revenues and one for discounting expenditures. These computations will typically discount all future investments (expenditures) of the project back to time zero using a low, "safe" discount rate of, say 8–12% to get a total present value expenditure.

This sum is then placed in the rate of return computation as the equivalent time-zero expenditure. Thus, the revenues will be discounted back to time zero using the rate of return discount rate which in nearly every case will be different than the low, "safe" time-value of money assumed for expenditures. This type of calculation violates our basic premises about the time value of money to a firm, and calculations of this type are not used widely for prospect analysis.

8. Another type of value criterion is the so called "book rates of return." Generally speaking, measures of value based on accounting methods (book rate of return, average annual rate of return) are not realistic measures of profitability. Such calculations are based on arbitrary schedules of depletion, depreciation, tax factors, etc. which usually do not relate directly to the true profit value of a given investment. Accounting books and tax books are not intended to be a direct measure of project profitability and usually are not used for investment decision making.

9. Although there is no explicit relationship between rate of return and payout time, they are somewhat correlative. If large cash flows are received early in the life of the project the payout time will be short and, all other factors equal, the rate of return high. For long extended cash flows payout time is usually longer and rate of return lower. It is possible to develop empirical correlations of payout versus rate of return for areas where the well costs and producing schedules are similar. Such a correlation can be used to approximate a value of rate of return based on a simple calculation of payout—without having to pursue the trial-and-error calculations of a rate of return computation. See Reference 2.6 for examples of such short-cut correlations.

10. In our discussion of the characteristics of the rate of return concept we mentioned that the computations included the implicit assumption that all cash flows are reinvested immediately upon receipt at an earnings rate equal to the rate of return. This assumption, in fact, applies to all discounting (NPV at i_0, etc.) but becomes especially critical as the rate of return increases. This is one of the principal reasons why rate of return, as a measure of true profitability, is falling into disfavor with many decision makers.

To amplify on this point, we give here a proof of the existence of the reinvestment assumption implicit with rate of return. Suppose we consider the following, simple investment project:

Initial Investment: $741,200
Future Series of Cash Flows Generated
  by the Investment:

Year	Year End Cash Flow
1	$500,000
2	$400,000
3	$300,000
4	$200,000
5	$100,000

Assume Annual Compounding

Assuming our firm's value system is based on rate of return the next step would be to compute the rate of return associated with investing $741,200 to secure the five year series of cash revenues. We'll leave the calculations to the interested reader and state only the answer—40%. Thus, by virtue of the meaning of rate of return, we can say that the investment of $741,200 to buy the five year series of cash revenues has an equivalent earnings rate of 40% per year compounded annually.

If the manager had the choice of investing in this project or any other alternative investment project yielding 40% per year he would presumably be indifferent regarding his available choices.

Following this line of reasoning, suppose he invested the $741,200 in some alternative project which yielded 40% per year compounded annually. His future net worth at, say, year 5 could be computed using the compound interest equation given earlier as equation 2.1:

$$C(1 + i)^n = S$$
$$C = \$741,200$$
$$i = 0.40 \qquad S = ?$$
$$n = 5 \text{ years}$$
$$S = (\$741,200)(1 + 0.40)^5 \cong \$3,986,400$$

His future net worth at year 5 from his investment of $741,200 would be $3,986,400.

Now we raise the "devil's advocate" question: "If he invested $741,200 to buy the future series of cash flows what would his future net worth be at the end of year 5?" The answer is that his net worth will be exactly the same *if, and only if,* he reinvests the 5 annual cash flows, when received, at exactly 40%, the rate of return. The proof is displayed pictorially in Figure 2.7 and computed in Table 2.7.

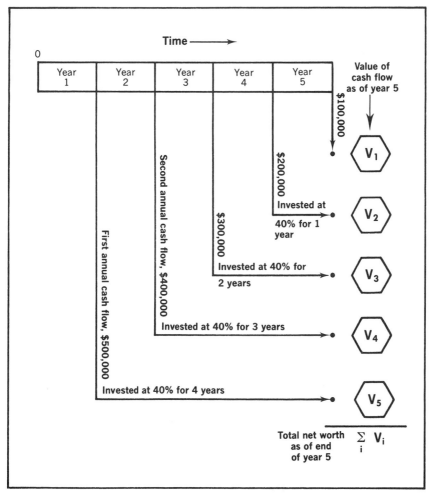

Figure 2.7 *Pictorial representation of reinvestment of the five annual cash flows to determine total net value of these cash flows as of the end of year 5. The appreciated values, V_i, of each cash flow as of the end of year 5 are computed in Table 2.7.*

From the computations of Table 2.7 we see that the total net worth as of year 5 from the five future cash flows is identical to the net worth resulting from investment of the $741,200 in an alternative project yielding 40% per year. This equivalence is true, however, only if the future cash flows are invested at the rate of return earnings rate as soon as they are received. If the cash flows are reinvested at any other rate (higher or lower) the initial investment will *not* have an equivalent earning rate (profitability) of 40%.

Table 2.7
Computation of Total Net Worth As of End of Year 5 of Example Cash Flow

Year	Year End Cash Flow	Number of Years Reinvested	Compound Interest Factor	Compounded or Appreciated Value of Cash Flow As of End of Yr. 5
1	$500,000	4	$(1 + 0.40)^4$	$1,920,800 = V_5$
2	$400,000	3	$(1 + 0.40)^3$	$1,097,600 = V_4$
3	$300,000	2	$(1 + 0.40)^2$	$588,000 = V_3$
4	$200,000	1	$(1 + 0.40)^1$	$280,000 = V_2$
5	$100,000	0	1	$100,000 = V_1$
				$3,986,400 = \Sigma V_i$

The reader should note two important points in the above proof. The common point in time to compare future net worth was chosen here as the end of year 5. But any future point in time could have been used to make the comparison. Also, note that the investment amount ($741,200) does not appear in the computations of Table 2.7. The reason is that this money was invested *in exchange* for receipt of the five future cash flows.

In drilling prospect analysis the initial investment would be in the drilling contractor's pocket — and any future earnings must be generated from reinvestment of future revenues from the well. This proof should also emphasize the statement given at the beginning of the chapter that the value of money received from a special project being considered is determined by the rate at which the revenues from the project can be reinvested to provide appreciation of the firm's total assets.

11. Finally, we should give emphasis to another important consideration in profitability analysis—the source of investment capital. Thus far we have not mentioned where the investment funds will come from, the cost of these funds, or how the funds will be repaid. We more or less implied that the funds would be available for the drilling prospects being considered. Further, our profitability computations did not consider the cost of these funds.

This practice is wide spread in prospect analysis—if for no other reason than the fact that the decision-making function (evaluating prospect worth, etc.) is normally separate and distinct from the financing function of the firm. In recent years, however, petroleum exploration investments in offshore and/or remote areas have involved extremely large capital outlays early in the life of the project, and sources of funds, as well as costs of funds have become an increasingly important consideration.

Under these conditions the decision criterion and profitability computations must include financing charges. In general, these are called "leveraged transactions," and the financial considerations get a bit more complex than the intent of this discussion on measures of profitability. Those wishing to pursue the topic of leveraged transactions would do well to consult reference 2.18.

ANALYSIS OF RATE ACCELERATION INVESTMENTS

A special type of investment in the petroleum industry is one intended to accelerate the rate at which a future schedule of revenues is received. These investments provide a somewhat different problem in capital budgeting analysis. The general approach is to evaluate the present worth of the existing cash flow schedule. This number is then compared to the present worth of the accelerated cash flow schedule, less the investment required to obtain the acceleration. If the latter net present worth exceeds the present worth of the original cash flow, the project becomes a candidate for consideration.

Examples of such investments include infill drilling in a field to accelerate depletion of an oil and gas reservoir, and installation of large volume lift equipment in producing oil wells. Such investments may increase ultimate recovery (equity) but this is not necessary for the project to be classified as a rate acceleration investment.

Consider a simple example:

Year	Present Future Cash Flow
1	$300
2	200
3	200
4	100
5	100
	$900

This cash flow already exists, and no future investment is required to assure its receipt. Now suppose that if we spent $50 we could receive the entire $900 in just two years; $500 the first year and $400 in the second year. We will assume mid-year cash flows, annual compounding, and an average opportunity rate of 10% per year. Should we spend the $50 to accelerate the project or not?

Solution:

1. First we find the net present value of the existing cash flow schedule.

Year	Present Cash Flow	10% Discount Factor*	10% Discounted Cash Flow
1	$300	0.953	$285.90
2	200	0.867	173.40
3	200	0.788	157.60
4	100	0.716	71.60
5	100	0.651	65.10
	$900		$753.60

* From Appendix A.

2. Next we find the present value of the accelerated cash flow:

Year	Accelerated Cash Flow	10% Discount Factor	10% Discounted Cash Flow
1	$500	0.953	$476.50
2	400	0.867	346.80
			$823.30

The *net* present worth of the accelerated cash flow is $823.30 less the $50.00 required to obtain the earlier cash flows, or $773.30.

3. Compute the net increase in present worth by investing in the project:

$$\$773.30 - \$753.60 = \underline{\$19.70}$$

Thus it *appears* that the acceleration investment is attractive and should be considered, together with other current investment opportunities.

The above steps can be combined into a single calculation using the technique of "incremental analysis."

Rate Acceleration Investment Analysis Using Incremental Analysis

Year	Present Cash Flow	Accelerated Cash Flow	Incremental Cash Flow*	10% Discount Factor	10% Discounted C.F. (Incremental)
0	0	−$50	−$50	1.000	−$ 50.00
1	$300	500	+200	.953	+ 190.60
2	200	400	+200	.867	+ 173.40
3	200	0	−200	.788	− 157.60
4	100	0	−100	.716	− 71.60
5	100	0	−100	.651	− 65.10
					+ $19.70

* (The plus sign means the accelerated cash flow exceeds the present cash flow in a given year.)

Notes on above incremental calculation:

a. Notice we get the same answer that we obtained previously by computing present values of each separate cash flow schedule. The positive sign in the incremental analysis was arbitrarily set as corresponding to an advantage of the acceleration. (If we had selected as positive those terms which were favorable to the present cash flow, the answer would have been of opposite sign, −$19.70, indicating that the accelerated cash flow is more desirable.)

b. Notice the sign changes in the incremental cash flow column.

When later cash flows become negative be alert to the possibility of multiple roots if you are attempting to calculate a true rate of return of the incremental cash flow. Perhaps the classic example of the trap you can get into in rate acceleration analyses:

> "A well is expected to produce $10,000/year for the next two years. What could we afford to pay for larger lift equipment to recover the entire $20,000 in just one year? By calculating a true rate of return on the incremental cash flow it can be shown that if we spend $827 for such a pump the rate of return is 10%. If we spend $1600 for it, the rate of return is 25% or 400% depending on how one evaluates which multiple rate of return is appropriate, and if we spend $2500 the rate of return is 100%! The more we spend the higher the rate of return!"

By discounting the incremental cash flows at a specified discount rate we of course eliminate the chance of multiple roots and/or absurd answers.

c. Should we take the decision option of acceleration or not? By the above calculations it appears desirable because the firm's overall present worth is increased by $19.70, even after spending the $50. The problem, though, is one of trying to balance acceleration investments with investments which add new equity to the firm, such as oil reserves. Obviously, if the firm accepted all of the rate acceleration projects having a positive present value it is conceivable that no additional capital would be available to replace reserves. The result would be a firm going out of business at an accelerated rate!

Unfortunately we do not have totally satisfactory ways of trying to strike such a balance. Some companies set a "minimum," such that the increase in present worth *per dollar invested* must exceed a specified number, such as 1.0. The problem in realistic implementation of such an approach is determining how to set the minimum.

d. Having computed the incremental cash flow column, virtually any other profit criterion can be utilized to evaluate the incremental worth from the acceleration investment. We used net present value above, and mentioned that we could have used rate of return. But we could also calculate a PGI value and an appreciation of equity rate of return. All measures would be different and suggest different degrees of desirability.

The ultimate question that must be answered is — "What could we do with the money received earlier and at what rate could it be reinvested?" The net present value calculation above answers the question, assuming the firm has a bountiful supply of investment opportunities which will yield a rate of return equal to i_0. The intangible part of the decision is whether accepting this acceleration investment will preclude accepting available investments which would increase the firm's equity.

3 The Expected Value Concept

. . . we ought always in the conduct of life to make the product of the benefit hoped for, by its probability, at least equal to the similar product relative to the loss.
— Laplace, 1814

THE measures of investment worth considered in the previous chapter were all "no-risk" parameters. That is, the criteria did not include explicit statements about the degree of risk and uncertainty associated with a given investment or drilling prospect. We are all aware, however, that petroleum exploration involves a great deal of risk and uncertainty. In fact, the element of risk is frequently the most critical factor in decisions to invest capital in exploration. In reality each time he decides to drill a well the decision maker is playing a game of chance in which he has no assurance that he will win.

When faced with decision choices under uncertainty the manager has the option of making an informal analysis of the prospect, or he can, in some manner try to consider the element of risk and uncertainty in a logical, quantitative manner. The informal approach utilizes the decision maker's intuition, judgment, hunches, experience (and luck?) to determine which prospects to drill. This subjective approach to decision making was an "art" which sufficed for many managers in earlier days of petroleum exploration when most wells were shallow and the drilling anomalies were numerous and easy to locate.

But as we pointed out in Chapter 1 the worldwide game of petroleum exploration has changed, and many managers now get a little uncomfortable basing their decisions completely on the informal approach to decision making.

The discipline of Decision Analysis considers the element of risk and uncertainty in a quantitative manner and provides a means to incorporate the dimension of risk into a logical and consistent decision strategy under conditions of uncertainty.

The cornerstone of Decision Analysis is the expected value concept—a method for combining profitability estimates with quantitative estimates of risk to yield a risk-adjusted decision criterion. The concept is not a substitute for managerial judgment, but rather a method of analysis whereby the various consequences of each decision option can be evaluated and compared. Virtually all formal strategies for decision making under uncertainty rest on the expected value concept.

In this chapter we will study the expected value concept and its logic, meaning, and application to drilling prospect decisions. We will attempt to show that the expected value concept is a remarkably versatile and simple approach that will provide an important summary parameter to the manager. But, as might be expected, implementation of a new concept such as this is not without its share of frustrations and misunderstandings, and we will discuss these misunderstandings as well.

Comments about Risk and Uncertainty

Before we begin we should pause to amplify a bit on risk. We will consider the words "risk" and "uncertainty" to be synonymous. We will use the words interchangeably to reflect the fact that, at the time of decision making we are not able to predict or measure specific values for all of the parameters contributing to overall prospect profitability. For example, we usually cannot measure exact pay thickness values, or porosities, or recovery factors before the well is drilled.

Similarly, we do not know exact drilling costs or the exact amount of reserves (if any) that the well will produce. Uncertainties about future events (crude sales price fluctuations, future tax and governmental policies, storm damage, etc.) also contribute to the overall risk of the drilling prospect. These uncertainties taken in combination imply that there can be more than one outcome to any decision alternative.

A decision to drill could result in a dry hole, or a marginal discovery, or a giant discovery. Decision making under uncertainty always involves at least two possible outcomes for each decision alternative. Each outcome has some likelihood of occurrence, but none is certain to occur.

By comparison, if we knew exact values for all the parameters which affect overall profitability we would be able to compute an exact value of project profitability. Such a calculation would be called a "deterministic" value of profitability. If we do not know exact values for each of the parameters (as is the case in exploration prospect analysis) the computation of profitability is said to be "stochastic."

If the consequences of all possible decision alternatives could be computed exactly the process of decision making would be much simpler. It is the unknown resulting from our inability to measure or predict values of the profit parameters before the well is drilled that makes the decision making process complicated.

Quantitative statements about risk and uncertainty are given as numerical probabilities, or likelihoods of occurrence. Probabilities are decimal fractions in the interval zero to 1.0. An event or outcome which is certain to occur has a probability of occurrence of 1.0. As the probabilities approach zero the events become increasingly less likely to occur. An event that cannot occur has a probability of occurrence of zero. The probability of a head on the flip of a fair coin is 0.5.

Assigning probabilities to the various outcomes of a drilling venture is generally called risk analysis and we will discuss this important topic in much detail beginning in Chapter 7. For the moment we will assume that reasonably accurate estimates of these probabilities can be obtained, and focus our attention on the concept which unites risk and economic factors into a meaningful decision criterion — the expected value concept.

As a matter of historical interest, the expected value concept is thought to have originated in 1654. At that time mathematicians began studying "the caprices of fate" in some of the popular gambling games of dice. An exchange of letters between the French mathematicians Fermat and Pascal in 1654 apparently set down the first formal statement of the theory of expectations — and in fact, the theory of probability itself.[1]

Many famous names appear in the developing history of the new theory: Leibnitz, Huygens, Bernoulli, D'alembert, Bayes, Laplace, Gauss, Poincare and Keynes, to mention just a few. The concept had various names in its early history — Laplace called it "mathematical hope," perhaps an appropriate name for its use in oil exploration.

Although the concept is over 300 years old, its use in the context of business decisions under uncertainty did not receive much emphasis until the 1950's. And it has only been in the last ten years or so that

---

1. Reference 1.7 has an interesting discussion on this historical exchange of letters, as well as a description of the dice games which prompted the letters. About 100 years prior to this time a controversial Italian mathematician, physician, lecturer, and ardent gambler, Gerolamo Cardano (1501–1576) developed much of the basic logic of probability theory. His book relating his discoveries was not published until 1663 and, strangely, received little attention in the rapid development of probability thereafter. Reference 3.9 contains a complete translation of his now famous *Liber de Ludo Aleae* (The Book on Games of Chance).

there has been much discussion of its use in petroleum exploration decision making. So it is probably safe to generalize that use of the expected value concept in petroleum exploration is a fairly contemporary idea. Other business sectors have been using the concept for some time (insurance business, gambling houses, etc.) but its application in drilling prospect analysis seems to represent a "radical departure" from traditional ways for many managers.

We mention this point to warn the reader that understanding the logic and use of expected value is only one-half of the battle. The other half is associated with the education of others who must work with expected values and the striking down of many misconceptions and misunderstandings which they have. The concept is rather simple to describe and use but it is not always clearly understood by those preparing or deciding on drilling prospects. More about this at a later point. First, let's define a few terms and see how the expected value concept works.

DEFINITIONS AND EXPECTED VALUE COMPUTATIONS

A *decision alternative* is an option or choice available to the decision maker. For example, his decision alternatives in petroleum exploration may include drilling the well, farming out the drilling rights, participation in the well with a reduced working interest, dry hole contributions, or various types of back-in privileges. Each decision alternative will have at least two possible *outcomes.*

An outcome can be thought of as a "state of nature," or an "event that could occur if a given decision alternative is accepted." If a decision maker selects the alternative *drill,* the outcomes might include, for example, a dry hole, or a completed well with 500,000 barrels reserves, or a completion having reserves of 1,000,000 barrels, etc.

Two definitions are important to the understanding of the expected value concept.

1. EXPECTED VALUE OF AN OUTCOME: The product obtained by multiplying the probability of occurrence of the outcome and the conditional value (or worth) that is received if the outcome occurs.

The value received if an outcome occurs can be expressed in various ways: monetary profits and losses, opportunity losses, preference or utility values based on associated profits and losses, etc. For the special case where the values received are expressed as monetary profits or losses the product is usually called *expected monetary value* of the outcome, or *EMV*. Most of our subsequent discussion of the expected value concept in this chapter will be for this special, but none the less common case of monetary values. For this special case monetary losses are usually expressed as negative profits.

The word "conditional" means that the value will be received only if that particular outcome occurs. That is, the value received is "conditional" upon the occurrence of the outcome. Many articles omit the word "conditional," and for simple decisions under uncertainty this causes no problem. In the analysis of complex decision alternatives it will become necessary to retain the adjective so as to be able to distinguish conditional profits and losses from profits and losses which are not conditional to a specified outcome.

The words "outcome" and "event" are interchangeable in the above definition. Since probabilities are dimensionless numbers, the product obtained by multiplying a probability by a conditional value received, will have the same units as the conditional value. That is, if we multiply the probability of a dry hole by the monetary loss from a dry hole, the product or "expected value of outcome dry hole" will have the same monetary units.

>   2. EXPECTED VALUE OF A DECISION ALTERNATIVE: The algebraic sum of the expected values of each possible outcome that could occur if the decision alternative is accepted.

The expected value of a decision alternative can be positive, zero, or negative. We will see in a moment that the expected value of a decision alternative is the numerical criterion used to compare competing decision choices available to the decision maker.

At this point perhaps it would be well to illustrate these important definitions with a few numerical examples. Suppose you are offered a wager (gamble) which consists of flipping a fair coin and the rewards of winning $2 if the flip results in a head and losing $1 if the flip results in a tail. Should you accept the wager?

This is certainly a decision under uncertainty—because if you accept the wager you have no assurance of winning $2. And it shouldn't take too much imagination to identify the decision alternatives: a) accept the gamble, or b) reject the gamble. If you accept the gamble the possible outcomes are that the coin flip will result in either a head or a tail. Having defined the problem in this way we could compute the expected value of the outcome "head" as shown on the following page:

$$0.50 \quad \times \quad (+\$2.00) \quad = \quad (+\$1.00)$$

$$\left(\begin{array}{c} \text{Probability} \\ \text{that outcome} \\ \text{will occur} \end{array}\right) \times \left(\begin{array}{c} \text{Conditional} \\ \text{value if} \\ \text{outcome occurs} \end{array}\right) = \left(\begin{array}{c} \text{Expected} \\ \text{value of} \\ \text{outcome "head"} \end{array}\right)$$

Similarly, the expected value of the outcome "tail" is computed as:

$$0.50 \times (-\$1.00) = (-\$0.50)$$

Expected
value of
outcome
"tail"

Notice that we
must keep track
of positive and
negative signs!

Finally, we could compute the expected value of the decision alternative "accept wager" by finding the algebraic sum of the expected values of all possible outcomes.

$$(+\$1.00) \quad + \quad (-\$0.50) \quad = \quad \underline{+\$0.50}$$

$$\left(\begin{array}{c} \text{Expected value} \\ \text{of outcome} \\ \text{"head"} \end{array}\right) + \left(\begin{array}{c} \text{Expected value} \\ \text{of outcome} \\ \text{"tail"} \end{array}\right) = \left(\begin{array}{c} \text{Expected value of} \\ \text{decision alternative} \\ \text{"accept wager"} \end{array}\right)$$

Pretty simple isn't it!

It is usually convenient to set up a table to calculate expected values of decision alternatives. For this illustration the table would appear as follows:

Outcome	Probability That Outcome Will Occur	Decision Alternative "Accept the Wager"	
		Conditional Value Received	Expected Value of Outcome
Head	0.50	+$2.00	+$1.00 ←— Col. 2 × Col. 3
Tail	0.50	−$1.00	− 0.50 ←
			+$0.50

Algebraic sum of fourth column: the expected value of the decision alternative of accepting the wager.

If you were to pursue the decision further you would compute a similar expected value for the alternative "reject the wager." In this little problem it would be zero because there would be no profit or loss if you rejected the gamble—your total cash position would be the same as if the gamble had never been offered. Note at this point we haven't resolved whether it is wise to accept the gamble, we've just computed the expected values for each of your decision choices. We will discuss the decision rules in just a moment.

Here is another simple numerical decision involving uncertainty:

We are evaluating a drilling prospect in which it is estimated that the probability of a successful well is 0.6, and the probability of a dry hole is 0.4. We have two decision alternatives: drill the well or farm the prospect out and retain an override. If we drill a dry hole our net loss will be $200,-000; if it hits our net profit (after drilling costs) will be $600,000. If we farm it out our income from a producer is $50,000. Of course there is no profit or loss to our overriding interest if the well is dry. What should we do?

Clearly this is a decision under uncertainty. There are two possible outcomes for each decision alternative, and neither outcome is certain to occur. We could probably all agree that if the probability of a completion is extremely remote the choice would be obvious—farm out rather

Table 3.1

Computation of Expected Monetary Values for the "Drill" and "Farm Out" Alternatives of Example Drilling Prospect

| Outcome | Probability Outcome Will Occur | DECISION ALTERNATIVES | | | |
| | | Drill | | Farm Out | |
		Conditional Monetary Value	Expected Monetary Value of Outcome	Conditional Monetary Value	Expected Monetary Value of Outcome
Dry Hole	0.4	−$200,000	−$80,000	0	0
Producer	0.6	+$600,000	+$360,000	+$50,000	+$30,000
			+$280,000		+$30,000
		EMV (drill)		EMV (farm out)	

than risk a $200,000 dry hole loss. Conversely, if it was highly probable that the well would be productive the choice would again be fairly obvious — drill. But what about situations between these obvious extremes?

Is a probability of 0.6 of a successful completion high enough to justify the gamble of drilling? What if the probability of production is only 0.3 — should we drill or farm out? How high does the probability of a dry hole have to be to make the farm out option more attractive? Well, this is where the expected value concept can help us to determine the "best" alternative under the conditions of uncertainty.

The specific computations for this example are given in Table 3.1.

Now that we know how to calculate expected values of outcomes and expected values of decision alternatives — what do we do with the numbers? What do they mean? How are they used to make a decision? The decision rule for the usual case where the conditional values received are expressed as monetary profits and losses is given in the following box.[1]

DECISION RULE FOR EXPECTED MONETARY VALUE CHOICES:
When choosing among several mutually exclusive decision alternatives select the alternative having the highest positive expected monetary value (EMV).

In this decision rule mutually exclusive means that acceptance of one alternative precludes acceptance of any other alternative. That is — the decision maker can select only one of the available alternatives.

By this rule we should accept the coin flip wager because the EMV of "accept wager" (+$0.50) exceeds the EMV of "reject wager" ($0). And in our example drilling prospect the EMV for "Drill" exceeds the EMV for "Farm Out" resulting in the choice of the drill option.

---

1. There are special types of analyses where the decision rule, as stated above, is not applicable. For example, if all the conditional values are expressed as costs, the decision maker would choose the alternative having the lowest expected value costs. Also, in some decisions it is useful to express values as "opportunity losses" and again the rule would be to take the alternative having the lowest expected opportunity loss. In most drilling prospect analyses, however, the values will be expressed as profits or losses, or some corresponding parameter related to profits such as preference values. For these more common situations the stated decision rule applies.

MEANING AND INTERPRETATION OF EXPECTED VALUES

The decision rule means that, all other factors being equal, the decision maker should accept the alternative which maximizes expected value. The expected value of a decision alternative is interpreted to mean *the average monetary profit per decision that would be realized if he accepted the alternative over a series of repeated trials.* The key words in this interpretation are "per decision" and "repeated trials."

As stated previously, the decision rule of maximizing positive EMV suggests that, given the choices of Table 3.1, the decision maker should elect to drill, rather than farm out. If the decision maker were presented with a repeated series of prospects having the same risk and profit values as those given in Table 3.1 his overall profit from the series of repeated trials would average $280,000 per decision (not per successful well) if, in each instance, he had accepted the drill option.

As proof of this point, suppose he had been presented 100 such prospects, and suppose further that he elected to drill each of them. The most probable results of the 100 well drilling series would be 40 dry holes and 60 producers.[1] His net revenue from the producing wells would be 60 × $600,000 = $36,000,000. His losses in 40 dry holes would be 40 × $200,000 = $8,000,000. His net profit, therefore, would be $36,000,000 − $8,000,000 = $28,000,000. His average profit *per decision* would be $28,000,000/100 = $280,000 — the expected monetary value of the decision alternative "drill."

If this meaning of the expected value of a decision alternative is accepted it should be obvious why the decision rule is to accept the alternative having the highest EMV. The average profit received *per decision* is not as much if any alternative with a lower EMV is accepted. The expected value of a decision alternative is really just a "weighted arithmetic average profit" that he would expect if the decision was repeated over a series of trials. The weighting factors in the actual computation of EMV are the probability numbers.

But suppose there is only one such prospect to consider — the usual case in the oil business. It should be clear that if he can only drill one well he will either receive a $600,000 profit or a $200,000 loss. How can he

---

1. Note that we said "most probable" results. This does *not* mean that 40 dry holes and 60 producers is the only result. Conceivably he could have drilled 41 dry and 59 producers, or 35 dry and 65 producers, etc. The important point here is that, of all the possible combinations of drilling results possible, the most probable is 40 dry holes and 60 producers. We'll prove this statement in Chapter 6.

make a net gain equal to the EMV (+$280,000) if he can only accept the gamble once? The answer is that he can not achieve a gain of $280,000. In this case the EMV number represents a "profit per decision" which is not even possible if he decides to drill.

What does this mean, if anything? Does the "repeated trial" clause rule out its use in business decisions where nearly every decision choice has different risk and profit levels? Have we just proved that expected value concepts might be useful for playing slot machines or flipping coins, but of no use in drilling decisions where every "trial" has different probabilities and profitabilities?

We are now at one of the central issues involved in using the expected value concept in business decisions. And, in answer to some of the above questions, it *is* possible to apply the concept to business (and petroleum exploration) decisions if it is recognized that the repeated trial aspect can be satisfied by continued investment decisions. The key is the following statement:

> If the decision maker consistently selects the alternative having the highest positive expected monetary value his total net gain from all decisions will be higher than his gain realized from any alternative strategy for selecting decisions under uncertainty. This statement is true even though each specific decision is a different drilling prospect with different probabilities and conditional profitabilities.

This statement is the absolutely essential element for any rational justification of the use of expected value in business decisions. The reader (and all others involved in trying to use and understand the expected value concept) must be convinced of its correctness. The statement suggests that as long as the decision maker continues to make investments, and as long as he adopts the consistent strategy of maximizing expected value, he will do better in the long run than any other strategy for selecting decisions involving uncertainty.

So we see that the expected value concept is more nearly a strategy, or philosophy for consistent decision making than an absolute measure of profitability. The strategy can only be applied to advantage if used consistently from day to day. The decision maker can not use expected value today, some other criterion tomorrow, and yet a third criterion on the third day. It is a guide for consistent decision making that is valid even if every decision alternative is different. He is, in essence, satisfying the "repeated trial clause" by continued investments.

It has been my experience in teaching courses on petroleum exploration decision analysis that at this point most "newcomers" to the world of quantitative decision theory are thoroughly bewildered. Assuming this will also be the case for some people reading this book let's see if we can back up and interpret the expected value concept in a slightly different manner. Hopefully this will clarify some of the issues involved.

We have said on several occasions that business decisions involving uncertainty are analogous to games of chance. The decision to "drill" in Table 3.1 could be paraphrased to the language of gambling by saying we have a 0.6 probability of winning the jackpot of $600,000 and a 0.4 probability of losing $200,000, as represented in Figure 3.1.

Now suppose a wealthy looking person approaches you and offers to give you x dollars cash in exchange for the right to play the gamble. How much cash would you accept in exchange for the gamble of Figure 3.1? (Incidentally let's assume for the moment that you are also wealthy and can afford to lose $200,000 without fear of bankruptcy.) Would you exchange your right to play the gamble for $100,000? For $300,000? For $50?

If you are totally impartial to money (that is, you have plenty of money and can afford the potential loss without any strong emotional

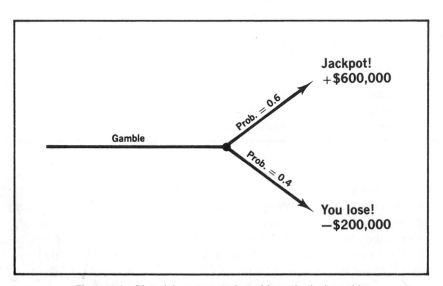

Figure 3.1 *Pictorial representation of hypothetical gamble.*

fears of its consequences) the expected value concept would say that you should be indifferent between the options of cash or the gamble if the cash offer was $280,000. If you accepted $280,000 you would have that much cash in hand. If you accepted the gamble you would realize an equivalent gain of exactly the same amount (assuming the gamble could be repeated so as to balance the losses with the gains.)

The gamble is said to have a certain "cash equivalent value" equal to its EMV of $280,000. Thus, in some textbooks and articles on the expected value concept the EMV is given the equivalent definition *"certainty monetary equivalent."* The reader would do well to note, however, that this idea is predicated on the assumption that you (the decision maker) are impartial to money. We'll speak about this extremely important assumption later in this chapter. The point to remember here is that this alternative definition of the meaning of EMV has a qualification which represents a major weakness of the concept in the view of some decision makers.

Another way to give meaning to the fact that the expected value represents the average gain per decision over a series of repeated trials is to return to our coin-flip wager mentioned earlier. We computed that your EMV for accepting the wager was +$0.50. And, by the interpretations given previously, this means that if you participated in a series of flips your cumulative winnings divided by the number of times the coin is flipped should be at or near +$0.50.

To illustrate this point the author began flipping a coin and observed the following actual sequence: tail, head, head, head, tail, tail, head, tail, tail, head, and head. It is of interest here to observe the parameter "cumulative gain per trial" as a function of the number of trials (flips) and compare it to the theoretical gain per trial—the EMV—which in this case is +$0.50. The computations are given in Table 3.2.

Now, if we plot "cumulative gain per trial" versus the number of trials, as in Figure 3.2, we can see how the cumulative gain per trial from the coin flip experiment is beginning to converge to the theoretical value given as the EMV.

If we had continued the coin flip experiment for many more trials the actual cumulative gain per trial function would converge ever closer and closer to the theoretical value—EMV. Thus, we see that over a series of repeated trials the EMV begins to assume a definite quantitative meaning, even though on any given trial the EMV represents only a theoretical expectation.

The reader is cautioned to not apply any quantitative conclusions

Table 3.2

Computation of Cumulative Gain Per Trial for Experiment Involving Eleven Flips of Fair Coin

(Monetary Values Involved: +$2 for Head, −$1 for Tail)

Trial No.	Actual Outcome	Gain or Loss from Outcome	Cumulative Gain	Cumulative Gain Per Trial
1	Tail	−$1	−$1	−$1/1 = −$1.000
2	Head	+$2	+$1	+$1/2 = +$0.500
3	Head	+$2	+$3	+$3/3 = +$1.000
4	Head	+$2	+$5	+$5/4 = +$1.250
5	Tail	−$1	+$4	+$4/5 = +$0.800
6	Tail	−$1	+$3	+$3/6 = +$0.500
7	Head	+$2	+$5	+$5/7 = +$0.715
8	Tail	−$1	+$4	+$4/8 = +$0.500
9	Tail	−$1	+$3	+$3/9 = +$0.333
10	Head	+$2	+$5	+$5/10 = +$0.500
11	Head	+$2	+$7	+$7/11 = +$0.637

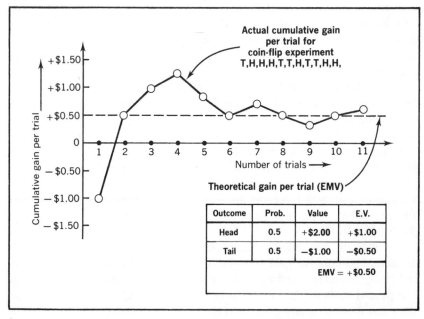

Figure 3.2 *Graph of cumulative gain per trial for coin flip experiment.*

from this illustration as to how many flips were, or would be required to approach the theoretical value. In general, the number of trials required will be dependent on the magnitude of the conditional monetary values, the probabilities of occurrence of the outcomes, the "closeness" to the theoretical value desired (that is ±2%, or ±5%, etc.), and the desired probability of being within the prescribed limits (that is, we must also specify that we want to be within ±2% of the EMV with a 95%, or 90%, or 99% certainty).

The point here is not the quantitative number of trials but rather the general observation of how the cumulative gain per trial function tends to converge to the EMV value as the number of repeated trials increase. This observation should help to give meaning to our earlier comment that the expected value of a decision alternative represents the average value per decision that would be realized if the alternative were accepted over a series of repeated trials.

CHARACTERISTICS OF EXPECTED VALUE COMPUTATIONS

The expected value concept has been described using two very simple numerical examples. The concept is, indeed, much more general, and the following list of characteristics and rules for expected value computations provide a basis for use of the concept in more complex decisions.

1. Any number of outcomes can be considered in an expected value computation. The only proviso being that the probabilities listed in the second column must add up to 1.0. This merely means that any number of possible outcomes can be listed, so long as *all* of the possible outcomes have been included. For our example drilling prospect of Table 3.1 the outcomes were only given in terms of producer or a dry hole. If several levels of reserves had been specified the outcomes might have been represented as a dry hole, reserves of 100,000 barrels, reserves of 200,000 barrels, etc.
2. Any number of decision alternatives can be considered.
3. The analyst has various options in expressing the conditional values received from the possible outcomes. A common situation is to express values in terms of monetary profits and losses. The profits can be expressed as "before-tax" or "after-tax" monetary

values. In addition, the monetary values can be undiscounted values or discounted net present value (NPV) profits.

If this latter case is selected the expected value of the decision alternative is usually called the *expected present value profit*. A positive expected present worth profit is the amount of expected NPV gain (over a series of trials) that would be realized *in excess* of an expected rate of return equal to the interest rate used in the discounting. If expected present value profit is zero it means that the alternative has an expected earnings rate equal to the discounting rate.

The conditional values received can also be expressed as costs, opportunity losses, preference values, or even conditional expected values. When values received are all expressed as costs the decision choice is to select the alternative having the lowest expected costs. In analyses of this type it is the usual practice to drop the negative sign from the cost (or expense) values.

Expressing values received as opportunity losses provides a useful insight into certain types of decision problems, and is described in further detail in Appendix E. Problems that lend themselves to the opportunity loss concept involve trying to quantify the cost of uncertainty and how much can be spent to obtain better (or perfect) information.

The notion of expressing values received in terms of the decision maker's relative preferences is an attempt to provide a more realistic quantitative decision parameter than expected monetary value. We will discuss this important, but controversial, idea in Chapter 5.

And finally, there will be certain types of decision problems in which it will be useful to express the values received as conditional expected values. That is, the numbers in Column 3 of the expected value computation will, themselves, be expected values. This will be clarified in Chapter 6 when we formally discuss conditional probabilities.

About the only measure of value which *can not* be used in the expected value computation is rate of return. As stated in Chapter 2 it is not correct to multiply a rate of return by a probability number to get an expected rate of return. Even if it was correct to do this what rate of return would we use for the conditional value in Column 3 for the outcome dry hole?

Of the options given above the most widely used measure of

value in prospect analysis is discounted net present values that have been computed on an "after-tax" basis using the anticipated reinvestment opportunity rate for the discounting.

4. The expected monetary value of a decision alternative does not have to be a possible monetary profit that could be received on a given trial. In fact in most discrete outcome analyses it will not be a value that could actually be received. (In our drilling example the EMV was $280,000, but the only possible profits we could receive if we drilled would be a $600,000 gain or a $200,000 loss!). In general it is incorrect to say that "the EMV is the *most probable* result" of selecting the alternative. Also, it is, in general, incorrect to say that we have a 50% chance of achieving a profit equal to or greater than the EMV.

These are common, but incorrect statements regarding EMV, and the reasons they are incorrect should become very clear when we get into our discussions of distributions and single point estimates of distributions in Chapter 6. When conditional values are expressed in monetary values the expected value of a decision alternative can be negative, zero, or positive.

5. The decision rule for the special case of the expected value concept in which the values are expressed as monetary values (i.e. EMV) implies that the decision maker is totally impartial to money and the magnitudes of money involved in the gamble. This means, among other things, that he has a very large amount of money and can afford any of the potential losses. This assumption may not be satisfied in some instances.

For example, suppose an independent oil operator having an annual drilling budget of $250,000 was considering the example prospect of Table 3.1. Even though the EMV for "drill" is higher than "farm out" he may very well decide *not* to drill if he is not willing to gamble 80% of his total annual budget in one well. He simply can't afford to lose $200,000 in a single well, even though the EMV decision rule indicates drilling to be the best choice.

Do we conclude, therefore, that EMV is a useless decision parameter unless we have an infinite supply of money? Is EMV a useful decision concept for the large major oil companies but of little or no use to the small operator with a limited asset position? Again we find ourselves at another of the central issues involved in applying expected value concepts to business decisions.

We'll try to speak to these questions later in the chapter and also in our discussion of preference theory (Chapter 5). The point we stress here is that for the special case of EMV the decision rule of maximizing expected monetary value profit implies that the decision maker is totally impartial to money.

6. If the decision maker is operating under a limited capital constraint it is meaningful to use the criterion of *EMV/Expected Investment Ratio*. This ratio has as its numerator the EMV computed in the usual manner. The denominator is the result of an expected value calculation in which the conditional values received are expressed as investments, or capital expenditures. The ratio provides a measure of the expected gain per unit of expected investment.

The decision maker would, presumably, wish to select alternatives which maximize this ratio if he does not have sufficient funds to accept all of the available projects. That is to say, he would select the projects which will yield the highest expected gain per unit of money invested. A numerical example to illustrate this will be given shortly.

7. There are several algebraic rules of operation related to the expected value concept which are useful in simplifying some of the computations. To explain these rules it will be necessary to be a bit more precise with symbols, and we define some symbols in a generalized expected value computation in Table 3.3.

a. If a constant amount, A, is added or subtracted to all of the conditional values received the EV is increased or decreased by a like amount. That is:

$$\text{If } V_i' = V_i \pm A, \text{ for } i = 1, 2, 3, \ldots, n$$

$$\text{then } EV' = EV \pm A, \text{ where } EV = \sum_{i=1}^{n} (P_i)(V_i)$$

b. If all the conditional values received are multiplied by a constant value, c, then the resulting EV will be equal to the original EV multiplied by c. That is:

$$\text{If } V_i' = (V_i)(c), \text{ for } i = 1, 2, 3, \ldots, n$$

$$\text{then } EV' = (EV)(c), \text{ where } EV = \sum_{i=1}^{n} (P_i)(V_i)$$

Table 3.3

Generalized Expected Value Computation

Outcome	Probability Outcome Will Occur	Conditional Value Received If Outcome Occurs	Expected Value of Outcome
O_1	P_1	V_1	$(P_1)(V_1)$
O_2	P_2	V_2	$(P_2)(V_2)$
O_3	P_3	V_3	$(P_3)(V_3)$
.	.	.	.
.	.	.	.
.	.	.	.
O_n	$\underline{P_n}$	V_n	$(P_n)(V_n)$
	1.000		

$$EV = \sum_{i=1}^{n} (P_i)(V_i) \leftarrow$$ Summation is, by definition, the expected value of the decision alternative, and is given the symbol EV.

(Note: The reader should note that we are considering the general expected value case, as distinct from the special case where the V_i terms are monetary values and the $\sum_{i=1}^{n} (P_i)(V_i)$ summation is called EMV.)

This is a useful rule to evaluate EMV's resulting from reduced working interests. For our example drilling prospect if we drilled the well with full ownership our EMV was +$280,000 (Table 3.1). If we had the option of drilling the well with only a 50% working interest ownership our computations of EMV would involve multiplying each of the conditional values received by $\frac{1}{2}$ (since we would now only own a $\frac{1}{2}$ interest in the well). But by the above rule we could compute the new EMV for 50% WI directly by multiplying the original EMV by $\frac{1}{2}$, i.e. the EMV for 50% ownership would be $(+\$280,000)\, (\frac{1}{2}) = +\$140,000$.

   c. If the decision maker selects k investment projects each having an EV computed in the usual manner, his expectation from the entire suite of projects is the sum of the individual EV's. That is:

Let EV_i be the expected value of the ith decision, where $i = 1, 2, \ldots, k$, then the aggregate, or total EV from all decisions, EV_{TOTAL}, will be

$$EV_{TOTAL} = \sum_{i=1}^{k} (EV_i) \qquad i = 1, 2, 3, \ldots k$$

In this rule the individual EV_i terms in the summation do not have to be equal. This rule says, in effect, that the expected value of a sum is equal to the sum of the expected values. *An example:*

Recalling our coin-flip example suppose we are offered the gamble ten consecutive times, and that each time we accepted. The expected value of the alternative "accept" was computed as +$0.50, and in this case the EV would also be the same for each of the subsequent wagers. To compute our expectation from accepting the wager ten times we simply find the sum of the EV's for each decision.

$$EV_{TOTAL} = EV_1 + EV_2 + EV_3 + \ldots + EV_{10}$$

or, $EV_{TOTAL} = \sum_{i=1}^{k=10} (EV_i).$

But since $EV_i = +\$0.50$ for all i

$EV_{TOTAL} = (10)(+\$0.50) = \underline{+\$5.00}$

In a more general sense each of our decisions in petroleum exploration will have different EV's. If we were interested in computing the expected value of all decisions made in a budget period we would simply sum the EV's of each decision alternative that had been accepted.

SOME ADDITIONAL NUMERICAL EXAMPLES

The numerical examples of Table 3.1 and the coin-flip gamble were very simple illustrations to use in our discussions of the expected value concept. We will examine here several numerical examples that more closely fit real-world decisions to invest capital in petroleum exploration. The first example will demonstrate how a decision maker could evaluate various decision alternatives with respect to single prospect. The second example illustrates how he might compare the relative merits of three different drilling prospects. And finally, we include an analysis of a player's expectations when playing roulette. (This last example was included on the premise that if we can't convince you on the use of expected value in drilling prospects we will at least be able to make you a better gambler!)

Example No. 1

Our exploration staff has been evaluating a proposed drilling prospect in the Anadarko Basin of western Oklahoma. The proposed well would be drilled as a 640 acre gas unit in Section 29. There are, at present, 256 net acres that are not leased, and our participation in the unit would be predicated on our purchase of the leasehold rights on the 256 acres (See Figure 3.3).

Basic data about the prospect include:

Gross Well Cost (Including Lease Equipment): $100,000
Gross Dry Hole Cost: $70,000
Estimates of Possible Outcomes and Their Probabilities of Occurrence:

Possible Outcomes	Estimated Probability of Occurrence of Outcomes
Dry Hole	0.35
2 BCF reserves	0.25
3 BCF reserves	0.25
4 BCF reserves	0.10
5 BCF reserves	0.05
	1.00

(BCF stands for *billion cubic feet*)

These probabilities have been obtained by a careful analysis of risk and uncertainty. We have not as yet described how to do this, but we will spend a great deal of time on this important part of the analysis in later chapters. For now we again ask the reader to accept these numbers as being representative.

Our staff has identified three possible alternatives with respect to

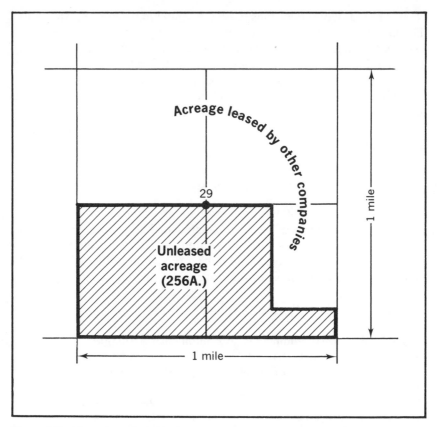

Figure 3.3 *Map of section 29 showing acreage status, proposed 640 gas unit in Anadarko Basin.*

participation in the proposed unit, given that we had secured the 256 acres.

    A. We could participate in the drilling of the well with a non-operating 40% working interest (256 Acres/640 Acres × 100% = 40%)

    B. We could farm out our acreage and retain a 1/8 of 7/8 ORI on 256 net acres[1]

    C. We could be carried under the penalty clause of the proposed operating agreement with a backin privilege (40%WI) after recovery of 150% of the investment by the participating parties.

    1. For those readers unfamiliar with some of the leasing terminology such as farm out, overrides, backin options, etc. refer to the Glossary of Terms given on page 570.

The next step in the analysis involved an engineering appraisal of the anticipated net present value profits and losses that would be realized from the possible outcomes for each of the above decision alternatives. The results of this analysis are given in Table 3.4.

Table 3.4

Table of Conditional Net Present Value (NPV) Profits and Losses That Would Be Realized from the Prospect for Each of the Decision Alternatives. All Revenues Are Net After Taxes, Royalties, Operating Expenses and Costs of the Well. The Discounting Rate Used Was 10%.

Possible Outcome	Decision Alternative		
	Drill with 40% Working Interest	Farm Out – Retain ORI on 256 Net Acres	Penalty Clause with 40% Backin Option
Dry hole	−$28,000	0	0
2 BCF	+$21,000	+$ 5,000	+$ 5,000
3 BCF	+$42,000	+$ 8,000	+$20,000
4 BCF	+$64,000	+$11,000	+$37,000
5 BCF	+$86,000	+$13,000	+$56,000

Given this information about the proposed drilling prospect in Section 29 we can respond to the following management questions:

1. How much can we afford to pay (if any) for the leasehold rights?
2. If we obtain the 256 net acres which of the three decision alternatives would maximize our expected net present value profit?

To solve this problem we must first make an expected value calculation for each decision alternative. The probabilities of occurrence to use with the conditional values of Table 3.4 were given on page 77. The expected value computations are given in Table 3.5.

From the expected value computations of Table 3.5 we see that the alternative "Drill with 40% WI" has the highest expected net present value profit and would be the preferred choice, given the management strategy of maximizing expected monetary profits.

Decision Alternative	Expected Net Present Value Profit
Drill	+$16,650 ← Highest EV
Farmout	+$ 5,000
Backin option	+$12,750

Table 3.5

Expected Value Computations for Anadarko Basin Drilling Prospect in Section 29 (Example No. 1)

Possible Outcomes	Probability Outcome Will Occur	Drill with 40% W.I.		Farm Out with ORI		Penalty with Backin	
		Conditional NPV Profit	Expected NPV Profit	Conditional NPV Profit	Expected NPV Profit	Conditional NPV Profit	Expected NPV Profit
Dry Hole	0.35	−$28,000	−$ 9,800	0	0	0	0
2 BCF	0.25	+$21,000	+$ 5,250	+$ 5,000	+$1,250	+$ 5,000	+$ 1,250
3 BCF	0.25	+$42,000	+$10,500	+$ 8,000	+$2,000	+$20,000	+$ 5,000
4 BCF	0.10	+$64,000	+$ 6,400	+$11,000	+$1,100	+$37,000	+$ 3,700
5 BCF	0.05	+$86,000	+$ 4,300	+$13,000	+$ 650	+$56,000	+$ 2,800
	1.00		EV = +$16,650		EV = +$5,000		EV = +$12,750

So the answer to the second question would be that participation with a 40% working interest would be the preferred strategy.

To determine how much we could afford to pay (if any) for the lease bonus we must recall an earlier definition that when conditional values received are expressed as NPV revenues discounted at a specified rate (in our problem 10%) the expected value of a decision alternative is interpreted as "the expected NPV gain that would be realized over a series of repeated trials *in excess* of a rate of return equal to the discounting rate."

This interpretation, together with the fact that the lease bonus to secure the drilling rights is a time-zero expenditure suggests that the maximum amount we could afford to spend for the leases (and still have an expected earnings rate equal to 10%) is equal in magnitude to the expected net present value profit.

Thus, in this problem we could afford to spend as much as $16,650 for the lease bonus and still have an expectation from the prospect of a rate of return equal to the NPV discounting rate, 10%. This corresponds to a maximum of about $65/acre for the 256 net acres that are unleased. Our lease broker would of course, attempt to get the acreage as cheaply as possible, and the $65/acre would be interpreted as the maximum amount he could offer as a lease bonus. If he secured the lease for $25 acre (total lease bonus: 256 acres × $25/acre = $6400) and the decision maker in turn accepted the "drill" option his overall expected net present value profit would be +$16,650 − $6,400 = +$10,250.

This example illustrates a very effective way to evaluate how much money can be spent at time-zero, or front-end costs and still yield a specified expected return from the project. These time-zero expenditures can include signature bonuses, bonus in offshore bids, lease costs, reconnaissance seismic data, etc.

The procedure is to include all subsequent cost factors exclusive of time-zero costs in the conditional values received (as in Table 3.4) and then determine the magnitude by which the expected present value profit exceeds zero. And of course, if the expected net present value profit, exclusive of time-zero costs, is zero or negative it means we can't afford to spend anything initially if we still want an expected rate of return equal to the NPV discount rate.

So, in answer to our proposed drilling prospect in Section 29 we have concluded the following.

1. We can afford to pay as much as $65/acre to secure the leasehold rights on the 256 acres in Section 29, and,
2. Given that we can secure the leases the decision alternative which maximizes expected net present value profit, discounted at 10% is "Drill with a 40% working interest."

Finally, we would like to use this numerical example to illustrate another very useful correlation that can be obtained from the expected value concept. It is a graph of expected value profit as a function of the probability of finding hydrocarbons. We will discuss the theory of the graph and how it is constructed in Chapter 6, and for now the reader's attention is directed only on the interpretation and use of such a correlation.

Specifically, in this example the probability of a dry hole was estimated to be 0.35. Therefore the probability of finding hydrocarbons (reserves of any amount) would be $1 - 0.35 = 0.65$. The expected values for each decision alternative for this probability of finding hydrocarbons were computed in Table 3.5. These values, together with the theoretical considerations to be discussed in Chapter 6 provide the data to prepare the correlation shown in Figure 3.4.

We shall see in later discussions that one of the parameters in risk analysis that is very difficult to quantify is the probability the well (or structure) will be productive of hydrocarbons. For an important parameter such as this it is useful to show its effect on expected value profit, and Figure 3.4 is an example of such a correlation. Even though the analyst may not be able to determine an exact value for the probability of finding oil or gas if he can estimate a general range he may be able to make his decision by use of a correlation like Figure 3.4.

In this example if the analyst was not certain of the probability of encountering gas reserves but felt it was probably in the range of 0.6 to

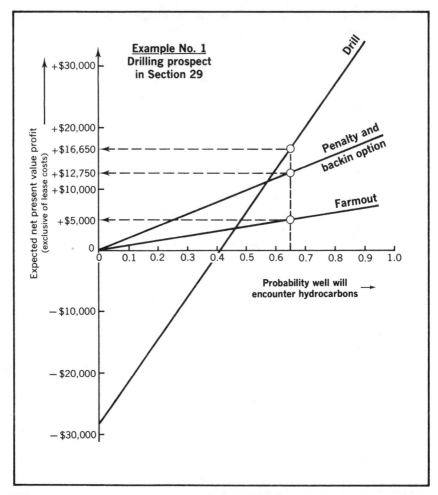

Figure 3.4 *Graph of expected net present value profit (exclusive of lease costs) vs probability well will encounter hydrocarbons for Example No. 1.*

0.8 for Section 29 the preferred choice would be to drill (since the "drill" strategy has the highest expected value in this range). If the analyst had determined that the probability of gas in Section 29 is probably in the range of 0.15 to 0.30 the preferred choice would be to take the penalty and backin option. In fact, we can read from this graph that for all values of probability of gas reserves less than about 0.57 the best choice would be the penalty. For probabilities of gas greater than 0.57 the preferred choice would be the drill option.

A correlation such as this will not tell the analyst what the probability of gas is, but if he can determine the probable range of the probability of finding gas his analysis may still indicate a firm (and correct) course of action to the decision maker. These graphs are very useful in exploration risk analysis and we will encounter them in later examples in the book. And again we stress that we will explain the theoretical basis for this correlation and how it was constructed in Chapter 6.

Example No. 2

This example will illustrate how the expected value concept can be used to compare the relative merits of three different drilling prospects. All revenues given in this example are net after taxes, royalties, operating expenses, and costs of the well. The same NPV discounting rate, i_0, was used in each case. The prospects are each in different sedimentary basins.

PROSPECT "A" – Our company would drill a fully owned development well having the following parameters: Dry hole cost = $80,000; Completed well cost = $100,000.

Possible Outcome	Probability of Occurrence	Discounted Net Revenue
Dry Hole	0.30	−$ 80,000
100M Bbls	0.30	+$ 25,000
200M Bbls	0.20	+$150,000
300M Bbls	0.10	+$250,000
400M Bbls	0.10	+$350,000
	1.00	

(M = one thousand, thus 100M = 100,000 bbls.)

PROSPECT "B" – Also a fully owned well, but one involving more expensive wells and a higher probability of failure. Dry hole cost = $200,000; Completed well cost = $250,000.

Possible Outcome	Probability of Occurrence	Discounted Net Revenue
Dry Hole	0.5	−$ 200,000
100M Bbls	0.1	−$ 100,000
400M Bbls	0.2	+$ 350,000
700M Bbls	0.1	+$ 600,000
1000M Bbls	0.1	+$1,000,000

PROSPECT "C"—We have the option to participate in the drilling of a gas prospect in which our company would own a 40% net working interest. Gross investments are: Dry hole = $70,000; Completed well = $100,000. Net, 40% WI investments are: Dry hole = $28,000; Completed well = $40,000.

Possible Outcome	Probability of Occurrence	Discounted Net (40%) Revenue
Dry Hole	0.35	−$ 28,000
2 BCF	0.25	+$ 25,000
3 BCF	0.25	+$ 50,000
4 BCF	0.10	+$ 80,000
5 BCF	0.05	+$100,000

Questions:

1. Based on the decision rule of maximizing expected present worth profit, which of these prospects would be the preferred choice?
2. If the decision maker had a limited capital constraint and wanted to maximize expected present worth profit per expected investment costs, which of these prospects would be the preferred choice?

In this example we will be using expected value to compare the merits of three dissimilar drilling prospects. The first step to resolve the questions is to compute the expected net present value profit for each prospect. These computations are shown in Table 3.6.

In answer to the first question all the prospects have positive expected present value profits and would, therefore, all be acceptable investments. The relative ranking of desirability would be in the order of descending expected net present value profit, i.e.

Prospect	Expected Net Present Value Profit
"B"	+$120,000
"A"	+$ 73,500
"C"	+$ 21,950

The second question imposes a capital constraint such that the decision maker would presumably be interested in investing in projects that will yield the highest expected value gain per unit of money invested.

Table 3.6

Expected Value Computations for the Three Drilling
Prospects of Example No. 2

PROSPECT "A"

Outcome	Probability of Occurrence	Conditional NPV Profit	Expected NPV Profit
Dry Hole	0.30	−$ 80,000	−$24,000
100M Bbls	0.30	+$ 25,000	+$ 7,500
200M Bbls	0.20	+$150,000	+$30,000
300M Bbls	0.10	+$250,000	+$25,000
400M Bbls	0.10	+$350,000	+$35,000
	1.00		EV = +$73,500

PROSPECT "B"

Outcome	Probability of Occurrence	Conditional NPV Profit	Expected NPV Profit
Dry Hole	0.50	−$ 200,000	−$100,000
100M Bbls	0.10	−$ 100,000	−$ 10,000
400M Bbls	0.20	+$ 350,000	+$ 70,000
700M Bbls	0.10	+$ 600,000	+$ 60,000
1000M Bbls	0.10	+$1,000,000	+$100,000
	1.00		EV = +$120,000

PROSPECT "C"

Outcome	Probability of Occurrence	Conditional NPV Profit	Expected NPV Profit
Dry Hole	0.35	−$ 28,000	−$ 9,800
2 BCF	0.25	+$ 25,000	+$ 6,250
3 BCF	0.25	+$ 50,000	+$12,500
4 BCF	0.10	+$ 80,000	+$ 8,000
5 BCF	0.05	+$100,000	+$ 5,000
	1.00		EV = +$21,950

At first glance Prospect "B" appears to satisfy this criterion because it
will yield the highest expected gain. But Prospect "B" also is the most
expensive well ($250,000 completed well cost, versus $100,000 and
$40,000 for Prospects "A" and "C" respectively) and has the highest
probability of failure.

We need a way to incorporate these factors into a decision criterion (they are included in the EV calculation but do not appear in a manner that provides direct comparison). And, as suggested earlier in this chapter the ratio of expected net present value profit divided by expected investment provides a criterion that can be used for decision making purposes.

The numerators for the ratios were computed in Table 3.6. The denominator terms, expected investment, are computed for each prospect as follows:

> PROSPECT A
> Dry Hole Cost: $80,000
> Completed Well Cost: $100,000
> Probability of Dry Hole: 0.30
> Probability of Completion: $(1.00 - 0.30) = 0.70$
> Expected Investment $= (0.30)(\$80,000) + (0.70)(\$100,000) = \underline{\underline{\$94,000}}$

> PROSPECT B
> Dry Hole Cost: $200,000
> Completed Well Cost: $250,000
> Probability of Dry Hole: 0.50
> Probability of Completion: $(1.00 - 0.50) = 0.50$
> Expected Investment $= (0.50)(\$200,000) + (0.50)(\$250,000) = \underline{\underline{\$225,000}}$

> PROSPECT C
> Dry Hole Cost (Net to 40% WI): $28,000
> Completed Well Cost: $40,000
> Probability of Dry Hole: 0.35
> Probability of Completion: $(1.00 - 0.35) = 0.65$
> Expected Investment $= (0.35)(\$28,000) + (0.65)(\$40,000) = \underline{\underline{\$35,800}}$

We can now summarize all of our expected value computations for Example 2 as follows:

Prospect	Expected NPV Profit (Table 3.6)	Expected Investment	Expected NPV Profit to Expected Investment Ratio (Col. 2 ÷ Col. 3)
"A"	+$ 73,500	$ 94,000	0.782
"B"	+$120,000	$225,000	0.533
"C"	+$ 21,950	$ 35,800	0.613

From this tabulation we see that, although "B" will yield the highest expected gain, it will *not* yield the highest gain per unit of expected investment. In fact, Prospect "A" yields the highest expected gain per unit of expected money invested (0.782 versus 0.533 and 0.613). So if the decision maker has a capital constraint his preferred selection (all other factors being equal) would be "A," then "C," and then "B." By setting a ranking such as this the decision maker can arrange the available investments in descending order, start at the top of the list, and work down the list until his available investment capital is expended.

This example should illustrate how expected value can be used to compare dissimilar drilling prospects, as well as showing the computations required to make the expected value criterion compatible with a limited capital constraint. We didn't say much about the specific geology of each of the three prospects, but it should be obvious that the expected value concept has given us a common basis upon which to compare the relative merits of prospects having greatly differing levels of investment and risks. In fact, it can be used to compare offshore investments versus onshore prospects, gas versus oil prospects, structural prospects versus stratigraphic trap prospects, etc. If you really think about it, a decision criterion this general is only a bit short of being an amazing breakthrough! What other decision criterion can you think of that is this general?

And by the way, if we had, in fact, selected Prospect "B" and drilled a dry hole did we make a wrong decision? Does getting a dry hole in Prospect "B" prove that the expected value does not work?

Finally, we use this Example to say a little more about the meaning of the term "Expected Investment." It has the same connotations as expected value profit, i.e., it is the expected investment *per decision* that the decision maker would spend over a series of repeated trials. If he participated in the drilling of a large number of drilling prospects the sum of the expected investments of each prospect *accepted* would be very close to the total investment capital expenditures for all the prospects. As such the summation of expected investments becomes a very good value to use as the basis for estimating future drilling budgets.

For example, if he had elected to accept all three prospects (they all had positive expected net present value profits) how much drilling capital would be required? The minimum amount of capital would be for the case where all three prospects were dry holes:

Prospect	Dry Hole Cost
A	$ 80,000
B	$200,000
C	$ 28,000
	$308,000 ← Minimum Capital Required

The maximum capital requirement would correspond to completing all three prospects:

Prospect	Completed Well Cost
A	$100,000
B	$250,000
C	$ 40,000
	$390,000 ← Maximum Capital Required

Budget controllers normally do not set their budgets on ranges, but rather, single values. What single value of capital required in the range of $308,000 to $390,000 would be most representative? The answer, assuming there is a fairly large number of prospects in the proposed budget, is the sum of the *expected investments* for each prospect. That is:

Prospect	Expected Investment
A	$ 94,000
B	$225,000
C	$ 35,800
	$354,800 ← Expected Capital Required

Thus, for budgeting purposes the best single value of capital to allow for these three prospects would be $354,800. Companies that have converted over to the expected value concept are finding this point to be particularly useful in preparing capital expenditure budgets for a fiscal budget period. Note that the denominator terms in the ratio, expected investment, consider only capital outlays, and it is the usual practice to drop the negative sign in expected investment computations.

Example No. 3

This illustration is for the gamblers. It doesn't have anything to do with petroleum exploration, but the example may prove useful to gain additional insights into the logic of expected value.

Suppose we are playing roulette at Monte Carlo and placing $100 wagers on a given number, say for example, the number 23. (The roulette wheel described here has 36 numbers plus a zero, or 37 compartments or slots around the edge of the wheel, as opposed to the U.S. type roulette wheels having 36 numbers plus a zero and a double zero, or 38 slots.) If the ball falls into number 23 the house gives us our wager back plus an additional $3500. If it falls into any other slot we lose our $100 wager. Assuming a mechanically perfect (unbiased) wheel and an honest croupier should we play? What is our expected monetary value profit or loss?

Again the first step is a standard expected value computation as given in Table 3.7.

Table 3.7
Expected Monetary Value Computation for Playing Roulette, Example No. 3

Outcome	Probability of Occurrence	Conditional Profit or Loss	Expected Monetary Value Profit
Number 23	1/37	+$3500	+$94.59
A number other than 23	36/37	−$ 100	−$97.30
		EMV =	−$ 2.71

Our expected value profit is a *loss* of $2.71! If we play the game 100 times our cumulative cash position will be approximately equal to a loss of −$2.71/trial × 100 trials = −$271! If we play 1000 times our cumulative position will be an approximate loss of $2710! Small wonder the gambling casinos have thick plush carpets and offer free drinks to its players – your loss is their gain!

Perhaps this also explains the reply given by a gambler who recently spent a weekend in Las Vegas. Upon his return a colleague asked him how he did in the casinos. His reply – "Well, Joe, I drove to Las Vegas in a $5000 automobile and came home in a $50,000 bus." Or another note regarding gambling – a service station along the Strip in Las Vegas displays a sign on its driveway saying "Free Aspirin and Tender Sympathy"!

But enough about Las Vegas or Monte Carlo. Our purpose is to learn about expected value. And this simple illustration regarding roulette wheels can be the basis for some interesting questions that you should consider:

1. Would your EV have been the same if you had placed your wager on number 6, rather than number 23? Does it matter what number you place your wager on?

2. Suppose you have been watching others play roulette over a long period of time and have noticed that the ball has not fallen into the number 14 slot as long as you've been there. Is it now a good time to place your money on number 14 (on the premise that number 14 is "overdue")? Does the EMV computed in Table 3.7 apply no matter when you choose to begin playing, or does it apply only after certain things have occurred before — such as no number 23, or 14, for the past 75 spins of the wheel?

3. You have been betting on a given number, say 23, and have lost 36 consecutive times. What is the probability of winning on the next trial? What is the probability of at least one win in 37 spins of the wheel? (Careful now — these are two questions that lead some people to all sorts of wrong conclusions!)

4. Is it correct to assume that each number will come up once in every 37 trials?

5. If a player's expectation in the game of roulette is negative why do people even play roulette? (Or any other game of chance for that matter!)

It is suggested that you jot down your answers to these questions at this time, and then compare your answers to the solutions that will be given in Chapter 6.

PROBLEMS OF INTERPRETATION AND APPLICATION OF EXPECTED VALUE CONCEPT TO DRILLING DECISIONS

Our discussions so far should illustrate the relative ease of making expected value computations and the versatility of the concept to compare investments having widely varying levels of risk and uncertainty. Yet, as we mentioned at the beginning of this chapter, despite the simplicity of the concept there still seems to be a pervasive air of misinterpretation and misunderstanding of the expected value numbers in the eyes of many business decision makers.

Some of the misconceptions are philosophical and some are caused by practical, real-world constraints. These misunderstandings are unfortunate, and have no doubt been one of the major reasons for the reluctance of many managers to accept the concept. Because the expected value concept is at the absolute center of any quantitative approach to decision making under uncertainty it is vital that we try to clarify some of the common misunderstandings.

Listed below are some of the attitudes and opinions expressed by petroleum exploration decision makers regarding expected value. Following the listing we will briefly respond to each point in hopes of casting a little more light on the problems of applying the concept to exploration prospect analysis.

Attitudes and Opinions Expressed by
Some Decision Makers about Use of
Expected Value Criteria in Drilling Decisions

1. The decision maker sees no need to quantify risk, hence no need to become involved with expected value criteria.
2. The decision maker rejects the expected value concept because he sees no apparent benefit from its use.
3. The decision maker accepts the expected value concept, but sees no way to get accurate probability numbers, and therefore concludes that it cannot be applied to drilling decisions.
4. The decision maker rejects the expected value concept because every drilling prospect is a distinct, unique consideration, thus there is no way to satisfy the "repeated trial" aspect implicit with expected value.
5. The small oil company manager (or independent operator) feels expected value criteria probably are useful to the big, fully integrated corporations but of little use to his small company.
6. The decision maker accepts the philosophy of expected monetary value (EMV) but feels there are other overriding concerns that he must consider which may lead him to reject alternatives having the highest EMV. These overriding concerns may include consideration of his asset position (can he afford the potential losses?), his or the firm's immediate and long term goals, and his attitudes about taking risks.
7. The decision maker accepts the expected monetary value criterion and believes it should be the dominant (or possibly the only) criterion to consider in selecting decisions under uncertainty.

Attitude No. 1:   At first glance most of us would probably agree that an exploration manager who does not choose to consider risk is extremely naive about the oil business. A dry hole should convince anyone of the risks involved. Of course failure to quantify risk does not necessarily imply that the decision maker has ignored the element of risk. In all probability he has chosen to account for risk in other ways. One way is to protect against risk by raising the "minimum" profit level necessary for project acceptance. For example he may have as his objective an overall rate of return of 12%, but since some prospects end up as dry holes he sets his minimum acceptable rate of return (computed on a no-risk basis, as in Chapter 2) at 25%, or 30%. Presumably, such a policy would yield a return commensurate with the degree of risk. These policies are often arbitrary and difficult to establish. How does one establish the "minimum cut-off" level of profitability? Is it right to reject a relatively certain project having a rate of return of 24% (relatively certain in the sense of having a high probability of obtaining the predicted cash flows) in favor of a high risk, rank wildcat which, if successful will yield a rate of return of 34%?

Apparently there are some business decision makers that choose to ignore the element of risk from a quantitative standpoint on the premise that their own superior experience, intuition and insight will remove the uncertainty involved. (See Reference 3.10 for further observations on this point). Looking at the petroleum industry in general one must conclude that if this latter approach was valid, and if the petroleum industry had good, competent decision makers (which it does) then we would no longer be drilling dry holes or investing in projects which return only marginal earnings. Yet these outcomes occur almost daily throughout the world, a fact which refutes this arbitrary approach at decision making.

The issue probably reduces to saying that there are other ways to account for risk but they are, at best, poor substitutes to a realistic appraisal of risk and uncertainty. The language of decision making involves numbers, and the quantification of risk and uncertainty is a vital key in trying to improve our decision making capabilities.

Attitude No. 2:   One of the ironies of the expected value concept is that it is difficult to judge whether it is working in the short term. Yet the obvious way to test any new idea is to try it for awhile and compare results with previous methods. Expected value is not an oil-finding tool, and a manager should not necessarily reject the concept if, after using expected value criteria for a week, he has not found a giant new field. Nor will its short term use cause an immediate change (up or down) in his firm's profit and loss statement. The proof of its value can only be measured in the long term.

Thus, the advocates of expected value criteria as a basis for decision making under uncertainty often find themselves in the position of trying to convince their management of its value but not being able to prove its value until the idea has been used consistently over a long series of investment decisions. And this situation is just as difficult from the reluctant manager's viewpoint, because he is being urged to adopt a new concept without any clear assurance that it will work to his benefit.

There isn't much we can do about this point, except to be careful to not sell the merits of expected value on the basis that it will cause immediate changes in our reserve ledgers or profit statements. I met a decision maker a few years ago who confided that his staff had urged him to try using EMV as a basis for selecting drilling prospects. He indicated that he had reluctantly agreed to try it for a few days—but that both wells he accepted to drill on this basis turned out to be dry holes.

This brief experience convinced him EMV and the oil business were incompatible and he went back to his older, less formal ways of choosing which prospects to drill. (No report on how he has been doing recently!)

Attitude No. 3: There is, of course, a sound basis for rejection of any concept if the input data are of no value. If the probabilities (or profitabilities for that matter) used in the expected value computation are meaningless certainly the EMV will have little or no meaning. Rejection of the concept in petroleum exploration because of the difficulties in obtaining the column 2 probabilities is certainly a frequently voiced opinion of many managers.

To be certain, the problem of assigning probabilities to the outcomes of drilling prospects is difficult. In fact, sometimes it is difficult to even state what the possible outcomes could be! But what alternative do we have? We could throw up our hands in despair and seek to devise a whole new system of describing risk and uncertainty. However, with the possible exception of using adjectives to describe the degree of risk, the inescapable fact is that there is no other system available.

If we are ever going to make effective use of decision methods that consider risk in a quantitative manner we will be faced with using probability numbers. The determination of probabilities—risk analysis—is still very much an "art" in petroleum exploration rather than a carefully defined science. We are just beginning to learn how to deal with probability concepts in prospect analysis.

Can the dilemma be resolved? Yes it can, so long as everyone working with risk analysis and making decisions under uncertainty recognize the limitations on the accuracy of the probability numbers and adopt a positive attitude toward trying to improve their expertise in dealing with probabilities. Everyone needs to recognize that answers to "what if" questions (sensitivity analyses) may prove to be extremely helpful in formulating decision strategies, even though the actual, or exact probabilities are not known. Certainly as our awareness of risk and our understanding of probability theories enlarges we will continue to improve our capabilities of estimating probabilities.

If we are operating in new exploration areas our data and experience will be very limited and the reliability of probabilities used in expected value computations will have to be viewed accordingly. If we are in an area which contains a large amount of historical data it is certainly useful to begin the analysis of risk by considering the statistical results of previous exploration efforts.

From this data base the analyst can compare the prospect under consideration to those which have been drilled. The comparison would seek to determine the ways in which the prospect is similar and the characteristics which may make the prospect unique. The past data are used as a starting point, with any changes in computing probabilities made after such a comparison. All of this will be discussed in much detail in later chapters.

So it seems that a realistic response to those who would reject expected value on the basis that there is no way to get accurate probabilities in petroleum exploration would be to agree that the probabilities are, indeed, the weakest part of the concept as applied to prospect analysis. But having recognized that there is really no alternative it is equally important that all persons concerned adopt a positive attitude about risk analysis and work together to improve their understanding and ability to estimate realistic values for the probabilities required to complete expected value computations. Hopefully this positive attitude will be apparent in our discussions of risk analysis in this book.

As a corollary to these comments some decision makers will take the position that the only time decision analysis techniques can be applied is in mature, drilled up areas having a great deal of statistical data available (success ratios, reserve distributions, and the like). They further state that in new, unexplored basins the techniques cannot be used because there is no data or experience upon which to base the probability and profitability numbers.

Do you agree? Or do you agree with those who take the opposite viewpoint—that in the undrilled areas it is all the more important to consider risk and uncertainty because these are the situations where risk is the greatest? That is, the less experience and data one has the more important it is to use decision analysis techniques such as the expected value concept, simply because one has no other way to evaluate the wisdom of investing capital to search for oil.

If you do not have a firm opinion about these opposing viewpoints you can "ride the fence" so to speak for a bit longer. But as we get into the complexities and frustrations of petroleum exploration risk analysis and the task of convincing others of the merits of decision analysis it will be quite obvious which side of the fence we'll have to be on. More about this in Chapters 7 and 8.

Attitude No. 4: Misunderstandings about the repeated trial aspect of the expected value concept are common. Almost everyone can grasp the idea of what expected value means when considering coin flipping experiments. It is an illustration involving risk that has the advantage that you can flip the coin many times and actually observe how the cumulative gain per trial converges to the EMV as suggested by Figure 3.2.

Having explained this the advocate of expected value will then typically state that the meaning of an "expected net present value profit of ____ million dollars" on Block 16 in an offshore area has the same connotations. Whereupon the skeptical decision maker will usually reply: *"Yes, but there is only one Block 16 and we can't continue to experiment with it many times to achieve some cumulative gain per trial. We can only drill it once and it either has oil or it will be dry. In the oil business you never get a chance to repeat the experiment (that is, to drill the prospect) more than once, therefore you can't satisfy the repeated trial condition."*

Pretty good point, right? The gambling houses don't worry too much about this, because they satisfy the repeated trial aspect by having lots of patrons and spinning the roulette wheel many, many times each evening. And the insurance companies (who, incidentally, base their premiums on expected value considerations) satisfy the repeated trial aspect by insuring as many people or cars or houses as possible. But, how can we satisfy the repeated trial condition in the oil business if we can only drill each prospect once?

The repeated trials must come from acceptance of many investment projects over a period of time. In essence, the "petroleum exploration game" has to be played enough times to average, or offset the losses with the gains. If the decision maker consistently seeks to maximize positive expected monetary value profit in these many projects it can be shown that he will have done better, in the long run, than he could have done with any alternative strategy for selecting decisions under uncertainty.

This statement holds even if every decision involves a different, unique investment project having its own specific levels of profitability and risk.

We mentioned on page 67 this statement is the absolutely essential link in the application of the expected value concept to exploration decisions. The decision maker must accept the validity of this statement, and there can be no compromise on this most essential point. If he rejects the statement, or believes it is correct "only to a certain degree," he may as well forget the whole idea of expected value.

The statement is, admittedly, a strong statement which carries no restrictions or conditions, and it is my opinion that the statement is quite difficult for many decision makers to grasp. How would you propose to prove or disprove whether the statement is valid? That is, what is implied by the words *it can be shown that . . .* ? Well one way, of course, would be to set up a series of hypothetical drilling decisions in which the decision maker could use various value systems (EV, rate of return, intuition, etc.) to choose drilling prospects.

Then by use of an analog which would duplicate the role of chance you could determine his cumulative gain over a long series of decisions made under uncertainty to see which criterion yielded the best results. The analog could be a computer model in which the element of chance is modeled using random numbers (more on this in Chapter 8). Or, as an alternative, you could adopt a more heuristic approach based on the logic and meaning of probabilities, much as we did with the numerical illustration showing the results of drilling 100 prospects having the probabilities and profits given in Table 3.1.

That is, by a very careful series of logical observations about the meaning of chance and the possible outcomes of a decision involving uncertainty one can generally conclude by purely logical reasoning that the statement is valid. Try it and see if you can rationalize it's meaning.

The above comments and the statement regarding how the repeated trial aspect of the expect value concept can be satisfied in exploration decisions are probably the most important points of this book. Ponder them very carefully.

Attitude No. 5: The problems of applying the expected value philosophy in small oil companies are the result of real-world constraints. The decision maker in the small oil company who suggests the approach is OK for the big companies but not for his is probably implying he doesn't have a large enough staff to justify such a sophisticated analysis.

But is the level of risk and uncertainty in petroleum exploration conditional on the asset position of the operator drilling the well? In general the answer is no. A small operator has just as much uncertainty about whether his wildcat will strike oil as the large companies. In fact, the small company probably must be even more aware of risk and uncertainty because of his smaller asset position and the ever-present possibility of a "gambler's-ruin" run of bad luck.

It is true that effective use of the full range of decision analysis methods requires persons on the staff conversant with probability theory, computers, statistics, and the like, and many small companies simply may not have people on their staffs who can do this type of work. But there is no reason why the decision maker and his geologist and engineer can't sit down and at least make some simple expected value calculations.

Often the decision maker in a small company may be involved in all phases of the evaluation of a drilling prospect (geology, engineering, leasing, organizing funds, following drilling operations) and there may be situations where it is difficult for him to separate all of the important dimensions of the prospect for explicit consideration. Expected value calculations assist in this aspect because it allows him to consider the dimension of risk separate and apart from the profitability considerations. After he carefully thinks about each part, the logic of expected value provides a means to combine all these pieces into a summary measure of investment worth. It would seem this approach may be better for him than his attempting to think about all aspects of the prospect simultaneously.

Decision analysis methods, including the expected value concept, have equal value and merits for large and small companies alike. Practically, the small company may not be able to use the more complex methods because of staff limitations. But even the simple concept of expected value should prove helpful to the small firm in improving its ability to devise decision strategies under uncertainty.

Attitude No. 6: This viewpoint says, in essence, that even though expected monetary value profit is a better decision parameter because it includes the dimension of risk, there are still certain overriding factors that may cause a manager to accept alternatives which may not have the highest positive EMV's. No argument here—and this viewpoint is not based on any misunderstandings about expected value.

It is simply a statement of the real-world fact of life that maximizing expected monetary gain is not the sole criterion to consider in making decisions under uncertainty. This management viewpoint only relates to the special (but usual) case of the expected value concept in which the conditional values received are expressed as monetary profits and losses.

To speak further to this issue let's go back and review the philosophy we have followed in trying to achieve more realistic decision criteria. In Chapter 2 we started out by describing the parameters payout and undiscounted profit-to-investment ratios. Then we observed that these two parameters were not totally sufficient, particularly because they did not include anything regarding the time-rate patterns of when cash flows enter and leave the treasury. So next we added the idea of the time-value of money to account for this important dimension of overall investment worth.

In essence, we put the dimension of time-rate cash flow patterns under a common denominator, so to speak, so that we would have a consistent parameter (rate of return) to use to compare dissimilar projects. Then we noted some of the disadvantages of rate of return, namely the reinvestment assumption and the fact it can't be applied to certain types of cash flows, and made a modification to alleviate these restrictions by using a single discount rate and making decisions based on net present value profit.

The next important dimension we introduced into the decision parameter was risk and uncertainty. We have shown in this chapter how the expected value concept incorporates risk factors (probabilities) into a parameter that can be used for decision making — the expected value of a decision alternative. As such, the expected value parameter represents the most comprehensive criterion of those we've discussed — simply because it includes more of the dimensions of overall investment worth. Thus everything we've tried to accomplish so far has a sound philosophical as well as practical basis or justification.

The problems suggested with this management attitude did not come into focus until we chose to express the conditional values received in the EV computations as *monetary* profits and losses. This choice, together with the decision rule to select alternatives having the highest positive EMV carry with them the implicit assumption that the decision maker is totally impartial to money and potential magnitudes of money associated with his available decision choices.

The simple fact is that people are *not* impartial to money. It is a relatively easy task to prove this point (and we will do so in Chapter 5). Earlier we suggested that one way to consider the meaning of EMV was to think about how much cash you would accept in lieu of the chance to play or accept a gamble. We called this the certainty monetary equivalent and suggested that this cash value for the gamble of Figure 3.1 was equal to $280,000.

If we were totally impartial to money the gamble (having an EMV of $280,000) and the cash payment of $280,000 would have precisely the same intrinsic value to us. Yet there is probably not one person in a thousand that would have this exact point of indifference. In fact most people, when offered cash in lieu of the gamble of Figure 3.1, would probably accept substantially less cash in exchange for the option to accept the gamble.[1]

All of this points up the fact that when using monetary profits in the expected value computations we must always remember that the decision rule to maximize EMV assumes total impartiality to money. It's that simple. We shouldn't read into the meaning of EMV anything more or anything less.

Yet we would all agree that there are still more dimensions to consider such as: a) Can we afford the potential losses involved?, b) How does the particular prospect fit in towards achievement of short and long-term goals?, c) How does the degree of risk in a given prospect relate to the amount of risk the decision maker is willing to take?, etc. These and other irreducible and intangible factors are ever-present in virtually all petroleum exploration decisions.

The point, then, that this particular management attitude is making is that the EMV is certainly a more representative decision parameter than the no-risk profit parameters discussed in Chapter 2, but there are still other factors which he must consider that may override the EMV decision rule. This view is well taken and we must remember these points when using EMV. It is a very useful parameter, but it is not an end in itself.

Scholars of expected value theory have long recognized this shortcoming of EMV. In fact, as early as about 1720 people were beginning to try to enlarge the concept to include the biases and preferences that people associate with money into a quantitative decision parameter. In essence they were trying to put the decision maker's intangible feelings

1. The author will readily admit that if he had the right to accept the gamble with EMV of $280,000 he would be more than happy to trade the gamble for as little as, say, $50,000 cash! If he had that amount of cash he'd probably stop writing books on decision analysis and catch the first plane to the beaches of Hawaii!

about money under the common denominator also. The goal would be, presumably, an even more representative quantitative decision parameter upon which to base his judgments.

This whole approach is generally called preference theory, or utility theory. We will explore this fascinating subject in more detail in Chapter 5, but for now you should at least be aware of the fact that people have been working on this improvement to the use of expected value criteria in business for many years. Unfortunately their efforts have not as yet been successful, and the application of preference value concepts in daily business decisions is more nearly a proposed way of making a decision than a reality at this point in time.

Until the problems with preference theory approaches can be resolved the decision maker will probably have to continue to deal with the expected monetary value criterion and its shortcomings. However it's not all that bad, because for the many decisions where the potential losses are reasonably small (relative to the firm's total assets) the EMV parameter will approximately coincide with the decision maker's preferences regarding money. If the decision maker feels justified in suppressing his preferences and biases with respect to money and analyzing the project on the basis of actual profits and losses the EMV criterion will be completely realistic and valid.

Attitude No. 7: Decision makers who believe EMV should be the dominant, or only parameter to consider in decision making under uncertainty represent almost the completely opposite viewpoint than those we've discussed thus far. They are saying, in principle, that EMV is great, and the decision rule to maximize positive EMV should be the main concern of the decision maker. For those of us advocating greater use of decision analysis methods it is always gratifying to get a convert who is completely sold on the techniques.

But we must be careful to avoid the potential danger of "over-selling" our decision makers on the concepts. As stated on numerous occasions in this chapter; EMV is not a cure-all, it is not an oil-finding tool, and it is not an end in itself, i.e., it is not the "ultimate" decision parameter. It is certainly true that EMV is a more representative parameter than a strictly economic criteria having no explicit consideration of risk and uncertainty. But as just stated above there are usually certain intangible factors that must be considered that are not included in EMV numbers. In fact the logic we used above to show the restrictions of EMV, if correct, serves to contradict the view expressed here that EMV should be used as the sole criterion.

As with most things in our world today some degree of moderation or balance usually produces the best end results. Moderation and judgment are essential to gain the best use of the expected value concept. The viewpoint expressed here is one extreme that is probably as unrealistic as the other extreme of rejecting expected value concepts outright.

SOME FINAL COMMENTS

At this point it should be quite clear that the expected value concept is the beginning point for any realistic appraisal of decisions under uncertainty. We've tried to keep our discussion of the logic and philosophy as clear and explicit as possible so that we could gain an insight about how the concept should be interpreted. In later chapters we'll direct our attention to much more complex exploration decision problems, as well as to spend considerable time discussing how to assess the probability numbers in the first place. But the end result of these more complex problems and risk analyses will always be an expected value number having the basic logic we've discussed here. Everything that follows has as its basis the simple expected value concept that was originated back in 1654.

All of this is to say that, while we've probably digressed into philosophical issues at times, most of our comments in this chapter will form the basis for virtually all the decision analysis concepts to be developed later in the book. It is hoped that you will have gained a reasonably good understanding of the expected value concept at this point. If nothing else you should now be able to determine whether to play the roulette wheels in Las Vegas. . . . !

A Note about the Case Study That Follows:

For the next few moments we digress from our discussions of decision analysis to consider a typical, real-world conversation relating to a wildcat drilling recommendation. The dialogue is intended to provide a comparison between what we have just discussed about the expected value concept and the less formal approach to decision making under uncertainty. At the end of the Case Study we have listed a few questions to focus your attention on other key points in the conversations.

Hopefully, this Case Study will be useful in contrasting the two approaches to evaluating exploratory drilling prospects.

A Case Study

Profit Oil and Gas Company
Exploration Department Recommendation

The Blackduck Prospect

Mr. Tom Miller, President of Profit Oil and Gas Company, has just entered the conference room for the bi-weekly Executive Committee meeting. Seated around the table are the other members of the Committee:

Joe Brown—Manager of Exploration
Ken Smith—Manager of Production and Engineering
David Jones—Chief Landman
Bill Davis—Manager of Accounting and Economic Analysis Group

(The Executive Committee authorizes all exploratory and development drilling recommendations. Recommendations are presented to the Committee by members of the exploration or production staffs. The Company has considerable production and undeveloped acreage in most of the major domestic basins. About a year ago Mr. Miller set a policy of active exploration, and the company has been very aggressive since that time with its exploration program.)

TOM MILLER: Good morning! What do we have for Committee consideration today?

JOE BROWN: Tom, we have a wildcat recommendation on that new 4000 acre tract on the Blackduck Ranch which we leased last November.

KEN SMITH: And producing department has two offset development well recommendations in our new Pinetree Field.

TOM MILLER: O.K., why don't we start with your well, Joe.

JOE BROWN: Fine. Henry Oilfinder of my geology staff has worked up the prospect and will be here in just a minute.

TOM MILLER: By the way, Ken, how are the two wells in Pinetree holding up?

KEN SMITH: So far, real well. The Johnson well, which you recall was the discovery well, has been flowing at a pretty steady rate of about 850 barrels per day. The offset, Smith No. 1, has only been on the line for about 3 weeks, but it's been averaging about 710 barrels per day. The two wells we plan to recommend today are the north and east offsets to the Smith well. The logs are in on the south offset to the discovery well and it looks like we only have about 35 feet of pay in it, so it probably isn't going to be as good a well as the discovery well.

JOE BROWN: See, that's what happens when the engineers start drilling! We find oil for them—65 feet of pay—high potential, the works, and then they start development drilling of offsets and things begin to turn to mud! *(Chuckles around the table)*

KEN SMITH: O.K., but if you guys would ever find a bigger field we'd be able to do better. Besides, who was it that finally made some oil from that sorry discovery of yours up in Plymouth County last—

The friendly rivalry between exploration and producing departments is interrupted as Henry Oilfinder, staff geologist, enters.

TOM MILLER: Good morning, Henry. What do you have today?

HENRY OILFINDER: Good morning, Mr. Miller. I'd like to recommend the drilling of an exploratory test on the Blackduck acreage in Woodbury County. It will be a 15,000' test of the Akron sandstone.

TOM MILLER: Just a minute, Henry, I've lost my bearings. Where is this prospect from the Pinetree Field?

HENRY OILFINDER: Sir, this prospect is about 25 miles north of Pinetree. It will be testing the same formation and we think the anomaly is right on strike with Pinetree.

TOM MILLER: O.K., thanks, I see it on the map now. Go ahead.

HENRY OILFINDER: Well, Sir, we estimate reserves of about 16 million barrels on the Blackduck prospect, which will give us a net profit, after development costs and operating expenses, of $12 million. The wildcat will cost about $1 million. Here is our interpretation of how the Akron sand extends north from Pinetree. . . .

(He presents several maps, log cross-sections, seismic interpretations and the specific geological details of the prospect. Later in his presentation . . .)

Here is a summary of the economics of the wildcat:

Estimated net pay—50 feet
Estimated size of anomaly—1600 acres
Estimated recovery—200 Bbls/Acre-ft
Recoverable reserves: 16 million bbls
Net Revenues (at $2.00/Bbl): $32 million
Estimated development costs: 20 wells on 80 acre spacing, at $1 million each, or $20 million
Estimated net profit: $12 million
Chance of success: 10%
Wildcat dry hole cost: $1 million

TOM MILLER: Thank you, Henry. Does anybody have any questions or comments?

DAVID JONES: I'd like to point out that the primary term on the Blackduck lease is only two years, so we are going to have to decide what we want to do with the six-section block pretty soon. If we don't drill we are going to have to get started trying to farm it out.

KEN SMITH: Henry, how did you get the numbers in your reserve calculation?

HENRY OILFINDER: Well sir, largely by analogy with Pinetree Field. The Johnson well had 65 feet, so I just used 50 feet as an average value of net pay. The 1600 acres of closure represents about an average of the seismic data interpretation. The recovery factor of 200 Bbls/acre-ft was based on porosities, water saturation, etc. of the Johnson and Smith wells in Pinetree Field.

KEN SMITH: Well the south offset to the Johnson well has only 35 feet of pay, so it certainly seems possible to me that 50 feet might be a little high. We're risking a million dollar well, and if reserves are reduced by, say 50%, we'd end up with just 8 million barrels. Let's see—that's $16 million revenue less development costs of $20 million—we'd end up losing $4 million even if we found oil.

JOE BROWN: That's true, Ken, but it can also go the other way. Henry and Walter Klemme of geophysics reviewed the "seis" data with me in detail yesterday, and it's possible to interpret the data to show closure of as much as 2400 acres. The economics that Henry has given to you are representative of the average values, in our opinion. Sure, we agree the values can vary either way, but we have to hang our hat on something, and we think these are the most probable values.

HENRY OILFINDER: I'd like to add that the nearest well, the Perkins Unit, is 18 miles from the prospect, so we really have very little control. It seems to me that these average values are about all we can really say about it.

TOM MILLER: Why was the Perkins well dry?

HENRY OILFINDER: Sir, it encountered the Akron sand low and drill-stem tested all water. The logs had good porosity, though, and it seems to fit the general picture we have about the Akron.

KEN SMITH: Henry, I hate to keep after this, but it still bothers me that we could get large variations in these numbers that would really kill the economics of this thing. What do you think the lowest possible values of reserves and economics are? And what do you think would be the highest value of reserves?

HENRY OILFINDER: I calculated these numbers this morning, and here is the summary:

	Minimum Case	Average or "Most Likely" Case	Maximum Case
Net Pay (feet)	30	50	70
Acres	400	1600	2400
Recovery (Bbls/acre-ft)	150	200	225
Reserves (MM Bbls)	1.8	16	37.8
Net Revenue ($MM)	$3.6	$32	$75.6
No. of wells	5	20	30
Dev. Costs ($MM)	$5	$20	$30
Net Profit ($MM)	−$1.4	$12	$45.6

KEN SMITH: Boy, that's quite a range! I might add that there's another uncertainty to consider. You've used $1MM as the average well cost. We've been doing some estimates of well costs down in engineering and we've determined that $800,000 is about as low as we can get on development drilling in Pinetree when everything goes well. But it is my understanding that they really had a bad lost circulation problem in the Mississippian at about 12,000 feet in the Perkins Unit (the dry hole 18 miles from Blackduck) and it ended up that they sunk $2 million in that hole.

HENRY OILFINDER: Yes sir, I realize that. All of these costs were based on an average of $1 million per well.

TOM MILLER: Well, let's see if I understand our decision options correctly. We are risking a million dollar wildcat against a 10 percent chance of finding what you estimate will be oil worth a net of $12 million dollars. Is that right?

HENRY OILFINDER: Yes sir, that's correct.

TOM MILLER: Well, the odds are 9 to 1 of a $1 million loss, so we'd have to get at least $9 million if we hit to break even on a risk basis wouldn't we?

HENRY OILFINDER: Yes sir, and we are estimating $12 million, so the odds for profit are in our favor.

TOM MILLER: The 10% chance of oil is pretty critical, too, isn't it! What did you base that number on and what would be your recommendation if we only had a 5% chance of hitting?

HENRY OILFINDER: Well sir, as you are aware, that is a very difficult number to get, especially since there has not been a great deal of exploration in the area to base a success ratio. I guess it just seems like a realistic number, and reasonably close to our overall wildcat success ratio. I really have nothing more concrete to base the number on

TOM MILLER: I see. (*He pauses to think a moment.*) Well, Henry, you've done a very good job on this prospect, and at first glance it seems like a good deal. But I share some of Ken's concern about possible variations in some of the key factors. Perhaps we ought to think about it for awhile and see if these variations might cause the entire deal to turn sour. Bill, you haven't said anything so far—do you have any comments to make?

BILL DAVIS: Yes sir, I do. I can't vouch for the accuracy of geology or engineering data, but I would like to make a proposal on how we could perhaps tie all these variations in possible economics together. I have a chap on my staff who has done some work with decision trees and expected value. It seems to me that there are really four things that could happen if we drill: a dry hole, a marginal field of 1.8 million barrels, a medium sized field of 16 million barrels, and a large field with 37.8 million barrels. Then, in addition, there is apparently a variation in possible drilling costs for each of these four outcomes. If we can get a handle on the probabilities of occurrence of all of these possible outcomes we could make a series of expected value calculations to see if the project has a chance of profit.

TOM MILLER: What do you mean by a chance of profit?

BILL DAVIS: Well, I mean we could calculate the expectation of the decision alternative of drilling, and if it is positive, it means that statistically we'd make a profit.

TOM MILLER: Well Bill, I'm not sure I completely understand what you just said, but it's worth a try. Could you work with Henry on this?

BILL DAVIS: Sure, Tom, I'd be glad to.

TOM MILLER: What do you think, Joe? (*turning to Joe Brown, exploration manager*)

JOE BROWN: I'm agreeable. We are going to have a tough time with those probability numbers, but we will be glad to work with Bill. We'll try to get these numbers together before Friday's meeting.

TOM MILLER: Fine. Thank you, Henry.

HENRY OILFINDER: Thank you, sir. (*Henry Oilfinder leaves room.*)

TOM MILLER: He's got a nice prospect there, Joe, but I believe we'd better look at economics a little closer. We would have a lot of money riding on that single well.

JOE BROWN: I agree, Tom. You know, this is a good example of one of the biggest problems we have—that is to try to quantify some of these unknowns when we explore out in the boondocks. Our geologists, I believe, recognize the risk and uncertainty, but it sure is hard to boil it down to simple numbers.

TOM MILLER: Well, I certainly understand. Maybe after we see Bill Davis' analysis we might be able to tie all these things together better. Ken, are you ready with those offsets?

KEN SMITH: Yes sir, he will be up in a minute.

. . . (To be continued) . . .

QUESTIONS

1. What is your overall reaction to the discussions of the Executive Committee?

2. Do you feel Mr. Miller was justified to ask for further consideration and study?

3. Mr. Miller made the comment that with 9 to 1 odds of a dry hole and a $1 million loss, the field would have to be worth at least $9 million net to "break even on a risk basis." Was this a correct statement? Was Henry Oilfinder's reply correct?

4. How do you suppose Bill Davis' economic analysis group will go about solving the problem of combining profitability numbers with risk numbers?

4 Decision Tree Analysis

The decision tree can clarify for management, as no other analytical tool that I know of, the choices, risks, objectives, monetary gains, and information needs involved in an investment problem

— Magee

MOST drilling prospects involve only a single decision that is made at time zero. This single decision would be, for example, "drill," or "farm out," etc. Once the decision is made there are no later contingencies or decision options with which the decision maker becomes involved. For these types of decision choices the procedures for making expected value computations given in Chapter 3 are completely adequate. The analysis would involve a simple expected value computation of the type shown in Tables 3.1, 3.5, and 3.6.

There are, however, certain decision alternatives of a more complex nature that can't be analyzed in a simple, four column EV computation. Consider the basic, initial decision of whether to bid on an offshore lease. The immediate decision choices are to bid or not bid. But if the decision maker bids (and wins the tract) he then has a decision regarding whether to drill immediately or run more seismic.

If he drills an exploratory well that is dry or only a "teaser" he has another decision whether to drill another wildcat test or drop the acreage, etc. It becomes fairly obvious that the initial decision is but a link in a chain of future decision options and contingencies. And all of these future options must be considered when evaluating the feasibility of bidding. The simple decision whether to bid has suddenly become quite complex and fraught with future decision options and "what if's" at nearly every step of the anticipated exploitation of the lease or tract.

These types of decisions require a slight modification in our thinking with respect to the expected value concept. The logic, philosophy, and decision rules remain the same, but the mechanical steps to determine expected values for the initial choices are slightly different.

107

The analysis involves constructing a diagram showing all the subsequent chance events and decision options that are anticipated. The "decision diagrams" look very much like a drawing of a tree, and early in the development of the analysis technique people began calling the diagrams "decision trees." This name seems to have captured the imagination of analysts and decision makers alike, with the result that the technique to evaluate these more complex, sequential decision choices is now given the name *decision tree analysis*.

In this chapter we will attempt to define decision trees as well as to show how to solve a decision tree. You should remember that the technique is merely an extension of the expected value concept, and no new theory or logic is involved here. The only change with respect to our discussions of Chapter 3 relates to the mechanical steps required to compute the expected value of each of the decision maker's initial alternatives.

We will first discuss a few terms and definitions using a very simple numerical problem. Then we will consider a more complex example which can only be solved using decision tree analysis. And finally, we will discuss briefly some general rules of algebra and comments relating to decision tree analysis.

DEFINITIONS

There are a few definitions that we must know in order to construct and solve a decision tree. We will illustrate these definitions using the very simple numerical example shown in Table 4.1. This problem is

Table 4.1
Data for Example Drilling Prospect

Possible Outcomes	Probability of Occurrence	DRILL		DON'T DRILL	
		Conditional Monetary Profits	Expected Monetary Values	Conditional Monetary Profits	Expected Monetary Values
Dry Hole	0.7	−$ 50,000	−$35,000	0	0
2 BCF	0.2	+ 100,000	+ 20,000	0	0
5 BCF	0.1	+ 250,000	+ 25,000	0	0
	1.0		EMV = +$10,000		EMV = 0

(BCF stands for billion cubic feet of gas reserves.)

simple because it involves only an initial decision to either drill or not drill. And, in fact, we can compute expected monetary values in the usual manner as shown in Table 4.1. Based on the decision rule of maximizing EMV the choice would be to drill. Nothing new here — it's the same four column computation we used in Chapter 3.

This simple problem could also have been solved using decision tree analysis, and the first step would have been to draw a diagram of the decision choices and the possible outcomes. The diagram, or tree, for this example is shown in Figure 4.1.

A decision tree is merely a pictorial representation of a sequence of events and possible outcomes. The decision tree in Figure 4.1 illustrates the decision alternatives "drill" and "don't drill." If the well is drilled there are three possible outcomes: a dry hole, reserves of 2 BCF, or finding reserves of 5 BCF.

There is no scale to a decision tree, so the lengths of the lines, or branches, have no significance. Also, the angles between the branches have no meaning so the analyst need not worry about being precise in drawing the tree itself. The trees normally read from left to right and are drawn in the same order as the actual sequence in which the decision choices and chance events occur in the real world.

The point from which two or more branches emanate is called a node. A node surrounded by a square denotes a *decision node,* a point at which the decision maker dictates which branch is followed. An encircled node is called a *chance node,* a point where chance determines the outcome.

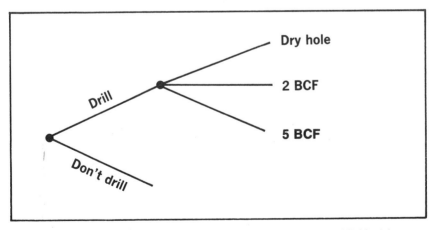

Figure 4.1 *Partially completed decision tree for example of Table 4.1.*

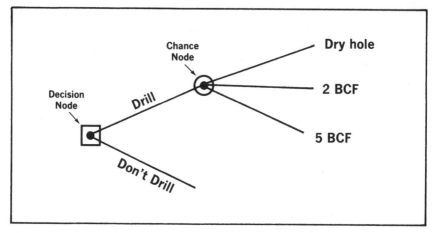

Figure 4.2 *Partially completed decision tree with chance and decision nodes indicated.*

Any number of decision alternatives or outcomes can emanate from a given decision or chance node. These nodes are indicated in Figure 4.2. It is important that all nodes be indicated as being either a chance node or decision node. (We won't be able to solve the tree if we don't!).

The convention of using circles and squares originated in some of the early papers on decision tree analysis and the tradition has followed through to this time. A decision tree can be drawn with two or more chance nodes in sequence as well as two or more decision nodes in sequence.

The next step in constructing a decision tree is that we must associate probabilities of occurrence to all the branches radiating from chance nodes and we must specify conditional values received at the endpoints of the tree. For our particular example, these figures were listed in Columns 2 and 3 of Table 4.1, and are shown on the tree in Figure 4.3.

In this example the conditional values received are expressed as monetary values, however we have the same options here that we discussed on page 71 relating to expected value. That is, we could use costs, preference values, opportunity losses, net present value profits, etc. at the endpoints. The decision tree as drawn in Figure 4.3 is interpreted as follows: There is probability 0.7 of drilling a dry hole with a resulting loss of $50,000; a 0.2 probability of finding 2 BCF reserves valued at a net profit of +$100,000; etc.

When indicating probabilities on a decision tree we must adhere to two important rules. First, the sum of the probabilities around a given

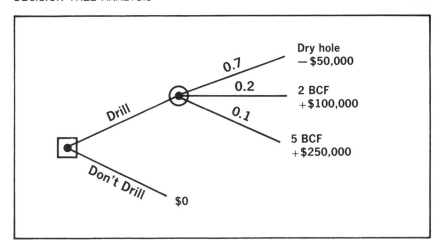

Figure 4.3 *Completed decision tree with monetary values and probabilities of occurrence shown for the outcomes of the chance node.*

chance node must add to 1.0. And second, no probabilities are shown on the branches emanating from a decision node. The probabilities around a chance node indicate the relative likelihoods as to which specific outcome (branch) will, in fact, occur. The specific path that occurs from a decision node, however, is not a function of chance, but rather the result of an explicit choice by the decision maker. (This assumes of course he doesn't make up his mind by throwing darts at a dart board!)

The ends of a decision tree are called *terminal points*. The adjective "terminal" means just that—there are no further contingencies, decisions, or chance events beyond that point. The terminal points of a decision tree are all said to be mutually exclusive points. This means that we will ultimately end up at one, and only one, terminal point on the tree. The cause of uncertainty is, of course, due to the fact that when the decision maker makes his initial decision he can't be sure at which of terminal points he'll end at. He can never escape the element of chance, or Lady Luck!

Once the tree has been drawn and we've indicated all of the probabilities and terminal point values received we are ready to solve the tree.

SOLVING A DECISION TREE

Decision tree analysis is one of the few things we solve by starting at the end and working backward. We start by making an expected value

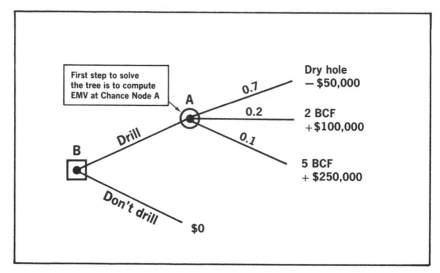

Figure 4.4   *Decision tree for example of Table 4.1 showing the first step in the solution of the tree.*

computation using the terminal points around the last chance node in the tree. For our simple example we first make an EMV computation at the chance node labeled A in Figure 4.4.

This EMV computation involves multiplying the probabilities of occurrence of each possible outcome (branch) by the corresponding terminal point values received. For the example of Figure 4.4 the computation is as follows:

$$\text{EMV}_A = (0.7)(-\$50,000) + (0.2)(+\$100,000) + (0.1)(+\$250,000)$$
$$\text{EMV}_A = +\$10,000$$

This expected value is written above chance node A as in Figure 4.5.

Next we proceed backward (to the left) in the tree to the next node, which in this case is a decision node. Now we make a decision as if the decision maker were actually standing in the square. His choices are summarized as having an expectation of +$10,000 if he drills and having an expectation of zero if he doesn't drill. That is, his choices at decision node B could be represented schematically as in Figure 4.6. Note that the EMV of +$10,000 at chance node A is used to represent, or replace all the tree beyond that point. It's as if we completely blank out the chance node and it's branches and replace them by their equivalent—the EMV at chance node A.

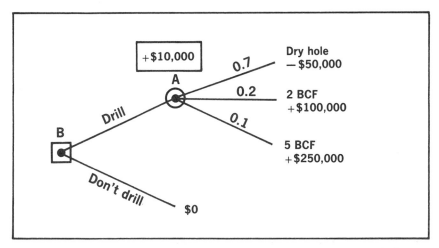

Figure 4.5 *Decision tree showing the expected value at chance node A.*

The decision rule at any decision node is to select the alternative (branch) which has the highest EMV. In this case it would be to drill. Hence the analyst crosses out the "don't drill" branch as being a sub-optimal strategy and moves the expected value for the alternative selected back to the decision node, as in Figure 4.7. Now everything to the right of decision node B is replaced by the expectation shown above the node, +$10,000. Next we would continue backward in the tree to the next node.

Each time we come to a chance node we make an EMV calculation, and each time we come to a decision node we make a hypothetical

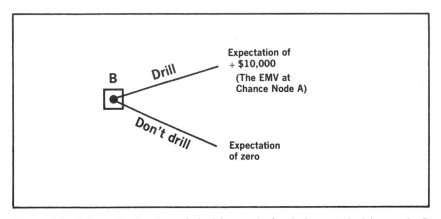

Figure 4.6 *Schematic drawing of decision maker's choices at decision node B.*

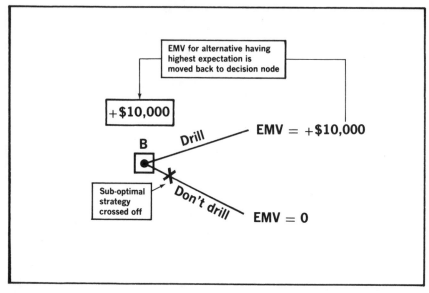

Figure 4.7 *Schematic drawing showing the decision that is made at decision node B.*

decision. This process is continued until the initial, time-zero decision node is reached, at which time the tree has been solved. In our simple illustration there was nothing to the left of decision node B, so our tree was solved at that point. The solution, by the way, was to "drill" — exactly the same solution we made on the basis of the EMV calculation of Table 4.1. In fact, we didn't really need a decision tree for this example because there were no subsequent decision points beyond the initial decision of whether or not to drill. We chose this simple example to illustrate the definitions and solution of a decision tree.

Decision tree analysis is pretty simple and logical, isn't it! As far as the definitions and mechanics of solution are concerned it is simple. The difficulties usually arise in being able to correctly organize and draw the trees for decisions of a more complex nature. In fact beginners with decision tree analysis, after trying to analyze real-world exploration problems that get a bit complex, frequently wonder when does a decision tree become a bushy mess!

My experience has been that it just takes a bit of practice, and with a little patience almost anyone can formulate and analyze some pretty complex decision strategies. When you finally stop to consider the whole technique it is all very logical, and the solution of a tree is just a series of repetitive EV calculations and hypothetical decisions. Let's practice with an example that is a bit more complicated. This example, by the way, can not be solved by any other method of analysis. We can not solve this problem with a simple four column EV calculation such as Table 4.1.

Example

Our management is considering the purchase of the leasehold drilling rights on a large block of acreage in Buena Vista County which, from available geologic control, appears to have a potential oil structure. The bonus required to secure the drilling rights is $3,000,000. We have no seismic coverage in the area, and a detailed survey is estimated to cost about $2,000,000. While there are many levels of reserves possible, we will consider for this example only two general field sizes, given that the structure has oil: a large reserve worth an estimated NPV profit of $40,000,000 and a smaller, marginal reserve worth an estimated NPV profit of $15,000,000. These profits are net after all operating expenses, taxes, royalties and subsequent development drilling costs. They do not include the wildcat discovery drilling costs. (The assumption of only two discrete field sizes was a simplification for this example. In Chapter 8 we'll see how we could have considered the many possible levels of reserves.)

From a preliminary analysis of the prospect our exploratory staff has suggested two possible exploration strategies if we had the leasehold drilling rights. Namely:

A. We could begin at once with plans to drill an exploratory test on the basis of present geologic interpretations and extrapolations.
B. Or, to be certain as to whether, in fact, a structure exists under the block of acreage, we could spend $2,000,000 for seismic and defer decisions about drilling until we've reviewed the seismic data. Seismic is thought to be quite reliable in this area, so the only uncertainty here involves whether or not a structure (and possible oil trap) exists. Our staff assesses the odds as being about 50–50 that their geologic lead is, in fact, a structure. As a further complication our staff has indicated it may even be feasible to consider drilling a second exploratory test if the first well encountered a structure but tested dry.

After a first meeting on the acreage purchase, our management seemed receptive to the prospect but felt that it would be wise to evaluate some of these alternate strategies and "what ifs" before committing $3,000,000 for leases. They stressed the need to consider these points initially by reminding us of the company's recent fiasco in the Hawarden basin in which we jumped in with $5.5 million dollars to secure a lease position before completely evaluating and anticipating the subsequent exploration and development costs and risks. (You can guess the results from management's use of the word fiasco!)

So we have been given the task of evaluating the feasibility of spending $3,000,000 for a block of acreage in Buena Vista County. And the inevitable question — what do we do now?

Well, we can all probably agree that this analysis falls into the general category of sequential decision problems because it involves management options beyond the initial "Buy" — "Don't buy" decision. So we will have to use a decision tree analysis if we have any hopes of computing expected values. Do you have any idea what the tree would look like? We've drawn the tree here, but before looking at it see if you can draw the tree for this example. Then compare your tree to the one shown in Figure 4.8.

Figure 4.8 is only partially completed because we still have not added the probabilities of occurrence around the chance nodes and the values received at the 15 terminal points of the tree. With regard to the interpretation of the tree we would mention that at terminal point (n) there probably should be a decision node with the choices of "drill" or "drop acreage." The tree was drawn without this option and shows only the terminal point "drop acreage" if the seismic shows that no structure exists. The implied assumption is that if no structure existed the option of drilling would be ruled out automatically.

Next we must add the monetary profits and losses associated with each terminal point. This is quite simple to do — merely add up all the costs and profits that had to occur to reach a given terminal point. For example, consider terminal point (k). In order for us to end at this point the following things would have had to happen:

a. We purchased acreage for $3,000,000.
b. We decided to run seismic for $2,000,000.
c. Seismic confirmed the presence of the structure and we decided to drill first wildcat costing $1,000,000.
d. The well was dry but we had decided to drill again costing another $1,000,000. (Total expenditures thus far, $7,000,000).
e. The second well was also dry and we abandon the lease and write off a $7,000,000 loss.

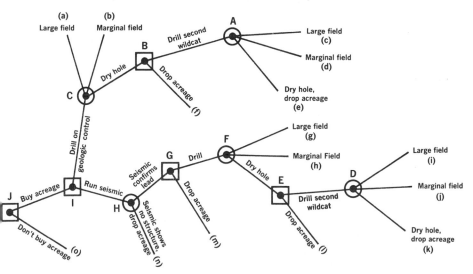

Figure 4.8 *Partially completed decision tree for example. Chance and decision nodes have been designated with capital letters (A, B, C, . . .), and the terminal points have been designated with lower case letters (a, b, c, ...). There are 15 mutually exclusive terminal points in this tree.*

This same approach is used for all the other terminal points. Table 4.2 summarizes the terminal point profit computations, and the values are shown on the tree of Figure 4.9.

The probabilities of occurrence associated with each of the chance node outcomes are determined by an analysis of risk of the type we will discuss in later parts of this book. For now we again ask your acceptance of the probabilities used in this example. We'll worry about how to get these most important numbers beginning with Chapter 6. The probabilities used are tabulated in Table 4.3 and shown on the appropriate branches of Figure 4.9.

The probabilities include the likelihood of the structure having oil as well as the likelihood we can locate the reserves. As noted from Table 4.3, these probabilities improve as more information is known to us — a fact that should be generally accepted by most explorationists. The probabilities around chance node (H) follow from the earlier observation of a 50–50 chance that the seismic will confirm the geologic lead.

The completed decision tree shown in Figure 4.9 is now ready to be

Table 4.2
Computation of Net Profits at Terminal Points (Units of Millions of Dollars)

Terminal Point Designation from Figure 4.8		Net Profit ($MM)
(a)	Large field less one wildcat and purchase: $40 - 1 - 3$	$36MM
(b)	Marginal field less one wildcat and purchase: $15 - 1 - 3$	11
(c)	Large field less two wildcats and purchase: $40 - 1 - 1 - 3$	35
(d)	Marginal field less two wildcats and purchase: $15 - 1 - 1 - 3$	10
(e)	Two dry holes and purchase: $- 1 - 1 - 3$	−5
(f)	One dry hole and purchase: $- 1 - 3$	−4
(g)	Large field, less wildcat, seismic, and purchase: $40 - 1 - 2 - 3$	34
(h)	Marginal field, less wildcat, seismic, and purchase: $15 - 1 - 2 - 3$	9
(i)	Large field, less two wildcats, seismic, and purchase: $40 - 2 - 2 - 3$	33
(j)	Marginal field, less two wildcats, seismic, and purchase: $15 - 2 - 2 - 3$	8
(k)	Two dry holes, seismic, and purchase: $- 2 - 2 - 3$	−7
(l)	One dry hole, seismic, and purchase: $- 1 - 2 - 3$	−6
(m)	Seismic and purchase: $-2 - 3$	−5
(n)	Seismic and purchase: $- 2 - 3$	−5
(o)	No action, (no profit or loss)	0

solved. As we indicated previously, the method of solution is to start at the terminal points and begin working backward through the tree. In this example there are only two places we can start: at terminal points (c), (d), and (e); or at terminal points (i), (j), and (k). (Why can't we start elsewhere in the tree, such as terminal points (a) and (b)?)

For convenience let's start at the points (c), (d), and (e). The first node we come to from these terminal points is chance node (A). Since it is a chance node this means we must make an expected monetary value computation at this point. The expected value at chance node A is computed as

$$EMV_A = (0.075)(+\$35MM) + (0.075)(+\$10MM) + (0.85)(-\$5MM)$$
$$= \underline{-\$0.875MM}$$

Moving backward (to the left) the next node is Decision Node B. Here we make a decision as if the state of events and choices had led us to this point. The choices are as shown in Figure 4.10. If he drills a second well, his expectation is a loss of −$0.875MM; if he drops the acreage at

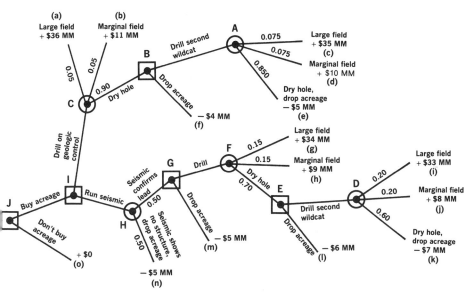

Figure 4.9 *Completed decision tree of Figure 4.8 showing terminal point monetary value profits (in units of millions of dollars, $MM) and the probabilities of occurrence around each chance node. The net profit computations are summarized in Table 4.2 and the probabilities used in the tree are summarized in Table 4.3.*

this point he writes off a −$4MM loss. The decision rule would be to drill a second well because the expectation is higher (less negative) than the alternative of dropping the acreage. So he crosses off the "Drop Acreage" strategy as sub-optimal and moves the EMV_A of −$0.875 back to decision node B.

Table 4.3
Probabilities Used in Figure 4.9

Outcome	One Wildcat Drilled on Geologic Control	Two Wildcats Drilled on Geologic Control	One Wildcat Drilled on Seismic	Two Wildcats Drilled on Seismic
Disc. of large field	0.05	0.075	0.15	0.20
Disc. of marginal field	0.05	0.075	0.15	0.20
Dry hole	0.90	0.850	0.70	0.60

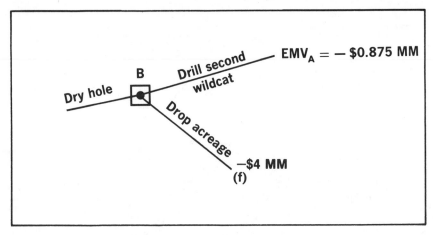

Figure 4.10 *Decision choices at B.*

Now proceeding backward again we come to chance node C. The expected value computation is as follows:

$$\text{EMV}_C = (0.05)(+\$36\text{MM}) + (0.05)(+\$11\text{MM}) + (0.90)(-\$0.875\text{MM})$$
$$= \underline{+\$1.5625\text{MM}}$$

Notice that originally there was no profit or loss associated with the "dry hole" branch at chance node C — but rather a decision node and subsequent decisions and chance events. By working backward, as we are doing, all the consequences which follow from the initial dry hole are represented in the −$0.875MM expectation at decision node B. Thus by using this value as an artificial terminal point on the "dry hole" branch, we are able to compute an EMV at chance node C.

Continuing our backward series of steps the next node we come to is decision node I. If the decision maker were at this point he would have two alternatives: drill on geologic control with an expectation of +$1.5625MM (the EMV we just computed at chance node C) or run seismic. But obviously we can't make this decision yet because we have not computed an expectation that summarizes everything that follows from the decision to run seismic. So this means we have to stop momentarily and work backward from terminal points (i), (j), and (k) on the lower portion of the tree. Before doing this we can summarize our calculations thus far by showing the upper portion of the tree, as in Figure 4.11.

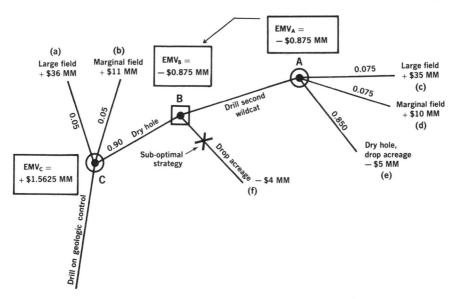

Figure 4.11 *Solution of the upper "Drill on geologic Control" portion of the tree.*

Solution of the lower main portion of the tree following the "run seismic" option at decision node I follows exactly the same series of steps. First we compute an EMV at chance node D as follows:

$$\text{EMV}_D = (0.20)(+\$33\text{MM}) + (0.20)(+\$8\text{MM}) + (0.60)(-\$7\text{MM})$$
$$= +\$4.0\text{MM}$$

Next we make a decision at node E from among the choices of: drill a second wildcat with an EMV of +$4.0MM, or drop the acreage and write off a −$6MM loss. Obviously we would choose to drill again so the EMV_D is moved back to decision node E and the branch "Drop Acreage" is crossed off as sub-optimal.

Now we can compute an EMV at chance node F as follows:

$$\text{EMV}_F = (0.15)(+\$34\text{MM}) + (0.15)(+\$9\text{MM}) + (0.70)(\$4\text{MM})$$
$$= +\$9.25\text{MM}$$

Next we find ourselves at decision node G with the choices of: drill the first wildcat with an expectation of +$9.25MM or drop the acreage with a −$5MM loss. The choice here is obvious — we would drill, so we move the EMV at F back to the decision node G and cross off the "drop acreage" option.

Chance node H, the next point in this backward sequence, requires another EMV computation as follows:

$$EMV_H = (0.5)(+\$9.25MM) + (0.5)(-\$5.MM) = \underline{+\$2.125MM}$$

With this computation we are back again at decision node I. Our computations and decisions on the lower main branch are summarized in Figure 4.12.

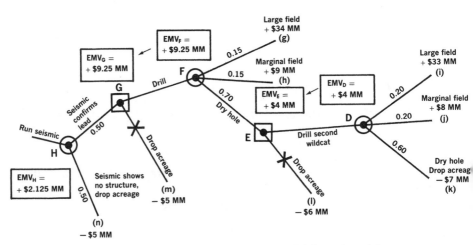

Figure 4.12 *Solution of the lower "Run Seismic" portion of the tree.*

Now we can make the decision at I based on the two choices: drill on geologic control with an expectation of +$1.5625MM or run seismic with an expectation of +$2.125MM as in Figure 4.13. By the rule of maximizing EMV the choice would be to run seismic, and the expectation of this choice is written at node I. The alternative of drilling on geologic control (without running seismic) is crossed out as an inferior strategy.

Finally the initial decision has been reduced to the choices shown in Figure 4.14. Presumably the "Buy Acreage" would be selected and the tree has been solved!

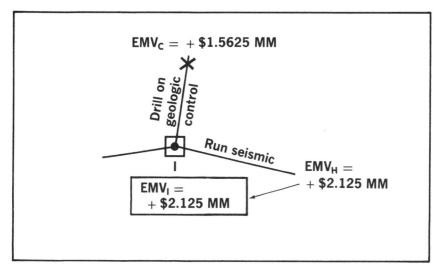

Figure 4.13   *Decision choices at decision node I.*

That took a little work, didn't it! But hopefully you will agree that we've been able to analyze a rather tricky set of options and alternatives in a systematic way. This is probably a good example supporting a comment we made in Chapter 1 that decision analysis provides a means to look at all the pieces and how they relate to the whole, rather than just looking at the whole and trying to decide if the prospect is feasible.

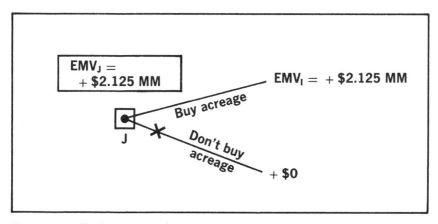

Figure 4.14   *Final resolution of the initial decision choices after solution of the entire tree.*

As a result of this analysis we have charted a very specific set of actions for the decision maker:

1. Buy the acreage and immediately run seismic.
2. If the seismic confirms a structure drill a wildcat.
3. If the initial well is dry drill a second exploratory well rather than releasing the acreage.

All these guidelines were developed *before* the money was spent for the leases, and this should certainly be a more rational approach than spending $3,000,000 for leases and then stop to figure out how (or whether) to proceed with the exploration of the tract.

ADVANTAGES OF DECISION TREE ANALYSIS

The advantages of this form of analysis include the following points.

1. All contingencies and possible decision alternatives are defined and analyzed in a consistent manner. The complex decision is broken into a series of small parts, then the parts are "reassembled" piece by piece to provide a rational basis for the initial decision.
2. Such an analysis provides a better chance for consistent action in achieving a goal over a series of decisions. That is, each step in the sequence has been analyzed ahead of time. This reduces the likelihood of a decision maker finding himself at some future option point (such as "should I drill a second wildcat or quit?") wondering how the sequence of events ever led him to that point and what should he do next.
3. Any decision, no matter how complicated, can be analyzed by the method.
4. The entire sequential course of action is set out prior to the *initial* decision. This is a good feature for delegating authority.
5. The decision tree can be used to follow the course of events. At any decision node, if conditions have changed, the remaining alternatives can be re-analyzed to develop a new strategy for that point forward.

My experience in teaching decision tree analysis to geologists, engineers, and managers has been that most people feel the idea is very logical, straightforward, and easy to use. It is, after all, really just a formal way of thinking in a clear and logical manner. Most also feel that an added benefit from decision tree analysis is that it requires that we think about all the possible future contingencies, options, and "what if" situations. Just drawing the tree will force us to think about the problem in more detail than most of us have been doing in the past.

To use decision tree analysis effectively in petroleum exploration decisions requires that the analyst have a degree of confidence in his ability to construct and solve decision trees. And this comes with a little practice. After you have solved two or three examples it should become an analysis technique that you can use effectively in analyzing the more complex exploration decisions.

A FEW GENERAL COMMENTS ABOUT DECISION TREE ANALYSIS

Before we close our discussion on decision trees we should mention several "rules of operation" or rules of algebra, so to speak, that apply to decision tree analysis.

1. The analyst has two options with respect to computing the monetary values received at the terminal points. One way is to carry all the costs and profits out to the ends of the tree and make a single accounting for them at each terminal point. This is what we did in the example problem above (Table 4.2). The other option is to indicate on the tree itself when any money is spent or received, and at the terminal points list the gains or losses of that single outcome or decision.

    Either way will lead to precisely the same answers and either method is perfectly correct. You are cautioned, however, that you must be consistent in the tree and use either the first or the second approach, but *not both!* You must be consistent and write off all gains and losses at the terminal points or you must enter the monetary values in the tree as they occur.

    The reason we mention this is that some people get a bit confused in situations such as the decision at node B in Figure 4.10. When we solved the tree we made the decision at B that he would proceed to drill a second wildcat because an EMV of $-\$0.875MM$ was better than an EMV of $-\$4.0MM$.

    The confusion arises because some people will argue that you don't have to spend $4 million dollars to quit—you can just walk off the location at no expense. Their point is that the $4 million dollars had already been spent (for leases and one wildcat), and it costs nothing further to quit.

In point of fact this is true, but if we had accounted for our expenditures as they occurred all the cost values that entered into the EMV computation at A would have been $4 million dollars higher and the EMV_A would have been +$3.125MM. Hence the choice at B would have been: drill again with an expected *additional* gain of +$3.125MM or quit with an expected *additional* gain of zero. The decision would have been the same.

If the second option had been chosen, the terminal values would all have been proportionately higher, and solving the tree all the way back to decision node I would have given us an expectation at that point of +$5.125MM. But then the decision maker at the initial time—zero point would have to make the final observation: "If I obtain the leases my expected gain is +$5.125MM, but the cost to obtain the leases is $3MM, thus my net expected gain is +$5.125MM − $3.MM = +$2.125MM." Which is the exact value we computed previously. With either option the final EMV will be the same and all sequential decision choices will be identical.

2. What would our strategy have been if the large field was valued at a net of only $30MM, rather than $40MM? Or, what would our strategy have been if the probabilities at chance node C were 0.01, 0.01, and 0.98 respectively? These questions are perfectly valid and they suggest the importance of making sensitivity analyses to test the effects of various parameters on the overall decision strategy. How is it done? Simple.

   Just change the appropriate numerical values and re-solve the tree. The tree itself doesn't change, just the terminal point values and/or the probabilities. So once the tree is drawn we have the option of making all sorts of sensitivity tests of various parameters without having to redraw the tree itself. We'll see a good example of this in a decision tree problem in a later chapter.

3. So far we have drawn the chance nodes with only three branches, or possible outcomes. As mentioned previously there can be any number of branches emanating from a chance node, providing the sum of all the probabilities add to 1.0. Conceptually there could even be an infinite number of branches on a chance node. For example the outcomes of drilling could be a dry hole plus an infinite number of specific reserve levels.

   In this case the pictorial representation of the chance node

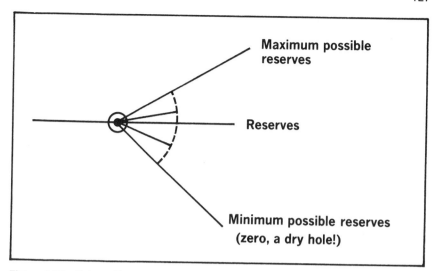

Figure 4.15 *Schematic representation of a chance node having an infinite number of branches.*

appears as in Figure 4.15. This schematic picture is sometimes called an "event fan." We will deal with these types of situations in Chapter 8 when we consider the outcomes from a decision under uncertainty as being a random variable having a distribution of possible values. Although we won't be drawing decision trees too often with a "fan" as in Figure 4.15 we will be using the concept of an infinite number of possible outcomes quite extensively in Chapter 8.

4. When all of the terminal point values received around a chance node are equal, the expected value at the chance node is equal to the terminal point value. This fact is probably obvious, and a simple illustration should suffice for proof. Consider a chance node with two branches as in Figure 4.16. The values received at each terminal point are equal, V_1. The probability of occurrence of the upper branch is p and the probability of occurrence of the lower branch is $1 - p$.

The computation of the expected value at chance node A would be:

$$EMV_A = (p)(V_1) + (1 - p)(V_1)$$

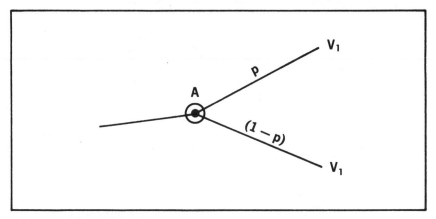

Figure 4.16 *Chance node having equal values received at both terminal points.*

But this is equivalent to

$$EMV_A = (p)(V_1) + (1)(V_1) - (p)(V_1)$$

or simply,

$$EMV_A = V_1 \qquad QED$$

5. Finally, in constructing trees having two chance events in sequence, the analyst can draw the tree in either of two ways. Let's consider some phenomenon consisting of two chance events in sequence.

First Chance Event	*Second Chance Event*
A	$\begin{cases} X \\ Y \\ Z \end{cases}$
B	$\begin{cases} X \\ Y \\ Z \end{cases}$

The first event would result in either A or B happening. The second event would result in either of three things happening, X, Y, or Z. An example might be a seismic survey. The chance events A and B could represent getting good quality records or poor quality records and the events X, Y, and Z could represent

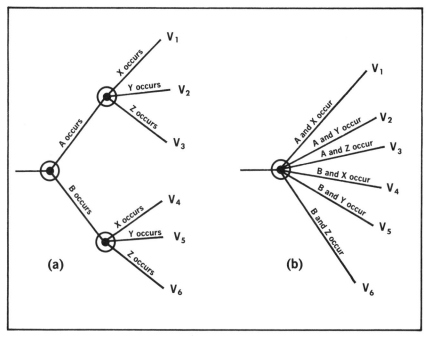

Figure 4.17 *Two ways to construct a tree having two chance events occurring in sequence. Either way is correct and both will result in the same EV at the initial chance node.*

the various possible geologic interpretations such as closed structure, structure but no closure, and no structure. Such an example might be a portion of an analysis to determine if a seismic survey is feasible from an expected value gain standpoint.

The two ways to draw situations involving two chance events in sequence are shown in Figure 4.17. We will give a formal proof in Chapter 6 as to why these two ways of representing two chance events in sequence are equivalent. The reason we are deferring the proof is that we must first define some probability terms and rules of algebra before we will be able to prove the equivalence, and it will be more convenient to discuss these points at that time.

The alternative (a) in Figure 4.17 is probably the way most people draw the trees when they are thinking in terms of one event (or decision) at a time. Alternative (b) consists, in essence, of combining the two separate chance events into one joint, or compound event (e.g. A *and* X occur in combination, A *and* Y occur, etc.).

Nothing really earth-shaking about this point other than to say either way is correct and both will result in exactly the same EV at the left (initial) chance node. Of course if a decision option exists between the first chance events (A or B) and the second chance events (X, Y, or Z) this equivalence no longer holds, and the tree must be constructed to show each event and decision step as it occurs in the sequence.

Before concluding this chapter it would perhaps be appropriate to return again to Henry Oilfinder and the Blackduck Prospect. This part concerns his presentation to the Executive Committee about a week later. During the interim he has been reassessing his initial recommendation at the request of the President, Mr. Miller. Let's pick up the story at that point. . . .

A Case Study

The Blackduck Prospect (Part II)

Friday—next meeting of Executive Committee. Recall that Mr. Miller, the president of Profit Oil and Gas Company, has asked Henry Oilfinder, staff geologist, to furnish additional economic analyses of his wildcat recommendation on the Blackduck acreage. Other members of the Executive Committee are:

Joe Brown—Manager of Exploration
Ken Smith—Manager of Production and Engineering
David Jones—Chief Landman
Bill Davis—Manager of Accounting and Economic Analysis Groups

TOM MILLER: Good morning, everyone. I understand we have a busy agenda today. What do you have, Joe?

JOE BROWN: We want to run the "Blackduck Prospect" that we had in the other day through again. We've also got a wildcat recommendation in Powder River Basin and an acreage purchase proposal in O'Brien County.

TOM MILLER: Ken, what are you up to today?

KEN SMITH: Producing has four development well offsets to bring in and a recommendation to plug our Scott Unit in Sioux County.

TOM MILLER: Is the rig still on the well?

KEN SMITH: Yes, sir. They are standing by. The drill-stem test we ran last night recovered 30 feet of slightly gas-cut mud; no sign of oil or gas.

TOM MILLER: O.K., we'd better get busy. Joe, what have you come up with on the Blackduck wildcat?

JOE BROWN: Well, with the help of some of the boys in Economic Analysis we've made a much more detailed study, including risk. It looks pretty marginal, and we will recommend farming it out. Henry will be up in just a minute to present the details.

TOM MILLER: David, what did we pay for that block?

DAVID JONES: $80,000 — 4000 acres at $20 per acre.

TOM MILLER: You mean we paid $80,000 for a lease a few months ago and now we turn around and farm it out?

JOE BROWN: Yes, sir. Of course, we didn't have the "seis" data at the time we leased it.

TOM MILLER: Maybe we'd better sharpen up our pencils from now on before we buy acreage instead of afterward. We'll never make any money that way.

Henry Oilfinder enters conference room.

TOM MILLER: Good morning, Henry. Are you ready to go over the Blackduck prospect?

HENRY OILFINDER: Yes, sir, I am.

TOM MILLER: Why don't you briefly review what you presented Tuesday to jog our memory.

HENRY OILFINDER: Certainly. (*He reviews his recommendation and presentation and summarizes the reasons he was asked to reconsider the recommendation.*) Exploration has worked closely with Mr. Davis and members of his staff, and after looking at some of the variations in reserves and development costs I'm afraid the deal doesn't look as good as before. In fact, we recommend that we farm the block out rather than risk our money in the wildcat. I'll go through the logic and numbers that we used in this analysis. The first thing we looked at was the possible variations in drilling costs. The engineers estimated that the minimum well cost in the "Blackduck Prospect" area would be $800,000, probable costs would be $1 million, and that costs could go as high as $2 million if we hit lost circulation problems. I indicated to them that we needed to associate a probability number with each of these three values of well costs. After lengthy consideration they felt that there was a 20% chance of costs about $2 million, a 10% chance of wells costing only $800,000, and a 70% chance that well costs would be about $1 million each. Using the concept of expected value, the expected cost for wells drilled in Blackduck will be about $1.18 million each.

KEN SMITH: Henry, how did you get that?

HENRY OILFINDER: Well sir, it is the sum of the expected values of the three outcomes. That is,

$$(0.1)(\$.8MM) + (0.7)(\$1MM) + (0.2)(\$2MM) = \$1.18 \text{ MM}$$

KEN SMITH: O.K. I agree, go ahead.

HENRY OILFINDER: The next step was to compute dollar profits and losses from drilling for each of the four possible outcomes. Remember that the oil is worth $2 per barrel net. These calculations are shown in this table:

Outcome	Reserves (MM Bbls)	Net Revenue ($MM)	No. of Wells	Total Well Costs @ $1.18MM Each	Net Profit ($MM)
Dry Hole	—	—	—	—	−$ 1.18
Minimum Case	1.8	$ 3.6	5	$ 5.9	− 2.30
"Average" Case	16.0	32.0	20	23.6	+ 8.40
Maximum Case	37.8	75.6	30	35.4	+ 40.20

KEN SMITH: But wait, Henry, you forgot to assign probabilities to each of these outcomes.

BILL DAVIS: Just hold your horses, Ken, he's coming to that. What he is describing
to you now is how he obtained the numbers that appear in the final decision tree.
He's going through the "data preparation" phase, in essence.

KEN SMITH: OK, sorry. I'll keep quiet for awhile.

HENRY OILFINDER: Now the next step, as you were suggesting, Mr. Smith, was to
come up with probability numbers. As I mentioned Tuesday, we feel there is a 90%
chance of a dry hole and 10% of finding oil. But, then if we found oil how much
did we find? That is, if we made a discovery which of the three field size categories
would it be? Well this again was a tough question to answer. After thinking about
it for a great length of time and discussing it with Mr. Davis, I felt there was a 10%
chance of the 1.8 million barrel field, a 75% chance of the "average," or 16 million
barrel field, and a 15% chance of finding the maximum reserve of 37.8 million
barrels.

DAVID JONES: Hold on a minute, Henry. You lost me there. First you said there was
only a 10% chance of hitting oil, but then you say there is a 75% chance that it
will be in the 16 million barrel category. That doesn't make sense to me. Do you
mean a 10% or 75% chance of 16 million barrels? Seems to me it has to be one or
the other, doesn't it?

HENRY OILFINDER: No sir, you misunderstood what I said. We feel that *if* we find oil
(of which there's only a 10% chance), then the chance that it's a 16 million barrel
field is 75%. I believe Mr. Davis called the 75% a conditional probability. It means
that something has to happen first. In this case it is finding oil and *then* we have a
75% chance of it being the average case. It means that two separate things have
to occur in order to get a 16 million barrel field: first, it has to be oil (10%), and then
at the same time it has to be 16 million barrels. The probability of both of these
things happening is 10% multiplied by 75%, or 7.5%. That number is an unconditional probability. Similarly the probability of hitting oil and finding only 1.8
million barrels is 1% and the probability of discovering the large field is 1.5%.

DAVID JONES: Great scott! The oil business will never be the same! Pretty soon we're
going to have so many probability numbers floating around we'll forget what the
original question was!

BILL DAVIS: Give Henry a chance. What he is going through is necessary to get risk
quantified in some manner. Sure, I agree it's difficult to understand at first, but
the language of economics is numbers, and we simply have to express these judgment factors as numbers. Otherwise we are back to the situation of basing our
decisions on hunches. Since we are risking a million dollars per well I sure feel
better about the deal if we try approaches such as what Henry is describing.

DAVID JONES: Well, maybe so, but I'm sure confused now. Guess I should be glad
that a landman's worst problem is just irritable royalty and landowners! I believe
they aren't as bad as all these probability numbers.

TOM MILLER: Let's let Henry get through his presentation and then you can ask questions about his approach. Go ahead, Henry.

HENRY OILFINDER: Well, to summarize that rather circular description of probabilities I have listed the outcomes and the probabilities of occurrence of each in
this table:

Outcome	Probability
Dry Hole	0.900
Minimum Case	0.010
Average Case	0.075
Maximum Case	0.015
	1.000

Notice that the probability numbers are expressed as decimal fractions instead of "% chance." Now the last step we had to consider before we drew a decision tree was the income from a farmout. Of course the outcomes are the same whether we drill it or someone else does, as are the probability estimates and reserve estimates. The only thing that changes is that we would get gross dollars, before operating expenses, etc. and there would be no investment charges. Engineering told us the oil would be worth $3 per barrel to the overriding royalty interest. This table summarizes the income received from a $1/8$ ORI:

Outcome	Reserves (MM Bbls)	Total Revenue @ $3/Bbl ($MM)	Net Revenue to ORI, $1/8$ of Total ($MM)
Dry Hole	—	—	0
Minimum Case	1.8	$ 5.4	$ 0.675
Average Case	16.0	$ 48.0	$ 6.0
Maximum Case	37.8	$113.4	$14.175

Now we can describe all of the outcomes and decision options in a graphical form called a decision tree. Each branch of the tree presents a possible outcome from Mr. Miller's initial decision as to whether we drill or farm out. At the end of each branch is the profit or loss we would realize if, in fact, we ended up at that point. These are the numbers that I just presented to you in the tables. Notice also that each branch has a probability of occurrence by it. These are the probabilities I showed you a few minutes ago. Now to solve this thing we simply multiply probabilities by profits for the possible outcomes of drilling. The sum of these products is the expected value of drilling. Then we do the same for the alternative of farming out. The results of these calculations are as follows:

Decision Alternative	Expected Value Profit ($MM)
Drill	+$0.148
Farmout	+$0.669

The decision rule with these expected value numbers is to take the alternative having the highest positive expected value, which in this case is the farmout.

So it appears that we should not risk the wildcat. That's about the sum and substance of the analysis. I hope I said everything right. I must confess I've had my nose in a probability book for the last few days trying to learn the logic of this expected value idea. Mr. Davis was a great help to me, I might add.

TOM MILLER: Thank you, Henry. Any questions?

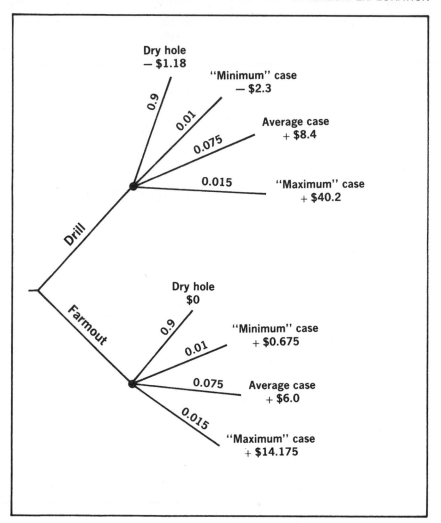

Decision tree for blackduck prospect—profits and losses are in units of millions of dollars. Prepared by Henry Oilfinder, Staff Geologist, Profit Oil and Gas Co.

DAVID JONES: Henry, you had probability numbers around some of the branches in that tree, but you didn't have any around the "drill" branch or the "farmout" branch. Why was this?

HENRY OILFINDER: (*A considerable pause, accompanied by a helpless look on Henry's face.*) Gosh, have I made a mistake? I can't seem to think of the—

BILL DAVIS: Henry, I'll answer that for you. It's because the decision maker dictates which branch we enter at the drill and farmout branch point. The outcomes with probabilities by them mean that chance determines which branch we finally enter.

HENRY OILFINDER: Oh yes, thank you. I recall that now. I believe I forgot to designate the decision nodes and the chance nodes by the squares and circles. Sorry. The squares, circles, nodes, etc. are part of the jargon of those who use decision trees.

TOM MILLER: Henry, would you state what the expected value numbers mean in words?

HENRY OILFINDER: Sir, they are the average profit we'd get over a series of such decisions. That is, if we drilled a whole bunch of these and added up our losses and gains, our average gain *per decision* would be the expected value.

TOM MILLER: I see. That sounds like the calculations they make in Las Vegas in figuring the house advantage.

HENRY OILFINDER: Yes sir, it's the same concept.

KEN SMITH: Well I'm pleased to see that we've considered these variations in our analysis, but I'm sort of like David in that I don't believe I fully comprehend all those probability calculations. But it just seems to me that his conclusion supports the feeling I have about the deal. That is, we are risking a whale of a lot of money to just barely break even. I'd much rather play it safe and take a farmout.

TOM MILLER: What do you think, Joe?

JOE BROWN: Well sir, I would like to take the deal. I think it's the type of gamble we have to live with in exploration, but I can't refute the numbers that Henry and Bill came up with. Everything they did seems quite logical.

TOM MILLER: Well, I guess we'd better see if we can farm it out to somebody. Maybe we can find someone willing to take a bigger gamble than I am. Thank you, Henry. By the way, what do you think of the approach you just described?

HENRY OILFINDER: Well sir, to be honest I'm not quite sure. The method seems very logical and I agree that we were able to consider more than just the "average," or most probable case. But the thing that bothers me is that it is so dependent on those probability numbers. And I just don't know how to get a handle on estimating all of those numbers.

JOE BROWN: I think Henry has a good point. The idea of expected value may be OK, but how in the world can we get all these probabilities?

TOM MILLER: Well I think this is a good case in point to show that we all need to do some further study on the idea, myself included. I guess this is the statistical decision theory stuff I read about in Harvard Business Review. It always seemed logical when they talked about expected values of bets or flipping coins, but I could never quite fit all the pieces together to see how it would apply to our work. Bill, why don't you continue working with the geologists and engineers on the use of these probability concepts. I'd like to see us look at more of the possible outcomes than just a dry hole or success. In fact, something has been bothering me about this. Henry described a dry hole and three possible field sizes if we hit oil. But isn't it possible that we could get something between 1.8 and 16 million barrels? Say 9 million barrels, or 9.2 million?

KEN SMITH: Yes, in fact I guess we could say there's an infinite number of possible sizes of the field. That means instead of a dry hole branch and three reserve branches in his decision tree there really should be a dry hole branch and, say, a hundred, or a thousand, or an infinite number of field size branches.

TOM MILLER: That could get awful complicated, couldn't it! How could we get probability numbers on all those thousand, or million branches? We have a difficult time with just four. But yet it seems completely logical that we could get reserves other than the 1.8, 16, and 37.8 million barrels. Hmm — an interesting problem.

BILL DAVIS: I read something the other day that I think is intended to solve that very problem. And amazingly enough it was real simple to do.

TOM MILLER: How does it work?

BILL DAVIS: Well I can't recall at the moment. But I'll dig it out and see if it might work for drilling prospects.

TOM MILLER: Please do. I think it might be worth looking into. This Blackduck well has raised my curiosity. If we hadn't started thinking about these variations we may have made a bad deal and drilled it. Thanks again, Henry. Joe, who is going to present the Powder River Basin well? . . .

(We leave the Committee as they continue their meeting.)

* *

QUESTIONS

1. If Profit Oil found a partner and drilled the wildcat on a 50–50 basis, would their expected value of profit be the same? Greater? Less?

2. If Profit had elected to drill, what is the probability of a monetary loss?

3. Was Henry's decision tree solved correctly?

4. Was Mr. Miller correct in thinking that there were many additional possible sizes of reserves which could occur but were not considered?

5. What technique do you suppose Bill Davis read about that would consider all these additional possible outcomes?

6. Do you feel it necessary to consider more than just the four outcomes which Henry considered? Or do you feel that they will have little, or perhaps no effect on the indicated decision option?

5   Preference Theory Concepts

A full purse is not as good as an empty one is bad.
— Marschak

IN this chapter we explore the fascinating, but controversial subject of preference theory. The theory relates to an extension of the expected value concept in which the decision maker's attitudes and feelings about money are incorporated into a quantitative decision parameter. The result will be, presumably, a more realistic measure of value than the EMV or expected present worth profit criteria that we discussed in Chapter 3.

Preference theory concepts are based on some very fundamental, solid ideas about decision making that are accepted by nearly everyone who has studied the theory. The real-world application of preference theory, however, is still very controversial and its value in the decision making context is questioned by some scholars and many businessmen.

In this Chapter we will discuss the issues involved with preference theory concepts and briefly describe the theoretical basis for preference theory and how it would be used in the decision making process. We will then discuss some of the practical problems that exist which make it difficult to apply to real-world decisions, and outline the research efforts that are being made to resolve the problems. And finally, we will discuss some specific applications of preference value concepts which can be of use in the petroleum industry, even though some of the basic philosophical issues remain unresolved.

In many articles and books on decision making the concept discussed in this chapter is sometimes called "utility theory." The word "utility" is somewhat like the word "average"—it can mean many different things. We will use the term "preference theory" for the simple reason that it is not as ambiguous as the term "utility theory."

THE ISSUES INVOLVED WITH PREFERENCE THEORY

In Chapter 3 we discussed the fact that the expected value concept is the fundamental basis for decision making under uncertainty. We also mentioned that the conditional values received could be expressed in several forms: monetary values, net present value profits, costs, opportunity losses, etc. We then proceeded to describe the meaning of the expected value concept by using examples in which the values received were given as monetary profits and losses.

This led to the special case of dealing with expected *monetary* values, or EMV. We pointed out that using EMV as the criterion for decision making implied that the decision maker (or his firm) was totally impartial to money and the magnitudes of potential profits and losses.

And therein lies the central issue—*people are not impartial to money!* Rather, they have specific attitudes and feelings about money which depend on the amounts of money, their personal risk preferences, and any immediate and/or longer term objectives they may have. A decision maker's attitudes and feelings about money may change from day-to-day, and may even be influenced by such factors as his business surroundings, the overall business climate at a given time, etc.

The net result of all this is that the simple, easy-to-use concept of EMV and its decision rule of maximizing EMV may not provide the most representative decision criterion. This is because the EMV parameter does not include (in a quantitative form) any consideration of the particular attitudes and feelings the decision maker associates with money.

Preference theory is an attempt to incorporate these attitudes and feelings regarding money into a quantitative parameter called *expected preference value*. This parameter would have all the features of EMV plus having the additional benefit of accounting for the decision maker's specific attitudes and feelings about money.

It is certainly not difficult to list situations in which EMV does not provide a clear-cut decision strategy. We mentioned several in Chapter 3 and here are some more.

1. Suppose the EMV (or expected net present worth profit) for a drilling venture is positive but the potential loss from a dry hole is greater than the decision maker can afford. Should he honor the EMV parameter and drill—with the hopes of not getting a dry hole?

2. An independent oil operator will frequently drill a farmout that a major oil company did not wish to drill. Assuming both had access to essentially the same data and both used EMV as a decision criterion why did they reach different conclusions?

3. Some decision makers are conservative in their preferences for

taking risks—others may be "riverboat gamblers." Given the same data both would probably reach the same decision using EMV. Yet if the drilling venture involved a great deal of risk the conservative decision maker may pass it up. Why?

Another way to bring the issues to focus is to consider the following questions as if your personal funds were involved.

1. On the flip of a fair coin you would receive $2 if a head occurs and lose $1 if a tail occurs. Would you accept this gamble?
2. Your entire fortune is $10 million. On the flip of a fair coin you would receive an additional $20 million if a head occurs and lose $10 million (your entire fortune) if a tail occurs. Would you accept this gamble?

What are your answers? Most people will answer "yes" to the first gamble and "no" to the second, which probably agrees with your answers. Of course if we based our decision to accept or reject on EMV the answers would have been to accept both gambles. (The EMV of gamble No. 1 is +$0.50, and the EMV of the second is a fantastically large +$5,000,000!) Why do most reject the second gamble, even though the EMV is positive and very large?

The obvious answer is that most people do not care to risk losing their entire $10 million fortune for an equal chance to gain another $20 million. In their view they feel they can afford the possibility of losing $1, but not $10 million. This suggests that the decision maker's asset position is one of the important "non-monetary" considerations in the decision process which may cause him to override the decision strategies (choices) suggested by EMV.

Now consider two different gambling situations. As before assume that your personal funds are being used.

3. You have saved $5 to spend for an evening of entertainment. You are then offered the chance to participate in a coin-flip gamble in which you receive an additional $5 if a head occurs and lose the $5 you have saved if a tail occurs. Would you accept this gamble?
4. You are desperate to see the big college football game. You have $5, but a ticket costs $7.50. You are then offered the chance to participate in a coin-flip gamble in which you receive an additional $5 if a head occurs, and lose your $5 if a tail occurs. Would you accept this gamble?

What are your answers to these two gambles? The usual reactions are that some would take gamble 3, some would not (probably depending on what the entertainment was!), but nearly everyone would accept gamble 4. On the basis of expected monetary values both gambles have an EMV of zero, which presumably means the decision maker is neutral, or indifferent to each gamble.

The strong preference to accept No. 4, relative to No. 3 is probably due to the realization that if you really wanted to see the game your only hope was to accept the gamble. Losing $5 in the coin-flip wouldn't be any more undesirable than not being able to see the game. This, of course, suggests that immediate (or long term) goals or objectives represent another "non-monetary" consideration in the decision process.

All of this is positive evidence that people simply are not impartial to money. The desirability and preferences we associate with various amounts of money are not proportional to the quantities of money. The proportionality is implied with EMV, but in reality under almost any decision context the proportionality between desirability and quantity of money is non-existent.

This observation about preferences and money is not particularly recent. In fact, soon after the development of the expected value concept in 1654 scholars became intrigued with the attitudes people associate with money. Perhaps the classic example was the famous Saint Petersburg Paradox. Briefly stated, the Paradox involves the following gamble:

Player A pays $1 to Player B for the privilege of playing the game. The game continues until the first tail appears on flip of a fair coin. Player A receives $1 for each head that occurs prior to occurrence of the first tail. Player A can repeat playing the game as long as he desires, but each time he must pay the $1 stake.

Question: Should A play the game? How much does he stand to gain if he does?

Solution: A reward table is constructed. (H = head, T = tail)

Outcome	Probability Outcome Will Occur	Reward to Player A, Dollars
T	1/2	0
HT	1/4	$1
HHT	1/8	$2
HHHT	1/16	$3
HHHHT	1/32	$4
. . . etc.	. . .	. . .

Player A's expected monetary profit, EMV_A, is computed as:

$$EMV_A = (1/2 \times 0) + (1/4 \times \$1) + (1/8 \times \$2) + (1/16 \times \$3) + (1/32 \times \$4)$$
$$+ \ldots - \$1$$

(the last term is the cost of playing the game)

$$EMV_A = (\$1/4) + (\$2/8) + (\$3/16) + (\$4/32) + \ldots - \$1$$

Excluding the last term, the above equation is an infinite series whose sum is +$1. Hence $EMV_A = +\$1 - \$1 = 0$. Player A's expected winnings are zero. Over a series of games he will neither win nor lose any money. In the terminology of mathematical game theory this is called a "fair game."

Now the rewards are changed slightly. Player A receives $1 for one head, $2 for two heads, $4 for three heads, $8 for four heads, $16 for five heads, etc., each time doubling the previous payoff. With this exception, the game is played as before. Naturally Player A should be required to pay a higher stake since his winnings are higher.

Question: How much can Player A afford to pay to play this revised game and still expect to at least break even?

Solution: A revised reward table is constructed.

Outcome	Probability Outcome Will Occur	Reward to Player A, Dollars
T	1/2	0
HT	1/4	$ 1
HHT	1/8	$ 2
HHHT	1/16	$ 4
HHHHT	1/32	$ 8
HHHHHT	1/64	$16
. . . etc.	. . .	. . .

Player A's expected monetary profit is:

$$EMV_A = (1/2 \times 0) + (1/4 \times \$1) + (1/8 \times \$2) + (1/16 \times \$4) +$$
$$(1/32 \times \$8) + (1/64 \times \$16) + \ldots - \text{stake}$$

or

$$EMV_A = (\$1/4) + (\$1/4) + (\$1/4) + (\$1/4) + (\$1/4) \ldots - \text{stake}$$

The sum of this infinite series is, of course, plus infinity. Therefore if Player A offered an infinitely large stake *each time he plays* his overall expectation would be zero. For any stake less than infinite Player A's EMV would be positive! And of course Player A's gain is Player B's loss.

Do you believe that? How much would you offer to Player B to play the game? If you were Player B would you have played if Player A offered you $100 each time? $1000? $1,000,000?

This simple coin-flip game seemed to imply to the early scholars that decision strategies based on expected monetary value may not be correct after all. The Paradox implied that Player A could offer any amount of money he could obtain and still win in the long run and Player B should never accept the gamble, because his EMV will always be negative for anything short of an infinite stake. And yet the scholars reasoned that for most people the reverse situation would apply: the more money Player A offered the more probable it was that A would lose in the long term, and conversely, the more Player A offered the more likely it would be that Player B would accept the money and offer to play the game!

The noted Swiss Mathematician, Daniel Bernoulli (1700–1782) studied this Paradox and about 1738 concluded that an individual's preference (utility, desirability, usefulness) for money is inversely proportional to the amount he already has. If he doesn't have much money he may associate more preference or desirability to receiving an additional $100 than he would receiving the same amount given that he already had $100 million dollars. Bernoulli's resolution to the Paradox thus made a distinction between "moral fortune" and "physical fortune."

The mathematical implication of Bernoulli's idea is that the correlation between the amounts of money and the corresponding preferences for money follows the shape of a logarithmic function. To solve the paradox he computed an expected value in which the rewards to Player A were expressed as the (natural) logarithm of $2, logarithm of $4, etc. With this modification the infinite series once again becomes convergent and has a finite sum. Thus one could compute the monetary value of the stake Player A should offer which would reflect his value system of diminishing preference with increasing wealth.

It should be noted that Bernoulli's ingenious resolution of the Saint Petersburg Paradox merely led mathematicians to devise new versions of the coin-flip game in which the expectation for outcomes of the coin-flip once again became infinitely large. This was accomplished by changing the reward schedule for one head, two heads, etc. With the proper modified rewards the expectation again became infinite, even though the individual rewards were expressed as the logarithms of money!

To summarize, our comments thus far do not prove or imply that expected monetary value (EMV) is invalid as a decision parameter. Rather, we have suggested that all of us have certain attitudes and feelings about money which EMV does not consider. These attitudes and feelings may be caused by many different circumstances which exist at the time of decision making, including such things as the decision maker's asset position, his attitudes for taking risks, and his immediate and long term objectives. We must agree that some, if not all, of these factors will be pertinent in *every* drilling decision. So the basic issues involved are the following:

1. Expected monetary value (EMV) is a completely valid decision parameter if the decision maker is totally impartial to money.
2. But very few (if any) decision makers are completely impartial to money. Most have specific attitudes and feelings about money that are caused by such factors as asset position, risk preferences, goals, etc.
3. Given these two statements the decision maker has two choices with regard to applying quantitative decision criteria to petroleum exploration investments:
   a. Use EMV as an indicator of relative value and include consideration of his attitudes and feelings about the money involved in an informal, non-quantitative (arbitrary? subjective?) manner.
   b. Incorporate his attitudes and feelings into a quantitative decision parameter having all the characteristics of the expected value concept and use the resulting numerical parameter as a basis for decision making under uncertainty.

The choices of Item 3 above basically involve whether to consider the attitudes and feelings about money in a quantitative manner or whether to consider these factors in a non-quantitative manner (as has been done in the past). You should have this distinction clearly in mind as we proceed through our more detailed discussions of preference theory. We all have specific attitudes and feelings about money—the only question is whether or not we should incorporate them into a quantitative decision parameter to use as a basis for decision making.

And we should point out the fact that our consideration of this question leads to a more basic question—can the manager make a better set of decisions over a period of time if he considers the emotional aspects of money in a quantitative manner? If the answer to this is yes (or even perhaps, or possibly) then we should be considering the implementation of preference theory concepts in our daily decision making.

I am taking considerable time to focus these key issues because it has been my experience that discussions about preference theory concepts themselves get very emotional. Often explorationists and decision makers, in the heat of an argument about whether or not preference theory has any value, lose sight of what it was intended to do. It is important that initially we have a clear understanding of the key issues.

What follows will be a discussion of a concept that was developed quite recently to consider a decision maker's emotions about money in a quantitative decision parameter. Proponents of this new approach claim that it provides the decision maker with the most representative decision parameter ever developed. They argue that its use will result in a more *consistent* decision policy as compared to using EMV and accounting for the "non-monetary" factors in an arbitrary way. On the other hand, some scholars (and probably many decision makers) feel that it is impossible to quantify one's emotions regarding money, and therefore the whole idea of preference theory is more nearly an exercise in frustration than a basis for rational decision making.

Now that the battle lines have been drawn let's explore this fascinating concept called preference theory in more detail. But don't forget that the main issue is whether or not the decision maker can do a better job of decision making under uncertainty by quantifying his attitudes and feelings about money.

THE MATHEMATICAL BASIS FOR PREFERENCE THEORY

Although Daniel Bernoulli made an attempt in 1738 to quantify an individual's emotions about money we have seen that his approach was only a first step. It was not until 1944 that a formal, mathematical theory was set forth to describe, in a quantitative sense, a decision maker's attitudes and feelings about money. This theory was developed by two Princeton University mathematicians, von Neumann and Morgenstern (Reference 5.1).

Their treatise concerned the theory of strategies against an opponent or adversary, or game theory as it is called in mathematical circles. In order to study this theory they first had to define what a person was trying to gain, or maximize. They reasoned that maximizing a monetary gain (or minimizing monetary costs) was a valid starting point.

But they were aware of these same issues we've just been discussing and concluded that a person's value system regarding money included various emotional attitudes and feelings. So, in the absence of any prior theory, they developed the mathematical basis for what we here call preference theory. This pioneering work, while merely a preface to their more immediate objective of considering game theory, was to have a significant impact on our understanding and approaches to decision making in the business context.

A mathematical theory has as its basis one or more axioms. An axiom is simply a statement, or starting point from which the theory is developed. The resulting theory has meaning only to the extent that the axioms upon which it was based have meaning. von Neumann and Morgenstern started with eight axioms which they considered to be the basic fundamental logic of rational decision making. We won't go into details of this, other than to mention two or three of them to illustrate their general nature. One of the axioms:

a. If an individual has two choices, A or B he either prefers A to B, or he prefers B to A, or he is indifferent between A and B.

That seems pretty logical doesn't it? It simply implies that an individual is able to establish a preference of one choice over another or determine that the two choices are of equal desirability to him. Another axiom:

b. If an individual has three choices, A, B, and C and if he prefers A to B and prefers B to C, then he prefers A to C.

This axiom is called the "transitivity" axiom. It means that he is able to order, or rank his available choices.

Still another axiom involves choice between a gamble and a no-risk alternative:

c. Given three choices A, B, and C such that A is preferred over B, and B is preferred over C, there is some combination of A and C where A occurs with probability p and C occurs with probability $(1 - p)$ such that the combination (gamble) with A and C as possible outcomes will be preferred over the no-risk alternative of B. Similarly, there is some value of p such that the no-risk alternative B, will be preferred over the combination of A and C.

This axiom concerns choices between accepting a gamble having a desirable outcome A and an undesirable outcome C or accepting a no-risk alternative of intermediate desirability, B. The axiom implies that a decision maker, when given the value of p (the probability of A occur-ring), is able to determine if the gamble is preferred over the no-risk alternative, or vice versa.

These three axioms are given here merely to suggest to you the type of assumptions (axioms) upon which the mathematical preference theory is based. Virtually all scholars of preference theory are in general agree-ment as to the validity and completeness of the eight axioms proposed by von Neumann and Morgenstern. (Readers wishing to study the axioms in further detail are directed References 5.1 or 5.3.)

So what does all this have to do with a new approach for decision making? Well, the truly dramatic breakthrough that was made by von Neumann and Morgenstern was their mathematical proof that if a deci-sion maker had a value system described by the eight axioms of decision making there existed a function which completely described his attitudes and feelings about money. The proof was long and extended, but nonethe-less its importance was noted immediately by persons studying the emo-tional aspect of how people make decisions under uncertainty.

Stated differently, they proved that if a decision maker accepts the eight axioms as the basis of rational decision it is possible to describe his attitudes about money in a simple function or curve. This curve might appear as in Figure 5.1 and is called a preference curve, or a utility curve, or sometimes a decision curve.

This curve (function) is mathematically unique up to a linear trans-formation. This means that the scales of the preference curve are arbitrary until numerical values of two points have been specified, at which time the curve is unique. For example, we could arbitrarily assign a preference value of zero to receiving or losing no money, and a preference value of plus ten to receiving $100,000. Having specified these two sets of numerical values the curve is now mathematically unique, as in Figure 5.2.

There are several important properties of the preference curve:

a. The vertical preference scale is dimensionless and reflects only the *relative* desirability of an amount of money. The magnitude of the scale is arbitrary until two points on the curve have been defined. The "zero" point on the vertical scale is generally inter-preted as a point of indifference, or neutrality about money. Positive, upper values on the vertical axis denote increasing desirability and the negative portion of the scale denotes an

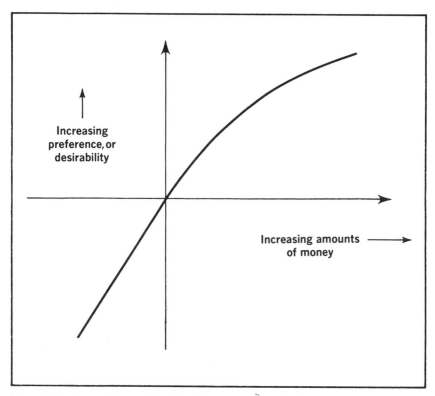

Figure 5.1 *General form of the function, or curve which relates a person's preferences for various amounts of money.*

increasing dislike for the corresponding amounts of money.
  b. The horizontal scale can be in any units of money, such as undiscounted monies, net present value monies discounted at a specified interest rate, incremental cash flows, current asset position, etc. In the context of decision analysis the horizontal axis is usually labeled as profits and losses, rather than the term "Increasing Amounts of Money."
  c. The curve is a monotonically increasing function. This means that the vertical parameter (preference) increases in numerical value as the amounts of money increase. This applies over the entire range of values of money included on the horizontal axis.
  d. The preference curve is based on an individual's preferences,

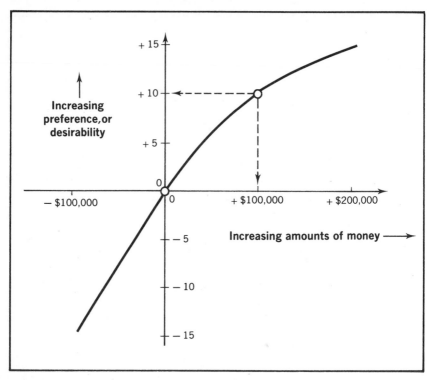

Figure 5.2 *Preference curve in which numerical values have been assigned to two points.*

and does not imply a comparison among individuals. The curve merely *describes* a person's preferences and attitudes for money, and does not imply he is wrong for having the attitudes, or that he should change his attitudes about money.

e. The shape of the preference curve reflects the attitudes and preferences of the decision maker. If he was totally impartial to money (the assumption with EMV) his preference curve would be a straight line passing through the origin. (More about this a bit later.)

f. The numerical preference values represent a measure of relative desirability for a given amount of money. For example, in Figure 5.2 the relative preference value for a gain of +$100,000 is +10. The preference value for a gain of +$200,000 is +14.75. These two numbers only have meaning when attempting to compare the *relative* desirability of receiving $200,000 versus $100,000. The +14.75 is greater than +10, hence receiving $200,000 is more desirable than receiving $100,000. This is, of course, nothing very earthshaking. The point we make here is that the numerical preference values only reflect *relative* preference or desirability.

g. Preference theory has the property of expectation. That is, we can multiply probabilities of occurrence by the relative preference values and compute an *expected preference value* for a decision alternative in the same manner that we previously computed expected monetary values. For example, suppose we have a drilling prospect with the following data:

Outcome	Probability of Occurrence	Conditional Monetary Value Received
Dry Hole	0.25	− $80,000
Producer	0.75	+$150,000

Next we could find the relative preference values for an $80,000 loss and a $150,000 gain from our preference curve. For the moment assume that Figure 5.3 represents our preferences and attitudes for money. The relative preference value for an $80,000 loss is −12.30, and the preference value for receiving a $150,000 gain is +12.75.

We can now make an expected preference value computation for the decision alternative of drilling as follows:

		DECISION ALTERNATIVE: DRILL		
Outcome	Probability of Occurrence	Conditional Monetary Value Received	Corresponding Preference Value (From Figure 5.3)	Expected Preference Value
Dry Hole	0.25	− $80,000	−12.30	−3.0750
Producer	0.75	+$150,000	+12.75	+9.5625

Algebraic Sum is called → +6.4875
the expected preference
value for decision
alternative "drill"

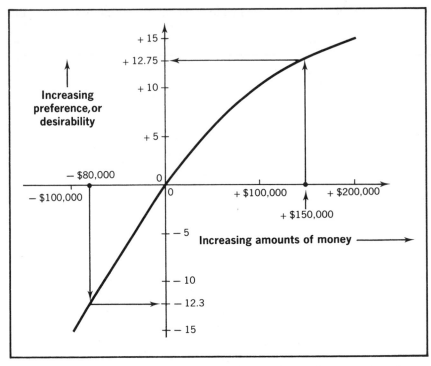

Figure 5.3 *Hypothetical preference curve. The relative preference values for an $80,000 loss and a $150,000 gain are −12.30 and +12.75 respectively.*

    h. The expected preference value of a decision alternative is the decision parameter used by the decision maker to accept or reject the alternative. The decision rule is to select the decision alternative(s) which has the highest positive expected preference value.

        In Chapter 3 we stated that the expected monetary value (EMV) of a decision choice represented a "weighted average profit." In preference theory the expected preference value of a decision alternative represents a measure of "weighted average desirability" or "weighted average preference." The decision rule thus suggests that the manager select projects which maximize this "weighted average desirability."

    The parallel between expected monetary value (EMV) and expected preference value should be quite obvious:

Steps in Decision Analysis Using EMV	Steps in Decision Analysis Using Preference Theory
1. Define decision alternatives and possible outcomes	1. Define decision alternatives and possible outcomes
2. Determine numerical probabilities of occurrence for each outcome	2. Determine numerical probabilities of occurrence for each outcome
3. Compute a conditional monetary profit or loss for each outcome	3. Compute a conditional monetary profit or loss for each outcome
4. Multiply probabilities of occurrence of each outcome by conditional monetary values	4. Read preference values corresponding to each conditional monetary profit or loss from preference curve
5. Compute algebraic sum of the expected monetary values of all possible outcomes to yield the decision parameter EMV of decision alternative	5. Multiply probabilities of occurrence of each outcome by corresponding preference values
6. Select alternative which maximizes EMV	6. Compute algebraic sum of the expected preference values of all possible outcomes to yield the decision parameter expected preference value of decision alternative
	7. Select alternative which maximizes expected preference value

The only difference between the two concepts is that in EMV we multiply probabilities by the monetary values to be received whereas in preference theory we multiply probabilities by *the preference values which correspond to the monetary values to be received.* The approach to decision analysis is exactly the same. The only difference is to find the preference values for each monetary profit and loss and use these preference values in the expected value computation. von Neumann and Morgenstern noted this parallel by observing

"We have practically defined numerical utility (preference value) as being that thing for which the calculus of mathematical expectation is legitimate."

And indeed we should remember that expected monetary value and expected preference value are just special cases of the general expected value concept developed in 1654.

To summarize our discussions thus far on preference theory we can probably all agree that people are not impartial to money. This was evidenced as early as the eighteenth century with the Saint Petersburg Paradox, and for each of you in your responses to the coin flips described earlier. The question was not whether or not people had specific attitudes and preferences about money but how could these attitudes be incorporated into the quantitative decision process.

The dramatic breakthrough came with von Neumann and Morgenstern's proof that if a decision maker acted in a manner consistent with eight axioms of decision making there existed a mathematical function (curve) which completely described his value system with respect to money. This was indeed a significant step forward in the theory of decision making and, for the first time, suggested a way to include his attitudes in the quantitative decision parameters used in the decision process.

Before leaving this discussion of the mathematical basis of preference theory we should note two subtle, but important properties of the preference theory concept. First, we must remember that the preference curve and resulting decision analysis techniques are *descriptive*, not prescriptive. That is, the curve is merely a description of the decision maker's attitudes and preferences for money. It does not prescribe what his attitudes should be, nor does it mean that he is wrong for having his particular preferences. The preference curve is merely a description of his attitudes and preferences.

This point is important to remember when trying to implement preference theory ideas into the decision process. Many decision makers, when first exposed to the ideas, adopt a defensive (negative) attitude because they fear that preference theory will show they've been making poor decisions or have had the wrong attitudes about money. A preference curve only describes attitudes and there is no "correct" curve. Its use is intended to provide a means to quantitatively account for a decision maker's attitudes about money in a consistent and unambiguous manner, and not to prescribe attitudes or decision policies.

The second point to note regards our comments in part g) above. In our computation of the expected preference value of the alternative "drill" we entered the preference curve (Figure 5.3) with the −$80,000 and +$150,000 monetary values and *not* the expected monetary values of each outcome. That is, we did not first multiply 0.25 by −$80,000 to get an expected monetary value of the outcome "Dry Hole" of −$20,000 and then enter the curve at −$20,000. Instead, we entered the curve with the actual profit or loss and then *later* multiplied the corresponding preference value by the probability. We must always follow this latter sequence. The horizontal scale of any preference curve is a monetary value, and not an expected monetary value.

To show why we do not enter the curve with expected monetary values merely requires that we consider another possible outcome with a potential loss of −$800,000 and a probability of occurrence of 0.025. The expected monetary value of this outcome is also −$20,000 (0.025 × −$800,000). But clearly the decision maker's attitudes about losing $800,000 will be different than losing $80,000 as in our example in part g)! So the point is to always enter the preference curve on the horizontal axis with the actual profits and losses, and not with expected profits and losses. The probabilities enter into the analysis when computing expected preference values, and are not considered with the preference curve itself.

APPLYING PREFERENCE THEORY TO
PETROLEUM EXPLORATION DECISIONS

Having discussed some of the theoretical basis of preference theory let's consider how the concept could be applied to exploration decisions.

In Chapter 3 we studied several numerical examples to illustrate the application of expected monetary value. Specifically, Example No. 1 (page 77) concerned whether or not to participate in the drilling of a proposed 640 acre gas prospect in the Anadarko Basin of western Oklahoma. Pertinent cost data were given in Table 3.4 and the EMV computations were shown in Table 3.5. Assume that our decision maker's preferences and attitudes about money can be represented by the preference curve given in Figure 5.4.

On the basis of maximizing EMV the preferred alternative was to drill. If he based his decision on maximizing expected preference value would drilling still be the preferred choice? To pose another possible management question—having determined that drilling results in the highest EMV is it possible that he would be better off to choose another strategy having lower possible monetary expectations but with no chance at losing money?

Well, of course, to answer these questions we have to go back to the anticipated profits and losses of Table 3.4 and find their corresponding

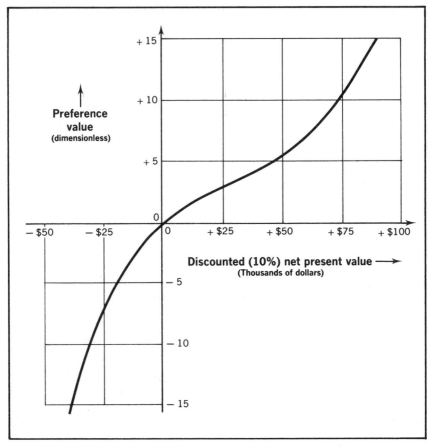

Figure 5.4 *Preference curve in which monetary values are given as NPV monies discounted at 10%.*

preference values from the preference curve of Figure 5.4. These parameters are summarized in Table 5.1. You should check some of the preference values so that you know how they were obtained.

The final step in the application of preference theory concepts to the analysis of the Anadarko Basin Prospect is to make an expected value computation using the preference values obtained from Figure 5.4 and tabulated in Table 5.1. The probabilities of occurrence are the same values used in the original EMV calculations for this prospect (Table 3.5). The expected preference value computations are summarized in Table 5.2.

Table 5.1

Table of NPV Profits and Corresponding Preference Values for Anadarko Basin Prospect (Example No. 1 of Chapter 3). The NPV Profits Are from Table 3.4 and the Corresponding Preference Values Were Read from the Preference Curve of Figure 5.4.

| | DECISION ALTERNATIVE | | | | | |
| | Drill with 40% Working Interest | | Farm Out, Retain ORI on 256 Net Acres | | Penalty Clause with 40% Backin Option | |
Possible Outcome	NPV Profit (Table 3.4)	Corresponding Preference Value (Figure 5.4)	NPV Profit (Table 3.4)	Corresponding Preference Value (Figure 5.4)	NPV Profit (Table 3.4)	Corresponding Preference Value (Figure 5.4)
Dry Hole	−$28,000	− 8.3	0	0	0	0
2 Bcf	+$21,000	+ 2.5	+$ 5,000	+0.9	+$ 5,000	+0.9
3 Bcf	+$42,000	+ 4.4	+$ 8,000	+1.2	+$20,000	+2.4
4 Bcf	+$64,000	+ 7.4	+$11,000	+1.4	+$37,000	+4.0
5 Bcf	+$86,000	+14.0	+$13,000	+1.7	+$56,000	+6.1

From the computations of Tables 3.5 and 5.2 we can summarize the relative feasibility of the three decision options as follows:

Decision Alternative	Expected Net Present Value Profit (EMV) (Table 3.5)	Expected Preference Value (Table 5.2)
Drill with 40% WI	+$16,650	+0.27
Farmout, Retain ORI	+$ 5,000	+0.76
Backin Option	+$12,750	+1.54

On the basis of maximizing EMV the preferred choice would be to drill with a 40% working interest. If the decision maker considers his attitudes and preferences about money (as represented by the preference curve of Figure 5.4) the preferred strategy is to exercise the backin option because this alternative has the highest expected preference value.

It is relatively easy to see why the decision strategies are different. In Figure 5.4 you will note that the preference curve is bending downward quite rapidly in the third quadrant. This is suggesting a strong dislike (negative preference) for losses greater than $20–$25 thousand dollars. The result is a large negative preference value (relative to other amounts of money) for a loss of $28,000 if the well is a dry hole.

The large negative preference value of −8.3 and the resulting large negative expected preference value of −2.90 for this outcome reduce the

Table 5.2
Expected Preference Value Computations for Anadarko Basin Drilling Prospect (Example No. 1. of Chapter 3)

Possible Outcome	Probability Outcome Will Occur	DRILL WITH 40% WI		FARM OUT WITH ORI		PENALTY WITH BACKIN	
		Preference Value Corresponding to Anticipated NPV Profit (Table 5.1)	Expected Preference Value for Outcome	Preference Value Corresponding to Anticipated NPV Profit (Table 5.1)	Expected Preference Value for Outcome	Preference Value Corresponding to Anticipated NPV Profit (Table 5.1)	Expected Preference Value for Outcome
Dry Hole	0.35	− 8.3	−2.90	0	0	0	0
2 Bcf	0.25	+ 2.5	+0.63	+0.9	+0.23	+0.9	+0.23
3 Bcf	0.25	+ 4.4	+1.10	+1.2	+0.30	+2.4	+0.60
4 Bcf	0.10	+ 7.4	+0.74	+1.4	+0.14	+4.0	+0.40
5 Bcf	0.05	+14.0	+0.70	+1.7	+0.09	+6.1	+0.31
	1.00		+0.27		+0.76		+1.54

EXPECTED PREFERENCE VALUE OF DECISION ALTERNATIVES

algebraic sum of the expected preference values. This results in the "drill" strategy being less desirable than the backin option involving no monetary losses but smaller profits from a successful well.

With the preference curve and the expected preference value computations of Table 5.2 the decision maker is able to determine a decision strategy that considers both his desire to maximize overall gain and his relative distaste for the possibility of losing $28,000. This analysis was objective and free of any arbitrary or subjective nuances or interpretations. And if he continued to use the same curve for future decisions he would be considering his specific attitudes and preferences for money in a logical, consistent way over a period of time. The end result would, presumably, be a more consistent series of decisions over time. This is the principal motive for including preference theory in the decision analysis process.

SHAPES OF PREFERENCE CURVES

At this point it would perhaps be well to comment briefly on the interpretation of the general shape of preference curves. In Figure 5.5 we have shown sketches of various preference curves for comparative purposes. In all of these sketches the horizontal axis is money of increasing amounts from left to right and the vertical scale is a numerical measure of preference, increasing in an upward direction. For this discussion we will assume that the numerical scales of each parameter are the same in all the curves.

First we can make some general observations about the shape of the curve in the third quadrant by comparing the two curves in Figure 5.5 (a) and (b). Generally, the shape of the curve in this quadrant is reflecting specific attitudes and preferences related to asset position or capital constraints. If a decision maker (or corporation) has a definite asset position limit as shown in Figure 5.5 (a) the preference curve will bend downward very sharply and approach the capital limit asymptotically.

Thus, a potential monetary loss of an amount at or near this limit would have a very large (perhaps infinitely large) negative preference value—which implies he simply is not willing to lose that amount of money in any investment decision. If the decision maker or firm has a much larger asset position and can afford larger potential losses the curve will not bend downward as sharply in the third quadrant, as shown in Figure 5.5 (b).

The general shape of the preference curve in the first quadrant is a

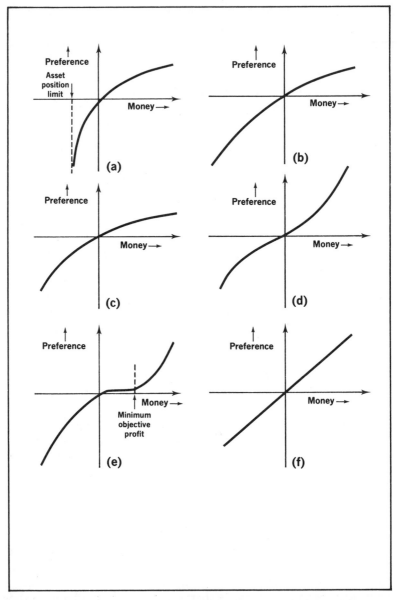

Figure 5.5 *Sketches of various shapes of preference curves. Figure (a) suggests a limited asset position or capital constraint, whereas Figure (b) does not imply a limit in the range of money shown. Figure (c) represents a conservative, or risk-averse attitude for taking risk. Figure (d) suggests the opposite of being conservative— "a risk-seeking" attitude. Figure (e) represents an attitude of being interested in only profits above a specified minimum level. Figure (f) is a preference curve for a decision maker who is totally impartial to money. This is the preference curve which is implied with the decision rule of maximizing expected monetary value (EMV).*

function, in part, of the decision maker's attitudes for taking risks. A preference curve which is concave downward in the first quadrant, as in Figure 5.5 (c) is representative of a conservative attitude for risk taking. This is sometimes called a "risk-averse" attitude, and a person having this attitude is called a "risk-averter." By contrast, a curve which bends sharply upward in the first quadrant as in Figure 5.5 (d) is representative of a "risk-seeking" attitude — the opposite of being conservative.

Recent research on the psychology of making decisions under uncertainty has suggested that most people, when given explicit information about uncertain choices, will exhibit a conservative, or risk-averse attitude for taking risks. We stress the words "explicit information" in the previous statement to mean they are given specific information about the probabilities and conditional profits and losses.

Frequently the same people may turn around and patronize the gambling activities of Las Vegas or Monte Carlo, and the only way one can rationalize gambling in the preference value context is with a risk-seeking curve of Figure 5.5 (d). Probably in the gambling context an otherwise conservative person is not aware of the specific probabilities of success and failure and/or he considers the thrill of a possible jackpot to be worth the inevitable fact that he is going to lose money in the long run!

The shape of the curve in the first quadrant can also be influenced by an immediate or long-term objective of being interested in only gains above a certain value. For instance, suppose the exploration manager has as his policy the objective of not being particularly interested in the small drilling prospects — but rather he is only interested in prospects where the potential gain is, say, greater than $5,000,000. In essence he's saying he's not after the little fish in the ocean, only the whales. The shape of the curve which relates an objective such as this is shown in Figure 5.5 (e).

In this curve his relative desirability or preference is small for profits less than a stated minimum objective profit. Beyond that point the curve rises sharply, indicating a much stronger preference for anticipated profits beyond the minimum objective level. While this management objective may be valid it can be shown that this policy implies that the decision maker is willing to take an inordinate amount of risk for the chance of a large potential profit. Some managers question whether this is, in fact, a good long-term policy for making decisions under uncertainty.

Finally, the preference curve shown in Figure 5.5 (f) represents a totally impartial attitude about money. This is the type of preference curve that is implied if we base our decisions solely on maximizing expected monetary value (EMV). If this linear preference curve represented our preferences and attitudes about money we would not have to use preference theory concepts — we could simply base all our decisions on EMV.

The slope of this curve is not important (since we are free to choose the magnitude of the vertical preference scale), only the fact that it is linear. With this curve a series of competing investment alternatives would be ranked in exactly the same order of desirability using EMV and expected preference value.

However, we must remember from our initial discussions in this chapter we concluded that very few, if any, people are totally impartial to money over the large range of potential profits and losses encountered in petroleum exploration. So this linear preference curve is probably the exception, rather than the rule.

It is possible for a decision maker's preference curve to have the characteristics of several of the types of curves shown in Figure 5.5. There is no "correct" curve and it is not appropriate to say that a given curve will yield more profit over the long term for a company. We must always remember that the preference curves merely *describe* attitudes and preferences for money. They do not prescribe or dictate what the attitudes should be. Everyone is slightly different and it is logical to presume that everyone's preference curve will be different. But we can't say that a person who is conservative is more correct or less correct than another person who is a "riverboat gambler" and willing to take gambles or decisions which the conservative person would reject.

PRACTICAL PROBLEMS IN APPLYING PREFERENCE THEORY

In our comments thus far we have been stressing preference theory as an improved basis for decision making under uncertainty. But, as is the case in any new concept, there are a number of problems and questions which arise when trying to apply the concept to the day-to-day real-world sector of decision making. We can summarize some of the principal problems and questions as follows:

1. How does the decision maker get his curve? Where does the curve come from?
2. Will the whole concept even work? Is anyone using it today? What has been their experience? If our firm begins to use pref-

erence theory in our daily decisions is there a chance of making a very serious error, or bad decision which could have serious consequences for the firm?

3. In a large corporation who's curve is used? Does the preference theory concept apply to corporations as well? Is there a corporate decision curve?

4. Will a decision maker's curve change with time? If so, how rapidly will it change? Would he have a different curve each month? Week? Day? Hour?

5. If subordinate staff members had access to the decision maker's preference curve why would they even need the decision maker any longer?

6. The objective of using preference theory is to be able to introduce our feelings about money into the decision process in a consistent and quantitative manner. In a competitive business environment is it even a good corporate strategy to be consistent?

7. When dealing with decisions under uncertainty the decision maker must consider potential profits and losses as well as the probabilities of occurrence of the possible outcomes. Is it possible to separate one's feelings about the monies involved from feelings or opinions about the probabilities of occurrence of the outcomes? Will a manager's feelings, prejudices, or biases he may have regarding the probabilities provided by his technical staff influence or cloud his feelings about the monies involved?

8. Will a decision maker's attitudes and preferences for money be influenced by his recognition that a given gamble or decision is just a single, one-time event that cannot be repeated? That is, will the repeated trial aspect of decision making under uncertainty (or lack thereof) influence his preference curve?

9. If our firm's exploration and production division uses a preference curve would the same curve be used for making refining or marketing capital investment decisions? If our company diversified and opened a new division to sell trucks would we use a different preference curve than the one used for exploration decisions? Or should a single preference curve be used for all corporate investment decisions?

10. When capital expenditure authority is delegated to lower management echelons (subject to specific expenditure limits at each level) is it possible that each management level uses a

different value system (preference curve)? If so, is this a good or bad situation?

A very imposing list of problems and questions! So imposing, in fact, that some are convinced that preference theory will never work in business decisions. Others feel that expected preference value is the best single numerical parameter that exists for selecting decisions under uncertainty. Certainly if the latter view is correct the above problems and questions must be resolved. So let's look at each of the issues raised in a bit more detail.

1. *Where does the preference curve come from?*

The first issue mentioned in the previous list is undoubtedly the most critical problem that presently hinders the widespread application of preference theory. von Neumann and Morgenstern proved that a preference curve exists for an individual if he makes his decisions consistent with the eight axioms. But once they proved that a unique preference curve exists the mathematicians went on to their study of game theory, and did not provide or suggest a method for you or I, or Tom Miller of Profit Oil and Gas Company to obtain our preference curves. And unless we can draw the curve on paper it is of little consequence to know that our curves exist in theory.

Since 1944 many researchers have studied this problem but so far their attempts have been only marginally successful, even to this date. The unfortunate truth is that we (as a business community) do not as yet have a satisfactory way to construct an individual's preference curve.

If we could ever describe a decision maker's curve we could then try using preference theory and resolve many of the questions on the previous list. Until that time trying to answer some of the interpretative questions regarding preference theory is similar to a dog chasing his tail around a tree—it's an endless circle!

Attempts to describe a decision maker's preference curve have generally consisted of obtaining his responses to a carefully designed set of hypothetical investment questions. In these tests the decision maker is offered a choice between a gamble having a very desirable outcome A and an undesirable outcome C, or a no-risk alternative B of intermediate desirability. For example, he might be asked the following question:

"Which alternative would you prefer—a gamble in which you would receive $100,000 with probability 0.6 and lose $50,000 with probability 0.4, or receive $20,000 cash?"

The $20,000 cash is considered the no-risk alternative in that he would receive it with certainty. Suppose the decision maker responded to the question by saying that he would prefer the gamble. Then the probabilities would be changed and the decision maker given the second question:

"Which alternative would you prefer—a gamble in which you would receive $100,000 with probability 0.5 and lose $50,000 with probability 0.5, or receive $20,000 cash?"

In this question the monetary values were held constant, the only change being that the probability of receiving the large gain was reduced from 0.6 to 0.5. Depending on his response to the second question the probabilities would again be changed until he reached a point of being indifferent between the gamble and the no-risk alternative. That is, the probabilities were adjusted (holding the monetary values constant) over a series of questions until the gamble and the $20,000 were of equal value, or preference to him. Assume for the moment that a probability of receiving the $100,000 of 0.55 (and the resulting probability of 0.45 of losing $50,000) resulted in his being indifferent between accepting the gamble or the $20,000 cash.

From this reply we could determine three points on his preference curve using this relationship:

$$[PF(A) \times p] + [PF(C) \times (1 - p)] = PF(B) \qquad (5.1)$$

where $PF(A)$ is the preference value for outcomes A, etc., and p is the probability of occurrence of outcome A. The left hand portion of the equation is simply the expected preference value for the gamble in which outcome A occurs with probability p and outcome C occurs with probability $(1 - p)$.

By equating the expected preference value of an uncertain decision (gamble) to the preference value of a no-risk alternative B we imply that each choice is of equal value to the decision maker. Equation 5.1 states that he is indifferent between the gamble and the no-risk alternative.

For the hypothetical questions $p = 0.55$, outcome A is receiving $100,000, outcome C is losing $50,000, and outcome B is receiving $20,000 on a no-risk, certain basis. Recall that we stated earlier that a preference curve was unique up to a linear transformation, and this meant that we could arbitrarily assign two preference values before the curve became unique.

For example, suppose we assign a preference value of −10 to a loss of $50,000 and a preference value of +5 to receiving $20,000. With these values we can solve equation 5.1 for the relative preference value of outcome A — receiving $100,000:

$$[PF(A) \times (0.55)] + [(-10)(1 - 0.55)] = (+5)$$

Preference value assigned to a loss of $50,000

Preference value assigned to receiving $20,000

Solving this equation gives a preference value for outcome A (receiving $100,000) of +17.3. Thus we have now defined three points on his preference curve:

Monetary Value	Corresponding Preference Value	
−$ 50,000	−10.0	Assigned values to select
+$ 20,000	+ 5.0	the preference value scale.
+$100,000	+17.3	Computed from decision maker's response.

The next step would be to ask him another series of hypothetical questions which involve two of the three monetary values having specific preference values. For example:

"Which alternative would you prefer — a gamble in which you would receive $300,000 with probability 0.75 and lose $50,000 with probability 0.25, or receive $100,000 cash?"

The probabilities would again be varied to determine his point of indifference. Assume the point of indifference corresponded to a probability of 0.7 of receiving $300,000. Now equation 5.1 is solved again as follows:

$$[PF(A) \times (0.7)] + [(-10)(1 - 0.7)] = +17.3$$

Preference values determined from previous questions

This equation results in a computed preference value for outcome A (receiving $300,000) of +29. Now we have an additional, or fourth point on the curve — a preference value of +29 for a $300,000 gain.

The test is then continued with another set of hypothetical gambles. In each successive set of questions two of the three monetary values are selected from the list of preference values previously determined. Each set of questions yields one additional point. In this illustration we started with the two preference values that we assigned arbitrarily and computed a third value corresponding to a gain of $100,000. Then using two of these three values in a second set of questions we were able to compute the preference value for receiving $300,000. The third set of questions would use two of the four values known to compute a fifth value, etc. Eventually the progressive series of questions will lead to enough points to plot his preference curve over the desired range of monetary gains and losses.

Researchers have developed various test procedures to construct preference curves but all of the approaches use the basic steps described. They all require that the decision maker identify his point of indifference between a gamble and a no-risk alternative. Equation 5.1 is then used to compute a preference value for one of the three monetary values involved in the test question. These tests have only been marginally successful, with the principle shortcomings being:

a. The test procedure involves hypothetical gambles and decisions rather than actual gambles involving an exchange of money.

b. Most decision makers are not used to making decisions on the basis of a precise discernment of probabilities required to justify a gamble. Rather, the probabilities are usually specified for a given investment and the parameter that he keys on is whether the successful gain is sufficient to justify the gamble.

Item a, of course, is a consequence of the testing being a series of abstract questions. Any testing procedure short of using actual real-world decisions will suffer this weakness. It could very well be that a decision maker's responses to a set of abstract questions may be different than his responses would have been if the actual monies were received or lost. The obvious question then is "Why not use actual decisions that he makes over a period of time to construct his preference curve?" The answer is that it sounds logical but such an approach leads to some computational "dead-ends."

The reason is that in actual decisions he either accepts or rejects a decision (gamble) and rarely ever stops to explicitly state what the probabilities would have had to be for him to have been indifferent between the gamble and a no-risk alternative. Thus the equal sign in Equation 5.1

would be an inequality sign—the expected preference value of the gamble was either greater than or less than the value of the alternative.

If he made, say, 100 decisions we would end up with 100 inequalities but no way to solve the relationships for explicit preference values. We'll say a bit more later in the chapter about an approach which could possibly resolve this dilemma. But at the present time the researchers have found they must rely on using hypothetical, rather than real-world decisions.

The second shortcoming above follows from the usual approach to exploration decisions in which risk factors (probabilities) and dry hole costs are specified. The manager then is asking how large do the oil or gas reserves (and resulting profits) have to be to justify the gamble of drilling. Instead of fixing the cost factors and varying the probabilities to determine whether to accept the gamble he is, in essence, fixing the probabilities and the costs for outcomes B and C and varying the magnitude of the successful outcome A in his thought process.

A test which incorporated this modification was developed in 1967 (Reference 5.3) which used actual drilling prospects in the testing. The test was given to 23 petroleum exploration decision makers in various companies. The results seemed to be encouraging in that the participants were able to better understand the testing procedure and the resulting preference curves generally agreed with their stated attitudes about money. The results, or applicability of this improved test were still clouded, however, by the uncertainty of not knowing whether their responses would have been the same if money was actually received or lost.

So in summary to the first problem listed about where do we get the preference curve the answer is that we really don't have a good, reliable way to draw a decision maker's curve at this point in time. This is one of the crucial missing links in the quantitative analysis of decisions under uncertainty. Research is continuing on this point, and a technique will no doubt be developed in time. At present there does not appear to be a completely suitable method to translate a person's feelings about money into a simple two-dimensional preference curve.

Some proponents of preference theory have taken the position that until such a method is developed there may be several reasonably acceptable alternative ways to obtain a preference curve. These alternatives are discussed in a later section of this chapter. Although they involve certain compromises, these alternatives may be very useful in the interim period, and, in my opinion, provide a way to make good use of many of the preference theory features. More about this a bit later.

2. Will preference theory work? Is anyone using it? Could preference theory lead to some very bad corporate decisions?

Formal use of preference theory in petroleum exploration decision making has been used only to a limited extent. Various companies have, or are experimenting with its use but no results or experiences have been publicly announced. From my contacts with industry representatives from around the world I am convinced that the various attitudes and preferences about money enter into the decision making process informally and/or subjectively, but I am not aware of any company that has been successful in applying the concept in a formal, quantitative sense.

The reason for this goes back to the problems of constructing a preference curve in the first place. Until we can find a way to describe a decision maker's preferences accurately we will be unable to try the concept to see whether it works. So these questions are still open at this point.

You should remember, however, that the fact that it still isn't being widely used does not necessarily imply that preference theory is wrong or of no value. Our motives for considering its use were that most people are *not* impartial to money, and the fact that theoretically such a curve must exist if we make decisions consistent with eight simple statements about the logic of rational decision making. And so far these motives remain unchallenged. It's just that we need to try it for awhile to see if it works and this hasn't yet been feasible because of the problems of defining a decision maker's preference curve.

3. Does preference theory apply to corporate decisions? Is there a corporate decision curve? If not, who's curve is used?

The von Neumann–Morgenstern derivation of preference theory was strictly based on individual preferences. There has been limited research on group or social choices in an attempt to define a similar concept for a corporation (or board of directors) but so far these attempts have failed. When decisions are based on the preferences of a group or corporation we get into some very sticky theoretical problems of synthesizing the diverse attitudes and preferences of each person into a single, composite representation of preferences. So strictly speaking, it has not been proved theoretically that a corporation has a preference curve.

However, one of the interesting observations from the tests that have been given to construct an individual preference curve is that a group of executives who make decisions as a board or committee generally have curves of about equal shape. This may be caused by the fact that each have modified their personal attitudes towards a common set of attitudes about money by virtue of having worked together as a board for a period of time. Or it could suggest that, in fact, there exists some overall, composite set of attitudes and preferences that might be reasonably close to what we might call a corporate preference curve.

An executive decision maker's preference curve is probably a mixture of some of his personal feelings and attitudes about money and the corporate asset position, policies, and attitudes as he perceives them. This leads to an interesting paradox. If there is, in fact, no such thing as a "corporate preference curve" does this mean that he must rely on his own personal value system (and preference curve) in order to ever be able to use the concept in drilling decisions? If a decision maker in a large oil company is of only modest personal wealth (measured in thousands of dollars) is it even possible for him to project his personal attitudes up to a level of the millions (billions?) of dollars that are involved in the corporate decisions he makes? It seems doubtful.

For example, try to describe what your preferences and attitudes would be if your personal asset position was increased to $100 million dollars. Can you do it? Most people do not have a personal value system that extends into the millions, so any feelings about dealing with monies as large as one hundred million would almost be speculative.

In fact, what probably occurs is that a decision maker may (and probably does) have a preference curve based on the corporate funds he deals with in his daily decisions. The shape of this curve is probably reflecting his perceptions and attitudes about what the corporate policies and objectives are, or should be. Thus to make preference theory work in drilling decisions the preference curve would more nearly represent his preferences regarding corporate funds than his preferences regarding his personal money. The theory and application are the same—the only difference is that the horizontal scale represents corporate money rather than personal wealth.

4. *Will a decision maker's curve change with time? If so, how rapidly will it change?*

Here is another point that is hard to answer until we can get some experience using preference theory concepts. Certainly most people will agree that our attitudes and preferences change with time. We get older and presumably wiser, and our asset position usually changes over a period of time. Also our objectives change with time. So it seems clear that a preference curve would change also. Just how fast it would, or should, change is not known at this point.

Some people have conjectured that a preference curve would change even in a single day. Others argue that if things change that rapidly there is even more need to introduce an element of consistency in the decision making process. Still others will suggest that attitudes about investing capital will change within a budget period or a tax period depending on the availability of budget money or tax writeoffs respectively.

The usual response to that is that the artificial constraints of a budget period, per se, may cause inconsistencies—a further reason to seek the consistency offered by preference theory. They reason that when budget money becomes in short supply we raise our acceptable "minimums" and begin to reject drilling prospects that would have been accepted earlier in the budget period. Or if we have too much money left at the end of the budget period we lower the "minimum" and begin accepting prospects that would have been rejected earlier.

My feeling is that it will not change as rapidly as some would suggest. To me it seems logical that a decision maker's attitudes will remain reasonably constant with time. The only significant changes would be caused by a sudden and dramatic change in asset position (you have had a run of dry holes and you are nearing bankruptcy—or your firm just finds a major new discovery that immediately doubles the firm's assets) or by a major change in the philosophy or objectives of exploration.

An example of the latter is probably represented by the dramatic changes in the world oil business that occurred in October and November 1973 with the oil embargos from Middle Eastern oil producing countries.

The question of how fast a curve will, or should, change will also remain an open question until we get some experience working with preference theory.

5. *If staff personnel have access to their decision maker's preference curve the decision maker would no longer be needed.*

The scenario would have the explorationist compute an expected preference value for each drilling prospect. If the parameter is negative he would not even present the prospect, on the presumption the decision maker would have rejected it anyway. And if the expected preference value is positive he would (presumably) accept the alternative. Since both of these options would be known at the completion of the economic and risk analyses why is the decision maker even needed any longer?

This viewpoint about preference theory reflects obvious tunnel vision. The expected preference value, while considered by many to be the best single indicator of value, is still not an end in itself. Even though a decision maker's attitudes and feelings about money are added to the quantitative consideration of risk and economic factors there are still many irreducible and intangible factors which management must consider that are not included or considered in the expected preference value parameter. There may be certain overriding factors which would justify acceptance of a project having a negative expected preference value, or rejection of a project having a positive expected preference value.

It would clearly be a mistake to screen projects using preference theory before they are presented to the decision maker. The intent of preference theory is to help the decision maker adopt a more consistent decision policy, not to override and/or replace his function and responsibilities.

6. *In a competitive business environment is it a good strategy to try to be consistent (the objective of using preference theory)?*

Some question whether consistency is a positive corporate strategy. They usually cite as examples an athletic contest such as football or soccer. If the offensive team consistently used the same play to advance the ball it would not be long until the defensive team can devise a strategy to thwart the offense. They continue by saying the element of surprise in these games is a strength, rather than a weakness to the offensive team.

The analogy to the oil exploration sector is, in general, weak. Probably the only situation where a truly competitive situation exists in our business is in the competitive bidding of offshore leases in the U.S. Conceivably, if Company A was using preference theory, over a long period of time, and *if* a competing company could "back-calculate" Company A's preference curve from its previous bids, then it might be possible for the competing company to anticipate Company A's bid on the next sale (because they would presumably bid in the same consistent manner), raise their bid a few dollars and outbid Company A!

But there are several big "ifs" in that scenario that make the possibility of it ever happening almost zero. Besides, those who have ever been a part of the determination of corporate bid strategies will realize there are many, many, factors considered in these decisions that could never be represented in a preference curve.

With this possible exception the business environment is not competitive in the same sense as a football game. In most decisions a firm is more nearly competing with itself to maximize profit or reserves, given the leases, concessions, etc. available for consideration. And in these situations most would probably agree that consistency would be a plus factor in their overall corporate investment policies.

7. *Is it possible to separate one's feelings about money from his opinions about the probabilities of occurrence of the outcomes? Will a manager's feelings, prejudices, or biases he may have regarding the probabilities provided by his technical staff influence or cloud his feelings about the monies involved?*

One of the gambling games offered at Las Vegas is keno. Casinos typically will have large marquees exhorting people to "Play Keno — Win $25,000 Jackpot!." How many people do you suppose decide to play with the feeling they have a good chance to win $25,000? Do you think as many people would play if they were advised that, in fact, the odds for winning the jackpot are only one chance in 8,911,710?

Is it possible that the emotions people associate with winning $25,000 could obscure, or cloud their opinions as to the likelihood of the event ever occurring? Or suppose we ask someone for his (subjective) opinion as to the likelihood of getting a head on twenty consecutive flips of a fair coin. Do you think his response would have been the same if we asked instead the following question: "If you flip a coin twenty times and get twenty heads you win $10,000,000. What do you think the chances are of this occurring?"

There is pretty good evidence in some of the research that has been done to define preference curves that decision makers frequently let their feelings about money influence their subjective opinions about probabilities, and vice versa. It is very possible that this happens in exploration decision analysis because exploration frequently involves large amounts of money and a lot of subjectively assigned probability numbers.

The decision maker is forced to think about two variables at the same time: (a) his judgments about the probabilities of occurrence of the possible outcomes, and (b) his likings or preferences for the monetary values to be received from each outcome.

Some scholars have even suggested that a preference curve is not a description of a person's attitudes about money, but rather a description of his subjective opinion as to the likelihood of occurrence of the outcomes yielding given amounts of money. In essence, they are suggesting that a preference curve describes subjective probability judgments of the decision maker, rather than his preferences for money. This suggestion is easily refuted, however, by a test to determine one's preference curve in which the gamble in each test question involves a coin-flip.

The respondent presumably accepts that the odds are 50/50, so his responses will be reflecting just his attitudes about money. In these tests it has been shown that the resulting preference curves all have the general shapes and characteristics expected of the von Neumann Morgenstern formulation. This evidence seems to refute the suggestion that a preference curve is but an expression of subjective probability.

In the application of quantitative decision analysis we should honor the step-by-step procedure as much as possible. That is, we must try to separate the analysis into specific steps of estimating probabilities of occurrence of the outcomes as one step and assigning preference values to the monetary values as a second step.

When solving a decision tree, for example, we could evaluate all the probability factors at the chance nodes before ever even writing a monetary value at the terminal points. The systematic sequence of decision analysis should be helpful in trying to separate our opinions about probabilities from our opinions and attitudes about money.

8. *Will the repeated trial aspect of the expected value concept influence a decision maker's preference curve?*

Suppose you were offered the choice of receiving $1 million cash or participating in a single coin-flip in which you win $3 million if it's a head and win nothing if it's a tail. Which would you take? Would your choice be the same if you were assured that the gamble would be repeated and you could participate as often as you wish?

If the choice is a single, one-time opportunity some people will take the $1 million cash without hesitation. But they will take the gamble if they can participate over a series of repeated trials. This viewpoint suggests that an additional factor which may influence a person's attitudes about money is whether or not the repeated trial aspect applies.

Of course the resolution of this goes back to our Chapter 3 discussion of the nature of the repeated trial condition and how it must be satisfied in order for expected value criteria to have meaning. We pointed out that this can only be accomplished by *consistently* maximizing an expected value criterion over a period of many business decisions.

The same logic applies to the interpretation and use of preference curves. The monies involved in decisions under uncertainty must be considered as resulting from repeated trials. The trials (decisions) do not have to be of the same odds and monetary values, but as long as a consistent decision strategy is used the repeated trial aspect is satisfied.

So the conclusion is that the preference curve should reflect the decision maker's attitudes about money assuming that many decisions will be made so as to achieve repeated trials. If the decision maker considered every decision as a unique, one-time opportunity he would have to have a preference curve for every separate prospect. This would be totally impractical to use and would contradict a basic tenet about the value of money in the corporate treasury—it loses it's identity as to its source.

9. *Should a preference curve that is used for petroleum exploration decisions also be used for the company's decisions in marketing, or refining? Or should a different curve be used for each of the various operating functions of a fully integrated oil company?*

Because preference theory concepts in exploration are only being used to a limited extent at this time most corporate policy makers have not yet had to face this question. Ideally, it would seem that a value system (preference curve) for corporate funds would be valid for all phases of its corporate investments. Presumably a dollar profit from marketing should have the same intrinsic value as a dollar profit from producing operations.

The basis for even suggesting different curves for the various functions probably stems from two considerations: (a) The types and magnitudes of risks are different in marketing investments than they are for, say, petroleum exploration; and (b) the corporate philosophy of taking the profit downstream (marketing, refining), and that exploration decisions should be directed to supplying and replacing the crude requirements of these downstream functions.

Regarding the first of these considerations we must remember that a preference curve does not consider or include probability statements. It merely provides a relative measure of desirability for various levels of money—without regard to the probabilities of achieving the monies. The risk factors are considered in the computation of expected preference values for each outcome.

If the analyst can realistically assess the levels of risk and uncertainty in each type of capital investment decision the variations in the types of investments will have been considered. Thus it would seem that a single preference curve would be suitable for all the various corporate functions.

If the firm has a stated objective of making its desired profit in downstream functions the objectives of exploration, per se, change somewhat. Under these conditions the value system for exploration decisions approaches one of minimizing crude oil replacement costs. The exploration decision process becomes a sub part of the overall integrated corporate investment policy, rather than being self-standing from a profitability standpoint. The preference curve would probably not be of much value in these types of exploration decisions.

For a fully integrated oil company it would seem logical that a single value system for corporate funds would be preferred. The problems in formulating such a unified corporate strategy seem rather imposing, however, and it will no doubt be a number of years before such an objective could be achieved.

10. *Does the delegation of authority for expenditures to lower management levels result in different value systems at each corporate level?*

Many larger oil companies have various subordinate management levels of authority located throughout their geographic areas of operations. Under the corporate headquarters the firm may have two or more second level divisions, or regions. Within each of these second level groups there may be two or more smaller, third level groups called districts or areas. Typically the manager or superintendent of the third level office will have authority to spend up to, say, $100,000 on drilling prospects without requiring approval of the next higher office.

The second level manager will usually have greater authority for expenditures — say up to $300,000. Any drilling prospect costing more than $300,000 would require headquarters (level one) approval. Does this organizational scheme lead to three distinct value systems regarding money?

Some say it does. Not so much by design, but rather due to the very human characteristic of striving for job security, approval, and promotion on the part of the lower management personnel. For example, the third level decision maker may express some reluctance in accepting prospects which require expenditures approaching his limit of $100,000. The reluctance may be caused by his desire to avoid having to possibly accept responsibility for a very expensive (as related to the limit of his authority) dry hole.

The same situation could be repeated at the second level when expenditures approach the $300,000 limit. These reluctances at the second and third levels of decision making would imply that their respective preference curves would bend downward sharply in the third quadrant near their limits of authority. An example is shown in Figure 5.6.

If this hypothesis about the differing value systems of each level of management is correct the inconsistency should be obvious. A prospect with a $99,000 dry hole loss may very well be rejected due to the large negative preference value the third level manager would assign to the loss. On the other hand, a prospect with a $101,000 dry hole loss would have to be authorized by the next level.

Assuming that potential profits are about the same, the prospect

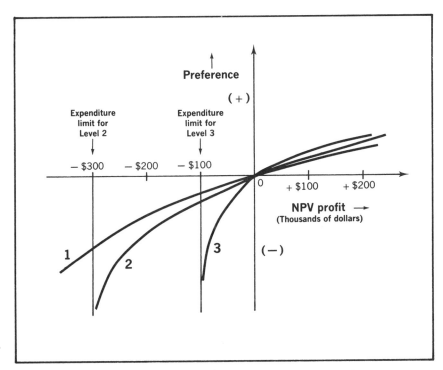

Figure 5.6 *Possible representation of different value systems within a given corporation. The third level manager having an expenditure limit of $100,000 may have a value system such as curve 3. The second level manager's curve may appear as in curve 2. The first level manager, having no expenditure limits in the range shown on this graph may have a preference curve similar to curve 1. The comparison here is in the third quadrant, and the shapes of the respective curves in the first quadrant were just drawn arbitrarily.*

could very well be accepted. This would be possible due to the much smaller negative preference value associated with losing $101,000, as read from the second level manager's curve. The same inconsistency could apply for prospects having potential losses near the $300,000 expenditure limit of level 2.

These inconsistencies clearly violate one of the mathematical properties of a preference curve which specifies that the curve is monotonically increasing function as the magnitudes of money increase. This property applies for negative money (losses) as well. With the value systems shown in Figure 5.6 as we proceed from losses of $400,000 to $300,000 the preference values suddenly become very large in the negative direction—in clear violation of the monotonicity property.

The organizational structure and delegated expenditure limits described are common in the petroleum industry. The hypothesis of three different value systems may also be fairly common. If so we may wish to reconsider whether such a policy is in the best interests of the company. I have no particular alternatives to suggest to remedy such a situation, other than the obvious need to modify the attitudes of the second and third level decision makers.

One possibility is to use the preference curve to disseminate corporate decision policies to lower levels so as to assure that the same value system is used throughout the corporation. A single curve would eliminate the inconsistencies at the lower level expenditure limits. At the very least this hypothesis suggests the need for complete and continuing communication of decision policies, attitudes, and values between corporate levels.

POSSIBLE APPLICATIONS OF PREFERENCE THEORY

It should be fairly obvious to you at this point that there are lots of problems related to trying to use preference theory concepts in petroleum exploration decisions at this point in time. The missing link is our inability to realistically draw or construct a decision maker's preference curve. Until a method or procedure is developed to accomplish this we cannot use the preference theory concepts as described in the quantitative analysis of decisions under uncertainty.

There are, however, several possible applications which could be used immediately in our decisions relating to petroleum exploration. These applications could be classified into three areas:

A. Alternative methods for defining a decision maker's preference curve,
B. Use of preference theory concepts as a descriptive learning aid for new management (decision making) personnel,
C. Use of a "certainty equivalent" parameter together with EMV to provide a dual, or parallel set of decision criteria to management.

A. Alternate Methods for Defining Preference Curves

Earlier in this chapter we described the usual approaches that have been developed to define preference curves—a series of test questions. It will be recalled that these questions were designed to locate the decision maker's point of indifference between accepting a gamble and a no-risk cash alternative. Researchers have not been completely successful in these tests, and recent research has been directed to finding different ways to accomplish the same objective.

One new approach is based on using actual, "real-world" decisions over a period of time (Reference 5.3). The obvious advantage is that such a procedure would eliminate the bias that is created by the fact that the tests involve only hypothetical investments, not real decisions involving exchanges of money. The disadvantage, as mentioned earlier in the chapter, is that the data points needed to plot the curve usually can't be computed from "real-world" decisions because the precise indifference point is rarely considered and recorded. This new approach is a compromise.

Suppose our decision maker accepted the following prospect several months ago:

Prospect A

Estimated NPV Profit from a Completion	$1,500,000
Estimated Dry Hole Cost	$ 500,000
Estimated Probability of Completion	0.6
Decision	Drill

Note that the costs and probabilities were those estimated at the time of decision—and not the costs which actually occurred. Since he accepted this over the alternative of rejecting the prospect with no gain or loss we could assume that the expected preference or desirability of the gamble exceeded the preference or desirability of "do nothing." That is:

$$[0.6 \times PF(+\$1,500,000)] + [0.4 \times PF(-\$500,000)] > PF(\$0)$$

where PF($1,500,000), PF(−$500,000) etc. stand for the preference value of gaining $1,500,000, the preference value of losing $500,000, etc.

As we have stated on several earlier occasions, we are free to arbitrarily define two points on the preference curve. Suppose we assign a preference value of −10 for a $500,000 loss and a preference value of 0 for a zero profit or loss. Substituting these values into the above inequality gives

$$[0.6 \times PF(+\$1,500,000)] + [0.4 \times (-10.0)] > 0.$$

For the inequality to hold the preference value for a gain of $1,500,000 must be greater than 6.67. That is,

$$PF(+\$1,500,000) > 6.67.$$

Now suppose that several days later he rejects another drilling prospect having the following parameters:

Prospect B

Estimated NPV Profit from a Completion	$2,000,000
Estimated Dry Hole Cost	$ 500,000
Estimated Probability of Completion	0.4
Decision	Do not drill

His decision to not drill Prospect B implies that the expected preference of the gamble was less than the "do nothing" option. That is

$$[0.4 \times PF(+\$2,000,000)] + [0.6 \times PF(-\$500,000)] < PF(\$0).$$

But having previously defined PF(−$500,000) = −10 and PF($0) = 0, we can compute that for the inequality to hold the PF(+$2,000,000) must be less than 15, or

$$PF(+\$2,000,000) < 15.$$

We now know the following information about his preference curve, based on his two previous "real-world" decisions:

$$PF(-\$500,000) = -10$$
$$PF(\$0) = 0$$
$$PF(+\$1,500,000) > 6.67$$
$$PF(+\$2,000,000) < 15.0$$

The last two inequalities could be thought of as constraints, or limits on the region in the first quadrant within which his preference curve must lie. As successive decisions are made additional constraints or limits on the region would be defined. And with time the region will decrease and (theoretically at least) converge to a single preference curve.

Graphically, we could illustrate this gradual converging process by using circles to plot the preference values on the lower edge of the region and squares to identify the upper limiting values. That is, PF(+$1,500,000) > 6.67 would be plotted as a circle to denote the fact that the theoretical preference value for a gain of $1,500,000 is greater than, or above 6.67. The inequality PF(+$2,000,000) < 15.0 would be plotted as a square to note that the actual preference value is less than, or below 15.0.

From the two decisions made thus far the preference graph would appear as in Figure 5.7(a). Subsequent decisions would yield additional constraints. After a sufficient number of decisions had been made the preference curve could be drawn in the first quadrant as the line which separates the circles from the squares, as in Figure 5.7(b). The curve would simply be drawn or faired by eye, subject to the requirement that the curve is rising monotonically as NPV quantities increase.

The approach just described would only work for the case where the dry hole costs were at, or very near $500,000 in each prospect. This is because we needed to know two of the preference values to compute the third value of each inequality. This would be possible, for example, if all the decisions were in the same general area having essentially the same drilling costs. This control area would be used to define the shape of the curve in the first quadrant, as in Figure 5.7(b).

Once the curve was fairly well defined for profits the third quadrant could be developed from decisions involving substantially different drilling costs. In this case the unknown in the inequalities would be the preference value associated with the (varying) dry hole losses. The preference value for a zero profit would, of course, still be 0, and the preference value for the gain would be read from the first quadrant of the just-constructed preference curve.

In this process of using real-world decisions it may develop that some of the points computed from the inequalities would be redundant or inconsistent, as shown in Figure 5.8. In the analysis of a decision maker's past decisions suppose it was established from one decision that the preference value for a $500,000 gain was less than +12. And suppose a separate decision that he made established that his preference value for a $500,000 gain was less than +20. The latter value is called a redundant point, for obvious reasons. The +12 value is more restrictive than the +20 value. Such a situation could arise from his rejection of the following drilling prospects:

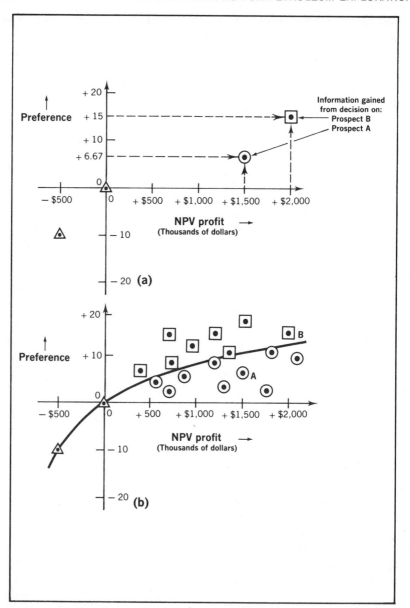

Figure 5.7 *Developing a preference curve from past "real-world" decisions. The preference values of −$500,000 and $0 are assigned initially and noted as triangles. Circled points imply actual preference value (and curve) is greater than, or above the value. The Squares represent points in which the true preference value is less than, or below the value.*

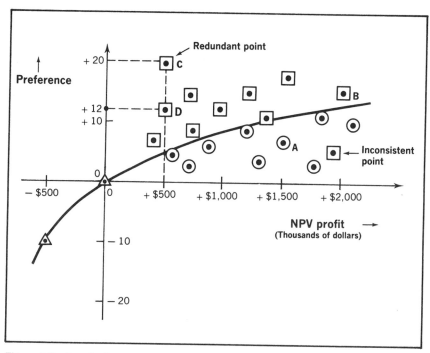

Figure 5.8 *Developing a preference curve from past "real-world" decisions, showing a redundant data point and an inconsistent data point.*

Prospect C

Estimated NPV Profit	+$500,000
Estimated Dry Hole Cost	−$500,000
Probability of Completion	0.33
Decision	Do not drill
Resulting Preference Value Data Point	PF(+$500,000) < 20

Prospect D

Estimated NPV Profit	+$500,000
Estimated Dry Hole Cost	−$500,000
Probability of Completion	0.45
Decision	Do not drill
Resulting Preference Value Data Point	PF(+$500,000) < 12

The inconsistent point in Figure 5.8 would represent a decision to not drill. The analyst would have to go back to the decision maker to determine his specific reasons why it was rejected. In all probability it will have been caused by other intangible factors such as a possible shortage of rigs, or a lack of investment capital. If it could be established that there were, in fact, no extraneous factors that could be pinpointed by the decision maker his decision would have to be judged as inconsistent, relative to the risk attitudes and preferences he used in the other decisions. This would, of course, present an even stronger argument for use of preference theory concepts in his decision making — to reduce or eliminate such inconsistencies!

A possible variation of this general approach to defining preference curves would be to construct an approximate preference curve initially using the usual test procedures involving hypothetical decisions. The decision maker would then continue making decisions using his former criteria. For each successive decision a "greater than" or "lesser than" point would be computed. By the successive steps described the preliminary preference curve would be modified from his subsequent actual decisions. When the decision maker felt that the curve reached a point of being as realistic as possible he could then convert over to a preference value system for his subsequent decisions.

A different approach to defining preference curves has been developed recently. This approach involves more of an analytic philosophy than methods that have been described thus far. It offers a possible alternative to the testing procedures involving hypothetical questions. The approach first seeks to determine several general characteristics about a decision maker's attitudes and preferences about money. From these general observations the analyst determines a mathematical function (usually called parametic functions) or equation of a curve which satisfies his general attitudes about risk taking and money.

The procedure might proceed as follows:

a. The analyst would first seek to confirm the monotonicity of the decision maker's attitudes and preferences about money. That is, he would clearly establish that the following condition applies: For any two levels of profit P_2 and P_1 where $P_2 > P_1$ then $PF(P_2) > PF(P_1)$. This would be accomplished by talking to the decision maker and confirming that for a given level of profit, his preferences for achieving that amount are greater than his preferences for a lesser amount.

This then implies that the analytic function finally selected as the preference curve must be a monotonically increasing function as the monetary gain increases. (Note: We use the word *profit* for the preference curve attribute plotted on the horizontal scale. Of course it can be any measure of monetary reward, including incremental profits and losses, net asset position, etc. When we speak of profit it is implied that negative profits are losses.)

b. The next step would be to characterize the decision maker's attitudes for risk taking—is he averse to taking risk, or is he a risk seeker, or neutral to risk-taking? Knowledge of this characteristic will determine the general shape of the curve. (Refer to Figure 5.5 for a summary of the shapes of preference curves.)

The general risk-taking preferences of the decision maker could possibly be determined in a discussion with the analyst, or it may require his response to several hypothetical questions of the following type:

> "You own the exclusive right to participate in a gamble in which you win $3 million dollars if a head comes up and you win nothing if a tail occurs on the flip of a fair coin. What would you sell this exclusive right to the gamble for in cash?"

Suppose he feels that he would be willing to sell it for $800,000 cash, or more. This means that if he were offered anything less than $800,000 he would prefer to retain (and play) the gamble. For offers over $800,000 he would sell his rights to the gamble in exchange for the cash. His response of $800,000 is said to be his *certainty equivalent* for the 50–50 gamble of receiving $3 million or receiving nothing.

To the analyst this reply is significant. It establishes that the decision maker is risk-averse (conservative). If he were risk neutral the "value" of the gamble would be its EMV, or $[(0.5) \times \$3,000,000] + [(0.5) \times 0] = \$1,500,000$. But to the decision maker the equivalent value is less than $1,500,000, because of his apparent reluctance to gamble. Stated more formally we could define a new term called risk premium as follows:

$$\text{Risk Premium} = \text{EMV} - \text{Certainty Equivalent}$$

If the risk premium is positive (Risk Premium > 0) the decision maker is risk-averse. If it is negative (i.e. Risk Premium < 0) he is a risk seeker, and if the risk premium is zero he is risk-neutral (in which case EMV completely describes his value system). For our example his risk premium for the coin-flip gamble is $1,500,000 − $800,000 = +$700,000.

By a series of such questions the analyst would establish his risk-taking attitudes over the range of monetary values encountered in his decisions. Knowing this characteristic would provide another important clue regarding the choice of analytic functions to represent his preferences.

Incidentally, the decision maker doesn't have to have the same risk-taking attitudes over the full range of monetary values. For example he may be a risk-seeker in the low ranges of money and risk-averse as the monetary values increase, or vice versa. As a further note of interest most researchers with this approach have observed that a conservative, risk-averse attitude for risk taking is much more common than the risk-seeking attitude.

c. Finally, the analyst would pose a series of questions to determine if the decision maker's aversion for taking or avoiding risks is a function of the magnitude of monies involved. For the illustration above, for instance, the analyst would seek to determine if the decision maker's aversion for risk increases or decreases as the monetary values increase. Technically this defines whether he is "increasingly risk averse" or "decreasingly risk averse," and this is another important clue from an analytic sense.

From this general series of steps a mathematician can usually determine an analytic function which satisfies the conditions and characteristics implied by the decision maker's responses and stated attitudes for taking risks. That is to say, from a very limited amount of information about his preferences the mathematician can, by some rather extended, but sound logic, develop an equation (or curve if its plotted on graph paper) which would reflect the decision maker's preferences.

This approach is not necessarily a replacement for the usual testing procedures, but rather an alternative in cases where a realistic curve can not be determined by the usual testing methods. It is beyond the intended scope of this book to delve into the mathematics of this approach. For those interested in exploring this in more detail References 5.7 and 5.9 are recommended.

As a possible third approach some managers have taken the viewpoint of not being too concerned about hypothetical tests or analytic functions, but rather to construct a curve which describes what they think the attitudes about money should be. In essence they are saying "Let's forget about the tests and just sit down and figure out what the shape of the curve ought to be."

From a theoretical standpoint this plays a little havoc with the von Neumann-Morgenstern theory, but the end result may be a very practical compromise. It certainly has the distinct advantage of providing a definite, precise curve that can be used immediately by the decision maker(s). It would probably be developed by a consensus agreement of the top corporate policy makers as to what the corporate attitudes, objectives, and preferences are, or should be.

Then having specified these general attitudes a preference curve would be drawn which represents their opinions. This curve would then be used as the basis for computing expected preference value decision criteria for each investment considered.

One consequence of such an approach was made known to me several years ago. The policy makers, after considering the various attitudes and opinions concluded that perhaps the only real constraint to a preference curve should be the firm's asset position. This is a tangible consideration that applies to all firms, including theirs. But they felt that, with this exception, perhaps the best long range attitude should be risk-neutrality. The representation of this in terms of a preference curve is as shown in Figure 5.9.

The curve is linear until the potential losses begin to represent a significant proportion of total corporate worth, say L_1. For losses greater than L_1 (further to the left on the curve) there would be an increasing reluctance to accept the project, as evidenced by the downward bending of the curve. For investments involving potential losses less than L_1 the curve would be linear—which means impartiality for money.

In these cases the EMV would be a completely valid and correct decision parameter. The only time management would override the EMV criterion would be for investments with potential losses exceeding L_1. For a large corporation having many investment opportunities such an approach would seem to have considerable merit.

The point here is that it may serve a useful purpose to draw a preference curve to represent what the decision maker feels the attitudes, objectives, and policies should be. This may be a useful compromise to hasten the application of preference theory until such time that we can develop a better way to describe preference curves.

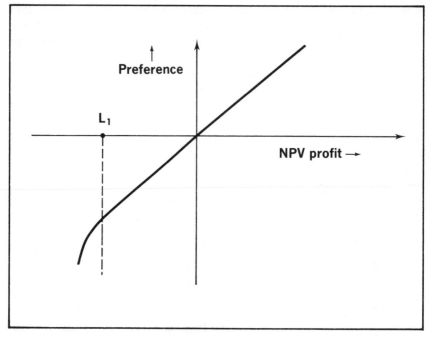

Figure 5.9 *A preference curve in which the only non-linear constraint is that of asset position. This type of curve may be representative for large corporations having large capital resources and many investment opportunities.*

B. Use of Preference Curves As a Learning Aid

Another practical application of preference theory falls in the general area of learning aids for management personnel. We all know, for example, that when technical personnel change to decision making functions they must change their viewpoints to a degree. The natural (and important) optimism of a geologist may have to be modified. Similarly, when an engineer changes to management positions he may have to modify his traditional conservative attitudes.

A useful way to evaluate such changes in attitude is to construct their preference curves. Although the testing procedures we've described may not be suitable to describe a curve that a decision maker is willing to use in real-world situations they are usually adequate to define general attitudes about money. To know this would be useful in determining what changes, if any, a new manager must make in his attitudes about money.

This possibility became very evident in some earlier research reported in Reference 5.3. The test developed in that effort was given to four top executives in a large integrated oil company. The same test was also given to about ten petroleum engineers of the same corporate office. Without exception the preference curves of the engineers were much more conservative (risk-averse) than the managers who acted upon their recommendations.

This is, possibly, as it should be to provide certain implicit "checks and balances" in decision making. As a learning aid the comparison would be extremely useful in preparing any of those engineers for the different attitudes which may be required when they are promoted to management positions.

Therefore, we see that the present test procedures could be very useful to simply describe attitudes. They could even be used to determine preference curves of experienced decision makers at a given level in the firm to compare the similarities and differences which may exist in their individual decision attitudes.

The hypothetical questions of the usual test procedures may not yet be adequate to give a curve which can be used with confidence in real decisions. But the test methods are usually sufficient to at least define the general shape of the curve. And this information may be valuable from the descriptive, or learning standpoint.

C. Use of Dual Criteria—EMV and Certainty Equivalent

A few moments ago when we were describing an analytic approach to determining preference curves we mentioned a new term *certainty equivalent*. It was defined as the no-risk, certain amount of cash the decision maker required in order to be willing to exchange, or sell his rights to a gamble. It was, in essence, the "cash value" of a decision alternative which involved uncertainty.

The adjective *certainty* implied that there was no risk to the receipt of the equivalent "cash value" of a gamble. All decisions under uncertainty have a certainty equivalent, and it may be very helpful to the decision maker to use this parameter in evaluating decision options.

To illustrate its use let's go back to the example we considered earlier in the chapter. The problem itself was first given in Chapter 3 and concerned the drilling of a 640 acre gas prospect in the Anadarko Basin.

The expected net present value profits (EMV) for each option were computed in Table 3.5. Then, using the preference curve of Figure 5.4 we reconsidered the options on the basis of expected preference value (Table 5.2). The results of each of these computations are summarized again as follows:

Decision Alternative	Expected Net Present Value Profit (EMV) (Table 3.5)	Expected Preference Value (Table 5.2)
Drill with 40% WI	+$16,650	+0.27
Farmout, Retain ORI	+$ 5,000	+0.76
Backin Option	+$12,750	+1.54

We earlier stated that the decision rule with preference theory was to maximize expected preference value. In this example this would mean to accept the backin option. We also mentioned the fact that preference numbers are actually relative indicators of desirability, and that the expected value computation using preference values is an attempt to get a "weighted average desirability" for each prospect—hence the rule to select the alternative having the highest positive expected preference value.

Some decision makers may feel that the abstract, dimensionless expected preference values in Column 3 are a bit difficult to grasp or comprehend. In this case it may be helpful to convert each of these expected preference values into their corresponding certainty equivalents so that the manager is using a criterion expressed in monetary units. This is simple to do. We merely enter the preference curve of Figure 5.4 on the vertical axis with the calculated expected preference values and read corresponding profits from the horizontal axis. These profit values are the certainty equivalents for each decision alternative. Note that this sequence is just the reverse of our usual way of using a preference curve.

For example, the expected preference value for drilling was computed as +0.27. We enter Figure 5.4 with this value and read a corresponding NPV profit of about $1,000. The corresponding certainty equivalents for expected preference values of +0.76 and +1.54 are about $4,000 and $13,000 respectively. These values are, by definition, the certainty equivalent for each option.

The decision maker, in consideration of his preferences as described by the curve of Figure 5.4 would, by this reasoning, be willing to sell his option to accept the backin for $13,000 cash. Or stated differently, the equivalent "value" of the backin option is $13,000. We have not added anything new to our preference theory concept by this approach — rather we have merely expressed the expected preference values in their corresponding certainty equivalent values.

With this approach we may have preferred to present our analysis to the manager in this form:

Decision Alternative	Expected Net Present Value Profit (EMV) (Table 3.5)	Certainty Equivalent Value, Based on Preference Curve of Figure 5.4
Drill with 40% WI	+$16,650	+$ 1,000
Farmout, retain ORI	+$ 5,000	+$ 4,000
Backin Option	+$12,750	+$13,000 ← Preferred Choice

This approach of presenting certainty equivalents to the manager has several advantages. It provides a way to express the abstraction of preference theory in equivalent monetary values. Presumably he can "relate" to being advised the gamble has an equivalent cash value of $13,000 better than being advised the option has an expected preference value of +1.54 (although each statement means the same thing).

Second, by listing EMV's in one column and certainty equivalents in the next column we focus his attention on the effects his preference curve is having on expectation. If he were completely impartial to money the certainty equivalents and EMV numbers would be identical.[1] When the two parameters for a decision option are substantially different the comparison raises a warning flag, so to speak.

For instance the drill option has the highest EMV and the lowest certainty equivalent. Why is that? Well, of course, we mentioned this earlier by observing that his curve drops rapidly in the third quadrant, and his relative preference for a possible $28,000 dry hole loss was a large negative number. This had the effect of lowering the "weighted average desirability," and hence the certainty equivalent. For him to reply to his question will force him to explicitly consider again his strong bias against losses of this magnitude.

1. Recall that in Chapter 3 we defined EMV as being the equivalent cash value we would be willing to take in exchange for the option to a gamble, assuming we were totally impartial to money. The certainty equivalents in this chapter consider specific preferences as described by the preference curve. The only time that EMV's and certainty equivalents will be equal is for the case of linear preference curve.

It may be that he will reaffirm that this is consistent with his feelings about money. Or it may be that this comparison may cause him to have second thoughts about whether he should be so adverse to losses of this amount. The important point is that the comparison calls his direct attention to the role his attitudes about money are playing in his decision making. This would seem to be a definite advantage in most exploration decisions.

In summary, it may be helpful to add one step to the usual sequence of steps used in preference theory. That is to find the corresponding certainty equivalent for each computed expected preference value. The certainty equivalents are read from the same preference curve used to determine preference values for the expected value computation.

With this approach both the EMV values and the certainty equivalents would be presented to management for each decision alternative. The decision rule would be to select the option having the highest (positive) certainty equivalent. Such a system of dual criteria would permit his direct comparison of the preferred choice based on being impartial to money and the preferred choice based on the specific preferences and attitudes about money represented in his preference curve.

This approach would be useful even if he did not as yet have full confidence in the reliability of his preference curve. The usefulness in that case would be that the two criteria would require his specific consideration of his attitudes about money but still give him an alternative criterion (EMV) to guide his thinking. This dual value system would be particularly useful in the early stages of working with preference theory.

SUMMARY

Perhaps the best way to summarize this chapter is to list a number of observations we have considered regarding the use of preference theory concepts in business decisions.

- There is overwhelming evidence that most people (decision makers) are not impartial to money. Rather, they have specific attitudes, preferences, and opinions which they associate with gaining or losing money and taking risks.

- The expected monetary value (EMV) criterion is based on being completely impartial to money. Although EMV is a very important decision parameter it does not include any information about attitudes for the monies involved or risk-taking preferences.

■ Preference theory provides a means to incorporate a decision maker's specific attitudes and preferences about money and risk-taking into a quantitative decision parameter. The parameter is called expected preference value and it is considered to be a more realistic measure of value than EMV.

■ Preference theory requires that a decision maker's attitudes about money be described by a two dimensional preference curve. The principal problem in applying the concept to exploration decisions is that we have not found a completely satisfactory way to draw, or define the curve. Present methods will give an indication of the general shape of the curve, but certain weaknesses of the tests reduce the reliability of the curves for "real-world" decisions. Research continues to find better ways of describing a decision maker's preference curve.

■ Some applications of the concept which can be used at this time include:

a. Constructing a curve to fit what management feels the corporate attitudes, preferences, and objectives about money are, or should be. Once these attitudes are defined it is a simple task to draw the appropriate curve.

b. Use of the present test methods as a descriptive, or learning aid for new management personnel. Can be used to compare attitudes to determine if any changes may be required to be more in line with corporate attitudes.

c. Use of approximate preference curves (defined by existing test methods) to compute certainty equivalents for each decision alternative. A dual value system is then presented for each prospect: EMV and certainty equivalents.

The key issue in the preference theory controversy seems to be whether or not there is an advantage to describing our attitudes about money and taking risks in a quantitative manner. Some feel our present methods which consider these factors in a subjective and arbitrary way (or perhaps not at all) are adequate.

They justify the view by observing that we have done a reasonably good job of decision making over the past 100 years of the oil business. So why change? Why bother with introducing a new and abstract dimension to the already complex decision making process, particularly when the scholars can't even agree among themselves as to whether the new dimension of preference theory will even work?

Many others, however, support the opposite viewpoint. They see evidence almost daily that most people are emotional about money. You no doubt proved that you had specific attitudes about money from your responses to the wagers given at the beginning of this chapter. And we can all probably agree that petroleum exploration itself involves a great deal of emotion.

Decisions to explore for petroleum have many similarities to games of chance, and the ever present hope for finding the huge new discovery, or the jackpot, as the case may be. Are we really able to temper our emotions about money and risk-taking to the extent that they are considered on a consistent, uniform basis in each decision? It seems doubtful.

The use of preference theory in business decisions can be a positive step forward. The concept forces a precise definition of attitudes toward money and risk-taking and permits explicit consideration of these attitudes in decision making. Use of the concept also has the important advantage of providing a way for the decision maker to separate his feelings about money from his subjective feelings about probabilities of occurrence of the possible outcomes. Preference theory, together with the quantitative analysis methods such as decision trees and expected value, provides a powerful new approach for the analysis of decisions under uncertainty.

6 Basic Principles of Probability and Statistics

What we need is probabilistic thinking, which is for so many practical and serious problems the only realistic kind of thinking. We must be trained in the technique of weighing alternative risks. We must face the grim fact that we cannot eliminate risk. The problem, therefore, is to be able to evaluate risks, to identify a reasonable risk, and to adopt it with courage, and care, and hope.

—Weaver

THIS chapter is the beginning of our consideration of the general topic of risk analysis—the assessment of the probabilities of occurrence for each possible outcome. Risk analysis is a vital part of decision making and is also probably the most complex and least understood part of decision analysis. Figure 6.1 provides a general picture of the decision analysis process which may be helpful in showing where we have been and where we are headed.

We briefly summarized the economic analysis portion of the overall decision process in Chapter 2. In Chapters 3, 4, and 5 we considered the expected value concept and its use as a decision parameter reflecting economic factors, risk, and uncertainty. Now we complete the process by discussing risk analysis. Essentially the remainder of the book will be dealing with this important subject.

We will see in later discussions that petroleum exploration risk analysis can be extremely complex, and the degree to which decision analysis can be used effectively is largely dependent on our abilities as geologists, engineers, and decision makers to think in terms of probabilities and statistics. For many of us it is a new way of thinking which will require that we consider many new terms and definitions.

But probability theory is a fascinating subject and I urge you to "jump in and get your feet wet"! Hopefully we'll be able to learn some

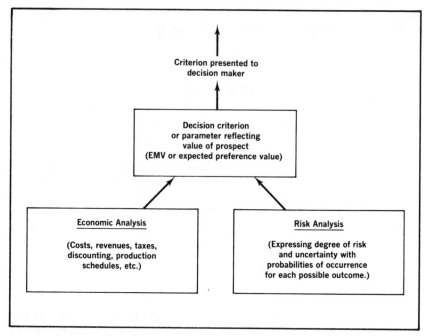

Figure 6.1 *A general picture of the analysis of a drilling prospect.*

new analysis methods that will be ever so much better than some of our earlier efforts at defining risk. Even if that objective fails we should be able to pick up enough about the subject to improve our poker game! So let's roll up our sleeves and, as the friendly Australians say, "Have a go at it!"

In this chapter we will be mainly concerned with defining terms and concepts which will be used in our later discussions of specific exploration risk analysis techniques. The chapter is divided into four major sections:

    I. Probability Definitions
   II. Probability Distributions and Single Value Parameters of Distributions
 III. Specific Distributions of Interest in Exploration Risk Analysis
 IV. Bayes' Theorem

Our objective in this chapter is not to make each of you a competent statistician or probability theory expert, but rather to introduce just enough important definitions to be able to study and understand exploration risk analysis methods. For this reason we will not cover some of the topics included in most textbooks on statistics and probability theory. Those wishing to dig deeper into the subject are referred to some of the new books pertaining to statistics in petroleum geology and engineering, such as References 6.5, 6.11, 6.12, and 6.16.

I. PROBABILITY DEFINITIONS

We have mentioned the word *probability* many times in the previous chapters without being very specific about its definition. It is important at this point, however, to be more precise in our meaning of *probability,* or a *probability of occurrence.* To do this we must first define a sample space and an event.

> *Sample Space:* A sample space is a set or list of all the possible things that can occur from an experiment, chance phenomenon, or a decision under uncertainty. It is sometimes called a population. The word *space* does not mean a dimension or volume; rather, a complete collection or list of objects or descriptions. A sample space can be finite as to the number of outcomes in the set or list, or there could be an infinite number of descriptions in the set.

As an example suppose we have a standard, 52 card deck of bridge cards and we are considering a wager involving the withdrawal of a card at random from the deck. The experiment, or chance phenomenon is drawing a card from the deck, and the sample space is the listing of all the possible cards which could be withdrawn. In this case there are 52 elements in the sample space.

There are various ways to define a sample space in petroleum exploration, depending on the specific analysis being made. Examples: all the possible values of recoverable reserves in a structure, the number of producing wells in a field, the number of structures which have oil in a basin, etc. Because of the many options in defining sample spaces it will be very important that we always clearly define the sample space in risk analysis.

> *Event:* An event is defined as a subset, or part of the sample space whose occurrence is of special interest to us. A word we've been using which means the same thing is outcome. An event (outcome) may be defined in any way we please, and it may contain one of the elements of the sample space, two or more of the elements of the sample space, all the elements of the sample space or none

of the elements of the sample space. An event is said to have occurred if the outcome of the experiment or chance phenomenon is
included in the subset defined as the event.

To amplify on this definition a bit, we could define an event as drawing a "7" of any suit from the deck of cards, or drawing a heart, or drawing
a five of diamonds, etc. In petroleum exploration an event might be
defined as finding at least 100 million barrels of reserves in a new structure, or finding 15 structures productive in a basin, etc.

We can now formally define what we mean by a probability number,
or a probability of occurrence. There are three main definitions which
can be given to define probability: relative frequency, or statistical;
a classical, or objective definition; and a subjective definition.

  1. *Relative Frequency Definition of Probability:* By this definition
     probability is defined as the long run ratio of the number of times
     the event has occurred divided by the total number of times the
     experiment or chance phenomenon has been repeated. An al-
     ternative definition is that probability is the ratio of the number
     of elements defined as the event divided by the total number of
     elements in the sample space. This latter definition applies for a
     finite sample space in which all the elements of the sample space
     are equally likely to occur. These definitions are sometimes
     called the statistical definition of probability.

The relative frequency definition of probability should be reasonably
straight forward. For our card deck if we define event A as drawing a 7
of any suit, the probability of A occurring, written as P(A), is 4/52. There
are four elements in the sample space defined as event A and there are a
total of 52 elements which are equally likely to occur in the sample space.
In essence, we find the "size" of the event and the "size" of the sample
space and make a ratio of the two numbers.

Or, if we are studying some repetitive experiment or phenomenon
we can compute a relative frequency by recording the number of times
the event occurs and divide this number by the total number of times the
experiment or process is repeated. An example of this might be a wildcat
success ratio. There are certain mathematical qualifications which we
must add to this relative frequency definition to be theoretically complete
and precise (considerations of whether the long-run ratio converges, the
law of large numbers, etc.). But for the purposes that we will be using this
definition we won't need to become involved with these special qualifications.

This first definition of probability can be used in many cases, but there are some chance phenomena of interest where a probability cannot be computed in the manner described as the relative frequency definition. For example, what is the probability of an earthquake occurring on June 2, 1995 twenty miles from Anchorage, Alaska? A rather far fetched question about something we hope never occurs. Nonetheless Anchorage lies in an area of known earthquake activity, and there may, in fact, be a finite probability of the event occurring. How would we determine its value?

By the relative frequency definition of probability we would determine this probability by observing the long-run frequency of occurrences of such an event, or alternately, evaluate the "size" of the event and the sample space. But neither of these approaches will apply in this case — principally because the phenomenon of interest is not a repetitive process that we can experiment with and study.

The stresses in the rock strata on that particular day and location are probably unique, and not directly comparable to previous stresses leading to the possibility of earthquake. Nor is it an easy matter to define the "size" of the sample space and the event. Thus, we must consider an alternative definition of probability, usually called the classical, or objective definition.

2. *Classical, or Objective Definition of Probability:* By this definition probability is defined as a measure of the degree to which available evidence supports a given hypothesis. This measure is determined by purely objective logic. Purely objective means apart from human opinion or bias.

The usual way we think about risk in oil exploration is an example of this meaning of probability. We hypothesize that a structure has oil (or a given level of reserves) and then look at all the evidence which supports the hypothesis: the nearby structure has oil, rock strata appear correlative, source rock immediately below objective zone, etc. Then, by purely objective logic, the probability is determined as the degree to which this evidence supports the hypothesis "Structure has oil." Or, we hypothesize that an earthquake will occur 20 miles from Anchorage on June 2, 1995 and then determine the degree to which the evidence supports this hypothesis.

From a practical viewpoint this definition represents a theoretical definition which may be difficult to use. The qualification 'by purely objective logic' may be difficult to achieve, because explorationists making the analysis are usually not free of human emotion about their hypotheses. This leads to a third definition of probability:

3. *Subjective Definition of Probability:* A personal opinion of the likelihood that an event will occur. A subjective probability estimate represents the extent to which an individual thinks an event is likely to occur—a "degree of belief," perhaps. Subjective probability estimates are sometimes used where past statistical data are not available and/or the information available is of an indirect nature. Subjective probability estimates are influenced by a person's biases, emotions, past experiences, etc.

In the view of the mathematician, subjective probability has no place in the theories relating to probability and statistics. From the practical, real-world viewpoint of oil exploration it may be the only definition which can be used—simply because we may not have the data or evidence to determine probabilities by either of the first two definitions.

In petroleum exploration risk analysis we will probably be using probabilities based on all three definitions. We normally won't have to concern ourselves about which of the three definitions is most appropriate to use, or which we should use in a given drilling prospect.

Probabilities are expressed as decimal fractions from 0 to 1.0. An event which is certain to occur has a probability of occurrence of 1.0. An event which cannot occur has a probability of occurrence of zero. When we say the probability of a dry hole is 40% we have merely expressed the probability of occurrence multiplied by 100. That is, a 40% chance of a dry hole means that the probability of occurrence is 0.4. Probability numbers are dimensionless and are always positive. There is no such thing as a negative probability of occurrence. For a given chance phenomenon the sum of the probabilities of occurrence of all possible events must be 1.0.

Sometimes we express judgments about the likelihood of occurrence of an event by saying "the odds for success are 6 to 5." This means that the probability of success is 6/11. If the odds were 8 to 4 the probability of success would be 8/12. To say an event has a 50–50 chance of occurrence implies its probability of occurrence is 0.5. If we say the chance of discovery is 1 in 10 we are implying that the probability of a discovery is 1/10, or 0.10.

Besides the basic definitions of sample space, event, and probability of occurrence we must also define several other important terms in probability theory.

Equally Likely Events: Two or more events are said to be equally likely if they have the same probability of occurrence.

Mutually Exclusive Events: Events are mutually exclusive if the occurrence of any given event excludes (or precludes) the occurrence of all other events. Mutually exclusive events have no points of the sample space in common.

Independent Events: Two or more events are independent if the occurrence of one event in no way affects, or is affected by the occurrence of the other events.

Conditional Probability: When two or more events are not independent they are said to be dependent. Conditional probability is the probability of an event *given* that some other event has already occurred.

The first of these definitions, equally likely events, is obvious and should require no further amplification. The terms mutually exclusive event and independent event are frequently confused. Both terms are important to exploration risk analysis, and we must keep them clearly defined and understood in our minds. The events or outcomes that we list in the first column of an expected value computation (e.g., dry hole, 2 BCF, 3 BCF, etc.) is an illustration of mutually exclusive events. If we get a dry hole we can't get 2 BCF reserves, or 3 BCF reserves, etc. The occurrence of one of the outcomes rules out the possibility of any of the other outcomes occurring.

The definition of independent events, on the other hand, relates to a sequence of events and whether or not the probability of occurrence on any given trial in the sequence is affected by what has occurred previously.

An example is flipping a coin five times. A coin has no memory, and each flip is a new experiment. Whatever has happened previously has no effect on the probability of a head or tail on the next flip. Thus each flip is said to be an independent event. Even though we may have tossed 100 consecutive heads the probability of a head on the next trial is 0.5, the same as it was on the very first flip.

We have used the notion of a sequence to describe independent events – but the definition is not restricted to a sequence. If we flipped 5 coins simultaneously each coin would represent an independent event. The test is whether or not the occurrence of one event has any bearing, or is influenced by any of the other events. If not, the events are defined as being independent events.

If events are *not* independent they are said to be dependent. This means that the probability of occurrence of an event is dependent upon what has already occurred. Such probabilities are called conditional probabilities. The probabilities are conditional upon, or dependent on something having occurred previously. We must be very precise about the distinction between independent and dependent events in exploration risk analysis. The reasons for this will become apparent later in this chapter and in Chapter 7.

Probability theory has a set of rules of operation in the same sense that our number system has a set of rules of operation (addition, multiplication, etc.). We will not take the time to define all the rules of algebra of probability, rather we will just mention two which lead to two very useful "rules of thumb." These two rules of operation are called the addition theorem and the multiplication theorem. We define the symbols used in these theorems as follows:

Symbol	Definition	Alternative Symbol and Definition Sometimes Used in Statistics Books
P(A)	The probability of event A occurring.	
P(A + B)	The probability of event A and/or event B occurring.	P(A ∪ B) The *union* of A and B.
P(AB)	The probability of A *and* B occurring.	P(A ∩ B) The *intersection* of A and B. The probability of A and B occurring is sometimes called the *joint probability* of A and B, or the *compound probability* of A and B occurring.
P(A | B)	The probability of A given that B has occurred. This is the symbol for a conditional probability.	

Addition Theorem:

$$P(A + B) = P(A) + P(B) - P(AB)$$

This theorem is interpreted as: The probability of event A and/or event B occurring is equal to the probability of A occurring plus the probability of occurrence of B minus the probability of both A and B occurring. To illustrate the use of this theorem suppose we were offered a gamble in which a card is drawn at random from a standard deck of bridge cards. We define event A as drawing a 4, 5, or 6 of any suit and we define event B as drawing a jack or queen of any suit. If event A and/or B occurs we win a prize. What is the probability of winning the prize, i.e., What is $P(A + B)$?

First we evaluate $P(A)$ as being 12/52 because there are 12 cards in the deck which are 4's, 5's, or 6's. Similarly, $P(B)$ is 8/52 (there are four jacks and four queens in the deck.) The last term in the theorem, $P(AB)$, is zero. This is because a card cannot be both A and B simultaneously — it is either A (a 4, 5, or 6) or B (a jack or queen), or neither. Therefore, probability of winning the prize is:

$$P(A + B) = P(A) + P(B) - P(AB)$$
$$P(A + B) = 12/52 + 8/52 - 0 = \underline{20/52}$$

Consider the same wager but with events A and B defined as follows: event A is a 4, 5, or 6 of any suit and event B is defined as a 6 or 7 of any suit. Now what's the probability of winning the prize? Again we must solve the addition theorem for $P(A + B)$. As before, $P(A) = 12/52$ and $P(B) = 8/52$ (there are four 6's and four 7's in the deck). The last term, $P(AB)$, is not zero, however. If a 6 is withdrawn it represents an element in the sample space which is defined as being part of event A *and* B. The probability of drawing a 6 of any suit is 4/52. Thus, the answer to this second wager becomes:

$$P(A + B) = P(A) + P(B) - P(AB)$$
$$P(A + B) = 12/52 + 8/52 - 4/52 = \underline{\underline{16/52}}$$

The reason that the last term, $P(AB)$, is subtracted in this theorem is to account for the fact that the elements in common were counted twice when we added the $P(A)$ and $P(B)$ terms. We can show this pictorially as in Figure 6.2. The rectangle represents the entire sample space. The large circle represents the elements of the sample space defined as event A, and the smaller circle represents the elements of the sample space defined as event B. The probability of event A can be thought of as being the area of the large circle divided by the total area of the rectangle. The same applies for event B. When we use the addition theorem to find the probability of A and/or B occurring we are, in effect, finding the combined area within the two circles as a fraction of the total area of the rectangle. The first two terms of the right hand side of the equation represent the sum of the areas of the two circles. But it will be noticed from Figure 6.2 that the circles overlap, or have common points. So we must subtract

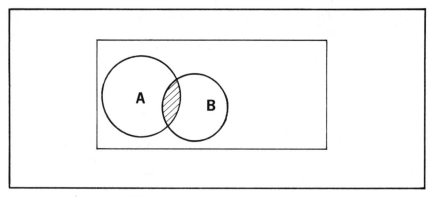

Figure 6.2 *Pictorial representation of two events A and B. The rectangle represents the entire sample space. Diagrams of this type are called Venn diagrams.*

the area of overlap (shaded in Figure 6.2) which was included twice in the summation. Diagrams of this type are called Venn diagrams, and are used extensively in a part of probability theory called set theory.

Referring back to the first of these two numerical examples we notice that the last term in the theorem was zero. Events A and B were *mutually exclusive,* and had no points in the sample space in common. The diagram for this situation is shown in Figure 6.3. To find the probability of A and/or B, if the events are mutually exclusive, we merely add the probability of event A to the probability of event B. This leads to a very useful "rule of thumb":

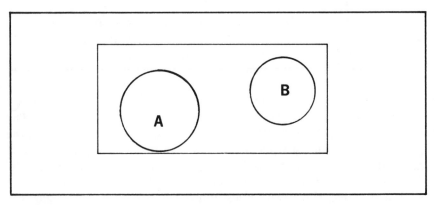

Figure 6.3 *Venn diagram showing two events which are mutually exclusive.*

For a number of events which are mutually exclusive, the probability of one of the events occurring is the sum of the individual probabilities.

If we want to know the probability of A, or B, or C, or D, . . . occurring and each of the events is mutually exclusive we simply add the individual probabilities. A useful way to remember whether this rule of thumb applies is the word *or;* the probability of A *or* B *or* C occurring.

This very brief description of the addition theorem considered only two events A and B. But the same logic and meaning is valid for 3 or more events. The only difference is that the theorem gets quite a bit longer to write. For example, the addition theorem for three events A, B, and C is:

$$P(A + B + C) = P(A) + P(B) + P(C) - P(AB) - P(BC) - P(AC) + P(ABC)$$

And as before, if A, B, and C are mutually exclusive the last four terms become zero and the theorem reduces to $P(A + B + C) = P(A) + P(B) + P(C)$. This is exactly what we would have obtained from our rule of thumb. The rule of thumb is valid for any number of mutually exclusive events.

Multiplication Theorem:

$$P(AB) = P(B \mid A) \, P(A)$$

This theorem is used to find the probability of two events occurring in sequence (or simultaneously). In words, the theorem is interpreted as: the probability of A *and* B occurring is equal to the probability of B, given that event A has occurred, multiplied by the probability of event A occurring in the first place. The term $P(B \mid A)$ is called a conditional probability. This theorem applies for both independent events and dependent events.

Let's illustrate this with an example using the deck of bridge cards. Define event A as drawing a jack of any suit on the first draw and event B as drawing a four of hearts on the second draw. Assume for the moment that the first card is *not* replaced in the deck prior to the second draw. What is the probability of both events occurring, i.e., What is P(AB)?

The likelihood of drawing a jack on the first draw, P(A), is 4/52. Since the card is not replaced in the deck there are only 51 remaining cards on the second draw, and the likelihood of drawing a four of hearts on the second draw, given that a jack had already been withdrawn is 1/51. This number corresponds to the conditional probability term of the multiplication theorem, $P(B \mid A)$. Hence, we can compute the probability of events A *and* B occurring as:

$$P(AB) = P(B \mid A)\, P(A)$$

$$P(AB) = (1/51)(4/52) = \frac{4}{(51)(52)} = \frac{4}{\underline{2652}}, \quad \text{or} \quad \frac{1}{663}$$

The sequence in this example is said to be dependent. The probability of a four of hearts on the second draw is dependent on whether or not a jack had been drawn on the first draw.

Now as a second example, suppose we are considering the same wager with the only change being that after the first card has been withdrawn and examined it is placed back in the deck and the deck thoroughly shuffled before drawing the second card. Now what is the probability of both events occurring, i.e., a jack of any suit on the first draw and a four of hearts on the second draw?

The probability of drawing a jack on the first draw, $P(A)$ is still 4/52. If the card is replaced before making the second draw the probability of a four of hearts on the second draw is 1/52 (as compared to 1/51 in the previous example). Using the multiplication theorem again we obtain:

$$P(AB) = P(B \mid A)\, P(A)$$

$$P(AB) = \left(\frac{1}{52}\right)\left(\frac{4}{52}\right) = \frac{4}{(52)(52)} = \frac{4}{\underline{2704}}, \quad \text{or} \quad \frac{1}{676}$$

It is important to note that in this last example the act of replacing the first card in the deck before making the second draw had the effect of making events A and B independent of one another. The probability of event B, drawing a four of hearts on the second draw, is independent of what happened on the first draw.

The second drawing is a completely new and separate experiment from the first. And in this case the conditionality of $P(B \mid A)$, in fact, becomes $P(B)$. That is, for independent events the probability of event B given that A has occurred reduces to just the probability of B. The occurrence of event B is not dependent on whether or not A has occurred. Thus, for the special case of independent events the multiplication theorem simplifies to the following: $P(AB) = P(B)\, P(A) \leftarrow$ Multiplication theorem for independent events.

This leads to a second "rule of thumb" which we will find very useful in our later discussions of risk analysis:

If a series of events are independent the probability that all the events will occur simultaneously (or in sequence) is the product of the individual probabilities of occurrence.

If we want to know the probability of A, *and* B, *and* C, *and* D, . . . , occurring, where A, B, C, and D are independent events, we simply multiply the individual probabilities. A useful way to remember this rule of thumb and to distinguish it from our previous one relating to mutually exclusive events is to note the word *and;* the probability of A *and* B *and* C, etc.

As with the addition theorem, this brief discussion of the multiplication theorem using just two events A and B can be generalized to three or more events as well. The multiplication theorem for three events is:

$$P(ABC) = P(A) \ P(B \mid A) \ P(C \mid AB)$$

and in general, for n events:

$$P(A_1 A_2 A_3 \ldots A_n) = P(A_1) \ P(A_2 \mid A_1) \ P(A_3 \mid A_2 A_1) \ldots$$
$$P(A_n \mid A_1 A_2 \ldots A_{n-1})$$

We will have occasion to use this generalized form of the multiplication theorem in our later discussions of figuring probabilities of various numbers of discoveries in a multi-well drilling program.

We have two final terms to consider in this section on probability definitions: sampling with replacement and sampling without replacement.

Sampling with Replacement: A phenomenon or experiment in which an observed sample (or outcome) is placed back into the sample space prior to the next trial or sampling. Sampling with replacement implies that the successive trials are a series of independent events. It also implies that the probability of a given event on successive trials remains constant with time. The sample space is the same for each successive trial or sample.

Sampling without Replacement: A phenomenon or experiment in which an observed sample (or outcome) is *not* replaced back in the sample space prior to the next trial or sampling. Sampling without replacement implies that successive trials are a series of dependent events. It also implies that the probability of a given event on the next trial changes after each successive trial. The sample space changes after every successive trial.

In the two examples used to explain the multiplication theorem the first wager was sampling without replacing. The first card was not replaced and we had to be concerned with the conditional probability term. The second wager was a sampling with replacement scheme, for the obvious reason that the first card was replaced in the deck before making the next draw.

In considering multi-well drilling programs it will be important, indeed imperative, that we establish whether the sequence of wells drilled is a series of independent events or dependent events. The reason is that the methods for computing probabilities of a given number of discoveries are not the same. Speaking very generally for the moment, exploring for oil in a basin is usually a sequence of dependent events — sampling *without* replacement. Once a structure has been drilled it is removed from the sample space of remaining undrilled locations. On the next structure tested the odds for discovery or failure will be different. But more about this in Chapter 7.

Examples and Illustrations of Probability Definitions

Before moving on to the second main section of this chapter it is perhaps worthwhile to illustrate how some of the terms we have just defined can be used. To do this we will discuss four examples or topics:

1. An exploration example involving conditional probabilities
2. Proof of the linear relationship between expected value and the probability of finding hydrocarbons
3. Proof of the two ways of drawing decision trees having two chance nodes in sequence
4. Answers to the questions in Chapter 3 relating to playing roulette

The first two topics relate to important facets of our later discussions on exploration risk analysis. For this reason it is suggested that you take the time to study them. Item 3 is a proof of a technique we mentioned in Chapter 4. Those not interested in proofs and a few equations can skip the discussions of Item 3. And Item 4 is for those of you who took the trouble to answer the questions in Chapter 3 about roulette wheels. The non-gamblers can skip this section if desired.

1. *An Exploration Example Involving Conditional Probabilities*

Tom Smith, a geologist for Profit Oil and Gas Company (Henry Oilfinder's Company) is analyzing a drilling prospect in the Northeast

Bryarwood Field of O'Brien County. There have been 20 successful gas completions drilled thus far in the field having reserves ranging from 2–5 BCF per well. The field appears to have some rather complex stratigraphic and lithologic variations within the field. These variations have been difficult to predict and the quality of the pay sand (and hence ultimate reserves) seems to be more directly related to these variabilities than pay thickness or structural position. So, as an attempt to determine probabilities of obtaining various levels of reserves in the proposed well, he tabulated the reserves from the 20 completed wells, as shown in Table 6.1.

Table 6.1
Tabulation of Gross, Per-Well Reserves
in Northeast Bryarwood Field

Gross Recoverable Gas Reserves, BCF	No. of Wells Having Given Reserves	Percent of Wells Having Given Reserves
2 BCF	7	35% ← ($7/_{20}$ × 100%)
3 BCF	7	35%
4 BCF	4	20%
5 BCF	2	10%
	20	100%

The next step in Tom's analysis was to associate monetary profits for each level of reserves for the two decision alternatives available to them — drill the prospect or farm it out and retain an override. With the help of the engineering staff and their economic analyses of wells in the field he obtained the data of Table 6.2.

Tom felt the principal uncertainty in the prospect regarded the likelihood the prospect would even be productive. After detailed study and comparisons to N. E. Bryarwood Field and nearby correlative areas he concluded the probability of even finding gas to be about 0.25. What is the best decision choice, based on maximizing EMV? What minimum probability of finding gas is required to even justify the risk of drilling?

Solution: To determine the best decision choice requires a straight-forward EMV calculation for each decision option. The difference, however, between Tom's problem and the EMV examples we studied in

Table 6.2

Revenues Associated with Various Levels of Reserves in N. E.
Bryarwood Field. All NPV Values Are Net Profits After Taxes,
Royalties, Operating Expenses, and Producing Well Costs.

Gross Recoverable Gas Reserves, BCF	NPV Profit If Profit Oil and Gas Co. Drills	NPV Profit If Prospect Is Farmed Out
2 BCF	+$ 40,000	+$ 9,000
3 BCF	+ 90,000	+ 12,500
4 BCF	+ 130,000	+ 15,000
5 BCF	+ 200,000	+ 18,000

(Dry Hole Cost = $70,000)

Chapter 3 relates to the probabilities of occurrence for each possible outcome. When Tom determined the relative percentages in Table 6.1 he was tabulating only the previous drilling outcomes which resulted in producing wells.

There are no dry holes in his listing of Table 6.1. So his conclusion of probability 0.35 of 2 BCF really means that, given the well finds production, there is a 35% chance of producing only 2 BCF. The probabilities of Table 6.1 are, in fact, conditional probabilities. What he needs for his expected value computation are probabilities of finding gas *and* given levels of reserves. To do this he must use the multiplication theorem we discussed previously. For example, let's first consider the probability of finding gas and that reserves are 2 BCF. We can define event A as finding gas and event B as 2 BCF reserves. We are interested in finding P(AB), the left side of the multiplication theorem.

$$P(AB) = P(B \mid A) \, P(A)$$

The likelihood of event A, finding gas, was estimated to be 0.25. The probability of 2 BCF, *given we find gas*, P(B | A), is 0.35, the value he computed in Table 6.1. Thus

Prob. of gas and 2 BCF = P(AB) = 0.35 × 0.25 = 0.0875.

The same procedure is applied for the remaining reserve levels. Each of the conditional probabilities of Table 6.1 must be multiplied by the likelihood of finding gas in the first place, 0.25. The resulting probabilities are summarized as follows:

Possible Outcome	Probability of Occurrence
Dry Hole	(1 − Prob. of gas) = (1 − 0.25) = 0.7500
2 BCF	(Cond. Prob. of 2 BCF)(Prob. of gas) = (0.35)(0.25) = 0.0875
3 BCF	(Cond. Prob. of 3 BCF)(Prob. of gas) = (0.35)(0.25) = 0.0875
4 BCF	(Cond. Prob. of 4 BCF)(Prob. of gas) = (0.20)(0.25) = 0.0500
5 BCF	(Cond. Prob. of 5 BCF)(Prob. of gas) = (0.10)(0.25) = <u>0.0250</u>
	1.0000

Now Tom can proceed with the usual EMV Computations, as summarized in Table 6.3.

Table 6.3
Expected Value Computations for Drilling Prospect in N. E. Bryarwood Field

Possible Outcome	Prob of Occur.	DRILL		FARM OUT	
		Conditional Values Received (Table 6.2)	Expected Value of Outcome	Conditional Values Received (Table 6.2)	Expected Value of Outcome
Dry Hole	0.7500	−$ 70,000	−$52,500	0	0
2 BCF	0.0875	+ 40,000	+ 3,500	+$ 9,000	+$ 787
3 BCF	0.0875	+ 90,000	+ 7,875	+ 12,500	+ 1094
4 BCF	0.0500	+ 130,000	+ 6,500	+ 15,000	+ 750
5 BCF	0.0250	+ 200,000	+ 5,000	+ 18,000	+ 450
	1.0000		−$29,625		+$3081

Conclusion: Decision Alternative	EMV
Drill	−$29,625
Farm Out	+ 3,081 ← Preferred Choice

From the EV calculations of Table 6.3 we see that the preferred decision choice, based on maximizing EMV, is to farm out. The new concept, illustrated by this example, is the notion that tabulations of reserve data from producing wells or fields result in relative frequencies or probabilities which are conditional upon having found oil or gas. These frequency data must be multiplied by the likelihood of finding oil or gas in the first place in order to compute the actual probability of finding a given level of reserves. By this multiplication process we are, in essence,

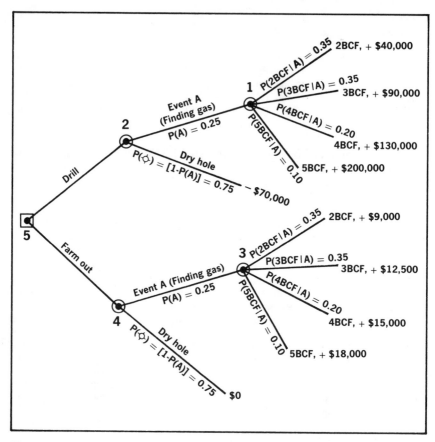

Figure 6.4 *Alternative solution to Table 6.3 using a decision tree. Event A is defined as finding gas. The probabilities around chance nodes 1 and 3 are conditional probabilities from Table 6.1.*

"normalizing" the reserve frequency data so that they sum to the probability of finding gas (0.25 in this example), rather than 1.0.

Before we consider the second question which Tom raised let's look at another way we could have analyzed the prospect. It is with a decision tree, as shown in Figure 6.4.

The solution to the decision tree of Figure 6.4 is as follows:

Step 1. Compute EV at Chance Node *1:*

$EV_1 = (.35)(+\$40,000) + (.35)(+\$90,000) + (.20)(+\$130,000)$
$\qquad\qquad\qquad\qquad\qquad\qquad\qquad\qquad + (.10)(+\$200,000)$

$\underline{\underline{EV_1 = +\$91,500}}$

The EV at Chance Node 1 is called *the conditional EV given that gas is found* for the strategy "drill."

Step 2.   Compute EV at Chance Node *2:*

$$EV_2 = (0.25)(EV_1) + (0.75)(-\$70,000)$$
$$EV_2 = (0.25)(+\$91,500) + (0.75)(-\$70,000) = \underline{\underline{-\$29,625}}[1]$$

Step 3.   Compute EV at Chance Node *3:*

$$EV_3 = (.35)(+\$9,000) + (.35)(+\$12,500) + (.20)(+\$15,000)$$
$$+ (0.10)(+\$18,000)$$

$$EV_3 = \underline{+\$12,325}$$

The EV at Chance Node 3 is called *the conditional EV given that gas is found* for the strategy "farm out."

Step 4.   Compute EV at Chance Node *4:*

$$EV_4 = (0.25)(EV_3) + (0.75)(\$0)$$
$$EV_4 = (0.25)(+\$12,325) + (0.75)(\$0) = \underline{\underline{+\$3,081}}$$

Step 5.   Make decision at Decision Node *5:*

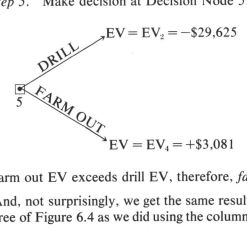

Farm out EV exceeds drill EV, therefore, *farm out.*   (QED)

And, not surprisingly, we get the same results from solving the decision tree of Figure 6.4 as we did using the columnar computation of Table 6.3.

We can extract another useful technique from the steps we performed to solve the decision tree. Figure 6.5 is a picture of the equivalent tree

---

1. Note that the EV calculation at Chance Node 2 involves multiplying probabilities by numbers which themselves are expected values. This perfectly valid operation was mentioned briefly on page 72 in Chapter 3, but we did not have the occasion to show an example until this point.

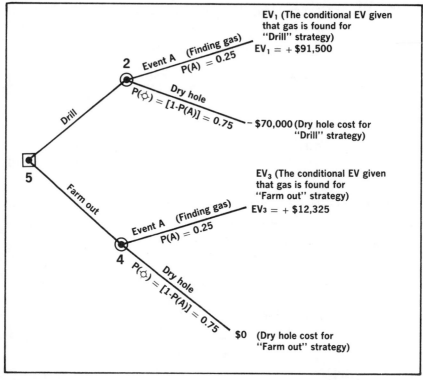

EV₁ (The conditional EV given that gas is found for "Drill" strategy)
$EV_1 = +\$91,500$

$-\$70,000$ (Dry hole cost for "Drill" strategy)

EV₃ (The conditional EV given that gas is found for "Farm out" strategy)
$EV_3 = +\$12,325$

$0 (Dry hole cost for "Farm out" strategy)

Figure 6.5 *Partially solved tree of Figure 6.4. EV_1 and EV_3 were computed by finding the expected values at Chance Nodes 1 and 3 respectively of Figure 6.4.*

after computing the EV at Chance Nodes 1 and 3. To find the EV for "drill" consisted of finding the EV at Chance Node 2. The computation was made previously under Step 2.

$$EV_2 = EV_{DRILL} = (0.25)(EV_1) + (0.75)(-\$70,000) = \underline{-\$29,625}$$

This computation can be generalized to a useful equation for computing the EV of a drill or farm out decision directly:

$$\begin{matrix} \text{EV of} \\ \text{decision} \\ \text{alternative} \end{matrix} = \begin{bmatrix} P(A) \end{bmatrix} \begin{bmatrix} \text{Conditional EV,} \\ \text{given that oil} \\ \text{or gas is found} \end{bmatrix} + \begin{bmatrix} 1 - P(A) \end{bmatrix} \begin{bmatrix} \text{Costs associated} \\ \text{with dry hole} \\ \text{failure} \end{bmatrix}$$

$$(6.1)$$

where

$$A = \text{Event "finding oil or gas"}$$
$$P(A) = \text{Probability of finding gas or oil.}$$

Equation 6.1 can be used to compute the EV for drilling or farming out – the only difference being that the conditional EV terms will be different for each strategy. Equation 6.1 leads us to another discovery. For a given strategy, such as drill, the conditional EV terms are constant. The only variable on the right side of the equation is P(A), the probability of finding hydrocarbons. P(A) only appears in Equation 6.1 to the first power (there are no $P(A)^2$, or $P(A)^3$ terms) which means that a graph of EV of the decision strategy versus P(A) will be linear. An important discovery! We can make a graph of EV versus P(A) for each strategy and be able to compare strategies for all values of P(A), not just 0.25. And this will probably lead us to the answer of Tom's second question regarding the minimum probability required to justify drilling.

Let's see how it works. First we must compute two expected values for drilling and two expected values for farm out.

Step 1: Compute two EV's for drill strategy using two different values of P(A)

$$\text{Conditional EV, given that gas is found} = EV_1 = +\$91,500$$
$$\text{Costs associated with dry hole} = -\$70,000$$

a. Let P(A) = 0.25.   Solving Equation 6.1 we get:

$$EV_{Drill} = (.25)(+\$91,500) + (1 - .25)(-\$70,000)$$
$$= -\$29,625$$

b. Select another value of P(A), say P(A) = 0. Compute EV_{Drill}:

$$EV_{Drill} = (0)(+\$91,500) + (1 - 0)(-\$70,000)$$
$$= -\$70,000$$

Step 2: We now have two sets of data:

$P(A)$	EV_{Drill}
0.25	−$29,625
0.00	−$70,000

These are plotted on coordinate graph paper as in Figure 6.6 and the points connected with a straight line to give EV_{Drill} as a function of P(A).

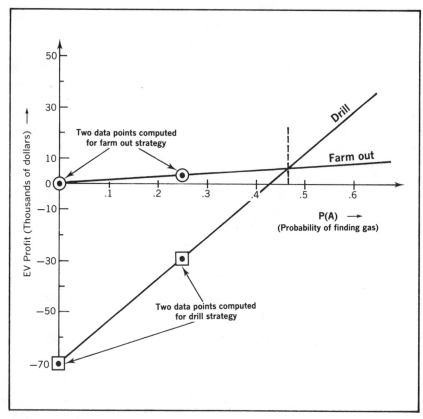

Figure 6.6 *Expected value as a function of P(A), probability of finding gas, for N.E. Bryarwood Field Prospect. The two functions cross at P(A) = .47. For values of P(A) <0.47 the farm out is the preferred decision strategy. For P(A) > 0.47 the drill strategy gives the highest EV. The intersection value of P(A) is the minimum probability of finding gas required to justify the risk of drilling.*

Step 3: Compute two EV's for farmout strategy using any two values of P(A).

Conditional EV, given that gas is found (farmout)
$$= +\$12,325$$
Costs associated with dry hole (farmout) $= 0$

a. Let P(A) = 0.25.   Solving Equation 6.1 we get:

$$EV_{Farmout} = (.25)(+\$12,325) + (1 - .25)(0) = +\$3,081$$

b. Select another value of P(A), say P(A) = 0.   Compute $EV_{Farmout}$:

$$EV_{Farmout} = (0)(+\$12,325) + (1 - 0)(0) = 0$$

Step 4:   We now have two sets of data for the farmout strategy.

$P(A)$	$EV_{Farmout}$
0.25	+$3,081
0.00	0

These are plotted on Figure 6.6 and the points connected with a straight line to give $EV_{Farmout}$ as a function of P(A).

From Figure 6.6 we see that the EV functions cross at a value of P(A) of about 0.47. For any value of P(A) less than 0.47 the farmout strategy will give a higher EV. For probabilities of finding gas greater than 0.47 the drill strategy is preferred. So we've answered Tom's second question graphically – the minimum probability of finding gas required to justify the risk of drilling is 0.47.

If he feels the probability of finding gas is in the lower range around 0.25 the prospect should be farmed out. Figure 6.6 is exactly analogous to Figure 3.4 in Chapter 3, and correlations of this type are extremely useful to portray to management the sensitivity on EV for various values of the probability of finding hydrocarbons. Incidentally, when using Equation 6.1 and the graphical approach just described any two values of P(A) can be used to get the necessary plotting points.

For the mathematicians reading this, the minimum probability of finding hydrocarbons required to justify drilling could have been computed directly, rather than going to the trouble of making a graph such as Figure 6.6. To do this we merely have to solve Equation 6.1 for P(A) for the case where EMV_{Drill} is set equal to $EMV_{Farmout}$. For this example the equation would be as follows:

$$EMV_{Drill} = EMV_{Farmout}$$

$$[P(A)][+\$91,500] + [1 - P(A)][-\$70,000] = [P(A)][+\$12,325]$$
$$+ [1 - P(A)][\$0]$$

The numerical constants are those shown in Figure 6.5. Now we merely solve this equation for P(A), the probability of finding gas which makes the EV of both strategies equal. With a bit of algebra you should be able to solve for a value of P(A) ≅ 0.47. This is the same value we read from Figure 6.6 corresponding to the intersection of the two functions.

Well, that answers Tom Smith's questions and the example should tell you everything you wanted to know about conditional probability (and perhaps some things you didn't want to know)! The notion of conditional probability is an important part of risk analysis, and hopefully this example has given you an insight about what it is and how it is used in expected value computations.

2. *Proof of the Linear Relationship between Expected Value and the Probability of Finding Hydrocarbons*

As we have stated on several occasions, graphs such as Figure 6.6 and Figure 3.4 are very useful to isolate the effects of the probability of finding hydrocarbons on expected value. This probability is sometimes very difficult to estimate, and to be able to show its effects on EV in graphical form can be very informative to the decision maker.

We have already suggested one way to prove that EV as a function of the probability of finding hydrocarbons is linear in our discussion of Equation 6.1. We noted that P(A), the probability of gas, was the independent variable and only had terms to the first power. But this proof involved two constants—the conditional EV's, given that hydrocarbons were found. Is it possible to prove linearity using a conventional four-column EV computation? Yes it is, and the proof is as follows.

Define the following constants:

CF = Cost of failure (dry hole)

$R_1, R_2, \ldots R_n$ = Various levels of reserves (such as 2 BCF, 3 BCF, etc.)

$V_1, V_2, \ldots V_n$ = Values received from the $R_1, R_2, \ldots R_n$ reserves. Usually expressed as monetary NPV, but could be any of the acceptable ways of expressing conditional values received for each outcome.

$P(R_1 \mid A), P(R_2 \mid A), \ldots$ = Conditional probabilities of each level of reserve given that gas or oil is found. (Example: the conditional probabilities tabulated in Table 6.1 of Tom Smith's drilling prospect).

Also define the Event A as finding hydrocarbons of any amount and P(A) as the probability of finding hydrocarbons. P(A) is considered the independent variable and EV is the dependent variable (in the sense of algebraic functions).

With the constants as defined above we can prepare a generalized EV computation in which P(A) is considered an independent variable. This generalized EV table is given in Table 6.4.

Table 6.4
Generalized Expected Value Computation in Which P(A), the Probability of Finding Hydrocarbons, Is the Independent Variable

Outcome	Probability of Occurrence	Conditional Value Received from Outcome	Expected Value of Each Outcome
Dry Hole	$[1 - P(A)]$	CF	$[1 - P(A)] \times CF$
R_1	$P(R_1 \mid A) \times P(A)$	V_1	$P(R_1 \mid A) \times P(A) \times V_1$
R_2	$P(R_2 \mid A) \times P(A)$	V_2	$P(R_2 \mid A) \times P(A) \times V_2$
R_3	$P(R_3 \mid A) \times P(A)$	V_3	$P(R_3 \mid A) \times P(A) \times V_3$
.	.	.	.
.	.	.	.
.	.	.	.
R_n	$P(R_n \mid A) \times P(A)$	V_n	$P(R_n \mid A) \times P(A) \times V_n$
	1.000		$\Sigma = $ EV of decision strategy

The summation of Column 4, EV of the decision strategy, can be written algebraically as follows:

$$EV = [1 - P(A)] \times CF + P(R_1 \mid A) \times P(A) \times V_1 + P(R_2 \mid A) \times P(A) \\ \times V_2 + P(R_3 \mid A) \times P(A) \times V_3 + \ldots + P(R_n \mid A) \times P(A) \times V_n$$

But we can factor out the common term P(A) in the products following the first product as follows:

$$EV = [1 - P(A)] \times CF + P(A)[P(R_1 \mid A) \times V_1 + P(R_2 \mid A) \times V_2 \\ + P(R_3 \mid A) \times V_3 + \ldots + P(R_n \mid A) \times V_n] \qquad (6.2)$$

All the terms in the second brackets are constants, and CF, the dry hole cost, is a constant. The only variable on the right hand side of the equation is P(A), and it only appears to the first power. Hence, we can conclude that the EV of a decision alternative expressed as a function of P(A), the probability of finding gas or oil will be linear.   *QED*

The analogy between Equation 6.2 and Tom Smith's solution to the decision tree of Figure 6.4 should be obvious. The conditional probability terms of Equation 6.2 correspond to the P(2 BCF | A) = .35, P(3 BCF | A) = .35, . . . terms on the branches around chance nodes 1 and 3. And the V_1, V_2 . . . terms of Equation 6.2 correspond to the terminal points of the branches around chance nodes 1 and 3, (+$40,000, + $90,000, etc.).

Multiplying the conditional probabilities by values received in the second bracket of Equation 6.2 is exactly analogous to finding the conditional expected values at chance nodes 1 and 3. By replacing the product terms in the second bracket of Equation 6.2 with the words "conditional EV given that gas is found" we see that Equation 6.2 is exactly the same as the Equation 6.1 we used previously. Amazing!

For any expected value computation for drilling prospects of the form of Table 6.4 the EV for the decision alternative will always be a linear function of the probability of finding hydrocarbons. Only two points are required to plot a straight line, and any two values of P(A) can be arbitrarily chosen to yield the plotting data.

From graphs such as Figure 6.6 or 3.4 the decision maker can determine the EV for each available strategy for *all* values of P(A). In addition, he is able to compare the relative desirability of competing strategies (such as drill versus farmout) for various values of P(A). Even if the analyst is uncertain of the exact value of P(A) and can only estimate a probable range it may still be possible to make a definite (and correct) decision. This will be a most useful analysis technique in petroleum exploration risk analysis.

3. *Proof of the Two Ways of Drawing Decision Trees Having Two Chance Nodes in Sequence*

Now that we have defined some basic terms of probability as well as the addition and multiplication theorems we can digress a moment to prove a statement we made on page 129 in Chapter 4. Recall at that point we were showing the two ways to construct a decision tree for the situation of two chance events occurring in sequence. The alternatives were shown in Figure 4.17 and are shown again as Figure 6.7. The V_i terms are the conditional values received at each terminal point. The probability terms follow the nomenclature we just developed in discussing the rules of algebra of probability theory. The proof we are about to give is valid regardless of whether the two chance events are independent events or dependent events.

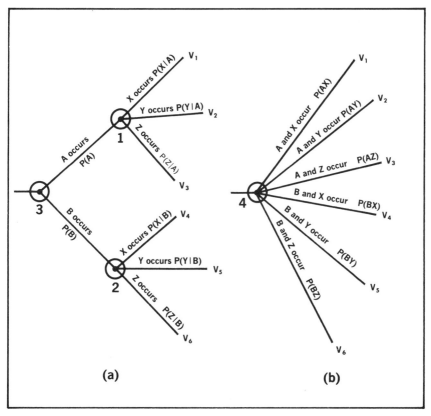

Figure 6.7 *The two ways to construct a decision tree having two chance events which occur in sequence.*

First, we compute the EV at chance node 1 of part (a):

$$EV_① = [P(X \mid A)][V_1] + [P(Y \mid A)][V_2] + [P(Z \mid A)][V_3]$$

Similarly the EV at chance node 2 is:

$$EV_② = [P(X \mid B)][V_4] + [P(Y \mid B)][V_5] + [P(Z \mid B)][V_6]$$

Moving backward in the normal manner for solving decision trees we can compute the EV at chance node 3 as:

$$EV_③ = [P(A)][EV_①] + [P(B)][EV_②]$$

or, by substitution we can obtain

$$EV_{\text{\textcircled{3}}} = [P(A)]\{[P(X \mid A)][V_1] + [P(Y \mid A)][V_2] + [P(Z \mid A)][V_3]\} + [P(B)]\{[P(X \mid B)][V_4] + [P(Y \mid B)][V_5] + [P(Z \mid B)][V_6]\}.$$

This relationship can be expanded to:

$$EV_{\text{\textcircled{3}}} = [P(A)][P(X \mid A)][V_1] + [P(A)][P(Y \mid A)][V_2] + [P(A)][P(Z \mid A)][V_3] + [P(B)][P(X \mid B)][V_4] + [P(B)][P(Y \mid B)][V_5] + [P(B)][P(Z \mid B)][V_6].$$

By virtue of the multiplication theorem the first two terms of each three-term product can be replaced by their equivalency. That is, for example, $[P(A)][P(X \mid A)] = P(AX)$ etc. Thus, we can write:

$$EV_{\text{\textcircled{3}}} = [P(AX)][V_1] + [P(AY)][V_2] + [P(AZ)][V_3] + [P(BX)][V_4] + [P(BY)][V_5] + [P(BZ)][V_6]$$

Now looking at alternative (b.) of Figure 6.7 we can compute the EV at chance node 4 as:

$$EV_{\text{\textcircled{4}}} = [P(AX)][V_1] + [P(AY)][V_2] + [P(AZ)][V_3] + [P(BX)][V_4] + [P(BY)][V_5] + [P(BZ)][V_6]$$

By comparing terms of this relationship to the one previously derived for $EV_{\text{\textcircled{3}}}$ you will note both relationships are identical, hence $EV_{\text{\textcircled{4}}} = EV_{\text{\textcircled{3}}}$. Thus we see that either way is correct and will yield the exact same EV at the left, initial node.

4. Answers to the Questions in Chapter 3 Relating to Playing Roulette

In Chapter 3 we gave an example of placing $100 wagers to play roulette (Example No. 3, page 88). At the conclusion of the discussion we mentioned some interesting questions about roulette and suggested that you try to answer them. Now that we have defined some probability terms we can speak to each question in a more specific manner. Here are the answers to the first four questions:

1. Would our EV have been the same if we had placed our wager on number 6, rather than 23? Does it matter what number we place our wager on?

Assuming a balanced roulette wheel and an honest croupier each of the 37 possible outcomes are equally likely to occur. The number 23 is not less likely or more likely to occur than any other number. So if we are placing our wager on a given number (as opposed to black or red) it doesn't matter which number we choose.

2. The ball has not fallen on number 14 for many, many spins. Is it now a good time to place your wager on number 14? Does the EMV computed in Table 3.7 apply no matter when you choose to begin playing, or does it apply only after certain things have occurred before—such as no number 14 in the past 75 spins?

The points raised by these questions are often misunderstood by the uninformed gambler. Each spin of a roulette wheel is an independent event. The wheel does not have a memory, and can not keep track of whether or not number 14 is "overdue." Therefore, it makes absolutely no difference when you begin. Your chances of winning on number 14 on the next spin are exactly $1/37$—no matter what has happened on the previous spins. Even if the number 14 had won for the last 50 spins the chances of winning on the 51st spin is just as good as it was on the first spin. (It's highly unlikely that the ball would land on the same number 50 consecutive times—but it is possible. And remember we are still talking about an honest wheel.) The EMV computation of Table 3.7 applies no matter when you choose to begin playing. It is not conditional on any specific previous sequences.

3. You have bet on number 23 and lost 36 consecutive times. What is the probability of winning on the 37th trial? What is the probability of at least one win in 37 spins?

We've already commented on the fact that each spin is an independent event. Therefore, the probability of winning on the 37th spin is the same as on any other spin—$1/37$. The second question is the one many people misunderstand. Their feeling is that they should get one number 23 every 37 spins, and since there have been 36 losses it's a certainty of getting a 23 on the 37th trial. *WRONG!* The probability of at least one win in 37 trials (at least one win means the same as one or more) is one minus the probability of no wins in 37 trials.

The probability of losing on any given spin is $36/37$, and the probability of losing on the first spin *and* the second spin *and* . . . *and* the 37th spin is the product of the individual probabilities (by virtue of our rule of thumb about independent events). Thus, the probability of 37 consecutive losses is $(36/37)^{37}$. In the sample space of outcomes of 37 spins one possibility is no wins, the only other possibilities are 1 win in

37, 2 wins in 37, . . . , or in other words—at least one win. Thus the probability of at least one win in 37 spins is:

$$\text{Prob. of at least one win in 37 spins} = 1 - \left(\frac{36}{37}\right)^{37} \cong \underline{0.637}$$

These two questions are entirely different. The first involves the sample space of outcomes on the 37th trial and the second involves the sample space of outcomes from 37 trials. We must be very careful how we define sample spaces in probability theory! If we are not careful about this we will end up with all sorts of misconceptions and errors of judgment. The impression that we are bound to have a win on number 23 after 36 losses is analogous to the exploration manager who comments "We've been having a long run of dry holes and we're due a discovery on the next well." (Have you ever heard your manager say this?)

4. Is it correct to assume that each number will come up once in every 37 trials?

The answer is no! While it is correct to say that one win in 37 spins is the most probable, or most likely outcome, the actual probability of *exactly* one win in 37 trials is probably a lot smaller than you would imagine. We can compute it using the second rule of thumb about independent events. Suppose the win came on the first spin followed by 36 losses. The probability of a win on the first *and* a loss on the second spin, *and* a loss on the third spin, *and* . . . *and* a loss on the 37th spin is the product of each of the individual probabilities. That is:

$$\text{Prob. of win on first spin, followed by 36 consecutive losses} = \left(\frac{1}{37}\right)\left(\frac{36}{37}\right)^{36} \cong 0.01008$$

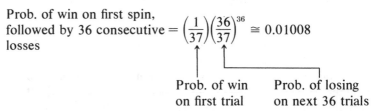

Prob. of win on first trial   Prob. of losing on next 36 trials

It is also possible that the win could occur on the second trial, rather than the first, and the probability of this outcome is:

$$\text{Prob. of the one win in 37 trials occurring on second spin} = \left(\frac{36}{37}\right)\left(\frac{1}{37}\right)\left(\frac{36}{37}\right)^{35}$$

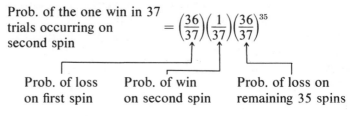

Prob. of loss on first spin   Prob. of win on second spin   Prob. of loss on remaining 35 spins

But this equation rearranges to $(^1/_{37})(^{36}/_{37})(^{36}/_{37})^{35} = (^1/_{37})(^{36}/_{37})^{36}$, which is the same as the probability of a win on the first trial. Winning on the first spin and winning on the second spin are mutually exclusive, and the probability of one or the other occurring is the sum of the two probabilities, i.e., 0.01008 + 0.01008. If we continue this line of reasoning it can be shown there are 37 mutually exclusive ways to have one win in 37 spins, and the (equally likely) probability of each way is 0.01008. Consequently, by use of our rule of thumb regarding mutually exclusive events the probability of any one of the 37 ways occurring is simply the sum of the individual probabilities. Thus the probability of just one win in 37 spins is $[37 \times 0.01008]$, or about 0.37. Surprised?

How many questions did you answer correctly before reading these answers? After finding out these facts are you still willing to play roulette? By the way, the fifth question on page 90 should have been answered in our Chapter 5 discussions on preference theory.

II. PROBABILITY DISTRIBUTIONS AND SINGLE VALUE PARAMETERS OF DISTRIBUTIONS

Another integral part of petroleum exploration risk analysis is the concept of probability distributions. Distributions are useful to describe in a very concise graph the range of possible values a variable can have as well as the probabilities of these values occurring. The new risk analysis technique called simulation is based on describing all the uncertain parameters in prospect analysis as distributions, including overall net present value profit. Therefore, it is important that we understand what distributions are, how to construct distributions from statistical data, and how to interpret distributions.

In this second main portion of the chapter we will define what a probability distribution is, the notion of relative frequency, and the meaning and uses of cumulative frequency distributions. We will also define several important single values from distributions which can tell us important information about the entire distribution. These numerical values are called single value parameters of distributions. The various definitions of this portion will be illustrated using core data from a well in southwestern Nebraska.

First we must define a random variable.

Random Variable: A parameter or variable which can have more than one possible value is called a random variable. Random variables may be thought of as those variables whose values cannot be predicted with certainty at the time of decision making. For each

possible value of the random variable there is associated a likelihood, or probability of occurrence.

Random variables are sometimes called stochastic variables to
denote the fact that the likelihoods of the values occurring are stochastic,
or probabilistic in nature. If the value of a variable is known or can be
predicted with certainty at the time of decision making the variable is
called a deterministic variable.

Examples of random variables in petroleum exploration include the
possible values of net pay thickness, ultimate reserves from a well or
field, drilling costs, ultimate net profit from a drilling prospect, etc. The
numerical values of a random variable can be positive or negative. Each
value (or range of values) will have a certain probability of occurrence.
Different values of the random variable can have the same probability
of occurrence. The adjective "random" implies that the values of a
variable have specific probabilities of occurrence, and it does not mean
or imply that the variable itself is random, or randomly distributed.

The reason we have taken the time to define random variables is
that random variables are the parameters being described when we deal
with probability distributions.

> *Probability Distribution:* A probability distribution is a graphical repre-
> sentation of the range and likelihoods of occurrence of possible
> values that a random variable can have. Probability distributions
> can be discrete or continuous, depending on the nature of the ran-
> dom variable. The horizontal scale of a probability distribution is
> the random variable and its appropriate units and range of values.
> The height of a probability distribution above the horizontal axis
> is proportional to the probability of occurrence of the values (or
> ranges of values) of the random variable.

An example of a continuous probability distribution is given in Figure
6.8. It is continuous in the sense that any value of recoverable reserves
within the range of $x_{minimum}$ and $x_{maximum}$ is possible. That is, there is a
continuum of possible values of x between the minimum and maximum
values. The parameter on the vertical axis, f(x), is called a probability
density function. It is a mathematical function such that when we determine the area under the distribution by integrating f(x) from $x_{minimum}$ to
$x_{maximum}$ the resulting area will be dimensionless 1.0.

The area under all probability distributions is by definition, *one.*
Another characteristic of all probability distributions is that the probabilities are always positive (or zero), but never negative. This means

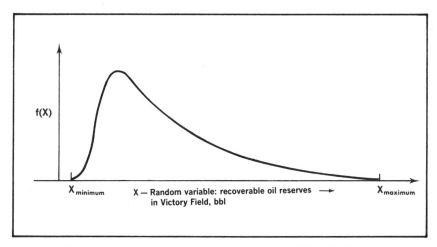

f(X)

$X_{minimum}$ X — Random variable: recoverable oil reserves ——→ $X_{maximum}$
in Victory Field, bbl

Figure 6.8 *An example of a continuous probability distribution.*

that the probability distribution curve, or function, never goes below the horizontal axis.

We will not become involved with defining and integrating the probability density functions on the ordinate of a probability distribution. This is for two reasons. First, by use of appropriate proportionality constants and/or some integration the vertical scale can be converted to a numerical scale that is proportional to the probability of occurrence of a value or range of values of the random variable. This leads us to the general statement given earlier that the height of the curve above the x-axis is proportional to the probability of occurrence.

Thus, in Figure 6.8 we can conclude that values of x (recoverable reserves) towards the low end of the range and under the high points of the curve are the most probable values. As x gets larger the curve gets lower and lower which implies that the larger ranges of x become decreasingly less probable. The probability of occurrence of a value of x which is less than $x_{minimum}$ or greater than $x_{maximum}$ is zero.

The second reason for omitting a formal discussion of probability density functions is that they are, in general, rather complex mathematical functions. And the mathematics and integral calculus required to define and use probability density functions is well beyond the intended scope of this book. (To which most of my students give a collective sigh of relief!) The mathematical operations dealing with probability density functions can be accomplished much easier by use of various graphical techniques and tables to be described later in the chapter.

Figure 6.8 is an example of a continuous distribution that describes a random variable which can have any of the infinite numerical values within a given range. There are some random variables, however, which can have only specific numerical values. The probability distributions which describe random variables of this type are called *discrete* distributions. Any random variable whose values result from counting is an example of a variable that would be described as a discrete probability distribution (examples: number of structures which have oil in a basin, number of producing wells in a field, etc.).

An example of a discrete distribution is given in Figure 6.9. The likelihoods of occurrence of each discrete value of the random variable can be read directly from the vertical scale of discrete probability distributions.

With discrete distributions the sum of the p(x) terms for each possible value of the random variable must be exactly one. That is, for the random variable x_i (i = 0, 1, 2, 3, ... n) the summation $\sum_{i=0}^{n} p(x_i) = 1.0$.

And as before, the height of the lines (or spikes, as they are sometimes called) is proportional to the probability of occurrence of the particular value of the random variable.

In Figure 6.9 the most likely number of productive structures in the basin is 15; there is a zero probability of 27 or more structures being productive, etc. Only integer values of the random variable are possible — we can't have 6.8 structures productive, or 14.1, etc. Just the integer values are possible — 0, 1, 2, 3, . . . This is, of course, why any random variable which results from counting is represented as a discrete probability distribution.

With regard to continuous distributions the theoretical limits of the probability density function are plus and minus infinity. For a random variable having limits such as in Figure 6.8 the probability density function is simply defined as being zero from minus infinity to $x_{minimum}$ and zero from $x_{maximum}$ to plus infinity. Another technical point with continuous probability distributions is that it is only meaningful to talk in terms of the probability of a *range* of values of x and not a specific value of x. For a continuous random variable the probability of occurrence of a specific, single value is, theoretically, zero.

When we work with probability distributions it is important that we

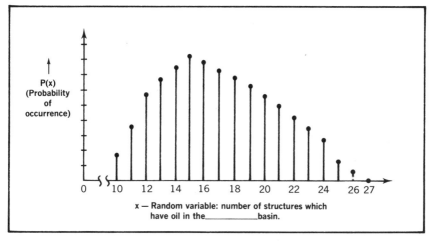

Figure 6.9 *An example of a discrete probability distribution. The probability of occurrence of each value of the random variable can be read directly from the ordinate.*

know how to interpret the meaning of the distributions. For instance, we may wish to know or determine the probability that the value of a random variable x will be between (within) two specified values, say x_1 and x_2. Referring to Figure 6.10 we are trying to find the likelihood of x being within the shaded area between $x = x_1$ and $x = x_2$. To do this we must find the area under the curve between x_1 and x_2. If we know the probability

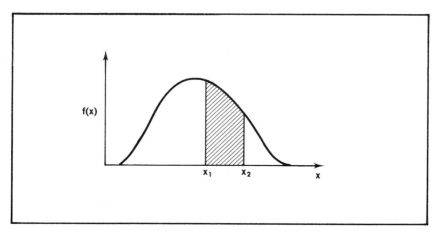

Figure 6.10 *A continuous distribution of the random variable x.*

density function we simply integrate the function from the limits $x = x_1$ to $x = x_2$. If we do not know the probability density function there are various graphical and tabular techniques for reading the area under the curve.

The point here is that if we want to know the probability that the value of a random variable will be within two specified limiting values we merely find the area under the probability distribution curve between the limits. To express this in symbols we would write $P(x_1 \leqslant x \leqslant x_2)$. Interpret this symbol as "the probability x will be within the limits x_1 and x_2."

As another example, we may wish to know the probability that some random variable, call it y, is *less than or equal to* a specified value, say y_1. In effect we would be trying to find the area under the curve to the left of y_1, the shaded area of Figure 6.11(a). Again, we could integrate the

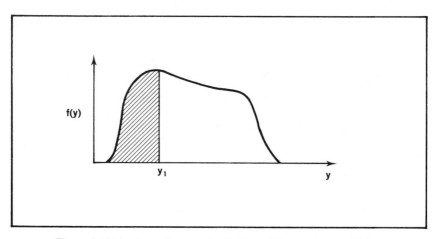

Figure 6.11(a)   *A continuous distribution of the random variable y.*

probability density function from minus infinity to $y = y_1$ or we could determine the area under the curve that is shaded by graphical means or tables. The shaded area in Figure 6.11(a) is sometimes called the left tail of the distribution. In symbols, the shaded area corresponds to $P(y \leqslant y_1)$. This symbol is interpreted as "the probability that y will be less than or equal to y_1."

Finally, we may be interested in knowing the probability that the value of some random variable, call it z, will be *greater than or equal to* a specified value, say z_1. To do this we would simply find the area under the probability distribution curve to the right of z_1 (the shaded portion of Figure 6.11(b)). In symbols we would write this probability as $P(z \geq z_1)$, and this is interpreted as "the probability that z will be greater than or equal to z_1." The shaded area of Figure 6.11(b) is sometimes called the right tail of the probability distribution.

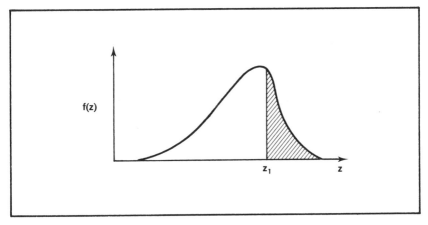

Figure 6.11(b)   *A continuous distribution of the random variable z.*

Relative Frequency Distributions

In most applied uses of distributions we will not be concerned with probability density functions in the sense just defined. Rather, we will usually be working with distributions obtained from statistical data. These are called relative frequency distributions and histograms. If we have a relative frequency distribution we can perform all the usual operations such as reading area under the curve, etc. without taking the final step of mathematically converting relative frequency distributions to probability distributions. The distinction between these two types of distributions will become evident with consideration of a numerical example.

Suppose a geologist is studying a new drilling prospect in an area in which twenty wells have been drilled. One of the unknown variables to be considered in the new prospect is net pay thickness. To get an idea of the possible likelihoods and ranges of possible values he has tabulated the net pay thickness values from each of the completed wells, as shown in Table 6.5. Tabulated data such as shown in Table 6.5 do not give a

Table 6.5
Net Pay Thickness (in Feet) of Twenty
Wells Completed in a Basin

Well No.	Net Pay Thickness, Feet
1	111
2	81
3	142
4	59
5	109
6	96
7	124
8	139
9	89
10	129
11	104
12	186
13	65
14	95
15	54
16	72
17	167
18	135
19	84
20	154

clear picture of ranges and probabilities of various values of the variable. It is desirable to be able to summarize the data in a clear, concise manner.

One way to do this is to divide the range of values of the random variable into a number of groups, or intervals and then tabulate how many values of the variable fall within each group or interval. This is shown in Table 6.6 for the net pay thickness data of Table 6.5. The numbers given in Column 2 of Table 6.6 are called *frequencies* and they are determined by counting how many of the original data points fall within each of the ranges listed in Column 1. The sum of the frequencies corresponds to the total number of data points or values. The decimal fractions in Column 3 are called *relative frequencies* and are computed by expressing each of the frequency values as a fraction, or proportion, of the total number of data points. Finally, it will sometimes be useful to express relative frequency data as a *percent* of the total number of points rather than a decimal fraction of the total. This, of course, is done by simply multiplying the relative frequency data by 100%, as shown in Column 4.

Table 6.6
Net Pay Thickness Data from 20 Wells Summarized As Relative Frequency Data

Range of Thickness (Feet)	Frequency (No. of Wells Having Thickness Values in the Range)	Relative Frequency (No. of Wells Having Thickness Values in Each Range, Expressed As a Fraction of the Total Number of Wells)	Relative Frequency Expressed As a Percentage
50–80	4	4/20 = 0.20	20%
81–110	7	7/20 = 0.35	35%
111–140	5	5/20 = 0.25	25%
141–170	3	3/20 = 0.15	15%
171–200	1	1/20 = 0.05	5%
	20	1.00	100%

We can represent the information of Table 6.6 in two graphical distributions known as a *histogram* and a *relative frequency distribution*. In Figure 6.12 we have shown a distribution called a histogram in which the vertical scale is the actual number of data points — the frequencies — in each range of net pay thickness. The width of the rectangles along the horizontal scale corresponds to the ranges or intervals given in Column 1 of Table 6.6.

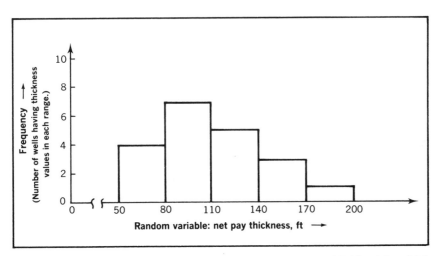

Figure 6.12 *Frequency distribution of net pay thickness data of Tables 6.5 and 6.6. This distribution is sometimes called histogram.*

As an alternative, we could have plotted relative frequency values on the vertical axis, and the resulting distribution would be called a relative frequency distribution. For the data of Table 6.6 the relative frequency distribution would appear as in Figure 6.13.

The number of intervals or ranges that are used to summarize statistical data will depend on the data and the degree of accuracy required. Generally the range of values of a random variable are divided into from 5 to 20 intervals, with 8–12 being probably the most common. The width or size of each interval does not have to be equal. Notice that as the number of intervals increase (for a given total range) the width of each interval will decrease. For the limiting case where the number of intervals becomes infinite the top points will define a continuous probability distribution.

This suggests one obvious difference between a relative frequency distribution such as Figure 6.13 and a probability distribution of Figure 6.8 – the relative frequency distribution is merely an approximation of a probability distribution using a limited set of observed sample data.

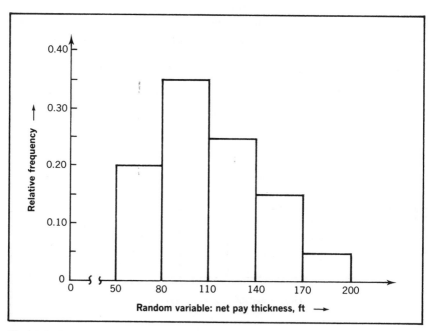

Figure 6.13 *Relative frequency distribution of net pay thickness data of Tables 6.5 and 6.6.*

Another difference between the two types of distributions is that the vertical scale of the relative frequency graph of Figure 6.13 is not scaled such that the area under the curve is precisely one. Also the units of the vertical axis must be the reciprocal of the units of the random variable in order for the area to be dimensionless, and clearly the relative frequency scales are dimensionless, rather than reciprocals of the random variable.

But even though we can't call a relative frequency distribution a probability distribution *per se* we can read areas under the curve as if it were a probability distribution. For example, the probability of a net pay thickness value less than or equal to 110 feet is simply the sum of the relative frequencies to the left of 110: $0.20 + 0.35 = 0.55$.

The way we interpret the probability within a range is to say there is probability 0.20 that thickness will be between 50 and 80 feet, a probability 0.35 that thickness will be between 80 feet and 110 feet, etc. Within each range the flat top of each rectangle implies that any value of the random variable within the range is equally likely to occur.

Cumulative Frequency Distributions

Every distribution can be expressed in an equivalent graphical form called a *cumulative frequency distribution,* or a cumulative frequency graph. There are two principal reasons for expressing distributions in their cumulative frequency form:

1. If we have a cumulative frequency distribution we can read any areas under the probability distribution desired without having to revert to integrating a probability density function.
2. In the mechanics of the risk analysis method called simulation it will be necessary that we convert distributions of possible values for each random variable into their equivalent cumulative frequency form. The reason for this will become evident in Chapter 8 when we discuss this important new technique in detail.

There are various ways to convert a distribution to its equivalent cumulative frequency, but probably the two most common ways that we will be using regard plotting relative frequency values from tabular data such as Table 6.6, or by an approximation method that will be described in a few moments.

In principle, each of these methods consists of moving from the left end of the distribution to the right end and computing the total area less than or equal to various values of the random variable within the range. These cumulative areas (probabilities) are then plotted on co-ordinate graph paper as functions of the values of the random variable corresponding to each cumulative area.

For example, we tabulated net pay thickness data from 20 wells in a basin in Table 6.6. The first three columns of Table 6.7 are repeated from Table 6.6. Column 4 is simply a cumulation of the relative frequency values as we proceed to larger values of the random variable. And Column 5 represents the cumulative relative frequencies expressed as percentages. We interpret columns 4 or 5 as follows: the area under the relative frequency curve less than or equal to 110 feet is 0.55 (or 55%), the area under the distribution less than or equal to 170 feet is 0.95 (or 95%), etc.

Table 6.7

Net Pay Thickness Data from 20 Wells in a Basin (From Data of Tables 6.5 and 6.6)

Range of Thickness (Feet)	Frequency	Relative Frequency	Cumulative Relative Frequency Less Than or Equal to Upper Limit of Col. 1 Range	Cumulative Percent Less Than or Equal to Upper Limit of Col. 1 Range
50–80	4	.20	.20	20%
81–110	7	.35	.55	55%
111–140	5	.25	.80	80%
141–170	3	.15	.95	95%
171–200	1	.05	1.00	100%
	20	1.00		

To represent the data as a cumulative frequency distribution we plot the cumulative frequencies of Column 4 versus the upper limit of each range, as shown in Figure 6.14. The cumulative frequency less than or equal to the minimum possible value of the random variable will always be zero, and the cumulative frequency less than or equal to the maximum value of the random variable will always be 1.0 (or 100% if a percentage scale is used).

Notice that the cumulative frequencies of Column 4 are plotted opposite the upper limit value of each range of pay thickness. The horizontal scale of a cumulative frequency graph will be the random variable itself.

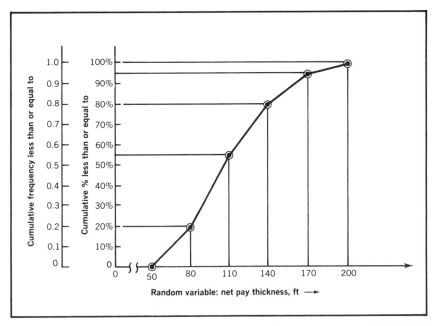

Figure 6.14 *Cumulative frequency distribution of pay thickness data of Table 6.7.*

There may be occasions when it is desirable to cumulate the areas under a distribution starting from the maximum value of the random variable and proceeding to smaller values of the random variable. In this case the cumulative distribution would appear as in Figure 6.15, and the vertical axis represents the cumulative frequencies *greater* than or equal to given values of the random variable.

The cumulative frequencies used in this graph are simply the results of subtracting each of the cumulative frequencies of Column 4 in Table 6.7 from 1.0. Also, please note that the cumulative frequencies *greater* than or equal to are plotted versus the lower limit value of each of the Column 1 ranges. Figures 6.14 and 6.15 are exactly equivalent, and either form can be used for interpreting areas under the curve or for use with some of the exploration risk analysis methods of Chapters 7 and 8.

Sometimes we will have occasion to talk about a given *percentile,* or *fractile.* A percentile is merely the value of the random variable corresponding to a specified "percent less than or equal to" value. The 20 percentile is 80 feet, as read from Figure 6.14; the 55 percentile is 110 feet, etc. A fractile is the equivalent definition for the case where the

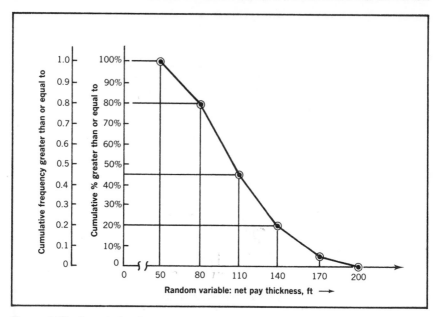

Figure 6.15 *Cumulative frequency distribution of pay thickness data of Table 6.7 in which area under the distribution was accumulated from right to left, rather than left to right as in Figure 6.14.*

ordinate scale is based on cumulative relative frequencies, rather than cumulative percentages.

An alternative, approximate method can be used if we have only the graphical representation of a distribution rather than the tabular data upon which the distribution may have been based. This method will provide a cumulative frequency distribution which will be approximately equal to the true cumulative frequency distribution that could be obtained from integration of a probability density function. The approximation is, for most purposes, satisfactory for our endeavors with exploration risk analysis, and this method will be most useful in our work with simulation.

Suppose we have been given a distribution of possible drilling costs in a new exploration area shown in Figure 6.16. The distribution may have been drawn from relative frequency data or it may represent the subjective judgment of the drilling engineer as to the range and likelihoods of anticipated drilling costs. (We'll talk more about subjective judgments expressed as distributions in Chapter 8.) Whatever its source

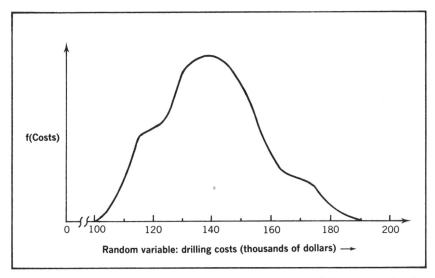

Figure 6.16 *A continuous distribution of drilling costs per well in a new exploration area.*

we have only the "picture" of the distribution and the numerical scale on the horizontal axis. The vertical coordinate is, of course, a probability density function, but this is also unknown to us. So the question we are trying to resolve is how to convert this picture of a probability distribution to its equivalent cumulative frequency distribution so as to be able to read probabilities directly.

The technique to accomplish this consists of a series of steps:

1. Define an arbitrary vertical scale. The magnitude of the scale is immaterial, and the scale has no units.
2. Divide the distribution into a series of geometric figures which approximate the shape of the distribution curve. The geometric figures can be triangles, trapezoids, and rectangles.
3. For each geometric figure read its vertical and horizontal dimensions from the arbitrary vertical scale and the given horizontal dimensions of the random variable respectively.
4. Compute the area of each geometric figure, and then find the sum of all the areas.
5. Compute the area of each geometric figure expressed as a decimal fraction of the sum of the areas. This fraction corresponds to

relative frequency, or the fractional area under the curve of each geometric figure.

6. Cumulate the fractional areas and plot the cumulative area data versus the upper limit value (of the horizontal scale) of each geometric figure. This graph will be an approximation of the cumulative frequency distribution of the original distribution.

These steps are illustrated for the distribution of Figure 6.16. The first step required that we define an arbitrary numerical scale on the ordinate. Figure 6.17 shows a scale which was chosen for this example. Figure 6.17 also shows the geometric figures used to approximate the shape of the original distribution. Once the figures have been defined the vertical and horizontal dimensions can be read from respective scales.

For example, Figure No. 7 is a trapezoid having parallel sides of 5.1 and 3.85, and the distance between the parallel sides, as measured along the horizontal scale is $(154 - 146) = 8$, in units of thousands of dollars.

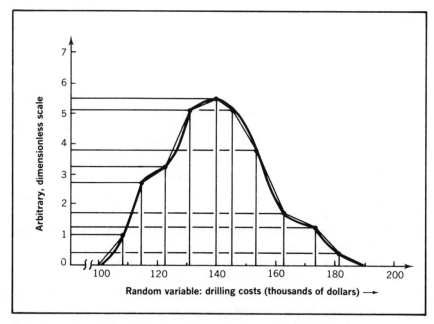

Figure 6.17 *Distribution of drilling costs of Figure 6.16 showing the arbitrary scale for the vertical scale and the division of the distribution into eleven geometric figures which approximate the shape of the original distribution.*

The number and type of geometric figures used to approximate a distribution will depend on the shape of the distribution and the desired accuracy of the cumulative frequency distribution. The more geometric figures used the greater the accuracy. Steps 4 and 5 relate to calculating areas for each geometric figure, and the calculations for the example of Figure 6.17 are tabulated in Table 6.8. A few sample calculations from Table 6.8:

Figure ①: A triangle. The base, as read from the horizontal scale is ($108.5M − $100.0M) = $8.5M. The height of the ordinate above $108.5M is 1.0 from the arbitrary vertical scale. The area of the triangle is $\frac{1}{2}$(base)(height) = $\frac{1}{2}$($8.5M)(1.0) = $4.250M, as listed in Column 8 of Table 6.8.

Figure ②: A trapezoid having the parallel bases standing vertical. The lengths of each of the parallel bases are the heights of the ordinates above $108.5M and $115.0: 1.0 and 2.8 respectively. The distance between the parallel bases is read along the horizontal scale as ($115.0M − $108.5M) = $6.5M. The area of trapezoid ② is thus: ($\frac{1}{2}$)(1.0 + 2.8)($6.5M) = $12.350.
etc.

The areas of each figure will have the same units as the random variable itself, $M in this example. Once the areas of each figure have been computed the areas are added to give a total area under the distribution, $238.625M in this case. Finally, each of the areas in Column 8 is divided by the sum of Column 8 to yield a dimensionless, fractional area. These fractional areas are equivalent to the relative frequency terms we discussed in Table 6.6.

Lastly, the fractional areas are accumulated, as shown in Column 10 of Table 6.8, resulting in the cumulative frequency data needed to plot the cumulative frequency distribution. Each cumulative frequency value is plotted versus the upper limit of each geometric figure. The result, Figure 6.18, is a close approximation of the cumulative frequency distribution of the original distribution of drilling costs of Figure 6.16.

To show how a cumulative frequency distribution can be used to read probabilities (areas under the distribution curve) directly let's pose a few questions relating to the drilling cost distribution of Figure 6.16.

a. *What is the probability that costs will be between $120M and $150M?*

To answer this we must find the area under the distribution curve between the limits of $120M and $150M. We can determine this area

Table 6.8

Calculations of Area of Each Geometric Figure of Figure 6.17. Numerical Dimensions of Each Figure Were Read from the Scales of Figure 6.17.

Figure No.	Type Figure	TRIANGLE ($M) Base	Height	TRAPEZOID Parallel Bases B_1, B_2	$\frac{1}{2}(B_1 + B_2)$	Distance between Parallel Bases ($M)	Area of Figure* ($M)	Area of Figure As Fraction of Total Area (Relative Frequency)	Cumulative Area (Cumulative Frequency)	Upper Limit of Each Figure ($M)
1	Triangle	8.5	1.0				4.250	0.0178	0.0178	108.5
2	Trapezoid			1, 2.8	1.9	6.5	12.350	0.0518	0.0696	115.
3				2.8, 3.3	3.05	8.0	24.400	0.1023	0.1719	123.
4				3.3, 5.1	4.2	8.0	33.600	0.1408	0.3127	131.
5				5.1, 5.5	5.3	9.0	47.700	0.1999	0.5126	140.
6				5.5, 5.1	5.3	6.0	31.800	0.1333	0.6459	146.
7				5.1, 3.85	4.475	8.0	35.800	0.1500	0.7959	154.
8				3.85, 1.7	2.775	9.0	24.975	0.1047	0.9006	163.
9				1.7, 1.2	1.45	11.0	15.950	0.0668	0.9674	174.
10	Trapezoid			1.2, 0.45	0.825	7.0	5.775	0.0242	0.9916	181.
11	Triangle	9.0	0.45				2.025	0.0084	1.0000	190.

TOTAL AREA → 238.625

PLOTTING DATA FOR CUMULATIVE FREQUENCY DISTRIBUTION, FIGURE 6.18

* Area of triangle = $(\frac{1}{2})$(Base)(Height).
Area of trapezoid = $(\frac{1}{2})(B_1 + B_2)$(Distance between parallel bases) *where* B_1 and B_2 are the parallel bases of the trapezoid.

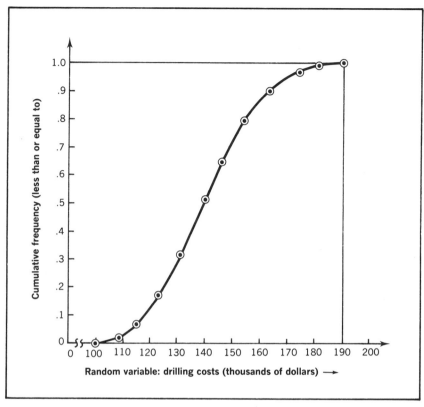

Figure 6.18 *Cumulative frequency distribution of drilling cost distribution of Fig. 6.16. Plotting data points for constructing this graph were computed in Table 6.8 and listed in Columns 10 and 11.*

directly using the cumulative frequency graph of Figure 6.18. First, enter the graph on the horizontal axis at the upper limiting value, $150M, and proceed up to the curve and across horizontally to read a cumulative frequency of about 0.72. This represents the *total* area of Figure 6.16 to the left of $150M. Then we repeat this step using the lower limiting value, $120M, and read a corresponding cumulative frequency of about 0.13. This value represents the *total* area of Figure 6.16 to the left of $120M. The difference represents the area between $120M and $150M, i.e. $(0.72 - 0.13) = 0.59$. The probability of costs being between $120M and $150M $= 0.59$.

What we have just done can be represented schematically in a

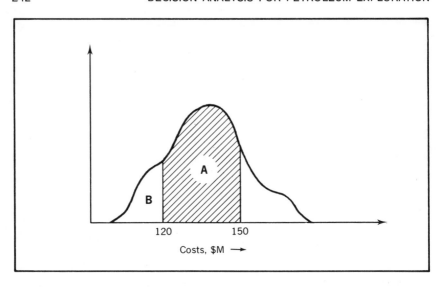

sketch. We are interested in finding the probability of costs being within the limits of $120M and $150M, and this probability is represented by the shaded portion A of the curve. The 0.72 we read from Figure 6.18 as the cumulative frequency corresponding to $150M is the total area to the left of $150M, Section A and Section B.

The cumulative frequency corresponding to $120M of 0.13 is the area to the left of $120M, Section B. The area we are interested in, Section A, is found by subtracting Section B from Section (A + B). The difference, $(0.72 - 0.13) = 0.59$, is the fractional area of Section A.

$$
\begin{array}{lcl}
\text{Area of Section A + B:} & & 0.72 \\
\underline{-\text{ Area of Section B:}} & & \underline{-0.13} \\
= \text{Area of Section A} & = & 0.59
\end{array}
$$

b. *What is the probability that costs will be less than or equal to $140M?*

We can read this directly from the cumulative frequency distribution as the cumulative frequency corresponding to $140M, about 0.51. We have a 51% chance that costs will be less than or equal to $140M.

c. *What is the probability that costs will exceed $160M?*

If we enter Figure 6.18 at $160M we read a corresponding cumulative frequency of 0.86. This means the area under the distribution curve to the left of $160M is 0.86. The total area under the curve is, by definition 1.0; therefore, the area under the curve to the right of $160M is $(1.0 - 0.86) = 0.14$. There is a 14% chance that costs will exceed $160M.

The method we have just considered for converting a probability distribution to its equivalent cumulative frequency distribution is not very elegant in the sense of mathematical theory and the calculus. But it gets the job done and provides an important way for the explorationist to work with, interpret, and use probability distributions without having to know the mathematical details such as integrating probability density functions to determine areas under the curve.

The important thing to remember about this approximate method of dividing the distribution into geometric figures is that its accuracy is dependent on how many intervals are used. The more intervals (geometric figures) used the better will be the approximation; the fewer intervals used the poorer will be the approximation.

The eleven intervals used in Figure 6.17 certainly provided a reasonably close fit to the original curve. If we had used only three or four geometric figures to represent the curve our approximation would have been much poorer.

Single Value Parameters of Distributions

Although a random variable can normally assume many possible numerical values there are certain specific values which can tell us important information about the entire distribution. We will only have need to consider two of the single value parameters of distributions: a) a parameter which describes central tendency, or "average" values of the random variable, and b) a parameter which describes the variability, or range within which the random variable is distributed. These, as well as other single value parameters all have a theoretical basis which relates to what is called a moment-generating function of a distribution. The parameters also have a physical significance that will be useful to us in our work with distributions.

a. *Measures of Central Tendency*

Knowledge of the distribution parameters which describe central tendency can provide information about "average" values of the random variable. The three measures of central tendency most commonly used are the *mean, median,* and *mode.*

Mean: The weighted average value of the random variable where the weighting factors are the probabilities of occurrence. It is synonymous to the expected value of a distribution and to the arithmetic average of a set of statistical data. If we are dealing with a distribution of net monetary value profit the mean value of this profit distribution is exactly equal to EMV.

All distributions (discrete and continuous) have a mean value. The units of the mean are the same as the units of the random variable. The symbol we will use for the mean value is μ.

From a statistical standpoint the mean value is, by far, the most important single value measure of central tendency. It is likewise very important in decision analysis because it is, by definition, equivalent to expected value, the basis for decision making under uncertainty. We'll discuss how to compute mean values in just a moment. As a matter of information, the symbol for the mean, μ, was chosen here because of its common acceptance in many statistical books. Occasionally the mean value is given as $\bar{x}$, or $E(x)$, but μ is the most widely used symbol for the mean.

Median: The value of a random variable which divides the area under the probability distribution into two equal parts. It is the value of the random variable which corresponds to the 50 percentile on a cumulative frequency distribution. The probability of the random variable being less than or equal to the median is 0.50. The probability of the random variable being greater than or equal to the median is also 0.50.

The median value is not as representative a measure of central tendency (averages) as the mean value. The reason is that the median is not affected by the magnitudes of the values of the random variable. It only tells us the value of the variable such that half of the possible values will be smaller and half of the values will be larger. If statistical data are listed in numerically increasing or decreasing order the median is the value half way down the list. The median of a lognormal distribution is the same as the geometric mean of the lognormally distributed variable. (The geometric mean is defined as the nth root of the product of n values of a random variable.)

Mode: The mode is the value of the random variable which is most likely to occur. It is the value of the random variable located under the highest, peak value of the distribution curve. Distributions can have more than one mode.

From the numerical computation aspect of risk analysis the mode does not have much value. However, the mode value is referenced frequently in the decision making process when we refer to the "most likely," or "most probable" values of the various random variables.

We stress again that of these three measures of central tendency the mean value is by far the most useful single value parameter. Another point to remember is that, in general, the mean, median, and mode are all different values for a given distribution. One exception to this are continuous distributions that are symmetrical in shape. In these cases the median and mean values will be coincident. Examples of these cases include the normal distribution and the uniform distribution (to be discussed in the next section of the Chapter).

The mean value of a distribution or a set of statistical data can be computed in several ways. If you have the actual data, one way to compute the mean value of the data is to add up all the numerical values and divide by the total number of data points. This is called an arithmetic average and is exactly the same as the mean value.

If the list of numerical data is large this can be very tedious, and it is frequently useful to compute the mean value by grouping the data into intervals or ranges of values and counting the number of values which fall into each range. With this procedure the equation that is used to compute an approximate value of the true mean is Equation 6.3:

$$\mu = \frac{\sum_i n_i x_i}{\sum_i n_i} \qquad (6.3)$$

where n is the frequency, or number of data points in each interval and x is the midpoint of each range or interval. i is the index to denote the various intervals (i = 1, 2, 3, 4, . . .)

The value of μ computed by Equation 6.3 is an approximation because it is based on the assumption that the values of the random variable within a range or interval occur at the midpoint. However, the approximation is very good and the value obtained from Equation 6.3 will be of sufficient accuracy for our work in exploration risk analysis. When analyzing statistical data in grouped intervals it is advisable to have at least eight intervals to get a representative value of μ. The size of the intervals do not have to be equal. Equation 6.3 can be used with all types of data that can have a continuum of values.

For random variables which have discrete values (such as something which is countable) the equation which is used to compute a mean value is given as Equation 6.4:

$$\mu = \sum_i (p_i)(x_i) \qquad (6.4)$$

where p_i is the probability of occurrence (relative frequency) of the x_ith value of the random variable.

Referring to the discrete distribution shown in Figure 6.9 the x_i terms are the possible discrete values that the random variable can have (10, 11, 12, . . .) and the p_i terms are the probabilities of occurrence of each x_i value as read from the vertical scale. Equation 6.4, when used for a discrete valued random variable, yields an exact mean value and not an approximation as with Equation 6.3.

For the mathematicians in the crowd the mean value of a continuous probability distribution is computed from the following integral:

$$\mu = \int_{-\infty}^{+\infty} x f(x) dx$$

where $f(x)$ is the probability density function of the random variable x.

Finally, it is possible to compute the mean value from a cumulative frequency distribution such as Figure 6.18 using the following steps:

1. Divide the range of possible values of the random variable into 8 or more intervals.
2. Determine the probability of the random variable having a value within each interval by subtracting the cumulative frequency associated with the lower end of the interval from the cumulative frequency associated with the upper limit of the interval. (Reverse these limits if the cumulative frequency graph is plotted as "greater than or equal to").

Table 6.9

Calculation of the Mean Value of the Probability Distribution of Figure 6.16 Using Data from Its Equivalent Cumulative Frequency Distribution of Figure 6.18

Range of Drilling Costs ($M)	Cumulative Frequency Value of Upper End of Each Range (From Fig. 6.18)	Cumulative Frequency Value of Lower End of Each Range (From Fig. 6.18)	Probability Within the Range (Col. 2 − Col. 3)	Midpoint of Each Range ($M)	Product of Probability Times Midpoint of Range, $M (Col. 4 × Col. 5)
100–115	0.07	0	0.07	107.5	7.525
115–125	0.20	0.07	0.13	120.0	15.600
125–135	0.40	0.20	0.20	130.0	26.000
135–140	0.51	0.40	0.11	137.5	15.125
140–145	0.62	0.51	0.11	142.5	15.675
145–155	0.80	0.62	0.18	150.0	27.000
155–165	0.91	0.80	0.11	160.0	17.600
165–175	0.97	0.91	0.06	170.0	10.200
175–190	1.00	0.97	0.03	182.5	5.475
			1.00		$\mu = 140.200$

3. Multiply the probabilities of Step 2 by the midpoint of each interval and sum the products for all the intervals to yield the mean value.

These steps are illustrated in Table 6.9 for the cumulative frequency distribution of Figure 6.18. The column headings should be sufficient to clarify how the computation proceeds. The sum of Column 6 is the approximate value of the mean: \$140.2M. The technique used to develop the cumulative frequency distribution of Figure 6.18 was perfectly applicable to any distribution. Knowing the cumulative frequency distribution we can use a computation of the type illustrated in Table 6.9.

These two techniques provide a way for us to be able to compute the mean value for virtually any type of continuous probability distributions. And as stated previously as long as we divide the distribution into at least eight ranges or intervals the approximate value of μ will be very close to the actual μ we would have obtained using the integral definition of the mean of a continuous distribution.

To summarize the various methods for computing mean values:

Type of Data or Information	Method to Use for Computing the Mean Value
Actual values of the random variable	Compute the arithmetic average by adding all the values and dividing by the total number of values
Values of the random variable listed as frequencies in each group or interval	Use Equation 6.3: $$\mu = \frac{\sum_i n_i x_i}{\sum_i n_i}$$
Discrete random variable in which relative frequencies are given for each value	Use Equation 6.4: $$\mu = \sum_i (p_i)(x_i)$$
The probability density function, f(x), of a continuous distribution is known	Evaluate the integral $$\mu = \int_{-\infty}^{+\infty} x f(x) dx$$
A graph of the probability distribution or its equivalent cumulative frequency distribution is available	Convert the distribution to its equivalent cumulative frequency distribution and use the calculation given in Table 6.9

b. *Measures of Variability*

The mean value of a distribution tells us important information about the "average," or expected value of the random variable, but it does not tell us anything about the spread, or variability of other possible values of the variable on either side of the mean. Will the range of possible values of the random variable be very small with all the values located at or near the mean, or will the range be very large with possible values much larger or smaller than the mean? Single value measures of variability can give us this information and are, therefore, useful parameters to describe distributions. The most important of the single value measures of variability, and the only one we will discuss here, is called *standard deviation*.

> *Standard Deviation:* Each possible value of a random variable is located a given distance from the mean, as measured along the horizontal axis. These distances are called deviations about the mean. The mean value of the squared deviations about the mean is called variance, and standard deviation is defined as the nonnegative square root of the variance. Or more concisely, standard deviation is the square root of the mean of the squared deviations about μ. All distributions have a standard deviation and its units are identical to the units of the random variable. The symbols for standard deviation and variance are σ and σ^2 respectively.

The physical significance of the standard deviation is that it tells us the degree of spread, or dispersion of the distribution on either side of the mean value. As standard deviation increases the spread of the distribution increases; the smaller the standard deviation the narrower the spread or dispersion about the mean. In Figure 6.19 the standard deviation of the narrow distribution A is less than the standard deviation of distribution B.

From a statistical point of view standard deviation is by far the most important measure of dispersion, or variability. In fact, certain distributions such as the normal and lognormal can be completely and uniquely defined by simply specifying the mean value and the standard deviation. Other measures of variability such as the range, mean deviation, 10–90 percentile range, etc. are much less useful to us in petroleum exploration risk analysis and will not be discussed.

As with the mean, the analyst has several options for computing standard deviation depending on the format of his information or data.

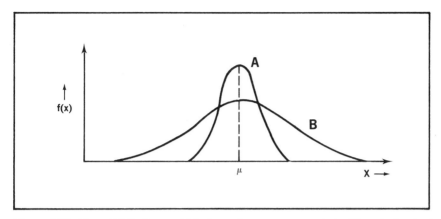

Figure 6.19 *Two distributions which have the same mean but different standard deviations. The standard deviation of distribution A is smaller than the standard deviation of distribution B. That is, $\sigma_A < \sigma_B$.*

If he has the entire list of N statistical data points, x_1, x_2, x_3, x_4, . . . x_N he can compute σ using the following equation:

$$\sigma = \sqrt{\frac{\sum\limits_{i=1}^{N}(x_i - \mu)^2}{N}} \qquad (6.5)$$

where x_i are the actual values of the random variable, μ is the mean value of the data, and N is the total number of data points or values.

Notice that to compute the standard deviation we must first compute the mean value of the data. The term $(x_i - \mu)$ is the distance, or deviation of the x_i value from the mean. When the term is squared, $(x_i - \mu)^2$, it is called the squared deviation about the mean. The mathematical operation to sum these squared deviations and then divide by N, the total number of data points, is simply computing the arithmetic average, or mean of the squared deviations. Finally, the square root is obtained of the mean of the squared deviations resulting in the value of σ, standard deviation.

If there are many values in the set of statistical data the arithmetic to solve Equation 6.5 can become very tedious. In these cases it is useful to divide the range of possible values of the random variable into smaller ranges, or intervals and count the number of data points which fall in each range. For grouped interval data such as this the equation to compute standard deviation is given as Equation 6.6.

$$\sigma = \sqrt{\dfrac{\sum_i \{n_i(x_i - \mu)^2\}}{\sum_i n_i}} \qquad (6.6)$$

Where n_i is the frequency, or number of data points in each range, x_i is the midpoint of each range or interval, and i is the index to denote the various intervals.

The value of σ computed by Equation 6.6 is an approximation because, as with Equation 6.3, it is based on the assumption that values of the random variable within a range occur at the midpoint.[1] However, this assumption is not critical as long as the total range of values has been divided into eight or more intervals. For this case, Equation 6.6 will be a most useful way to compute σ. With a little bit of algebra it can be shown that Equations 6.5 and 6.6 can be rearranged into the following form:

$$\sigma = \sqrt{\mu(x^2) - (\mu_x)^2} \qquad (6.7)$$

The first term under the square root, $\mu(x^2)$ is interpreted as the mean value of the squares of the random variable (that is the actual values squared or the midpoint values of each range squared). The second term is simply the mean value squared. Thus, Equation 6.7 reads that σ is the square root of the mean of the squares minus the square of the mean. For some types of computations Equation 6.7 involves a little less arithmetic.

For a random variable which has only discrete values, such as Figure 6.9, the equation to compute σ is as follows:

$$\sigma = \sqrt{\sum_i (p_i)(x_i - \mu)^2} \qquad (6.8)$$

1. To be technically correct the theory of statistics distinguishes between the standard deviation of sample data and the standard deviation of the total population or sample space from which the sample data was selected. In this context, Equation 6.6 is the equation to compute the standard deviation of the sample data. The corresponding equation for computing the standard deviation for the population is the same except that the denominator term under the square root is $(\sum_i n_i - 1)$, rather than $(\sum_i n_i)$. The reason for the -1 term is a bit complex and will not be derived here. As long as the number of data points, $\sum_i n_i$, is large (say greater than about 30) the terms are essentially equal and Equation 6.6 is acceptable to use. For statistical work requiring precise estimates of σ the -1 term cannot be ignored, and the analyst should consult a standard text on statistics for further explanation and definitions of equations for estimating population standard deviations.

where p_i is the probability of occurrence (relative frequency) of the x_ith value of the random variable and μ is as defined by Equation 6.4.

Use of Equation 6.8 for a discrete valued random variable yields an exact value of σ, rather than an approximate value.

If a probability density function, $f(x)$, of a probability distribution is known σ is computed from the following integral:

$$\sigma = \sqrt{\int_{-\infty}^{+\infty} (x - \mu)^2 f(x)dx}$$

Finally, if we only have a cumulative frequency distribution available to us (such as Figure 6.18) we must make use of the data used to compute the mean value from a cumulative frequency distribution. To illustrate this we will use the cumulative distribution of Figure 6.18 and the data of Table 6.9 to compute the standard deviation of the drilling cost distribution of Figure 6.16. Recall that from Table 6.9 we computed a mean value of $140.2M for the distribution.

Table 6.10

Calculation of Standard Deviation of the Probability Distribution of Figure 6.16 Using Data from Its Equivalent Cumulative Frequency Distribution of Figure 6.18 and data from Table 6.9 Used in the Computation of the Mean Value, μ, of $140.2M

Interval Number	Range of Drilling Costs ($M)	(p_i) Probability Within Range (from Col. 4 of Table 6.9)	(x_i) Midpoint of Each Range ($M)	Midpoint Deviation from the Mean, $M $(x_i - \mu)$	Squared Deviations $(x_i - \mu)^2$ ($M)^2	$(p_i)(x_i - \mu)^2$ Col. 3 Times Col. 6 ($M)^2
1	100–115	0.07	107.5	$(107.5 - 140.2) = -32.7$	1069.29	74.8503
2	115–125	0.13	120.0	$(120.0 - 140.2) = -20.2$	408.04	53.0452
3	125–135	0.20	130.0	$(130.0 - 140.2) = -10.2$	104.04	20.8080
4	135–140	0.11	137.5	$(137.5 - 140.2) = -2.7$	7.29	0.8019
5	140–145	0.11	142.5	$(142.5 - 140.2) = +2.3$	5.29	0.5819
6	145–155	0.18	150.0	$(150.0 - 140.2) = +9.8$	96.04	17.2872
7	155–165	0.11	160.0	$(160.0 - 140.2) = +19.8$	392.04	43.1244
8	165–175	0.06	170.0	$(170.0 - 140.2) = +29.8$	888.04	53.2824
9	175–190	0.03	182.5	$(182.5 - 140.2) = +42.3$	1789.29	53.6787
		1.00				317.4600

$$\sum_i (p_i)(x_i - \mu)^2 = \sigma^2 = \$317.4600M^2$$

$$\sigma = \sqrt{\sigma^2} = \sqrt{317.4600} = \underline{\$17.8174M}$$

The single value parameters of the drilling cost distribution of Figure 6.16 are: $\mu = \$140.2M$, and $\sigma = \$17.8174M$, as computed from Tables 6.9 and 6.10 respectively.

The various methods for computing the standard deviation of a distribution or set of data are summarized in the following table:

Type of Data or Information	Methods to Use for Computing Standard Deviation, σ
Actual values of the random variable	Equation 6.5 or Equation 6.7
Values of the random variable listed as frequencies in each group or interval	Equation 6.6 or Equation 6.7
Discrete random variable in which relative frequencies are given for each value	Equation 6.8
The probability density function f(x), of a continuous distribution is known	Evaluate integral $$\sigma^2 = \int_{-\infty}^{+\infty} (x - \mu)^2 f(x)dx$$ then compute σ from $\sigma = \sqrt{\sigma^2}$.
A graph of probability distribution or its equivalent cumulative frequency distribution is available	Convert the distribution to its equivalent cumulative frequency distribution and use the calculation given in Tables 6.9 and 6.10

When working with large quantities of statistical data, probably the most useful method for computing the mean and standard deviation of the data is to arrange the data into groups or intervals and use Equations 6.3 and 6.6. To illustrate an example of this approach let's consider the core data listed in Table 6.11.

Table 6.11
Porosity and Permeability Data from a Well in Denver-Julesburg Basin of Southwestern Nebraska

Depth	Permeability, md.	·	Porosity %
4805.5	0.0		7.5
06.5	0.0		12.3
07.5	2.5		17.0
08.8	59		20.7
09.5	221		19.1
10.5	211		20.4
11.5	275		23.3

Table 6.11 (*Continued*)

Depth	Permeability, md.	Porosity %
4812.5	384	24.0
13.5	108	23.3
14.5	147	16.1
15.5	290	17.2
16.5	170	15.3
17.5	278	15.9
18.5	238	18.6
19.5	167	16.2
20.5	304	20.0
21.5	98	16.9
22.5	191	18.1
23.5	266	20.3
24.5	40	15.3
25.5	260	15.1
26.5	179	14.0
27.5	312	15.6
28.5	272	15.5
29.5	395	19.4
30.5	405	17.5
31.5	275	16.4
32.5	852	17.2
33.5	610	15.5
34.5	406	20.2
35.5	535	18.3
36.5	663	19.6
37.5	597	17.7
38.5	434	20.0
39.5	339	16.8
40.5	216	13.3
41.5	332	18.0
42.5	295	16.1
43.5	882	15.1
44.5	600	18.0
4845.5	407	15.7

NO RECOVERY

Depth	Permeability, md.	Porosity %
4847.5	479	17.8
48.5	0.0	9.2
49.5	139	20.5
50.5	135	8.4
51.5	0.0	1.1

(Note: Consider the intervals with no permeability as non-productive.)

To compute μ and σ of the porosity and permeability data we first define ranges or intervals for each random variable, as shown in Column 2 of Table 6.12 and 6.13. The number of values of porosity and permeability in each range was determined and listed as the frequency values, n_i, in columns 3. The remainder of each table is merely the step-by-step computations that are required to solve Equations 6.3 and 6.6.

Note that after completing Column 5 you must solve Equation 6.3 for the mean, μ, before proceeding to Column 6. In the next section of this chapter we will see how we could have obtained the same mean value and standard deviation of the porosity data using a very simple graphical technique. And we will also discuss the fact that for these two specific random variables we can define the entire distribution with just the two single value parameters of μ and σ.

Table 6.12
Computation of Mean Value and Standard Deviation of Porosity Data of Table 6.11

i	Porosity Interval, Percent	Frequency n_i	Interval Midpoint x_i	$n_i x_i$	$(x_i - \mu)$	$(x_i - \mu)^2$	$n_i(x_i - \mu)^2$
1	$7.0 \leq \times < 10.0$	1	8.5	8.5	-9.2	84.64	84.64
2	$10.0 \leq \times < 12.0$	0	11.0	0	-6.7	44.89	0
3	$12.0 \leq \times < 14.0$	1	13.0	13.0	-4.7	22.09	22.09
4	$14.0 \leq \times < 16.0$	10	15.0	150.0	-2.7	7.29	72.90
5	$16.0 \leq \times < 18.0$	12	17.0	204.0	-0.7	0.49	5.88
6	$18.0 \leq \times < 20.0$	8	19.0	152.0	$+1.3$	1.69	13.52
7	$20.0 \leq \times < 22.0$	7	21.0	147.0	$+3.3$	10.89	76.23
8	$22.0 \leq \times < 25.0$	3	23.5	70.5	$+5.8$	33.64	100.92
		42		745.0			376.18

From Equation 6.3 the mean is computed as:

$$\mu = \frac{\sum_i n_i x_i}{\sum_i n_i} = \frac{745.0}{42} = \underline{\underline{17.7\%}}$$

From Equation 6.6 the standard deviation is computed as:

$$\sigma = \sqrt{\frac{\sum_i \{n_i(x_i - \mu)^2\}}{\sum_i n_i}} = \sqrt{\frac{376.18}{42}} = \sqrt{8.957} = \underline{\underline{2.99\%}}$$

Table 6.13

Computation of Mean Value and Standard Deviation of the Permeability Data of Table 6.11

i	Permeability Interval, md.	Frequency n_i	Interval Midpoint x_i	$n_i x_i$	$(x_i - \mu)$	$(x_i - \mu)^2$	$n_i(x_i - \mu)^2$
1	0–50	2	25	50	−295	87,025	174,050
2	51–100	2	75	150	−245	60,025	120,050
3	101–150	4	125	500	−195	38,025	152,100
4	151–200	4	175	700	−145	21,025	84,100
5	201–250	4	225	900	− 95	9,025	36,100
6	251–300	8	275	2200	− 45	2,025	16,200
7	301–350	4	325	1300	5	25	100
8	351–400	2	375	750	55	3,025	6,050
9	401–450	4	425	1700	105	11,025	44,100
10	451–500	1	475	475	155	24,025	24,025
11	501–700	5	600	3000	280	78,400	392,000
12	701–1000	2	850	1700	530	280,900	561,800
		42		13,425			1,610,675

From Equation 6.3 the mean is computed as:

$$\mu = \frac{\sum_i n_i x_i}{\sum_i n_i} = \frac{13,425}{42} = 319.6 \text{ md.} \cong \underline{320 \text{ md.}}$$

From Equation 6.6 the standard deviation is computed as:

$$\sigma = \sqrt{\frac{\sum_i \{n_i(x_i - \mu)^2\}}{\sum_i n_i}} = \sqrt{\frac{1,610,675}{42}} = \sqrt{38,349} \cong \underline{196 \text{ md.}}$$

III. SPECIFIC DISTRIBUTIONS OF INTEREST IN EXPLORATION RISK ANALYSIS

Thus far we have discussed the meaning of distributions without reference to any special shapes or types of distributions. There are several specific distributions, however, which appear rather frequently in exploration risk analysis and we will briefly describe the characteristics of each of these special types here.

The uses of these distributions will be explained in detail in Chapters

7 and 8. The distributions which we will describe in this section include the *normal, lognormal, uniform, triangular, binomial, multinomial,* and the *hypergeometric* distribution.

Normal Distribution

The normal distribution is probably one of the most common and widely used distributions in statistics and probability. It is a continuous probability distribution having a symmetrical shape similar to a bell, Figure 6.20. It is sometimes called a Gaussian distribution, after the German Mathematician Karl Friedrich Gauss who developed the mathematical basis of the distribution.

Some examples of random variables that can usually be represented by normal distributions include core porosity, percentages of abundant minerals in rocks, and percentages of certain chemical elements or oxides in rocks.

Specific characteristics of the normal distribution include:

a. The distribution is completely and uniquely defined by the two single value parameters μ and σ.

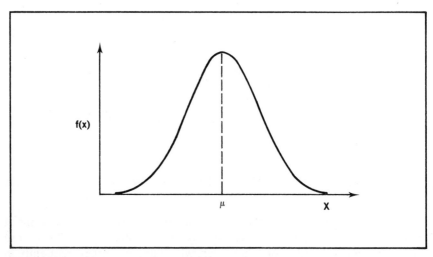

Figure 6.20 *An example of a normal distribution of a random variable x. The distribution curve is symmetrical and the mean, median, and mode values are all coincident and occur under the highest point of the curve.*

b. The mode (most likely value), median (value of the random variable which divides the distribution into two equal parts), and the mean are all equal.
c. The distribution curve is symmetrical and the inflection points of the curve occur at values of the random variable corresponding to $\mu + \sigma$ and $\mu - \sigma$. All normal distributions (regardless of

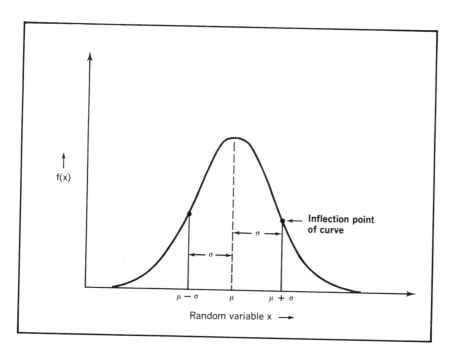

the units of x or the type of parameter described as x) have 0.6826 of the total area under the curve within the limits of $\mu \pm \sigma$.

That is, the probability of any normally distributed random variable having a value within the limits of $\mu + \sigma$ and $\mu - \sigma$ is 0.6826. The area under the curve between the limits of $\mu \pm 2\sigma$ is 0.9544 for all normal distributions, and the area within the limits of $\mu \pm 3\sigma$ is 0.9974.
d. The theoretical limits of all normal distributions are $-\infty$ and $+\infty$. In practice, however, we normally truncate the distribution at values of the random variable corresponding to 4 or 5 standard deviations ($\mu \pm 4\sigma$ to $\mu \pm 5\sigma$) on the basis that probabilities for

values of the random variable outside 4 or 5 standard deviations become infinitesimally small.

e. The cumulative frequency graph of a normal distribution, when plotted on coordinate graph paper, has the shape of a symmetrical "s," as shown in Figure 6.21(a). When a special graph paper called normal probability paper is used the cumulative frequency distribution plots as a straight line, as in Figure 6.21(b).

An illustration of the use of normal probability graph paper will be given in a moment. If statistical data are used to plot the cumulative frequency distribution, the parameters of the normal distribution can be read directly from the cumulative frequency graph. The mean value of the distribution will be the value of the random variable corresponding to the 50 percentile. That is, $\mu = x_{0.50}$. The standard deviation is determined by subtracting the value of the random variable corresponding to the 50 percentile from the value of the random variable corresponding to the 84.1 percentile. That is, $\sigma = x_{0.841} - x_{0.500}$. These graphical techniques for determining μ and σ apply only to the special case of a normal distribution.

f. In addition to the usual methods for reading probabilities (areas) from a normal distribution (reading probabilities from the cumulative frequency curve, integrating the probability density

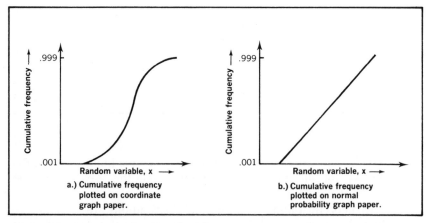

Figure 6.21 *General shapes of the cumulative frequency graphs of normal probability distributions when plotted on coordinate graph paper and a special type of graph paper called normal probability.*

function, etc.) we can also read areas under the curve using a special table given as Table 6.14. To use this technique you must compute a value of a standardized, dimensionless variable t:

$$t = \frac{x - \mu}{\sigma}$$

where x is the specific value of the random variable of interest, μ is the mean of the normal distribution, and σ is the standard deviation of the normal distribution.

Table 6.14 gives the total area to the left of the value of x used to compute the t value used in entering the table. An example of how to use Table 6.14 will be given later. The table can be used to read areas under the curve for any normal distribution, regardless of the type and units of the random variable.

To illustrate how to use normal probability graph paper we can refer to the porosity data from the Denver-Julesburg Basin Well, Table 6.11. The first step is to arrange the data in intervals, count frequencies, and compute cumulative frequencies just as we did in Table 6.7. For the porosity data this information is summarized in Table 6.15. Columns 1 and 2 are repeated from Table 6.12. The cumulative frequencies expressed as percentages are plotted versus the upper limit of each porosity interval on a special graph paper called normal probability paper, as shown in Figure 6.22. The abscissa of Figure 6.22 is scaled so that the cumulative frequency of a normally distributed variable will plot as a straight line. The ordinate of the normal probability graph paper of Figure 6.22 is a standard coordinate scale on which the random variable is plotted.

As we mentioned previously, if the cumulative frequency curve is plotted from statistical data we can read the mean and standard deviation of a normal distribution directly, rather than having to solve Equations 6.3 and 6.6 as we did in Table 6.12. To read the mean value of porosity enter the abscissa at the 50 percentile and read the corresponding value of porosity of about *17.6%* — which compares favorably to the $\mu = 17.7\%$ computed in Table 6.12.

The value of porosity corresponding to the 84.1% percentile from Figure 6.22 is about 20.6%. To compute σ we subtract the value of porosity corresponding to the 50 percentile from the value of porosity corresponding to the 84.1 percentile; i.e. $\sigma = x_{84.1\%} - x_{50\%} = 20.6\% - 17.6\% = 3.0\%$. This also agrees with the value of σ computed in Table 6.12. These steps are shown in the schematic diagrams of Figure 6.23.

DECISION ANALYSIS FOR PETROLEUM EXPLORATION

Table 6.14

Areas Under the Normal Distribution Less Than or Equal to the Value of x Used to Compute t. The Standardized Variable t Is Defined As

$$t = \frac{x - \mu}{\sigma}$$

t	.00	.01	.02	.03	.04	.05	.06	.07	.08	.09
−3.5	.0002	.0002	.0002	.0002	.0002	.0002	.0002	.0002	.0002	.0002
−3.4	.0003	.0003	.0003	.0003	.0003	.0003	.0003	.0003	.0003	.0002
−3.3	.0005	.0005	.0005	.0004	.0004	.0004	.0004	.0004	.0004	.0003
−3.2	.0007	.0007	.0006	.0006	.0006	.0006	.0006	.0005	.0005	.0005
−3.1	.0010	.0009	.0009	.0009	.0008	.0008	.0008	.0008	.0007	.0007
−3.0	.0013	.0013	.0013	.0012	.0012	.0011	.0011	.0011	.0010	.0010
−2.9	.0019	.0018	.0018	.0017	.0016	.0016	.0015	.0015	.0014	.0014
−2.8	.0026	.0025	.0024	.0023	.0023	.0022	.0021	.0021	.0020	.0019
−2.7	.0035	.0034	.0033	.0032	.0031	.0030	.0029	.0028	.0027	.0026
−2.6	.0047	.0045	.0044	.0043	.0041	.0040	.0039	.0038	.0037	.0036
−2.5	.0062	.0060	.0059	.0057	.0055	.0054	.0052	.0051	.0049	.0048
−2.4	.0082	.0080	.0078	.0075	.0073	.0071	.0069	.0068	.0066	.0064
−2.3	.0107	.0104	.0102	.0099	.0096	.0094	.0091	.0089	.0087	.0084
−2.2	.0139	.0136	.0132	.0129	.0125	.0122	.0119	.0116	.0113	.0110
−2.1	.0179	.0174	.0170	.0166	.0162	.0158	.0154	.0150	.0146	.0143
−2.0	.0228	.0222	.0217	.0212	.0207	.0202	.0197	.0192	.0188	.0183
−1.9	.0287	.0281	.0274	.0268	.0262	.0256	.0250	.0244	.0239	.0233
−1.8	.0359	.0351	.0344	.0336	.0329	.0322	.0314	.0307	.0301	.0294
−1.7	.0446	.0436	.0427	.0418	.0409	.0401	.0392	.0384	.0375	.0367
−1.6	.0548	.0537	.0526	.0516	.0505	.0495	.0485	.0475	.0465	.0455
−1.5	.0668	.0655	.0643	.0630	.0618	.0606	.0594	.0582	.0571	.0559
−1.4	.0808	.0793	.0778	.0764	.0749	.0735	.0721	.0708	.0694	.0681
−1.3	.0968	.0951	.0934	.0918	.0901	.0885	.0869	.0853	.0838	.0823
−1.2	.1151	.1131	.1112	.1093	.1075	.1056	.1038	.1020	.1003	.0985
−1.1	.1357	.1335	.1314	.1292	.1271	.1251	.1230	.1210	.1190	.1170
−1.0	.1587	.1562	.1539	.1515	.1492	.1469	.1446	.1423	.1401	.1379
−0.9	.1841	.1814	.1788	.1762	.1736	.1711	.1685	.1660	.1635	.1611
−0.8	.2119	.2090	.2061	.2033	.2005	.1977	.1949	.1922	.1894	.1867
−0.7	.2420	.2389	.2358	.2327	.2296	.2266	.2236	.2206	.2177	.2148
−0.6	.2743	.2709	.2676	.2643	.2611	.2578	.2546	.2514	.2483	.2451
−0.5	.3085	.3050	.3015	.2981	.2946	.2912	.2877	.2843	.2810	.2776
−0.4	.3446	.3409	.3372	.3336	.3300	.3264	.3228	.3192	.3156	.3121
−0.3	.3821	.3783	.3745	.3707	.3669	.3632	.3594	.3557	.3520	.3483
−0.2	.4207	.4168	.4129	.4090	.4052	.4013	.3974	.3936	.3897	.3859
−0.1	.4602	.4562	.4522	.4483	.4443	.4404	.4364	.4325	.4286	.4247
−0.0	.5000	.4960	.4920	.4880	.4840	.4801	.4761	.4721	.4681	.4641

Table 6.14 (Continued)

t	.00	.01	.02	.03	.04	.05	.06	.07	.08	.09
+0.0	.5000	.5040	.5080	.5120	.5160	.5199	.5239	.5279	.5319	.5359
+0.1	.5398	.5438	.5478	.5517	.5557	.5596	.5636	.5675	.5714	.5753
+0.2	.5793	.5832	.5871	.5910	.5948	.5987	.6026	.6064	.6103	.6141
+0.3	.6179	.6217	.6255	.6293	.6331	.6368	.6406	.6443	.6480	.6517
+0.4	.6554	.6591	.6628	.6664	.6700	.6736	.6772	.6808	.6844	.6879
+0.5	.6915	.6950	.6985	.7019	.7054	.7088	.7123	.7157	.7190	.7224
+0.6	.7257	.7291	.7324	.7357	.7389	.7422	.7454	.7486	.7517	.7549
+0.7	.7580	.7611	.7642	.7673	.7704	.7734	.7764	.7794	.7823	.7852
+0.8	.7881	.7910	.7939	.7967	.7995	.8023	.8051	.8078	.8106	.8133
+0.9	.8159	.8186	.8212	.8238	.8264	.8289	.8315	.8340	.8365	.8389
+1.0	.8413	.8438	.8461	.8485	.8508	.8531	.8554	.8577	.8599	.8621
+1.1	.8643	.8665	.8686	.8708	.8729	.8749	.8770	.8790	.8810	.8830
+1.2	.8849	.8869	.8888	.8907	.8925	.8944	.8962	.8980	.8997	.9015
+1.3	.9032	.9049	.9066	.9082	.9099	.9115	.9131	.9147	.9162	.9177
+1.4	.9192	.9207	.9222	.9236	.9251	.9265	.9279	.9292	.9306	.9319
+1.5	.9332	.9345	.9357	.9370	.9382	.9394	.9406	.9418	.9429	.9441
+1.6	.9452	.9463	.9474	.9484	.9495	.9505	.9515	.9525	.9535	.9545
+1.7	.9554	.9564	.9573	.9582	.9591	.9599	.9608	.9616	.9625	.9633
+1.8	.9641	.9649	.9656	.9664	.9671	.9678	.9686	.9693	.9699	.9706
+1.9	.9713	.9719	.9726	.9732	.9738	.9744	.9750	.9756	.9761	.9767
+2.0	.9772	.9778	.9783	.9788	.9793	.9798	.9803	.9808	.9812	.9817
+2.1	.9821	.9826	.9830	.9834	.9838	.9842	.9846	.9850	.9854	.9857
+2.2	.9861	.9864	.9868	.9871	.9875	.9878	.9881	.9884	.9887	.9890
+2.3	.9893	.9896	.9898	.9901	.9904	.9906	.9909	.9911	.9913	.9916
+2.4	.9918	.9920	.9922	.9925	.9927	.9929	.9931	.9932	.9934	.9936
+2.5	.9938	.9940	.9941	.9943	.9945	.9946	.9948	.9949	.9951	.9952
+2.6	.9953	.9955	.9956	.9957	.9959	.9960	.9961	.9962	.9963	.9964
+2.7	.9965	.9966	.9967	.9968	.9969	.9970	.9971	.9972	.9973	.9974
+2.8	.9974	.9975	.9976	.9977	.9977	.9978	.9979	.9979	.9980	.9981
+2.9	.9981	.9982	.9982	.9983	.9984	.9984	.9985	.9985	.9986	.9986
+3.0	.9987	.9987	.9987	.9988	.9988	.9989	.9989	.9989	.9990	.9990
+3.1	.9990	.9991	.9991	.9991	.9992	.9992	.9992	.9992	.9993	.9993
+3.2	.9993	.9993	.9994	.9994	.9994	.9994	.9994	.9995	.9995	.9995
+3.3	.9995	.9995	.9995	.9996	.9996	.9996	.9996	.9996	.9996	.9997
+3.4	.9997	.9997	.9997	.9997	.9997	.9997	.9997	.9997	.9997	.9998
+3.5	.9998	.9998	.9998	.9998	.9998	.9998	.9998	.9998	.9998	.9998

Reprinted with permission from *Engineering Statistics and Quality Control*, by Irving W. Burr, McGraw-Hill, 1953.

Table 6.15
Cumulative Frequency Computations for the Porosity Data of Table 6.11

Porosity Interval (%)	Frequency	Cumulative Frequency Less Than or Equal to Upper Limit of Interval	Cumulative Frequency Expressed as Percentage
$7.0 \leq x < 10.0$	1	1	2.4%
$10.0 \leq x < 12.0$	0	1	2.4%
$12.0 \leq x < 14.0$	1	2	4.8%
$14.0 \leq x < 16.0$	10	12	28.6%
$16.0 \leq x < 18.0$	12	24	57.1%
$18.0 \leq x < 20.0$	8	32	76.2%
$20.0 \leq x < 22.0$	7	39	92.9%
$22.0 \leq x < 25.0$	3	42	100.0%
	42		

From this illustration we see that if we have statistical data which are normally distributed we can determine μ and σ graphically from a cumulative frequency graph rather than having to solve Equations 6.3 and 6.6. Both result in essentially the same values (the values will be exactly equal if the data are exactly normally distributed). But remember also, that this neat graphical method to find μ and σ of the porosity data only worked because of the fact that the core porosities were close to being normally distributed.

If you have a set of data how do you know whether it's normally distributed so as to be able to read μ and σ graphically? Simple. Just arrange the data in intervals, count frequencies, compute cumulative frequencies as in Table 6.15 and plot the data on normal probability paper. If the data points are approximately linear the data can be represented as a normal distribution. If the data points are not linear then the data is not normally distributed, and the technique for reading μ and σ can not be used.

Is the straight line shown on Figure 6.22 a good fit of the core porosity data of Table 6.11? Technically, no. The data points for low values of porosity do not fall on the straight line, so we cannot say that the 42 values of porosity from the Denver-Julesburg well *are* normally distributed.

But it is a reasonably good fit, and assuming porosity was not an extremely critical parameter in a drilling prospect (as related to ultimate profitability), the linear approximation shown in Figure 6.22 is probably

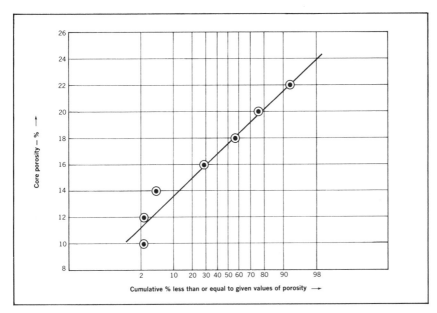

Figure 6.22 *Cumulative frequency graph of the core porosity data of Table 6.11. Specific plotting points were computed in Table 6.15.*

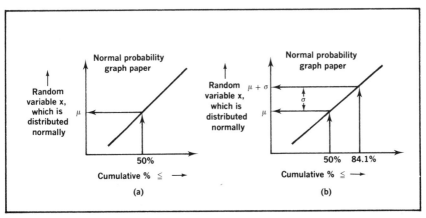

Figure 6.23 *Schematic diagrams of how to read μ and σ from the cumulative frequency graph of a normally distributed random variable. This graphical method to determine μ and σ applies only to random variables which are, or can be represented by normal distribution.*

satisfactory. Whether or not it is a good fit is largely judgmental, and we can give no prescribed rules for how closely a set of statistical data can be represented by a smoothed, standard distribution such as the normal.

Another point supporting the "goodness" of Figure 6.22 is that the 42 values of porosity used to determine the plotting points represent a fairly small sample from the statistical sense. Conceivably if we would have had twice as many values of porosity it may have been a better fit. But we must realize that lots of things in the real-world don't always agree with the carefully designed textbook examples, and this just happens to be a real-world example of a set of porosities which are not exactly distributed as a normal distribution.

Suppose we had computed μ and σ of the porosity data using the approach given in Table 6.12. How could we get the information about the porosity distribution to an analyst located 10,000 miles away? Again it's very simple. We merely tell him that the random variable porosity is normally distributed having $\mu = 17.7\%$ and $\sigma = 3.0\%$. When he receives this information he locates plotting points on a piece of normal probability paper at the following coordinates:

Values of Porosity	Corresponding Cumulative Percentile
17.7%	50%
17.7% + 3.0% = 20.7%	84.1%

These two points define the straight-line cumulative frequency graph of the normal distribution when plotted on normal probability graph paper. The cumulative frequency graph he would obtain from the above two plotting points would be identical to the graph we obtained in Figure 6.22 using all the available statistical data. Being able to draw the entire cumulative frequency curve from the numerical values of μ and σ only is an example of what we meant earlier when we said that the two single-value parameters μ and σ completely define the normal distribution.

When we were discussing the characteristics of a normal distribution we mentioned in Section f that Table 6.14 could also be used to read areas under a normal distribution curve, in lieu of using a cumulative frequency graph. To determine probabilities from Table 6.14 we must first compute μ and σ, using a method such as the computations of Table 6.12. For the porosity data the computed μ and σ were about 17.7% and 3.0% respectively.

Assuming these are the single value parameters of a normal distribution which approximately represents the statistical data we can determine areas under the curve using the method illustrated by the following example.

What is the probability that porosity will be within the range of 15% to 22%? (The shaded portion of the distribution in the sketch).

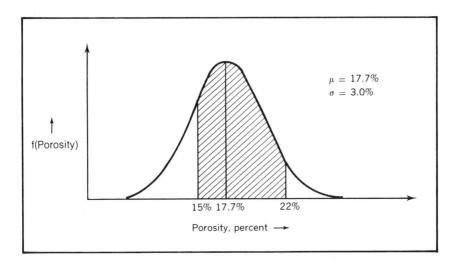

First we must compute a value of the standardized, dimensionless variable t corresponding to each of the limiting values of the porosity interval of interest.

1. $x =$ the specific value of the random variable $= 15\%$
$\mu = 17.7\%$
$\sigma = 3.0\%$

$$t_{15\%} = \frac{x - \mu}{\sigma} = \frac{15\% - 17.7\%}{3.0\%} = -0.90$$

Similarly, for $x = 22\%$:

$$t_{22\%} = \frac{22\% - 17.7\%}{3.0\%} = +1.43$$

2. Next, we use these values of t to read the cumulative area to the left of the value of x used to compute t from Table 6.14.

Value of t	Corresponding Cumulative Area Read from Table 6.14	Interpretation
$t_{15\%} = -0.90$	0.1841	The total area under the normal distribution to the left of 15% porosity.
$t_{22\%} = +1.43$	0.9236	The total area under the normal distribution to the left of 22% porosity.

3. Finally, the area under the curve between 15% and 22% is the difference between the two cumulative areas read from Table 6.14. That is:

Probability of porosity values between 15% and 22% =
$0.9236 - 0.1841 = 0.7395 \cong \underline{0.74}$

These steps and Table 6.14 provide a means to read any areas under a normal distribution, regardless of what the random variable is. The only data you need to use this approach are μ and σ of the normal distribution. You can check the value just computed using the cumulative frequency graph of Figure 6.22 by subtracting the cumulative percentage less than or equal to 15% from the cumulative percentage less than or equal to a porosity value of 22%. Within the accuracy of the graph the results are the same.

In summary, if the random variable of interest is normally distributed we can read areas under the distribution curve (probabilities) using Table 6.14 or graphically from the cumulative frequency graph. To use the table you must first know the numerical values of μ and σ. To plot the cumulative frequency graph you can use either the values of μ and σ or the actual statistical data that are used to compute μ and σ.

Lognormal Distribution

The lognormal distribution is a continuous probability distribution that appears similar to a normal distribution except that it is skewed to one side, as in Figure 6.24. The distribution can be skewed in either direction. The lognormal distribution skewed as shown in Figure 6.24 describes a random variable which has a small chance of large numerical

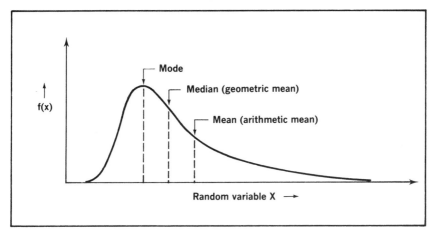

Figure 6.24 *An example of a lognormal distribution. When the curve is skewed to the right the median lies to the right of the mode, and the mean lies to the right of the median, as shown above. This order is reversed if the distribution is skewed to the left.*

values and a large chance of smaller numerical values of the variable. Some examples of random variables that can sometimes be represented by the lognormal distribution include core permeability, thicknesses of sedimentary beds, oil recovery (barrels per acre-foot) in a given formation producing by a common reservoir mechanism, and reserves per field in a sedimentary basin.

Specific characteristics of the lognormal include:

a. The distribution is completely and uniquely defined by the two single value parameters μ and σ.

b. If a random variable, x, is lognormally distributed, the logarithms of the numerical values of x are normally distributed. That is, if we make the transformation $y = \ln x$, where x is a lognormally distributed random variable, the distribution of the transformed variable, y, will be normally distributed.

c. The theoretical limits of a lognormal distribution are $\pm\infty$. In practice the distribution is usually truncated at values of the random variable where the distribution curve is nearly asymptotic with the abscissa.

d. The cumulative frequency graph of a lognormal distribution, when plotted on a special graph paper called lognormal probability graph paper, will plot as a straight line. This graph paper is similar to normal probability paper except that the scale used to plot values of the random variable is a logarithmic scale, rather than a coordinate scale.

The mean and standard deviation of a lognormally distributed variable *can not* be read graphically from such a graph (as was the case with the normal distribution). It is a common mistake to enter a lognormal cumulative frequency graph at the 50 percentile and assume the corresponding value of the random variable is μ. The value we read in this manner is, by definition, the median. But as we see from Figure 6.24 the median and mean of a lognormal distribution are not coincident. A numerical example showing how to use lognormal probability graph paper will follow shortly.

e. To compute μ and σ of a lognormally distributed variable we can use Equations 6.3 and 6.6 in a computation such as given in Table 6.13. If statistical data are plotted as cumulative frequencies on lognormal probability paper μ and σ can be computed using the following relationships:

$$\mu_x = e^{\mu_y + 1/2\sigma_y^2}$$
$$\sigma_x^2 = \mu_x^2(e^{\sigma_y^2} - 1)$$

where μ_x, σ_x are the single value parameters of a lognormally distributed variable x, and μ_y, σ_y are the single value parameters of the transformed variable y = ln x. The numerical values of μ_y and σ_y are the logarithms of values read from a lognormal probability graph of cumulative frequency.

This calculation can become a bit tedious, so the most practical way to determine μ and σ of a lognormal distribution is by use of Equations 6.3 and 6.6.

f. The meaning of standard deviation of a lognormal distribution is the same as with any other distribution — the larger the value of σ the greater the range (variability) of the variable. We cannot, however, say some of the things about σ that applied to a normal distribution (i.e. the area under the curve of a lognormal distribution between $\mu \pm \sigma$ may or may not be 0.6826, the value depending on the degree to which the distribution is skewed.)

To illustrate how to use lognormal probability graph paper we can refer to the permeability data of Table 6.11. First we must arrange the

Table 6.16

Cumulative Frequency Computations for Permeability Data of Table 6.11

Permeability Interval (millidarcies)	Frequency	Cumulative Frequency Less Than or Equal to Upper Limit of Interval	Cumulative Frequency Expressed As Percentage
0–50	2	2	4.8%
51–100	2	4	9.5%
101–150	4	8	19.0%
151–200	4	12	28.6%
201–250	4	16	38.1%
251–300	8	24	57.1%
301–350	4	28	66.7%
351–400	2	30	71.4%
401–450	4	34	81.0%
451–500	1	35	83.3%
501–700	5	40	95.2%
701–1,000	2	42	100.0%
	42		

data in intervals and compute cumulative frequencies. This is summarized in Table 6.16. Columns 1 and 2 are repeated from Table 6.13. The cumulative frequencies expressed as percentages of the total number of data points, Column 4, are plotted versus the upper limiting value of each permeability interval on lognormal probability graph paper, as shown in Figure 6.25.

If the original permeability values were exactly distributed as a lognormal distribution the cumulative frequency curve, when plotted on lognormal probability paper, would be a straight line. From Figure 6.25 we see that the actual core permeability data is reasonably close to a lognormal distribution, but not an exact fit.

From the cumulative frequency graph such as Figure 6.25 we can read any areas under the lognormal distribution desired. And it should be clear at this point that the analysis of statistical data which are log-normally distributed is exactly the same as the analysis of data which are normally distributed. The only difference is in the random variable scale of the probability graph papers.

We should also stress again that we cannot read the mean value of permeability directly from Figure 6.25. If we enter Figure 6.25 at the 50 percentile we read a corresponding permeability value of about 265md.

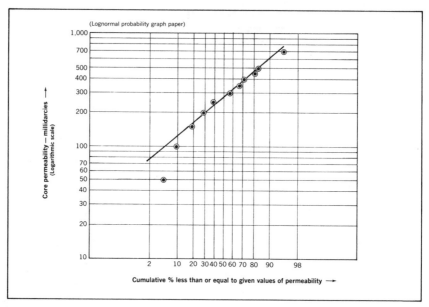

Figure 6.25 *Cumulative frequency graph of the core permeability data of Table 6.11. Specific plotting points were computed in Table 6.16.*

This value is the median value (geometric mean) of the permeability data but not the mean value.

This is a good opportunity to stress another point that was made in Chapter 3 regarding the fact that the EMV isn't necessarily the most probable outcome (mode), nor is it an outcome such that there is a 50% chance of exceeding the EMV. Recall that EMV and mean value are synonymous, and the above statement can be restated to say that the mean value of a distribution isn't necessarily the most probable value, nor is it a value such that there is a 50% chance of the random variable having a value exceeding the mean.

For this set of permeability data the most probable value (mode) is in the interval 251md–300md because it has the highest frequency (see Table 6.16). The mean value of this permeability data was computed using Equation 6.3 as 320md (Table 6.13). And the probability of a permeability value of at least 320md is only about 38%. This is determined by observing that the chance of permeability values ≤320md from Figure 6.25 is about 62%; thus the chance of permeability values ≥320md would be 100% − 62% = 38%.

As a matter of information for the reservoir engineers who might have occasion to read this the best single value of permeability to use in a fluid flow (Darcy's) equation is the median, or geometric mean. For this permeability data the median value is 265md. The reason this is a better single value of permeability to use in fluid flow equations relates to the geometric configuration of the heterogeneities of permeability and is fairly well documented in the early reservoir engineering literature.

The "average permeability" normally listed on core analysis reports is usually an arithmetic average value which is synonymous to what we have been calling the mean value. But this is not equal to the median value and its use in flow equations will result in computed flow rates which will be higher than can be expected under actual flow conditions. Reservoir engineers should be alert to this distinction when trying to represent a set of permeability data by a single, representative "average" value.

There are also some pitfalls to be aware of when using the lognormal distribution to represent or describe the possible values of reserves per field in a sedimentary basin. We will discuss these pitfalls in detail in Chapter 7.

To summarize the lognormal distribution we can conclude that statistical data can be analyzed in much the same way as normally distributed data with the exception that the cumulative frequency graph is plotted on lognormal probability graph paper. To avoid having to solve exponential equations the safest and easiest way to compute μ and σ is by use of Equations 6.3 and 6.6. The most common application of lognormal distributions in our subsequent discussions of exploration risk analysis will be to describe reserve data, recovery data (barrels per acre-foot), and formation thickness data.

Uniform Distribution

The uniform distribution is a continuous probability distribution describing a random variable in which any numerical value of the variable is equally likely to occur within an upper and lower limit. An example is shown in Figure 6.26(a) for a random variable x. All values of x between $x_{minimum}$ and $x_{maximum}$ are equally likely to occur. The uniform distribution is sometimes given the synonymous names of a rectangular distribution (for obvious reasons), or a random distribution.

The mean value and the median of a uniform distribution are coin-

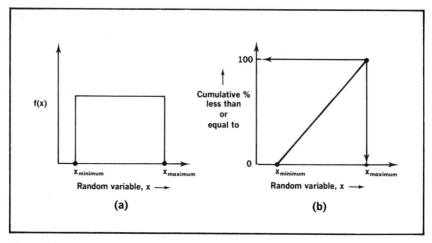

Figure 6.26 *An example of a uniform distribution of a random variable x is shown in part (a). The cumulative frequency distribution of a uniform distribution plots as a straight line on coordinate graph paper, part (b).*

cident and occur at the midpoint value of the random variable between $x_{minimum}$ and $x_{maximum}$. That is:

$$\mu = \frac{x_{minimum} + x_{maximum}}{2}$$

Equation for Mean Value of a Uniform Distribution

The standard deviation of a uniform distribution can be computed from the following equation:

$$\sigma = \sqrt{\frac{(x_{maximum} - x_{minimum})^2}{12}}$$

Equation for Standard Deviation of a Uniform Distribution

The cumulative frequency graph of a uniform distribution plots as a straight line on coordinate graph paper, as shown in Figure 6.26(b). The zero percentile (less than or equal to) corresponds to $x_{minimum}$ and the 100 percentile corresponds to $x_{maximum}$. The only two numerical values required to specify a uniform distribution are the upper and lower limiting values of the random variable.

The principal use of this distribution in exploration risk analysis is with the method called simulation (to be discussed in detail in Chapter 8). The simulation method of risk analysis permits the analyst to express uncertainty about the possible values of a parameter in the form of a distribution. In this context the uniform distribution is sometimes used to represent uncertainty about a variable in which the analyst can only specify a minimum and maximum possible value.

For instance, suppose it was the best judgment of the analyst that platform operating costs could vary anywhere from $20,000–$30,000 per day, and any value within the range is equally likely. This judgment about the uncertainty of platform operating costs could be represented by a uniform distribution having $20,000 per day as the minimum value and $30,000 per day as its maximum value. But more about this in Chapter 8.

Triangular Distribution

The triangular distribution is a continuous probability distribution which has the shape of a triangle, Figure 6.27. The triangle can be symmetrical or skewed in either direction, as in Figure 6.27. The mode value, x_2, can also be located at the minimum or maximum values of the random variable.

The triangular distribution is completely defined by specifying the minimum, most likely, and maximum values of the random variable. The

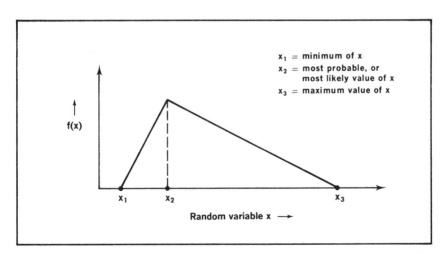

Figure 6.27 *An example of a triangular distribution of the random variable x. The most probable value of x, x_2, can be located anywhere in the interval $(x_1 \leq x_2 \leq x_3)$.*

mean value and standard deviation of a triangular distribution can be computed from the following relationships:

$$\mu = \frac{x_1 + x_2 + x_3}{3}, \text{ where } x_1, x_2, \text{ and } x_3 \text{ are the minimum, most likely, and maximum values of the random variable respectively.}$$

$$\sigma = \sqrt{\frac{(x_3 - x_1)(x_3^2 - x_1 x_3 + x_1^2) - x_2 x_3 (x_3 - x_2) - x_1 x_2 (x_2 - x_1)}{18(x_3 - x_1)}}$$

There are probably very few, if any, random variables which are, in fact, distributed as a triangular distribution. The frequent use of triangular distributions arises, again, in the risk analysis method called simulation. It is commonly used to represent a distribution of possible values of a random variable when the only information that is known or can be estimated is a minimum, most likely, and maximum value. We will discuss the use of triangular distributions in simulation in Chapter 8. We will also later point out a few interpretative problems to avoid when using triangular distributions.

There are several approaches which can be used to convert a triangular distribution to its equivalent cumulative frequency form, but perhaps the easiest way is to use the following two equations:

Equations to use to convert a triangular distribution to its equivalent cumulative probability distribution:

For a triangular distribution having the parameters:

$$x_1 = \text{minimum value of x}$$
$$x_2 = \text{most likely value of x}$$
$$x_3 = \text{maximum value of x}$$

For values of x less than or equal to x_2 (i.e. $x \leq x_2$):

$$\frac{\text{Cumulative probability}}{\text{less than or equal to x}} = \frac{(x')^2}{m} \tag{6.9}$$

For values of x greater than or equal to x_2 (i.e. $x \geq x_2$):

$$\frac{\text{Cumulative probability}}{\text{less than or equal to x}} = 1 - \left\{ \frac{(1 - x')^2}{(1 - m)} \right\} \tag{6.10}$$

where:

$$x' = \frac{x - x_1}{x_3 - x_1} \quad \text{and} \quad m = \frac{x_2 - x_1}{x_3 - x_1}$$

To find the cumulative area under the triangular distribution curve to the left of a value x you must solve Equation 6.9 if x is to the left of the mode and you must solve Equation 6.10 if x is to the right of the mode. Either equation can be used to compute cumulative probability to the left of x_2, the most likely value. The parameter m is a constant for a given triangular distribution, and the value of x' varies, depending on the value of x.

When we learn more about simulation in Chapter 8 we will find that we must convert the expressions of judgment about the distributions of possible values of the uncertain parameters into their equivalent cumulative frequency forms. For this reason we will have occasion to convert triangular distributions to their cumulative frequency form on a fairly regular basis. This fact, plus my observations about the confusion which sometimes arises when using Equations 6.9 and 6.10 perhaps justify a numerical example.

EXAMPLE: Our drilling engineers have estimated that drilling costs for the No. 1, DECISION ANALYSIS well could vary from a minimum value of $100,000 to a maximum value of $200,000. They further estimated that the most probable value of drilling costs for the well will be $130,000. If we chose to represent this expression of judgment about the possible values of drilling costs by a triangular distribution it would appear as in this sketch. Our problem is to convert this distribution to a cumulative probability distribution for use with a simulation analysis and/or to be able to read areas under the curve.

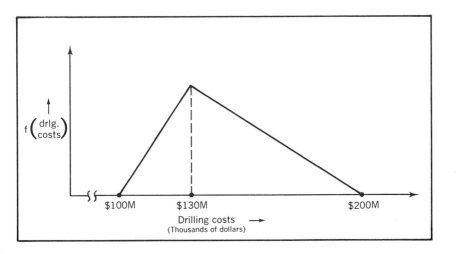

For this distribution

$$x_1 = \$100,000 = \$100M \quad \text{(M stands for thousands)}$$
$$x_2 = \$130,000 = \$130M$$
$$x_3 = \$200,000 = \$200M$$

and

$$m = \frac{x_2 - x_1}{x_3 - x_1} = \frac{\$130M - \$100M}{\$200M - \$100M} = \frac{\$30M}{\$100M} = 0.3$$

Next, we set x equal to various values of costs and compute cumulative probabilities less than or equal to x. By computing 9 or 10 of these cumulative probabilities we will have a sufficient number of data points to construct the cumulative probability distribution. For all values of x to the left of the peak value ($130M) we'll have to use Equation 6.9, and for values to the right of the mode we must use Equation 6.10. And, of course, the cumulative probability less than or equal to $x_1 = \$100M$ is zero and the cumulative probability less than or equal to $x_3 = \$200M$ is 1.0.

Here are a few sample calculations:

x = $110M:

$$x' = \frac{x - x_1}{x_3 - x_1} = \frac{\$110M - \$100M}{\$200M - \$100M} = 0.1; \quad x = \$110M \text{ is } < x_2 = \$130M$$

So we use Equation 6.9:

$$\text{Cum. prob. less} \atop \text{than or equal to } x = \$110M = \frac{(x')^2}{m} = \frac{(0.1)^2}{0.3} = \underline{0.033}$$

x = $120M:

$$x' = \frac{x - x_1}{x_3 - x_1} = \frac{\$120M - \$100M}{\$200M - \$100M} = 0.2; \quad x = \$120M \text{ is } < x_2 = \$130M$$

So we use Equation 6.9:

$$\text{Cum. prob. less} \atop \text{than or equal to } x = \$120M = \frac{(x')^2}{m} = \frac{(0.2)^2}{0.3} = \underline{0.133}$$

.
.
.

x = $160M:

$$x' = \frac{x - x_1}{x_3 - x_1} = \frac{\$160M - \$100M}{\$200M - \$100M} = 0.6; \quad x = \$160M \text{ is } > x_2 = \$130M$$

So we use Equation 6.10:

$$\text{Cum. prob. less} \atop \text{or equal to } x = \$160M = 1 - \left\{ \frac{(1 - x')^2}{(1 - m)} \right\} = 1 - \left\{ \frac{(1 - 0.6)^2}{(1 - 0.3)} \right\} = \underline{0.771}$$

.
.
.

etc.   (example continues on page 278)

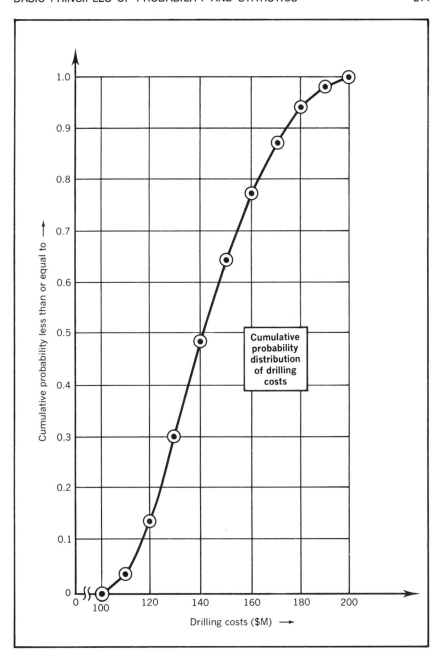

Cumulative probability distribution of drilling costs

The results of similar calculations for various other values of x are summarized in the following tabulation:

x, Value of Random Variable	Cumulative Probability Less Than or Equal to x	Equation Used to Compute Cum. Prob.
$x_1 \to$ \$100M	0	—
\$110M	0.033	6.9
\$120M	0.133	6.9
$x_2 \to$ \$130M	0.300	6.9 or 6.10
\$140M	0.486	6.10
\$150M	0.643	6.10
\$160M	0.771	6.10
\$170M	0.871	6.10
\$180M	0.943	6.10
\$190M	0.986	6.10
$x_3 \to$ \$200M	1.000	—

The values of drilling costs in Column 1 and their corresponding cumulative frequencies of Column 2 are plotted to yield the cumulative probability distribution of the original triangular distribution of anticipated drilling costs shown on page 277.

Binomial Distribution

The binomial distribution is a discrete probability distribution that describes the probabilities of a given number of outcomes in a specified number of trials. The distribution is widely used in quality control work to determine probabilities of a given number of defective pieces in a large quantity of manufactured pieces. In the petroleum exploration context the distribution can be used, under certain conditions, to compute the probabilities of a given number of discoveries (or completions) in a multiple well drilling program. Stochastic phenomena of this type are called *Bernoulli processes* and the repetitive trials are called *Bernoulli trials*.

The binomial probability equation is a special case of the general Bernoulli process in which only two outcomes can occur on any given trial. The outcomes can be called success or failure, dry hole or discovery, head or tail (from the flip of a coin), but the usual practice is to define the equation in terms of "successes" and "failures." The parameters required to specify a binomial distribution are n, the total number of trials, and p, the probability of a success on any given trial. The random variable is the number of successes in n trials, and is here given the symbol x. The binomial probability equation is given as:

Binomial probability of x
"successes" in n trials $= (C_x^n)(p)^x(1 - p)^{n-x}$ (6.11)

where x = number of "successes" $(0 \leq x \leq n)$, n = total number of trials, p = probability of success on any given trial $(0 \leq p \leq 1.0)$, and

$$(C_x^n) = \frac{n!}{x!(n - x)!}, \qquad \begin{array}{l} n! = 1 \times 2 \times 3 \times 4 \times 5 \times \ldots \times n \\ 0! \equiv 1.0 \end{array}$$

The term (C_x^n) is a mathematical notation which represents the number of mutually exclusive ways that x successes can be arranged in n trials. It is usually interpreted as "the combination of n things taken x at a time." The terms n!, x!, and (n − x)! are called factorials and are the product of the integers from 1 to n, 1 to x, and 1 to (n − x) respectively. Zero factorial, 0!, is defined as 1. A number raised to the zero power is one, so if x or (n − x) = 0, the corresponding terms $(p)^0$ or $(1 - p)^0$ are one. Zero raised to any power is also one, so if p or (1 − p) = 0, the corresponding terms $(0)^n$ or $(0)^{n-x}$ are one.

The binomial probability equation of 6.11 describes a stochastic process which has three important characteristics:

1. Only two outcomes can occur.
2. Each trial is an independent event.
3. The probability of each outcome remains constant over repeated trials.

We must always be certain that when we use the binomial probability equation the process we are considering can be characterized by these three conditions.

Let's try an example to show how the binomial equation is used. Suppose we are drilling five wildcats in a new basin where the chance of a discovery (wildcat success ratio) is 0.15 on each well. Assuming each well is a Bernoulli trial,[1] what is the probability of only one discovery in the five wells drilled? Simple. We merely solve Equation 6.11 for the case n = 5, p = 0.15, and x, the number of successes (discoveries) = 1.

---

1. This is a very critical assumption which we will discuss in detail in Chapter 7. For the moment just consider this example as an illustration of binomial probabilities, rather than a valid, real-world situation.

Binomial probability
of 1 discovery $= (C_1^5)(0.15)^1(1 - 0.15)^{5-1}$
in 5 wells drilled

$$= \frac{5!}{1! \, (5 - 1)!} \, (0.15)^1 (0.85)^4$$

$$= \frac{1 \times 2 \times 3 \times 4 \times 5}{1 \times 1 \times 2 \times 3 \times 4} \, (0.15)^1 (0.85)^4$$

$$= (5)(0.15)^1 (0.85)^4 = \underline{0.3915}$$

We could have similarly solved this example for the cases of $x = 0$, 2, 3, 4, and 5 discoveries in the $n = 5$ trials, resulting in the following binomial probabilities:

No. of Discoveries, x, in n = 5 Wildcats, Where p = 0.15 on Each Trial	Binomial Probability of x Discoveries, from Equation 6.11
0	0.4437
1	0.3915
2	0.1382
3	0.0244
4	0.0021
5	0.0001
	1.0000

From the results of these calculations we can draw the discrete binomial probability distribution of the number of discoveries possible in the five well wildcat drilling program, as shown in Figure 6.28.

If we had been interested in finding the probability of *at least* one discovery in the 5-well drilling program (rather than the probability of exactly one discovery which we computed as 0.3915) we would have to solve Equation 6.11 five separate times using $x = 1, 2, 3, 4,$ and 5. The sum of these five individual probability terms would be the probability of one or more discoveries. That is:

Binomial probability $= (C_1^5)(0.15)^1(1 - 0.15)^4$
of *at least* 1 discovery $\quad + (C_2^5)(0.15)^2(1 - 0.15)^3 + (C_3^5)(0.15)^3(1 - 0.15)^2$
in 5 wells drilled $\quad + (C_4^5)(0.15)^4(1 - 0.15)^1 + (C_5^5)(0.15)^5(1 - 0.15)^0$

or, more generally for this example:

Binomial probability
of at least 1 discovery $= \sum_{x=1}^{x=5} (C_x^5)(0.15)^x(1 - 0.15)^{5-x}$
in 5 wells drilled

In general, the probability of at least c successes in n trials, where the probability of success on any given trial is p is computed from the following equation:

Binomial probability
of *at least* c successes $= \sum_{x=c}^{x=n} (C_x^n)(p)^x(1 - p)^{n-x}$
in n trials

This equation is sometimes called the cumulative binomial equation.

The arithmetic solution of Equation 6.11 for this example was quite easy and can be done by hand calculation or slide rule. But as n becomes large, the solution can become extremely tedious. Fortunately, very complete tables of cumulative binomial probabilities have been developed, and tables for n = 2 − 20, 25, 50, and 100 trials are given in Appendix F. Instructions on how to use the table are given at the beginning of Appendix F.

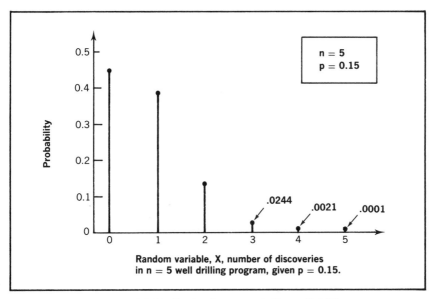

Figure 6.28 *Binomial distribution for example five well drilling program.*

There are also certain conditions when binomial probabilities can be approximated using other distributions. As n becomes large and p becomes small, such that the product np is constant, the binomial probabilities can be approximated by the Poisson distribution. The Poisson distribution is easier to solve than the binomial, and gives a good approximation to the binomial for values of n greater than 20 and values of p less than 0.05. In the limit as $n \rightarrow \infty$ and $p \rightarrow 0$ (such that np is a constant) the Poisson distribution and binomial distribution are equal. We will not discuss the Poisson distribution here, but the interested reader can find further details in any standard text on statistics.

The binomial distribution can also be approximated by the normal distribution if n is large and if p is not too close to zero. The approximation becomes better as p approaches 0.5. In general, the normal distribution gives a good approximation to the binomial if $[np \times (1 - p)] \geq 25$. When using the normal distribution as an approximation the parameters of the normal distribution, μ and σ, are computed as np and $[np(1 - p)]^{1/2}$ respectively.

The mean and standard deviation of a binomial probability distribution are given as:

$\mu = (n)(p)$ Mean Value of Binomial Distribution

$\sigma = \sqrt{(n)(p)\,(1 - p)}$ Standard Deviation of Binomial Distribution

A variation of the type of calculation represented by the binomial probability concept is the calculation of the probability of n consecutive failures (dry holes) before the occurrence of the first success (discovery). This computation is given as Equation 6.12. This equation applies for the case where each trial is an independent event and the value of p remains constant over repeated trials. This equation is useful in exploration areas having extremely low probabilities of large discoveries where the decision maker must be concerned about a "gambler's ruin" run of dry holes.

$$\begin{array}{l}\text{Probability of success}\\ \text{on } (n + 1) \text{ trial after} = (1 - p)^n \, p \\ \text{n consecutive failures}\end{array} \qquad (6.12)$$

where p = probability of success on any of the n + 1 independent trials and n = number of consecutive failures.

To summarize, the binomial probability equation will be of use to evaluate probabilities of the various number of discoveries in multi-well drilling programs. With use of binomial probability tables it is a simple, quick method for determining probabilities of 1, 2, 3, . . . discoveries. There are, however, only a limited number of exploration "scenarios" in which this simple calculation can be realistically applied, primarily due to the constraint of independent events. In general, the drilling of a sequence of wells is a series of dependent events, not independent events. We'll discuss all this in detail in Chapter 7.

Multinomial Distribution

The discrete distribution describing the general Bernoulli process in which any number of outcomes can be considered is called the *multinomial probability distribution*. The binomial distribution only considers two possible outcomes, whereas the multinomial is not restricted to just two outcomes. The equation to compute multinomial probabilities is given as Equation 6.13.

$$p_{x_1, x_2, x_3, \ldots, x_r} = \frac{n!}{x_1! \, x_2! \, x_3! \ldots x_r!} \, (p_1)^{x_1}(p_2)^{x_2}(p_3)^{x_3} \ldots (p_r)^{x_r}$$

(6.13)

where

n = total number of trials
r = number of possible outcomes
x_1 = number of times outcome 1 occurs in n trials
x_2 = number of times outcome 2 occurs in n trials
.
.
.
x_r = number of times outcome r occurs in n trials
p_1 = probability of occurrence of outcome 1 on any given trial
p_2 = probability of occurrence of outcome 2 on any given trial
.
.
.
p_r = probability of occurrence of outcome r on any given trial

and

$$(x_1 + x_2 + x_3 + \cdots + x_r) = n$$
$$(p_1 + p_2 + p_3 + \cdots + p_r) = 1.0$$

Equation 6.13 applies to processes in which each trial is an independent event and the probabilities of occurrence of each outcome remain constant over repeated trials. The binomial probability equation of 6.11 is a special case of the multinomial equation of 6.13 in which $r = 2$.

The following example will illustrate how the multinomial probability equation could be used.

From a detailed study of a new oil play it has been determined that the probabilities for finding various levels of reserves are as follows:

Field Size MM Bbls	Probability of Discovery, Fraction
0 (Dry Hole)	0.85
1–2	0.08
2–4	0.04
4–8	0.02
8–12	0.01
	1.00

In the drilling of 10 exploratory wells, what is the probability of obtaining 7 dry holes, 2 fields in the 1–2 MM Bbl range and 1 field in the 8–12 MM Bbl range?

For this example the following parameters apply:

Outcome	No. of Times Outcome Occurs in n = 10 Trials	Probability of Occurrence of Outcome
1 → Dry Hole	$x_1 = 7$	$p_1 = 0.85$
2 → 1–2 MM Bbls	$x_2 = 2$	$p_2 = 0.08$
3 → 2–4 MM Bbls	$x_3 = 0$	$p_3 = 0.04$
4 → 4–8 MM Bbls	$x_4 = 0$	$p_4 = 0.02$
5 → 8–12 MM Bbls	$x_5 = 1$	$p_5 = 0.01$
	$n = 10$	1.00

Substituting these values into Equation 6.13:

$$p\begin{Bmatrix} x_1=7 \\ x_2=2 \\ x_3=0 \\ x_4=0 \\ x_5=1 \end{Bmatrix} = \frac{10!}{7!2!0!0!1!}(0.85)^7(0.08)^2(0.04)^0(0.02)^0(0.01)^1$$

$$= \underline{0.00737}$$

The probability of 7 dry holes, 2 discoveries in the 1–2 MM barrel range and 1 discovery in the 8–12 MM barrel range in the 10-well exploration program is 0.00737.

The multinomial probability distribution can sometimes be more useful than the binomial distribution because we can consider various levels of reserves, rather than just the two-outcome case of dry hole or discovery. However, it also has the limitation that each trial is an independent event and the probabilities of occurrence of each of the r outcomes remain constant over repeated trials. These two conditions imply that the multinomial and binomial probability distributions are describing a multi-outcome process of sampling *with* replacement, whereas the drilling of sequence of wells is, in general, a sampling *without* replacement process.

Another practical limitation of a calculation such as this is that we are normally interested in the probabilities of finding, say, at least 4 MM barrels of oil in n trials rather than the specific combination of outcomes specified in the example. To do this would require similar solutions of Equation 6.13 for all combinations of possible outcomes of n = 10 wells which would result in at least 4 MM barrels of oil. And that is an extremely tedious job, to say the least! Fortunately, there are other approaches to determine the probability of finding at least ____ reserves in a multiwell drilling program which are much easier to evaluate.

Despite these qualifications the multinomial probability distribution is occasionally of use in exploration risk analysis, and the analyst should at least be aware of its value in expanding the binomial equation to more than two outcomes.

Hypergeometric Distribution

The binomial and multinomial probability distributions are based on the notion of independent trials, which in turn implies a sampling with replacement process (and/or an infinite sample space). The analogous discrete outcome distribution which is based on a finite sample space and sampling *without* replacement is called the *hypergeometric distribution*.

It is a discrete distribution which does not presume the independence of each trial as the Bernoulli process does. This important distribution will be useful to us in computing probabilities of various outcomes of a multiwell exploration program when there are only a limited number of prospects available.

The hypergeometric probability equation for the case of two possible outcomes is given as Equation 6.14.

$$\text{Probability of x ``successes'' in a sample of n trials} = \frac{(C_x^{d_1})(C_{n-x}^{N-d_1})}{(C_n^{N})} \qquad (6.14)$$

where

n = size of sample (number of trials)
d_1 = number of "successes" in sample space before the n trials were made
N = total number of elements in sample space before the n trials were made (N consists of d_1 elements identified as "successes" and $N - d_1$ elements identified as "not-successes")
x = the random variable of the hypergeometric probability distribution — the number of "successes" in the n trials. Obviously x cannot exceed, numerically, the value of n.

and

$$(C_x^{d_1}), \ (C_{n-x}^{N-d_1}), \ (C_n^N) = \text{the combination of } d_1 \text{ things taken}$$
$$x \text{ at a time, etc.}$$

To use this form of the hypergeometric probability distribution we must specify d_1, n, and N. The random variable is x. Note that d_1 and N represent the number of "successes" in the sample space and the total number of elements in the sample space *before* the sample of n trials is made. The ratio d_1/N corresponds to the proportion of "successes" in the total sample space before the first trial and is analogous to the parameter p of the binomial probability equation. The mean and standard deviation of a hypergeometric distribution are given as:

$$\mu = \frac{nd_1}{N}$$ Mean value of hypergeometric distribution

$$\sigma = \sqrt{\frac{nd_1(N - d_1)(N - n)}{N^2(N - 1)}}$$ Standard deviation of hypergeometric distribution

Equation 6.14 may, at first glance, look rather complicated and imposing, so let's work a simple example to show how it could be applied in exploration risk analysis.

EXAMPLE: Our company has identified ten seismic anomalies of about equal size in a new offshore operating area. In a nearby area of approximately equal geology 30 percent of the drilled structures were oil productive. If we test five of the anomalies (five exploratory wells) in our new area what is the probability of two discoveries?

First, we must define the terms of the hypergeometric probability equation 6.14. The number of trials, n, is five. The elements of the sample space are the ten anomalies so $N = 10$. We can define a "success" as a discovery, and the number of successes of interest is $x = 2$. Finally, the number of "successes" in the sample space before the $n = 5$ trials begin, d_1, is estimated as 30 percent of the total, by analogy with the nearby area. Thus, $d_1 = (0.30)(N) = (0.30)(10) = 3$. This is interpreted as "of the

ten anomalies in the operating area three are expected to have oil."
Now we can solve Equation 6.14 using the values $N = 10$, $n = 5$, $d_1 = 3$, and $x = 2$:

$$\text{Probability of 2 discoveries in 5 exploratory tests} = \frac{(C_2^3)(C_{5-2}^{10-3})}{(C_5^{10})} = \frac{(C_2^3)(C_3^7)}{(C_5^{10})}$$

$$= \frac{\left(\dfrac{3!}{2!1!}\right)\left(\dfrac{7!}{3!4!}\right)}{\left(\dfrac{10!}{5!5!}\right)} = \frac{30}{72} = 0.417$$

The computation we have just made is based on sampling without replacement, that is, once an anomaly is tested it is removed from the sample space of possible outcomes on the next well. This sequence results in the probabilities changing after every trial, depending upon what has occurred previously. We'll explore the importance of this sampling without replacement scheme in petroleum exploration, as compared to the sampling with replacement condition of the binomial probability equation in Chapter 7.

We will also reconsider this example at that time and solve it using logic, rather than Equation 6.14. At that time it should become apparent that the hypergeometric probability equation is a much more realistic model to analyze multiwell drilling programs than models based on Bernoulli processes.

As the values of N and d_1 become large the solution of Equation 6.14 can become extremely tedious. This is because we are having to evaluate the factorials of large numbers. In these cases it is helpful to use tables of the logarithms of factorials (such as Table G of Reference 6.13). Another alternative is the binomial approximation to the hypergeometric equation. If N is at least ten times greater than n, ($N \geq 10n$), the hypergeometric can be approximated using the binomial equation (6.11) for values of $p = d_1/N$.

The parameters n and x would be the same as defined for the hypergeometric. The larger the value of N (relative to the sample size n) the better is the approximation. In fact, the binomial equation of 6.11 can be regarded as the limit of the hypergeometric equation 6.14 as the sample space, N, becomes infinitely large.

The hypergeometric equation of 6.14 is for the case where the sample space is classified into only two outcomes (success or failure; discovery or dry hole, etc.). Equation 6.14 can be generalized for the case where we wish to classify the sample space into r outcomes ($r \geq 2$) using the following equation and nomenclature.

Outcome	Number of Elements in Sample Space Designated As Each Outcome before the n Trials	Number of Each Outcome Which Occur in the n Trials
1	d_1	x_1
2	d_2	x_2
3	d_3	x_3
.	.	.
.	.	.
.	.	.
r	d_r	x_r
	$\sum_{i=1}^{r} d_i = N$	$\sum_{i=1}^{r} x_i = n$

$$\text{Probability of } x_1, x_2, x_3, \ldots x_r \text{ outcomes in a sample of n trials} = \frac{\left(C_{x_1}^{d_1}\right)\left(C_{x_2}^{d_2}\right)\left(C_{x_3}^{d_3}\right) \ldots \left(C_{x_r}^{d_r}\right)}{\left(C_n^N\right)} \quad (6.15)$$

where

n = size of sample (number of trials), and $(x_1 + x_2 + x_3 + \ldots + x_r) = n$

N = total number of elements in sample space before the n trials are made, $(d_1 + d_2 + d_3 + \ldots + d_r) = N$

x_i = the number of outcomes in the n trials classified as the ith outcome. The x_i terms are the random variables of this multi-outcome hypergeometric distribution.

Comparison of Equations 6.15 and 6.14 will indicate that Equation 6.14 is a special case of Equation 6.15 for the value of r equal to 2.

We can summarize the applications of the binomial, multinomial, and hypergeometric equations in the following table:

No. of Possible Outcomes on Any Given Trial	For Sampling with Replacement Process (Bernoulli Process, Independent Trials) Equation to Use:	For Sampling Without Replacement Process (Finite Sample Space, Probabilities on Each Trial Dependent on Previous Outcomes) Equation to Use:
2	Binomial Probability Equation 6.11	Two-outcome Hypergeometric Probability Equation 6.14
More than 2	Multinomial Probability Equation 6.13	Multi-outcome Hypergeometric Probability Equation 6.15

Finally, I would call your attention to a subtle point regarding sampling without replacement processes – do the n trials have to occur simultaneously? The answer is no. Most statistics books will illustrate how to use the hypergeometric equation by considering a jar of colored red and white marbles. They will use the equation to compute the probability of x red marbles in a sample of n marbles obtained by reaching into the jar and simultaneously withdrawing a handful of (n) marbles.

Petroleum explorationists sometimes interpret this to mean that when we apply the hypergeometric equation to drilling wildcats that the wells must all be drilled simultaneously. This is not required of the hypergeometric equation, and in our example of testing five anomalies it does not matter whether the five wells are drilled simultaneously or sequentially. The only condition which must hold during the n trials is that the initial values of $d_1, d_2, d_3, \ldots d_r$ cannot change.

In the five-well drilling example this means that our estimate of $d_1 = 3$ productive anomalies can't be changed during the sequential drilling of the five wells. If, for example, we drill two wells and determine from geological or geophysical reassessments that the number of productive structures is probably zero, rather than three, the calculation we made to solve Equation 6.14 was not valid. Our estimates of d_1 can, of course, change as new information becomes available, but if d_1 changes the hypergeometric equation cannot be used to compute probabilities of outcomes of n trials.

IV. BAYES' THEOREM

In this and succeeding chapters we direct our attention to the determination of the probabilities of occurrence of the various possible outcomes of a decision choice. These probabilities, together with economic and profitability factors are used to compute EMV (or expected preference), the basis for accepting or rejecting the project. At the time of making the decision there may have been little or no information available upon which to base the probability estimates. Because of this it may be important to revise, or reassess these initial probability (risk) estimates as new information becomes available.

The statistical method to revise probability estimates from new information is called *Bayesian Analysis*. The fundamental basis of Bayesian analysis is a theorem developed by the English philosopher and clergyman, Thomas Bayes, which is appropriately called *Bayes'*

Theorem. In this section we will discuss the theorem, what it means, and the mechanics of solving the theorem.

Thomas Bayes (1701–1761) developed the theorem in his study of the theory of logic and inductive reasoning. The theorem provides a mathematical basis for relating the degree to which an observation (or new information) confirms the various hypothesized causes or states of nature. His major mathematical works, including the theorem, were published in 1763, two years after his death. The theorem was later proved independently by Laplace in 1774. The theorem is easily derived from the addition and multiplication theorems we discussed in the beginning of this chapter and we will give here only the final form of the theorem (for a derivation of the theorem see Reference 6.14).

Bayes' Theorem

Let E_1, E_2, ... , E_N be N mutually exclusive and exhaustive events, and let B be an event for which one knows the conditional probabilities $P(B \mid E_i)$ of B, given E_i, and also the absolute probabilities $P(E_i)$. The conditional probability $P(E_i \mid B)$ of any one of the events E_i, given B, can be computed from the following equation:

$$P(E_i \mid B) = \frac{P(B \mid E_i)\, P(E_i)}{\sum\limits_{i=1}^{N} [P(B \mid E_i)\, P(E_i)]} \qquad i = 1, 2, \ldots N \qquad (6.16)$$

This formidable looking theorem can be explained by supposing that the explorationist has defined an inclusive list of possible events, or states of nature: E_1, E_2, E_3, ... , E_N. Suppose further that he has initially assigned (or estimated) probabilities or likelihoods of occurrence for each of these hypothesized states of nature: $P(E_1)$, $P(E_2)$, $P(E_3)$, ... , $P(E_N)$. In other words, he feels there is a probability $P(E_1)$ of E_1 occurring, a probability of $P(E_2)$ that E_2 will occur, etc. The $P(E_i)$ terms are the original risk estimates and are sometimes called the *a priori* probabilities.

Next, suppose we define event B as the occurrence of some new bit of information that would cast additional light on the validity of his original probability estimates. Bayes' Theorem is used to compute new assessments of the likelihoods of occurrence of the states of nature, given the new information of event B. That is, we are trying to determine a new value of $P(E_1)$ given B — this conditional probability is written as

$P(E_1 \mid B)$; or a new value of $P(E_2)$ given $B - P(E_2 \mid B)$; etc. Equation 6.16 is a formula for computing these new assessments (revised probabilities, or *a posteriori* probabilities) based on knowledge of the conditional probabilities of the evidence occurring, given the hypothesized states of nature.

So the $P(E_1)$, $P(E_2)$, . . . terms are the original risk estimates and the $P(E_1 \mid B)$, $P(E_2 \mid B)$, . . . terms are the revised risk estimates after the evidence of event B has been considered. The fact that the original risk estimates are not zero or one reflects the fact that initially we cannot determine the true state of nature. That's the reason our exploration decisions are called decisions under uncertainty. One of the hypothesized states of nature exists, but we don't know which one it is initially. The whole purpose of using Bayes' Theorem is to try to identify the true state of nature from new information as quickly as possible so that management can make any modifications that may become necessary to corporate investment strategies.

Let's consider a simple numerical example. Suppose we have made a geological and engineering analysis of a new offshore concession containing 12 seismic anomalies of about equal size. But we are uncertain about how many of the anomalies will contain oil. So we hypothesize several possible states of nature as follows:

States of Nature
(E_i Terms)

E_1	⎡7 anomalies contain no oil ⎣5 anomalies contain oil
E_2	⎡9 anomalies contain no oil ⎣3 anomalies contain oil

Now suppose that, based on what little information is available, our initial judgment is that state of nature E_2 is twice as probable as state of nature E_1. This statement represents our initial risk estimates, i.e. $P(E_1) = 0.33$ and $P(E_2) = 0.67$ (rounded to two decimal places).

Note at this point that *if* we had perfect information to indicate that E_1 was, in fact, the distribution of dry and productive anomalies, then $P(E_1) = 1.0$ and $P(E_2) = 0$. Or, if our perfect information indicated that E_2 was, in fact, the true state of nature then $P(E_1) = 0$ and $P(E_2) = 1.0$.

We decide to drill a wildcat on one of the twelve seismic anomalies and it turns out to be a dry hole. The question is — "How can this new information be used to revise our original estimates of the likelihood of each of the hypothesized states of nature?" We can use Bayes' Theorem to gain an insight about the question.

Thus far we have the following information:

1. Definition of the possible states of nature, E_1, E_2
2. Original risk estimates, $P(E_1) = 0.33$, $P(E_2) = 0.67$
3. Outcome of first well drilled—a dry hole. We'll call this outcome event B.

The next step is to calculate the conditional probabilities that the evidence, B, could have occurred, given the states of nature. These are the $P(B \mid E_i)$ terms on the right hand side of Equation 6.16. To compute these conditional probabilities we answer the following questions:

a. What is the probability that the first anomaly drilled will be dry if, in fact, E_1 is the true state of nature?

Recall that E_1 was the state of nature in which 7 anomalies are dry and 5 oil productive. Assuming that the anomalies are all about equal size (and hence equally likely to have been selected on first trial), the probability of the first well being dry is 7/12. Thus $P(B \mid E_1) = 7/12 = 0.58$.

b. What is the probability that the first anomaly drilled will be dry if, in fact, E_2 is the true state of nature?

By the same reasoning the probability of a dry hole if the concession contains 9 dry anomalies out of 12 (and assuming each anomaly is equally likely to be selected for the first test) is 9/12. Thus $P(B \mid E_2) = 9/12 = 0.75$.

We now have all the terms we need to solve Bayes' Theorem.

c. What is the *revised* likelihood that E_1 is the true state of nature, given the evidence of one dry hole? That is, what is $P(E_1 \mid B)$?

Expanding equation 6.16 we have:

$$P(E_1 \mid B) = \frac{P(B \mid E_1)\, P(E_1)}{P(B \mid E_1)\, P(E_1) + P(B \mid E_2)\, P(E_2)}$$

$$P(E_1 \mid B) = \frac{(0.58)(0.33)}{(0.58)(0.33) + (0.75)(0.67)} = \underline{\underline{0.28}}$$

d. What is the *revised* likelihood that E_2 is the true state of nature, given the evidence of one dry hole? That is, what is $P(E_2 \mid B)$?

Again expanding Equation 6.16 we have:

$$P(E_2 \mid B) = \frac{P(B \mid E_2)\, P(E_2)}{P(B \mid E_1)\, P(E_1) + P(B \mid E_2)\, P(E_2)}$$

$$P(E_2 \mid B) = \frac{(0.75)(0.67)}{(0.58)(0.33) + (0.75)(0.67)} = \underline{\underline{0.72}}$$

We now have a set of revised, updated risk estimates based on the information received from the first well drilled.

State of Nature		Original Risk Estimate	Revised Risk Estimate, Based on the New Information, Event B
E_1	7 dry anomalies 5 oil anomalies	$P(E_1) = 0.33$	$P(E_1 \mid B) = 0.28$
E_2	9 dry anomalies 3 oil anomalies	$P(E_2) = \dfrac{0.67}{1.00}$	$P(E_2 \mid B) = \dfrac{0.72}{1.00}$

In general, the steps required in a Bayesian analysis can be listed as follows:

1. First, we must describe the possible states of nature which might exist. In applied decision situations the states of nature can be defined in many different ways. The Bayesian analysis computations will use the new information to identify which of the hypothesized states of nature is actually the source, or cause, of the observed information.
2. We must assign a likelihood of occurrence of each state of nature, the $P(E_i)$ terms. These probabilities are our original, or *a priori,* risk estimates.
3. Having gained or observed some new information or evidence we must compute the conditional probabilities that the evidence could have occurred, given the various states of nature.
4. Finally, we can compute the revised risk estimates using the theorem and the probabilities of Steps 2) and 3).

The entire solution of Bayes' Theorem can be performed in a standard, five column computation. Using the general nomenclature of the theorem, the columnar table would appear as in Table 6.17.

In some situations we may wish to obtain additional information before final selection of a management decision strategy. Bayes' Theorem can also be used in this instance. The cyclic process of obtaining information, revising probabilities, obtaining more information, again revising probabilities, . . . , is called sequential sampling.

Table 6.17

Generalized Solution of Bayes' Theorem, Equation 6.16

Possible States of Nature	Original Risk Estimates	Conditional Prob. That Evidence, B, Could Have Occurred, Given the States of Nature	Joint Prob. (Column 3 Multiplied by Column 2)	Revised Risk Estimates
E_1	$P(E_1)$	$P(B \mid E_1)$	$P(B \mid E_1)\,P(E_1)$	$P(E_1 \mid B)$
E_2	$P(E_2)$	$P(B \mid E_2)$	$P(B \mid E_2)\,P(E_2)$	$P(E_2 \mid B)$
E_3	$P(E_3)$	$P(B \mid E_3)$	$P(B \mid E_3)\,P(E_3)$	$P(E_3 \mid B)$
.	.	.	.	.
.	.	.	.	.
.	.	.	.	.
E_N	$P(E_N)$	$P(B \mid E_N)$	$P(B \mid E_N)\,P(E_N)$	$P(E_N \mid B)$
	1.000		$\displaystyle\sum_{i=1}^{N} P(B \mid E_i)\,P(E_i)$	1.000

Notes about Table 6.17

1. Column 2 sums to 1.0 because we have (presumably) listed all the possible mutually exclusive states of nature.

2. The sum of Column 3 has no meaning, so we need not bother to add the conditional probabilities.

3. The sum of Column 4 is the denominator of Bayes' Theorem. It will be a decimal fraction less than 1.0. The interpretation of the joint probabilities in Column 4 is as follows:

   "Consider the first term involving E_1 terms. In Column 3 we computed the probability of the evidence occurring if, in fact, E_1 is the true state of nature, $P(B \mid E_1)$. But originally we stated that the likelihood of E_1 being the true state of nature was only $P(E_1)$. The joint probability of the evidence occurring, weighted by the likelihood of E_1 existing in the first place is the product of these two probabilities."

4. All the revised probabilities in Column 5 are computed by dividing each term in Column 4 by the sum of Column 4.

To illustrate, recall our example in which the unknown states of nature were distributions of productive and dry anomalies.

State of Nature		Original Risk Estimate
E_1	7 dry anomalies / 5 oil anomalies	$P(E_1) = 0.33$
E_2	9 dry anomalies / 3 oil anomalies	$P(E_2) = 0.67$

Suppose as a result of a separate economic analysis it was determined that overall profitability from the concession would be acceptable if E_1 exists and unacceptable if E_2 is the distribution of dry and productive anomalies. By virtue of our initial risk estimates it would appear, *before drilling,* that the concession is probably unprofitable. But it may be desirable and/or feasible for management to defer a final decision about development of the concession until the results of one or two wells are known.

Previously we computed revised probabilities after the first anomaly tested dry of $P(E_1|B) = 0.28$ and $P(E_2|B) = 0.72$. Suppose it was decided to again defer the final decision regarding complete development until an additional anomaly had been tested. So a second exploratory well is drilled on one of the remaining eleven untested seismic anomalies, and for our example let's assume it is also dry. (Curse the luck!) Now management is faced with the question of whether to proceed with full scale development of the concession or surrender the drilling rights and try to find oil someplace else.

The results of the second well, of course, represent new information. We can use Bayes' Theorem to compute new, updated risk estimates in the context of sequential sampling. The only difference is that the revised risk estimates computed after the first anomaly was tested now become the "original" probabilities for the second Bayesian computation. The conditional probabilities of the second anomaly being dry are computed as follows:

a. $P(B \mid E_1)$: State of nature E_1 originally had 7 dry anomalies and 5 productive anomalies. One anomaly had already been tested dry; thus when the second anomaly was drilled six of the remaining 11 untested anomalies were expected to be dry. Assuming each of the anomalies was equally likely to have been selected for the second well, the probability of a dry hole was $6/11$. $P(B \mid E_1) = 6/11 = 0.545$. (Note: We are now defining event B as the outcome of the second exploratory well.)

b. $P(B \mid E_2)$: By the same reasoning, state of nature E_2 contained 11 untested anomalies when the second well location was selected, of which 8 were expected to be dry. Therefore, $P(B \mid E_2) = 8/11 = 0.727$.

The second sequential Bayesian computations are made as follows:

State of Nature		Original Risk Estimates*	Conditional Prob. That 2nd Anomaly Tested Dry, Given the States of Nature	Joint Probability	Revised Risk Estimates
E_1	7 dry anomalies 5 oil anomalies	0.28	0.545	0.153	0.226
E_2	9 dry anomalies 3 oil anomalies	0.72 1.00	0.727	0.523 0.676	0.774 1.000

*The original risk estimates of Column 2 are the revised probabilities computed previously after only one anomaly had been tested.

The revised probabilities of Column 5 represent the updated risk estimates based on the information gained from the two wells drilled thus far and our initial judgments of the likelihood of occurrence of each state of nature. These are the new risk estimates management would use to make the final decision about whether to develop the concession.

In each sequential computation the revised risk estimates from the previous sample become the original risk estimates for the next calculation. The Bayesian analysis of a sequential sampling process can be continued until a clear-cut management strategy becomes apparent and/or the costs of additional sampling exceed the costs of uncertainty. A "clear-cut" strategy would, of course, follow from knowledge of the true state of nature.

This occurs when the revised probabilities for one of the states of nature becomes sufficiently close to one. (The revised probabilities for the other hypothesized states of nature will be approaching zero simultaneously.) Evaluating the costs of additional sampling relative to the costs of uncertainty can be done with an expected opportunity loss computation. (See, for example, Reference 1.3, Reference 1.4, or Appendix E).

A logical question at this point would be to ask how many anomalies would have to be evaluated in the example problem before we could identify the true state of nature. In general, questions of this type are difficult to answer because many factors bear on the issue:

   a. Our initial risk estimates prior to first sample,
   b. The type of process generating the sample outcomes (or size of the sample space if it is a sampling without replacement process),
   c. Our description of the states of nature (how many E_i states are defined, whether the states are fairly similar in description or not,

size of the sample relative to the size of the sample space, etc.), and

d. The sequence of observed outcomes.

We should hasten to add at this point that if we are to use the sequential Bayesian analysis to determine how long to stay in an exploration play we will want to formulate the model in a manner more representative of the geologic province. For instance, we will undoubtedly want to describe the possible states of nature in terms of dry prospects and prospects with several levels of recoverable reserves. This would be to account for the obvious fact that not all "oil productive" anomalies produce the same amounts of oil. We will also probably want to define more than just two possible distributions of how oil is distributed in the identifiable prospects.

The application of Bayes' Theorem in the manner just described has been the source of much controversy between the "theorists" and the "practitioners" of statistical concepts and statistical inference. The theorem itself is completely valid and correct and is not the basis of controversy. Rather, the arguments arise when the theorem is used in practical situations in which the initial risk estimates may be subjective judgments (and/or guesses?).

The "theorists" do not recognize as valid the concept of subjective probabilities. Recall that the revised probabilities (Column 5) were computed using, in part, the original risk estimates (Column 2). The "theorists" reject the conclusions of Bayesian analysis because the original subjective risk estimates had no value or meaning. They challenge whether meaningful revised probabilities can be computed from a series of sample information when the numbers used in the initial step were useless.

The "practitioners" counter with one or more of the following points:

1. There is often at least some prior information or data upon which to base reasonably approximate risk estimates.
2. If a sufficient amount of sample information has been obtained the final revised probabilities are relatively insensitive to errors in the initial probability estimates.
3. Other statistical tests of hypotheses (such as confidence interval approaches) also include certain tenuous assumptions and/or technical problems of implementation, so what better way is there to revise or update risk estimates?

The statistical methods advocated by the "theorists," among other weaknesses, do not permit the analyst to introduce explicit judgments he may have about the states of nature. The "practitioners" maintain that the analyst (or decision maker) may, in fact, have some feelings about the states of nature, and that this information should be incorporated into the analysis. The interested reader is directed to Reference 6.15 or Chapter 10 of Reference 3.8 for a more detailed discussion of the two views about statistical inference.

On final balance, most proponents of statistical decision theory concepts feel the insights gained from Bayesian analysis far outweigh in value the theoretical points of the controversy. In petroleum exploration we base practically every decision on estimates of risk which involve a degree of subjectivity. This is because we often have no alternative method to quantify or measure risk and uncertainty. Consequently, we are usually willing to incorporate subjectivity in our decision making analysis. The net result is that the theoretical questions about the validity of subjective risk estimates become somewhat academic.

This brief description of Bayes' Theorem should be helpful to at least outline the mechanical solution of the theorem. We will discuss some specific, real-world exploration strategies in Chapter 10 which will require that we use Bayes' Theorem. These strategies involve whether to purchase imperfect information.

7 Petroleum Exploration Risk Analysis Methods

> *Our passions, our prejudices, and dominating opinions,*
> *by exaggerating the probabilities which are favorable to*
> *them and by attenuating the contrary probabilities, are*
> *the abundant sources of dangerous illusions.*
> — *Laplace*

HAVING mastered (?) the principles of probability and statistics of the previous chapter we now turn to the application of these principles to petroleum exploration risk analysis. In this context *risk* is somewhat of a catchall term, in that we may have the need to quantify or assess many types of risks:

Risk of an exploratory or development dry hole

Political risk, economic risk

Risk relating to future oil and gas prices

Risk of storm damage to offshore installations

Risk that a discovery will not be large enough to recover initial exploration costs

Risk of at least a given number of discoveries in a multi-well drilling program

Environmental risk

Risk of gambler's ruin

etc.

In this chapter we will discuss some of the methods that can be used to evaluate these various types of risks.

The problems relating to making exploration decisions under conditions of risk and uncertainty have been with us since the oil business began. Early attempts to define risk were pretty informal and usually involved adjectives rather than probabilities. In 1934 Hayward made the following analysis of the situation (Reference 7.1):

"It is surprising, but none the less true, that in everyday practice many deals are discussed and transacted without any concrete figure for the

299

chance or probability being arrived at. The cost of the well, the probable depth, and the profit to be realized from a successful result will be very carefully figured out; but the probabilities of success will be left at a 'sporting chance,' 'a sure shot,' 'a fairly good chance,' etc."

Several decades later the new discipline of statistical decision theory (decision analysis) began to emerge, and exploration decision makers began to take note of the potential benefits of decision analysis. As we have stated on numerous occasions, however, the problem involved in using decision analysis has been (and still is) where do we get all the probabilities required to solve a decision tree or compute an EMV? Risk analysis is certainly the weakest link (at least in the petroleum exploration context) of the overall decision process shown in Figure 6.1.

But what is our alternative? Ignore risk? In 1968 the industry began drilling in the Santa Barbara Channel off the coast of California. At the time of the initial lease sale it seemed to be the general consensus within the industry that the oil potential of the Channel was a virtual certainty.

Many of the oil productive onshore structures and trends extended out into the federal waters of the Channel, providing reasonably good geologic control. But when drilling began the story changed as summarized in a series of trade journal reports given in Table 7.1. These comments are not given here to cast judgment on the bids that were made or the reporting of the trade journals—but rather to illustrate that the "sure thing" may not be so sure afterall. Failure to recognize (and account for) risk can be disastrous when playing the high-stakes game of petroleum exploration.

I don't believe we can afford to ignore the element of risk. Even though risk analysis is tough, the alternative of ignoring it is untenable. So my feeling is that we must accept the realities of exploration risks and adopt a positive approach to learn all we can about risk analysis. This requires that we attempt to improve our expertise at evaluating risk and communicating our findings in a clear, concise manner to the decision maker.

There are no "handy-dandy" formulas which yield probability estimates nor does there appear to be much chance of ever finding such formulas. But the methods discussed in this chapter, together with the new, important method called simulation (Chapter 8) should provide the basis to use decision analysis more effectively in exploration decisions.

We will first give a brief summary of the basic problem and issues involved in exploration risk analysis. The remainder of the chapter will

Table 7.1

Sequential Reports of Santa Barbara Channel Exploration

Dec. 11, 1967 — Reserves under these (Santa Barbara Channel) federal waters have been termed greater than those of the Los Angeles Basin. (OGJ)

Jan. 22, 1968 — For the first time they (operators) have the chance to bid on a large offering of high potential oil lands. . . . It (the sale) involves the most thoroughly explored wildcat acreage in the deepest water ever offered. . . . The channel is much more a known quantity than, for example, the Gulf of Mexico.

No one believes there is anything less than a vast quantity of oil to be found off Santa Barbara. (OGJ)

March, 1968 — Santa Barbara bid broke all records. High bidding was due in part to the widespread knowledge of the area which is already producing from wells closer to shore. (PE)

May, 1968 — California's Santa Barbara Channel, with its ultradeep waters, ultra-high price tags and super potential, isn't going to disappoint anyone. (PE)

May 27, 1968 — Deep oil found on Tract _____. (OGJ)

June 17, 1968 — Third dry hole added to list of Santa Barbara Channel failures. (OGJ)

Aug. 12, 1968 — After $602.7 million in leases and $10.5 million for drilling of 12 wells in the Channel — the scoreboard:

    1   multizone discovery, Tract _____
    1   indicated discovery, 700' of oil sand, Tract _____
    4   officially announced dry holes
    4   tight holes abandoned — believed to be non-commercial

Another well suspended after hole difficulties . . . if the next 6 months of drilling doesn't produce better overall results than the first 6 months there'll be several Channel operators wishing that the federal leases had a money-back-guarantee. (OGJ)

Aug. 19, 1968 — _____ plugs third test on Tract _____, which already has two dry holes. _____ abandons initial wildcat on Tract _____. (OGJ)

Aug. 26, 1968 — Another abandonment was added to the Santa Barbara Channel's lengthening list last week. . . . The fifteenth well to be abandoned or suspended. (OGJ)

Sept. 9, 1968 — Operators have drilled five more holes in Santa Barbara Channel — and, so far, all are big question marks. Although initial results in the Channel are admittedly disappointing, the brisk drilling pace continues. (OGJ)

Sept. 23, 1968 — Biggest surprise so far, . . . has been the scarcity of productive oil sands. Only two out of thirteen tracts drilled have found commercial oil. (OGJ)

Sept. 30, 1968 — _____ abandons Tract _____ wildcat. _____ spuds No. ____ well on Tract _____ after abandoning No. ____. (OGJ)

(References: OGJ — Oil and Gas Journal; PE — Petroleum Engineer.)

be a compendium of risk analysis methods that can be used in petroleum exploration. Most of these methods are fairly well documented in the petroleum literature. Our purpose will be to discuss the strengths and weaknesses of each method, so that you will be in a better position to judge the conditions for which each method can be used.

We should continually remind ourselves as we study these methods that there are very few absolutes in petroleum exploration risk analysis. The value of these methods is the insight we can gain from sensitivity analyses. Answers to "what if" questions are very important and may provide the basis for a rational decision, even though some of the specific probabilities are not known to three decimal place accuracy.

The Basic Problem

The problem in risk analysis can be stated quite simply. For the drilling prospect (or series of prospects) being considered what are the probabilities of occurrence of all possible levels of profitability? (What are the chances of a dry hole? What are the probabilities of discovering a five million barrel field? What are the probabilities that estimated future crude prices will in fact occur? etc.)

Since an acceptable investment must have a reasonably good chance of making a profit the various outcomes of interest are expressed in levels of profitability, rather than just reserves such as 2 BCF, 3 BCF, etc. But of course, the uncertainties relating to possible reserve levels are, in many cases, a major unknown in the decision.

As we consider solutions to the problem we must bear in mind certain characteristics which are unique to the petroleum exploration decision process.

1. With regard to reserve level probabilities we cannot explicitly describe the process which originally generated the distribution of petroleum accumulations. This is a handicap to us because it means that we will probably never be able to develop an exact probabilistic model (equation) to serve as an analog to the exploration process.

   In probability theory the first step to solve any problem dealing with chance phenomena is to determine the basic process that is generating the outcomes of interest to us. Once the process is defined the mathematicians can usually develop an equation describing the process that can be used to compute probabilities of the outcomes on future trials. Perhaps the best illustration of this approach is in quality control of a product being manufactured by an automatic machine.

Suppose the machine operator has observed that periodically the machine produces a defective piece, and that the occurrence rate of these defectives is about one in every one thousand. Suppose he has further observed that there seems to be no pattern as to when the defectives occur. His problem is to be able to compute the probability of, say, less than five defective pieces in a batch of 10,000 pieces.

The mathematician will probably conclude from this evidence that the occurrence of defectives from the machine is equivalent to a Bernoulli process: only two outcomes, the probability of a defective seems to be constant over time, and the random occurrence of defectives all imply a Bernoulli process. From this conclusion he writes the binomial probability equation as an exact analog of the chance phenomenon, and he solves the equation using values of $p = 0.001$, $n = 10,000$, and $X < 5$ to answer the machine operator's question.

In petroleum exploration we do not know the basic process which controlled the origin, accumulation, migration, and entrapment of petroleum reserves. And we probably never will know this. Thus we can't say that reserve distributions in a basin are the result of a lognormal generating process, or a Poisson process, or any other process for that matter. The net effect is that we will be trying to determine probabilities of various levels of reserves in an undrilled prospect without having any kind of model to use as an analog.

Any "quasi-theoretical" model we develop will only be an approximation to what we think or perceive as the basic process — but we probably will never have an exact probability analog. This is the principal reason for the earlier statement that there is no "handy-dandy" formula for exploration probabilities.

2. As a related issue to item 1. the drilling of a sequence of wells is a series of dependent events based on a "sampling without replacement process." The few models that have been proposed in exploration risk analysis (e.g. binomial equation) are usually based on the notion of independent events. Because successive wells are a series of dependent events the probabilities for various levels of reserves on successive wells continually change. This

makes the problem much more complex, and models to approximately represent this process will be much more complicated than models based on independent trials.

3. Probability estimates must often be made on the basis of very little or no statistical data or experience. Additional data in petroleum exploration are usually from additional wells or seismic, and we normally can't afford to delay decisions until there is a sufficient amount of data upon which to base our probability estimates. In other industries it is frequently possible to obtain relatively inexpensive statistical sample data prior to major investment decisions (opinion polls, market surveys, etc.). But in our business we pretty much have to live with probabilities based on meager information.

4. With regard to the economic factors and the prediction of probabilities we have an equally complex situation to deal with. As evidenced by the never-before-experienced instability of the world crude prices of late 1973 and 1974, the economic factors relating to development and exploitation of a discovery can be difficult to predict. It is no longer valid to assume that future crude prices can be predicted with certainty, or that they will only fluctuate due to local market demand and supply patterns.

It is now clear that crude prices are subject to a much larger, worldwide set of political and economic factors and conditions. And we have had no previous precedent for predicting future crude prices in such a complex network of pushes and pulls on the system. So we are here faced with another instance of having to predict or estimate probabilities without the benefit of a realistic model or analog.

The effects of inflation, shortages of steel for platforms, casing for wells, etc. must be considered when determining ultimate value of a discovery. These factors are also part of a larger, more complex pattern of world trade and economic conditions. The future is an unknown and predictions of future costs can also be extremely difficult.

All these comments probably signal the end of the simple, straight-forward evaluation of a drilling prospect in which the "evaluation system" ended at the outlet side of a tank battery in the sale of crude at a stable, easy to predict price. The "evaluation system" now must, of necessity, encompass a much larger spectrum of economic and political considerations.

5. Geology is a very abstract science. An explorationist must try

to recreate in his mind how ancient river deltas existed, how the seas regressed, basins subsided, erosion occurred, etc. in his consideration of an exploratory project. But at the end of his analysis he is now being asked to quantify his (intuitive?) feelings about the prospect in terms of probabilities. Not an easy task, to say the least.

On the one extreme is the view that geology cannot (and should not) be quantified into probabilities. On the other extreme is the view that you can't sell intuition — exploration is a business decision which must be made on the basis of hard business data-profitability criteria and probabilities to quantify risk and uncertainty. Probably each extreme is partially valid, and we must seek to find a balance between the extremes.

A former student and good friend pictured this as a seesaw having a fulcrum which can be moved as necessary to achieve a realistic balance between the "pure geology" of a prospect and the need to express value in terms of profits and probabilities. It is not easy to quantify uncertainty relating to the abstract and non-quantifiable factors which an explorationist must consider in prospect analysis. Yet we must try if we are to make effective use of decision analysis in petroleum exploration.

These considerations certainly make the task of estimating probabilities in exploration uniquely complex and difficult. However, the methods discussed in the remainder of this chapter can often be used effectively to at least begin to evaluate sensitivities of the various unknowns on ultimate profitability. The technique of simulation to be discussed in Chapter 8 provides another significant step forward in analyzing the effects of risk and uncertainty.

Even though the above list of characteristics suggests many strikes against us we can still gain significant insights about how to best make our decisions in petroleum exploration by using these concepts. When overwhelmed by what seems to be the ultimate futility of estimating probabilities just ask yourself — "What's the alternative?"

METHODS FOR ESTIMATING PROBABILITIES IN PETROLEUM EXPLORATION

Subjective Probability Estimates

Subjective probability estimates are a common way of expressing the degree of risk and uncertainty relating to the uncertain parameters in exploration. The estimates represent a personal opinion as to the likelihood that a given outcome will occur. The opinion can be based on available statistical data, or it may simply be based on the analyst's opinion and feelings.

EXAMPLE: "I have studied all available geologic and well data and have constructed this isopach map. A well drilled in the center of Section 14 should encounter 150 feet of net pay and yield an ultimate net profit of $1,200,000. I estimate the chance of this occurring is 60 percent."

Such a statement probably represents his "degree of belief," or a "reliability of interpretation" that he associates with the prospect.

But what does a probability judgment of this type really mean? Does it mean, for example, that we have a 60 percent chance of 150 feet of pay and a 40 percent chance of no pay? Or a 40 percent chance of slightly less pay, say 145 feet? Does it mean that we have a 60 percent chance of a $1,200,000 profit? If so, what level of profit will occur the other 40 percent of the time?

These questions point out one of the problems of a so-called "single point estimate" of risk. With proper qualification it can tell us the chance of one outcome (such as achieving $1,200,000 profit and/or 150 feet of pay) but it tells the decision maker nothing about other outcomes which can occur. A single value estimate of risk about how many factors will occur does not give much information about the range of other possible outcomes. Other implications of subjective probability estimates include:

■ Subjective probability estimates are based on individual judgment. Two people, given the same basic data, may not reach the same conclusions regarding the degree of risk. Personal bias, emotions, and experience influence subjective judgments.

■ A subjective probability estimate of the type illustrated in the example is usually made at the end of the analysis and is meant to represent the likelihood of how the many uncertain parameters will, in fact, occur to yield the stated level of profitability. Such a probability estimate may be more difficult to assess than some people realize. For example, suppose the favorable outcome of a $1,200,000 profit depends on the simultaneous occurrence of four independent factors, each of which is subject to a degree of uncertainty.

The factors might be that net pay thickness must be 150 feet, the drilling costs must be as predicted in his analysis, etc. If each of these four factors has a 0.60 probability of having the single values he used to compute the profit of $1,200,000 the probability that all four occur simultaneously to yield the profit is *not* 0.60, but rather $(0.6)^4 = 0.1296$! Is that what he really meant when he gave the example recommendation? It's doubtful.

The point here is that if there are very many separate random variables it may be rather difficult to subjectively estimate the likelihood that favorable values of all the variables will occur simultaneously.

■ Subjective probability judgments are sometimes biased by the analyst's feelings about the consequences of the outcome. For example, if the potential monetary gain or loss is very large his preferences and attitudes about money may influence his estimate as to the likelihood of the event occurring. The complexities of our value system and our judgments about probabilities can become enmeshed without our even being aware of it happening.

The literature relating to the psychology of how we make judgments contains numerous examples of how value systems can bias probability estimates. To the extent possible, subjective probability estimates should be made without consideration of resulting monetary consequences of the outcome occurring. (Refer back to page 171 in Chapter 5 for more on this point.)

Despite the obvious limitations of subjective probability estimates we probably will have to use subjective probabilities in exploration risk analysis to some extent—simply because we may have no alternative ways to determine the probabilities. There may be better ways, however, to apply subjectivity to the analysis than the example recommendation (simulation, for example), and we must continually strive to improve our skills at making subjective probability estimates. This will require that we always be conscious of the implications and limitations of subjective probability estimates whenever we must use them.

Arbitrary Decision "Minimums" to Cover Uncertainty

In some instances risk is treated simply by raising the "minimum" no-risk standards required to accept a project.

EXAMPLE: A company may be earning 10% on its invested capital, but arbitrarily specify a minimum rate of return for drilling prospects of 25%. The logic is that the prospects which are accepted and which result in successful

completions will be profitable enough to carry the unsuccessful ventures
and still balance out with a 10% overall rate of return for the total drilling
program.

By this approach management is attempting to protect against uncertainty by raising the investment "cut-off" rate. Such a procedure
would presumably reflect the need to have the return commensurate with
the degree of risk. The reasoning is sound up to a point, but the weakness is that the arbitrary minimum (no-risk) economic rate does not explicitly consider the varying levels of risk between competing investments.

Who's to say, for example, that a high risk drilling prospect which
has a 27% rate of return, if successful, is better than a 23% return prospect which has a very high chance of success? If the minimum acceptable
no-risk rate of return is 25% the latter prospect would have been rejected when, in fact, it may have been a much more desirable investment
in terms of trying to balance the objectives of maximizing profits and
minimizing exposure to risk and uncertainty.

Another subtle weakness to the use of arbitrary minimums is the unfortunate fact that the minimums are sometimes used to regulate the
flow of drilling capital. When drilling capital is scarce the minimums are
sometimes raised so as to be even more selective in choosing the limited
number of prospects that can be accepted. Or, if funds are plentiful and
the budget period is about to end sometimes the minimum rate is lowered
so that more prospects are selected.

This, of course, defeats the whole purpose of using the arbitrary
minimums in the first place. Varying the minimum no-risk rate of return
to regulate flow of capital obviously leads to inconsistent decision
strategies—projects which were rejected earlier are now accepted, or
projects are rejected now that would have been accepted earlier.

On final balance this approach to incorporate risk and uncertainty
into the decision making process is a poor substitute for a careful,
quantitative analysis of risk.

Modifications to Rate of Return to Account for Risk

In Chapter 2 it was pointed out that probabilities cannot be used in
the strict definition of rate of return. However we further pointed out that
an approximation of the effects of risk could be obtained by use of Equation 2.7 to compute a "pseudo" investment for the rate of return computation. The approach is used by several companies as a means of partially

accounting for the possibility of a dry hole in a normal rate of return computation. Rather than repeat the discussion of the method here you are directed to pages 42–45 in Chapter 2 for details.

Allowable Dry Holes, Profit/Risk Ratios

Some explorationists feel that a valid measure of the relative degree of risk in exploratory drilling prospects is the parameter defined as the *allowable dry holes*. By this approach the analyst computes an estimated NPV profit that would result from the prospect if it was successful. This NPV profit is then divided by the cost of an exploratory dry hole. The result is a multiple of how many times the profit from a discovery exceeds the dry hole costs.

EXAMPLE: Estimated NPV profit from prospect being productive (computed after consideration of taxes, operating costs, development drilling costs, write-off of initial geophysical and leasehold expenses):

$$\$5,000,000$$

Exploratory dry hole cost: $500,000

$$\text{Allowable dry holes} = \frac{\$5,000,000}{\$500,000} = 10$$

In this example the potential profit is 10 times greater than the dry hole cost. Or stated differently, we could afford to drill as many as 10 dry holes to find the discovery and still have a rate of return equal to the discounting rate used in the NPV computations.

The multiple which has here been defined as allowable dry holes has also been called the Profit/Risk Ratio in some references.

Use of this approach does not give us any information about the probability of discovery, but rather it gives management an insight of how many exploratory wells he could afford to drill to locate the reserves. As such, it provides a *relative* indicator of desirability to management of competing exploration areas. Presumably, the higher the multiple the better the area.

For example, suppose management is considering exploration in two different basins, A and B, having approximately the same chances for discovery. If the allowable dry holes multiple in Basin A is 10 and in Basin B is 35 the latter would be the preferred choice. In Basin B he could drill more wells to locate a discovery and, all other factors equal, the more wildcats drilled the higher the chance of finding at least one discovery.

The approach can be useful in certain cases, but again we are faced with several limitations: it does not explicitly account for probability of discovery, and it does not provide a specific decision "go/no-go" criterion. The second limitation follows from the fact that this approach does not tell us that the allowable dry holes multiple must be greater than ____ in order to achieve an objective level of profitability.

Numerical Grading Techniques

Another method for evaluating risk is a technique which uses an arbitrary grading scale. The result is another relative indicator of risk, rather than an explicit statement of the degree of risk using probability numbers. References 7.2 and 7.3 are examples of this approach.

The technique consists of first listing all the factors which would have a bearing on possible presence of oil in a prospect. Examples of some of these factors: proximity to source rocks, structural closure, adequate permeability, etc. Next, the analyst considers each factor in his particular prospect and assigns a relative scale, or grade to the factor. The scale might range from one to four, with four being excellent and one being poor.

For example, if there is an indication of a source rock in close proximity to the objective pay zone he may assign a grade of 3 or 4. If he has indications the permeability may be very low he may assign a grade of only 1 or 2 to the permeability factor. After assigning a grade or score to all the possible factors (one of the references includes a list of factors which covers about six pages!) he then assigns a weighting multiplier factor to those factors which are particularly important to having an oil accumulation. The weighted scores are then added and compared to alternative prospects. The one having the highest score would (presumably) be the preferred prospect.

This approach certainly has one distinct advantage in that it forces the explorationist to think about each factor individually, rather than assigning some sort of risk factor at the end of the analysis on how all the factors will simultaneously occur. However, the limitations are also rather obvious. The numerical grading scale is quite arbitrary, as are the weighting factors used to increase the grade for some of the factors.

From comments which have been made to me by those who have tried this approach it would appear that the whole scheme is so long and tedious that it simply collapses under its own weight. After one has had to make 20 or 30 arbitrary judgments about a grade to assign to some of the more esoteric factors relating to petroleum geology he begins to wonder if it wouldn't be simpler to just guess at the answer and save all the bother.

In my judgment the insights to be gained by numerical grading techniques became pretty much overshadowed by the new risk analysis method called simulation (Chapter 8). If an analyst must assign a grade to porosity he'll probably consider the porosities from nearby wells in the equivalent formation.

This will involve looking at statistical data to find minimums, maximums, average values, etc. Having looked at all this data he must assign an arbitrary grade or score. Why not, as an alternative, express the available statistical data as a distribution of possible values of porosity and include the actual distribution in the analysis, rather than the arbitrary score? Simulation permits us to do this, as we shall see in the next chapter.

Estimating Probabilities of Discovery of Petroleum

The probability that a well (or wells) will result in a discovery or completion is, of course, an important link in computing expected values of drilling prospects. It is, in my opinion, one of the more difficult probabilities to estimate and we should recall the importance of being able to express the sensitivity of this parameter on ultimate expected value profit graphically such as in Figures 6.6 and 3.4.

Probably one of the most commonly used methods to estimate a discovery probability is to compute a wildcat success ratio (or a development well success ratio if considering additional drilling within a field). The ratio is computed by dividing the number of previous discoveries by the total number of exploratory wells drilled in the basin. Success ratio data are frequently published in trade journals and other professional society publications.

The important assumption that we make when using the statistical results of past wells drilled to estimate the probability of discovery on the next well is that the discovery probability does not change with time. This further implies independent events and a sampling *with* replacement process. We are in effect saying what has happened in the past is a mirror image of what will happen on subsequent wells to be drilled.

The problem is that this does not recognize the finiteness of the sample space of prospects in a basin and the fact that once a structure (or prospect) has been drilled it is removed from the sample space of undrilled prospects on the next well to be drilled. The sample space changes after each successive wildcat has been drilled and this results in the discovery probability continually changing.

Past success ratios are certainly useful "first approximations," but we must be aware that blind use of success ratios implies a probabilistic phenomenon that does not strictly apply to the real-world of petroleum exploration. Examples of where the use of past success ratios may lead us astray include:

1. Fifty wells have been completed in Field A, with average per well reserves of 80,000 barrels. Fifteen wells have been dry and abandoned, resulting in a well completion success ratio of 77%. Is the chance of completing the next well 77%? If so, how many more wells could we drill at a 77% success ratio? What happens when development approaches unknown productive limits?

2. Two hundred thirty exploratory wells have been drilled in geologic province B, of which thirty five (15.2%) have been successful in discovering new fields. Is it correct to assume the odds for finding a new field with the two hundred thirty first well are 15.2%? What if there were only thirty five total fields in the province — how many more wells would be required to prove this? Do the exploration success ratios make a distinction between total fields found and total fields which will be economic successes?

3. Field ABC was discovered six months ago. Since that time two successful offsets have been completed. Since all three wells have been successful, is the probability of success for the next development well 1.0? (This implies a certain event having no risk at all!)

We must remember that we cannot drill forever on a success ratio. Just because 10% of the previously drilled wildcats found oil or gas it does not necessarily follow that we can continue drilling with the expectation that 10% of the future wildcats will also result in discoveries. Sooner or later all the productive prospects will have been found and the discovery probability on subsequent wells will be *zero*, regardless of what the historical success ratio has been. Success ratios can be a useful approximation of discovery probabilities but use with caution!

A qualifying remark. There may exist certain areas where stratigraphic factors are so complex or inter-related that each well is, in essence, a separate entity. A possible example is the Morrow play in the Anadarko Basin of N.W. Oklahoma. Each well drilled is nearly an independent event, i.e. the outcome is nearly independent of what the offsets were.

In these cases the probabilities may be fairly stable with time, in which case the past frequency data may provide a realistic basis for risk analysis. To conclude that a play is truly random requires careful statistical analysis, however, and from the very nature of the oil exploration game this situation of randomness will not occur very frequently.

Some explorationists prefer to use an alternate approach for estimating discovery probabilities. They rationalize that a successful discovery requires that a trap be present and in the position indicated from geologic and geophysical evidence, that the objective formation must have sufficient thickness and the formation must contain hydrocarbons which can be readily produced into the wellbore. All these factors must be present to yield the successful discovery, and assuming each of the contributing factors is independent, the discovery probability is the product of the probabilities of each of the individual factors occurring. That is:

$$
\begin{array}{l}
\text{Probability} \\
\text{of Wildcat} \\
\text{Discovery}
\end{array}
\qquad (7.1)
$$

$$
= \begin{pmatrix} \text{Probability} \\ \text{of reservoir} \\ \text{trap} \end{pmatrix} \times \begin{pmatrix} \text{Probability} \\ \text{that trap} \\ \text{is in the} \\ \text{position} \\ \text{projected} \\ \text{from} \\ \text{seismic} \end{pmatrix} \times \begin{pmatrix} \text{Probability} \\ \text{of pay} \\ \text{thickness} \end{pmatrix} \times \begin{pmatrix} \text{Probability} \\ \text{of} \\ \text{hydrocarbons} \\ \text{in} \\ \text{formation} \end{pmatrix}
$$

To show how Equation 7.1 might be used suppose we are considering a structural prospect having the following parameters:

Probability of a reservoir trap	= 0.90	From experience of having good interpretative success with seismic in the area.
Probability of trap being in the position indicated	= 0.70	(A seismic reliability factor?)
Probability of sufficient pay thickness	= 0.80	Possibly from geological extrapolation from nearby structures.
Probability of hydrocarbons in the pay formation	= 0.25	Perhaps from analogy to nearby fields or perhaps a guess?

From these factors we could solve equation 7.1 to yield a discovery probability:

$$\text{Probability of Wildcat Discovery} = (0.90) \times (0.70) \times (0.80) \times (0.25) = \underline{0.126}$$

Conceptually, this approach has merit. One of the problems in applying it, however, relates to determining or estimating the individual terms on the right hand side of equation 7.1. If valid data are available it poses no problem, but if the prospect is in a new basin it may be very difficult to estimate probabilities for the four factors. In these situations one must consider whether four guesses multiplied together is better than one guess of the discovery probability itself.

Another situation which sometimes arises is computing a discovery probability in a multi-pay prospect. Suppose a wildcat is being considered in which there are three separate geologic sections which could be productive, A, B, and C. What is the probability that at least one of the zones will be productive? What is the probability that all three zones will be productive?

In this situation the analyst must first try to determine whether the occurrence of petroleum in a given zone is independent or dependent on whether or not the other zones are productive. To do this requires that he ask questions such as: If the shallowest zone A is productive does that mean that the chances of production in the deeper zones B and C are better? If zones A and B are dry is it still just as likely that zone C is productive? etc.

If the probability of occurrence in a zone is *dependent* on whether or not the other zones are productive the decision tree showing the possible outcomes of drilling three pay sections in a well is shown in Figure 7.1. The bar above the symbol for an event or probability of the event (e.g., $\overline{A}$, $P(\overline{A})$) means the event *not* occurring, and the probability of the event not occurring. To solve the tree the analyst must specify or estimate seven probability terms — the probabilities enclosed in rectangles.

The terms used to compute each of the terminal point probabilities are given in the right hand column of Figure 7.1. The probability that none of the zones will be productive is simply the probability of terminal point 8. The probability that all three zones will be productive corresponds to the probability of occurrence of terminal point 1.

The probability of two or more zones being productive is the sum of the probabilities for terminal points 1, 2, 3, and 5. (The four probability terms were summed because each was a mutually exclusive event, or

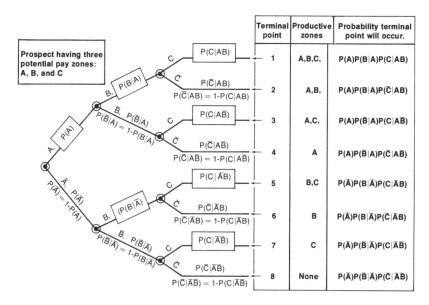

Figure 7.1 *Decision tree showing possible outcomes in a drilling prospect having three potential pay zones, A, B, and C. To use this method the explorationist must define or estimate the seven probabilities enclosed in rectangles. The bar above the event symbol ($\bar{A}$, $\bar{B}$, etc.) means the zone is not productive. $P(\bar{A})$ means the probability that the A zone is not productive. With this tree the likelihoods of production in given zones are dependent on whether or not the other zones in the prospect are productive. The same tree is shown for the case of independent events in Figure 7.2.*

outcome. Remember our Chapter 6 rule of thumb about mutually exclusive events?)

If the likelihood of production in a zone is *independent* of whether or not the other zones are productive the analysis is much simpler to complete and requires that the analyst specify only three probabilities, rather than seven. The tree for the case of independent events is shown in Figure 7.2. The interpretation and use of this tree is the same as for the case of dependent events. The three probabilities which must be specified are the probabilities of hydrocarbons in each individual zone.

This approach for analyzing probabilities in a multi-pay prospect is relatively straight-forward once the required probabilities are specified. But this is probably an example of where the probability model is more sophisticated than our ability to describe the prospect we are considering.

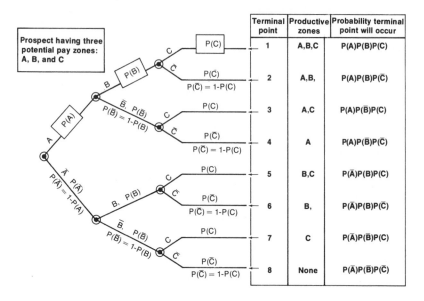

Terminal point	Productive zones	Probability terminal point will occur
1	A,B,C	P(A)P(B)P(C)
2	A,B,	P(A)P(B)P(C̄)
3	A,C	P(A)P(B̄)P(C)
4	A	P(A)P(B̄)P(C̄)
5	B,C	P(Ā)P(B)P(C)
6	B,	P(Ā)P(B)P(C̄)
7	C	P(Ā)P(B̄)P(C)
8	None	P(Ā)P(B̄)P(C̄)

Figure 7.2 *Decision tree showing possible outcomes in a drilling prospect having three potential pay zones, A, B, and C in which the probability of production in any given zone is independent of whether or not the other zones are productive. The assumption that A, B, and C are independent results in this special case of the more general tree of Figure 7.1. Only three probabilities must be specified to solve this tree — the probabilities enclosed in the rectangles. If A, B, and C are not independent events this tree does not apply, and the tree given in Figure 7.1 must be used.*

For example, to solve the dependent case shown in Figure 7.1 we'd have to specify or estimate such terms as:

P(C | AB̄) — The conditional probability that Zone C is productive given that Zone A is productive and Zone B is non-productive.

P(C | ĀB̄) — The conditional probability that Zone C is productive given that Zones A and B are non-productive.

etc.

And these terms may be pretty difficult to specify, particularly in new exploration areas where there is little or no past drilling information upon which to base the judgment. In multi-pay areas which have been extensively drilled (such as the Anadarko Basin of western Oklahoma) there may be enough information to determine these probabilities, in which case the methods of Figure 7.1 or 7.2 can be most useful.

Yet another approach for determining discovery probabilities was proposed in some recent research of the Kansas Geological Survey (Reference 7.14). Their approach was called a "re-experience" technique and attempted to relate post-drilling results to the pre-drilling perception of the geological structure in a densely drilled area of central Kansas.

When exploring for oil or gas in an area we normally prepare or estimate the geological structure interpretation of the prospect based on the available well control and seismic data. Suppose we perceive the interpretation to be a closed anticline. When trying to assess the likelihood of the anticline being productive a logical question we could ask is "What has been the results of drilling closed anticlines (as perceived before drilling) in this formation in nearby, similar areas?"

The research attempted to develop answers to questions of this type by studying a 24-by-24 mile area in northern Stafford County of central Kansas. Over the period of time from 1930–1970 this control area had been extensively explored (765 wildcats in an area of 576 square miles) and extensive information on formation tops, reserves etc. were available.

The first step in the technique was to prepare structure maps (using computer contouring algorithms) for each year based upon data available as of the particular year. (The map for 1935, for example, included tops from all wells drilled prior to 1935 but none of the wells drilled subsequently). These structure maps obviously changed with time as new wells were drilled and more control was available.

The exploratory wells were indexed by the year in which they were drilled, and an attempt was then made to define the structural interpretation (anticline, nose, homocline, syncline, etc.) at it's location *before* the well was drilled by computer and visual means. This "pre-drill" perception was then compared to the post-drill results of the exploratory well (dry, oil, etc.).

By relating this information for each wildcat drilled during the 40 year period they were able to establish frequency ratios of "post-drilling" results as related to "pre-drilling" perceptions of the structure. The idea was then to use these results (probabilities) of the densely drilled control area in other new areas of comparable geological conditions in the same formation as approximations of the probabilities of discovery.

This "re-experience" technique is, of course, dependent on having a densely drilled control area available upon which to derive the discovery probabilities for a new area. This limitation restricts it's application to mature exploration areas. It would probably be very difficult to use with any confidence in new, rank wildcat exploration areas such as offshore Greenland.

The research on this technique was also clouded somewhat by problems of finding consistent methods of classifying "pre-drill" structural interpretations around each wildcat well. We must also remember that such an approach implicitly requires that the new area of exploration interest be in the same formation and geologically similar to the conditions of the control area.

The control area probabilities couldn't be applied, for example, to a contour map of a shallow water sand! As the old adage goes, "Multiplying anything by zero still results in zero!"

To summarize, probably the most common method of estimating discovery probabilities is by use of past success ratios. When we use a success ratio to determine probability of discovery on the next trial we infer that the sequence of wells drilled in a basin is a series of independent events. While past success ratios are useful "first approximations" to estimate discovery probabilities we must use them with caution because the inference of independent trials is not strictly realistic of the real world context of petroleum exploration.

Discovery probabilities can be computed directly using Equation 7.1 if sufficient information is known about the prospect to be able to specify or estimate the probability terms of each contributing factor of the equation. If a prospect has multiple pay sections the discovery probability can be computed using the method shown in Figures 7.1 and 7.2.

Figure 7.1 is used if the probability of occurrence of oil or gas in one of the multiple pay zones is dependent on whether or not any of the other zones are productive, and Figure 7.2 is used if the probabilities of production in each zone are considered to be independent of one another. The dependent case requires specification of seven probability terms and its use may be limited due to the problems of estimating these values.

Finally we should mention a subtle nuance about interpreting the meaning of a discovery probability. If we say the probability of a discovery is 0.3 in a ten well drilling program we normally would interpret this to mean that the most probable outcome is that thirty per cent of the wells drilled would be productive – i.e., three discoveries in the ten well program. But what if we are just considering one well – what does $p = 0.30$ mean? Most would probably interpret this as the likelihood that the well will discover oil or gas, which is the correct interpretation.

This interpretation is troublesome to some, however, because they argue that the single wildcat is either going to be dry ($p = 0$) or productive ($p = 1.0$) and that $p = 0.3$ is really a meaningless number. We can't have 0.3 discovery – it either *is* or it *isn't* oil! My experience in responding to questions of this type is that one can quickly lose sight of the basic definitions of probability by following this series of logic.

I like to return to the old trusted example of flipping coins. When we flip a fair coin we say the probability of a head is 0.5 and we all agree. But when we actually flip the coin it is either a head or it isn't. So what's the difference between this and the meaning of p = 0.3 in the single well? None at all! When we interpret a discovery probability for a single well we must interpret it as the likelihood of discovering petroleum.

If our analysis involves multiple wells we can interpret the discovery probability as the fraction of the total of many trials which will be discoveries. These two interpretative statements must be given in that order – the logic does not work in the reverse direction. (That should be enough to even confuse those who thought they had it straight in the first place!)

Risk Analysis Models Based on Ultimate Reserve Distributions

There have been several attempts to define reserve level probabilities at various stages of exploration in a basin using an ultimate reserve distribution (e.g., References 6.3 and 7.4). These models are based on the hypothesis that the probability density function of reserves remaining to be discovered at any time in a basin is proportional to the area between the ultimate reserve distribution and the distribution of reserves discovered to date. Ultimate here is taken to mean after all prospects have been drilled. The random variable of these reserve distributions is reserves per field.

Let's back up and look at the rationale that precedes the hypothesis. Suppose we were able to estimate a probability distribution of reserves per field in a geologic basin. This distribution will be called the ultimate reserve distribution of the basin. It could be any shape and have whatever limits are applicable for the magnitudes of reserves per field.

Now suppose further that, by use of a suitable proportionality constant, we convert the ultimate probability distribution to a frequency distribution in which the vertical axis is now the number of fields in each reserve range, rather than the probability density function of the distribution. This frequency distribution might appear as in part a of Figure 7.3.

The shape of the ultimate distribution in Figure 7.3(a) is approximately lognormal, but remember again that the frequency distribution can have any shape we feel is representative. From this frequency dis-

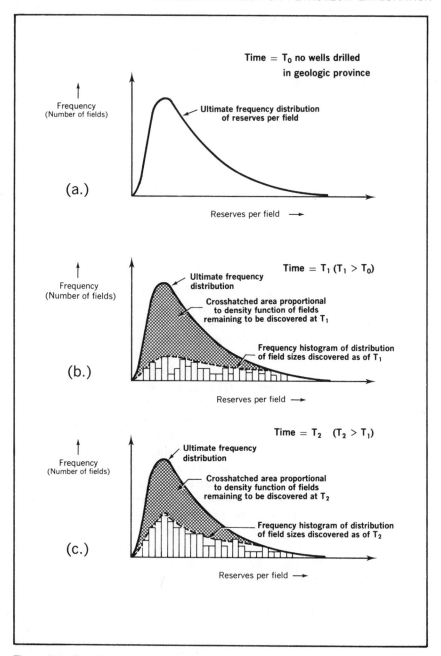

Figure 7.3 *Graphs of the frequency distribution of reserves discovered in a geologic province at various points in time.*

tribution we would be able to read that there are N_1 fields in the reserve range of R_0 to R_1 barrels per field, N_2 fields in the reserve range R_1 to R_2 barrels per field, etc.

Now suppose some exploration has occurred in the basin and as of time T_1 there have been fields discovered of various sizes. These data can be plotted as a histogram on the same graph, as shown in part b of Figure 7.3. By the method being described, the probabilities for finding various sized fields yet to be discovered are proportional to the area between the frequency histogram of what's been found to date (Time T_1) and the ultimate frequency distribution.

The area between these two curves is shown crosshatched in Figure 7.3(b). Conceptually this is equivalent to saying that if there are (ultimately) twenty fields in reserve range R_i to R_{i+1} and eight fields have been found as of T_1 there are twelve left to find on subsequent exploratory wells.

As time progresses, say to T_2, more fields are found and the frequency histogram of reserves found to date moves closer to the upper, ultimate distribution. If all the fields in a particular reserve range have been discovered the frequency histogram and the ultimate frequency distribution are coincident. The area between them for the reserve range is zero, thus the probability of finding more fields in that range is zero. Eventually, as (and if) all the prospects are drilled the two curves will be coincident over the entire range of reserves per field.

From T_0 the upper ultimate frequency distribution is constant and remains fixed on the distribution graph paper. The only curve that changes with time is the lower frequency histogram of discoveries to date. And at any point in time the probability density function (and hence the probabilities) of fields remaining to be discovered is proportional to the area between the two frequency curves.

This explanation of the rationale for this approach is a bit simplified. The original proposals for this method of determining reserve level probabilities did not make the extra step of using a proportionality constant to convert from the probability distributions to the frequency distributions of Figure 7.3. Rather they were formulated using the probability density function notations of the ultimate reserve distribution and the distribution of discoveries to date. The description here in terms of frequencies was used to describe the logic of the method while avoiding the horrendously complex mathematics involved.

Is the method realistic? Will it work? I believe the answers are yes and no respectively. Conceptually, the hypothesis seems valid. It recognizes the "sampling without replacement" nature of exploring for oil. It accounts for the obvious fact that once a discovery is made there is one less field in that reserve range left to be found on subsequent exploratory wells.

Using this method at any point in time we would be able to compute how many fields are left to find in each reserve range, express these values as conditional relative frequencies, and multiply the chances of finding oil in the first place by the conditional relative frequencies to yield the probabilities of various-sized discoveries. These probabilities would be used in a standard EMV calculation, and exploration would continue until such time that the EMV for continued exploration becomes negative and/or is less than the EMV for alternative exploration plays. That's the good news.

Now the bad news. It will probably not work for us in the real world because of the obvious question—How do you estimate the ultimate frequency distribution in the first place? If the upper curve can't be defined at time zero (or very early in the exploration of a new basin) the whole scheme is of little avail. The simple fact is that it is virtually impossible to estimate this ultimate distribution until the basin has been completely drilled—at which point reserve probabilities become moot anyway. If we knew the ultimate distribution of reserves per field at time zero many of our tough exploration investment decisions would be considerably simplified!

So we see that this method of risk analysis requires that we specify an ultimate reserve distribution, and unless this can be done the method is of little practical value. One researcher proposed that the ultimate reserve distribution would be lognormal and would be defined by the analyst by estimating its two parameters μ and σ (Reference 6.3). Not much help.

Others have attempted to re-study the sequential exploration of highly drilled basins to try to determine if there is some clue or trend by which an explorationist could estimate the ultimate distribution of a new play at a fairly early point in time. The prospects for a breakthrough here are very slim indeed.

Probably the only real-world situation where this method can be used effectively is in a new play in which the geology, stratigraphy, etc. are correlative to a nearby (or similar) basin which has been extensively explored. A case in point is the present (1974/1975) southwest extension of the Denver-Julesburg basin. It represents a new active play which appears to be correlative to the older (1950's and early 1960's) activity to the north and east.

In the older portion over 700 fields had been found by early 1970 and it appears that the distribution of reserves per field from this area can be applied to estimate the probable results of the new activity in the southwest portion of the basin (near Denver). Of course such a "by analogy" extrapolation implies that both areas have similar depositional and geologic histories.

My feeling is that this method of risk analysis is conceptually valid but extremely difficult, if not impossible, to apply because of our inability to define the ultimate reserve distribution in a new basin. My view, however, is biased by an even more critical issue. Even if we could, by some brilliant bit of perception, determine the ultimate probability distribution of reserves per field at or near time zero would that solve our problem? Would it be sufficient to know for example, that the probability of a field greater than one billion barrels is 0.10 (the area in the right tail of the reserve distribution for reserves greater than one billion barrels per field)?

The answer is *no*. (Sorry about that!) To just know the probabilities of various levels of reserves as read from the ultimate probability distribution is only one half of the problem. The missing link is to know that 0.10 of *how many fields* are greater than one billion barrels. That is, we need to know the parameter by which we can get from probabilities to the actual number of fields in each reserve range. This parameter is the proportionality constant we spoke of earlier in converting from a probability distribution to a frequency distribution.

The parameter is the total number of prospects in the basin that have oil or gas. We'll call this constant N, the total number of productive prospects in the basin. *It is an absolutely vital link in exploration risk analysis!* Without specifying this parameter we will be unable to determine probabilities of what's left to find at a point in time. To just say that 0.10 (10%) of the productive fields will have reserves greater than one billion barrels is not sufficient.

We must be able to say 10% of N fields will be at least one billion barrels. If N is 10 this means there is $0.10 \times 10 = 1.0$ field of this size in the basin, and if one has already been discovered we'd better quit drilling—there are no more left to find. If $N = 100$ this means there are $0.10 \times 100 = 10$. fields of this size in the basin. If one has been found we should keep drilling because there are still nine left to be discovered.

So the point here is that knowledge of the ultimate distribution of reserves per field in a basin will provide an important step forward in determining probabilities of what is left to find at a point in time. But in addition to knowledge of the ultimate distribution we must also know (or be able to estimate) the total number of productive prospects which exist in the basin. This important proportionality constant is, in most cases, almost as difficult to estimate early in the play as the ultimate reserve distribution. One or the other of these terms will probably lead us to a dead-end with this approach for determining probabilities. We'll talk about estimating N a bit more in the next section of this chapter.

Before leaving this brief discussion about reserve distributions I should like to draw your attention to several subtle points about analyzing reserve data. A very common statement I hear from explorationists is that reserves per field in a sedimentary basin are lognormally distributed. This statement is, in the strict sense of the word, probably incorrect. Perhaps what we should be saying is that in many basins it appears that the distribution of reserves per field *can be approximated by* a lognormal distribution.

The statement that reserves *are* lognormally distributed is simply too strong. In point of fact we do not know whether the origin, accumulation, migration, and entrapment of oil and gas reserves was controlled by (or the result of) a lognormal generating process. While the lognormal distribution appears to be a reasonably good approximation of reserves per field in many instances there are some basins where the distribution of reserves is just simply not lognormal. (For example, see References 6.3 and 6.7.) So we must be very careful not to assume that reserves *must* be lognormally distributed.

Another disquieting part of such a statement is that the usual parameter used as the random variable for these reserve distributions is recoverable reserves—the only type of reserves of value to us. Recoverable reserves are a function of how we, today, operate our leases and the price we get for the oil and gas. Recoverable reserves are a function of today's economics of producing oil and gas.

Is it reasonable to assume that nature distributed the oil and gas deposits in a nice consistent lognormal fashion millions of years ago knowing how we, today, are defining recoverable (economic) reserves? Extremely doubtful. Perhaps oil-in-place, or perhaps hydrocarbon pore volume might very well be defined in a lognormal manner, but recoverable reserves are a function of how we operate the oilfields and it seems doubtful that the influences of today's taxes, oil prices, costs, etc. would be so well behaved.

And to even carry the argument several steps further, when we plot recoverable reserves per field how do we account for the volumes of associated gas produced with the oil? Do we convert the produced gas volumes back to equivalent reservoir volumes on a pressure basis? Or on an equivalent dollar-value basis?

Or do we just ignore the produced gas volumes, and plot just the oil reserves as determined by cumulative production plus estimated remaining reserves? What if an oil field has an initial gas cap — how do we treat the gas volume and the produced gas in determining reserves for the field?

Another problem when dealing with reserve data relates to what is called truncation of data. For example if we had a list of all the fields in a basin having reserves greater than 100 million barrels each it might very well be that these reserves, when plotted on log probability graph paper would not fit a lognormal distribution. The reason is that by virtue of the 100 million barrel limit the data we had before us would represent only the right tail of the distribution.

The lower values of reserves were not included in the data. These lower values of reserves, when included with those greater than 100 million barrels may in fact be lognormally distributed, but if we initially only used the data of the large fields it is not surprising that the data did not fit a lognormal distribution. The lower values were truncated, or deleted from the set of data.

So the moral here is that when analyzing reserve data we must be sure that we have the smaller reserve values to the left of the mode so as to fill out the left tail of the lognormal distribution. But this statement then leads us to another dilemma. How small do the values of reserves per field have to be so as to insure that we have avoided the "truncation of data" problem? Fields with reserves as low as 100,000 barrels? As low as 10,000 barrels? 1,000 barrels?

Some explorationists take the view that if we included all the deposits of oil in the basin down to the limit of zero barrels per field that the distribution curve would never bend downward to the left of the mode (as in a lognormal distribution). Rather, the distribution would continue upward as reserves per field become smaller and smaller. Such a distribution, they claim, would more nearly be described as an exponential distribution (or a gamma distribution having the appropriate parameters), and that we would never see a lognormal distribution.

Such a distribution might appear as in Figure 7.4(b). The proponents of this view are saying, in essence, if we were ever able (and interested) in tabulating reserves per field for the very small fields less than R_{min} that instead of there being no fields in this range (as in Figure 7.4(a)) there would be a very large number of them, as shown by the area in Figure 7.4(b) between zero reserves and reserves of R_{min}.

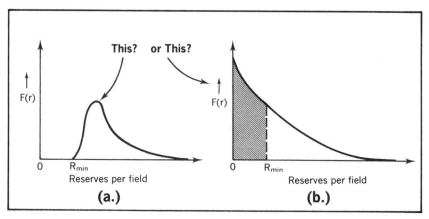

Figure 7.4 *Possible shapes of the distribution of reserves per field in a basin. R_{min} represents the minimum reserves required to justify completion of the discovery well.*

And of course this view would clearly imply that reserves would *never* be lognormally distributed if we did not artificially truncate our reserve data at some minimum value, R_{min}. Do you agree or disagree?

The obvious practical consequence of such a hypothesis is that we'll never know, because once the reserves per field get less than some minimum economic level we're not interested in it. We can't make money on a show of oil, or a deposit of oil having only one foot of pay zone. In practice everything smaller than R_{min} is classified as an exploratory dry hole and we move on and look for oil elsewhere. But the hypothesis is hard to refute. And the fact that we can make such a hypothesis with some degree of validity should make us wary of saying that reserves *must* be lognormally distributed.

The theoretical issues are good topics for an "after-hours" meeting at a bar someplace—but from a practical standpoint we can side-step the issues in our risk analyses. For reserves less than R_{min} we probably won't complete the discovery well, so we can classify these outcomes as exploratory dry holes. For discoveries greater than R_{min} we would compute an associated economic value, and the distribution of these reserves (lognormal or otherwise) would be one of the inputs to the analysis. This eliminates the need to define the reserve distributions for values of reserves less than R_{min}.

To summarize, in many basins of the world reserves per field can be reasonably approximated by a lognormal distribution—but there are some areas where the discovered reserves simply do not fit a lognormal distribution. So the moral is that we must avoid the statement reserves *are* lognormally distributed.

Probabilities of Outcomes of Multiwell Drilling Programs

In Chapter 6 we briefly mentioned several probability distributions that can be used in multiwell drilling programs. These included the binomial, multinomial, and hypergeometric distributions. We now will discuss in substantial detail how these methods can be applied to petroleum exploration. In particular, we'll show how these distributions (analogs, models) can be used to compute probabilities of outcomes resulting from the drilling of a series of exploratory wells.

The first issue that we face in analyzing multiple-well exploration programs is whether the drilling of the wells is a series of independent events or a series of dependent events. The reason we must speak to this issue is that the computations are different in each case. The issue is vital—we must specify one or the other as applying to the set of wells we're considering. There is no neutral position astride the fence, so to speak.

If it is the judgment of the explorationist that the sequence of wells is a series of independent events he is implying that the probabilities of various outcomes remain constant with time, and that the probability of any given outcome is not affected by or does not affect the probability of an outcome on any other trial. The notion of independence implies a sampling *with* replacement phenomenon, and hence a sample space which remains unchanged over all of time.

Conversely, if it is his best judgment that the proposed wells are, in fact, a series of dependent events he implies that the probabilities of various outcomes change after each successive exploratory well is drilled, depending on what has been found thus far. Dependent events imply sampling *without* replacement, and a sample space having only a finite number of elements.

When we define our sample space as consisting of all the geologic prospects in the basin the dependency condition will be the applicable representation of the sequence of drilling multiple wells. There are certain instances, however, when the notion of independent events will be approximately correct, and in some instances we may choose to analyze the decision options on this basis so as to gain advantage of simpler computations.

We summarize the two types of probabilistic phenomena and their applications in Table 7.2. The Table shows the two parallel, but dis-

Table 7.2

Summary of Analysis Methods for Computing Probabilities of Outcomes of Multiwell Drilling Programs

Dependent Events		Independent Events
The wells are considered to be a series of dependent events. Probabilities of outcomes on any given well are dependent on what has already been discovered. These probabilities change after each successive prospect is drilled. A sampling *without* replacement process. A finite number of prospects in the basin (or exploration area), N, and after each prospect is drilled there is one less element in the sample space of remaining undrilled prospects. The sample space changes after each well is drilled.	GENERAL CHARACTERISTICS OF THE PROBABILITY ANALOG	The wells are considered to be a series of independent events. Probabilities of outcomes on any given well are independent of anything that has happened before or will happen thereafter. These probabilities remain constant after each successive prospect is drilled. A sampling *with* replacement process. Implies an infinite number of prospects in the basin – and as long as we continue drilling we'll have a constant proportion of successes and failures. Does not recognize the finiteness of the number of prospects in a basin (or exploration area). Sample space remains the same after each prospect is drilled, even though, in fact, there is one less prospect left to drill after each successive wildcat is completed.
When the sample space of outcomes is defined as all the exploration prospects in a basin (or exploration area) this condition of dependency applies in all cases. It is the true representation of the drilling of a sequence of wells. It is particularly important to account for the "sampling without replacement" aspect of exploration if the total number of prospects in the basin or area, N, is small (of the same order of magnitude as the number of wells being drilled in the exploration program, n.). If the number of prospects is very large compared to the number of wells being drilled (for values of N at least $\geq 10n$) the probabilities computed using the notion of dependency can be approximated by equivalent "independent event" methods which have the advantage of being simpler to compute.	APPLICATION TO MULTIWELL EXPLORATION (OR DEVELOPMENT WELL) DRILLING PROGRAMS	Because of the above characteristics and the realities of exploration being a process of sampling without replacement (rather than sampling with replacement) this method of analysis has direct application in only a few special situations: • Each of the wells in the multiple-well program is in a different basin. • In a basin where the geologic strata of interest exhibit such rapidly varying and complex stratigraphy that every well drilled is, for all intents and purposes, in a separate reservoir. This is a very special case which only arises in a few basins in mature stages of exploration. The analysis method based on the condition of independent events provides an approximation of the probabilities derived using dependent event methods if the number of wells in the drilling program. n. is small compared to the total number of undrilled

	Hypergeometric Distribution	Binomial Distribution	Multinomial Distribution
PROBABILITY DISTRIBUTION MODEL APPLICABLE	this approximation to have much value N should be at least ten times the value of n (i.e. $N \geq 10n$). As N becomes very large the probabilities obtained by the independent event methods will be approximately equal to the corresponding probabilities based on a series of dependent events. Hypergeometric Distribution (Equation 6.15)	Only 2 outcomes being considered. (e.g. success or failure, producer or dry hole) Binomial Distribution (Equation 6.11 or Appendix F)	More than 2 outcomes are being considered. (e.g. a dry hole or discovery of reserves of 4 or 5 different levels) Multinomial Distribution (Equation 6.13)
FACTORS WHICH MUST BE SPECIFIED BY ANALYST TO USE THE METHOD	N = Total number of undrilled prospects in the basin or exploration area before the drilling program begins n = Number of trials (wells) in the multiple-well program r = Number of possible outcomes on any trial $d_1, d_2, d_3, \ldots, d_r$ = Number of elements in sample space designated as each outcome before the n trials with the condition that $(d_1 + d_2 + d_3 + \ldots + d_r) = N$ $x_1, x_2, x_3, \ldots, x_r$ = Number of outcomes which occur in each of the r categories, with the condition that $(x_1 + x_2 + x_3 + \ldots + x_r) = n$	n = number of trials (wells) in the multiple-well program x = number of successes of interest $(0 \leq x \leq n)$ p = probability of success on any given trial $(0 \leq p \leq 1.0)$	n = number of trials (wells) in the multiple-well program r = number of possible outcomes on any trial $x_1, x_2, x_3, \ldots, x_r$ = number of times the outcomes occur in n trials, with the condition that $(x_1 + x_2 + x_3 + \ldots + x_r) = n$ $p_1, p_2, p_3, \ldots, p_r$ = probability of occurrence of the outcomes on any given trial, with the condition that $(p_1 + p_2 + p_3 + \ldots + p_r) = 1.0$

tinctly separate analysis methods for analyzing multiple-well drilling programs – one based on the condition of dependent events and one based on the condition of independent events.

A. *Analyses Based on the Condition of Dependent Events*

We have hinted at several points in this discussion that the drilling of a series of wells is a situation of dependent events. What does this really mean? How can we tell if a real-world series of wells are independent or dependent events? Does it even matter that we make such a distinction?

To answer the first of these questions let's consider an experiment in which the 52 cards of a deck of bridge cards are thoroughly shuffled and then laid face down on a table. For this analog we'll assume that each card represents a geologic prospect in a basin or exploration area. Hence there are 52 "prospects" or elements in the sample space.

Now let's suppose further that all the diamonds represent oil productive prospects and all the hearts, spades, and clubs correspond to prospects having no oil or gas. Finally, to complete the analog suppose we assume that selecting a card (which is face down initially) and turning it face up is equivalent to drilling a prospect with an exploratory well. If a diamond is turned up the well found oil. If any of the other three suits are turned up the prospect is considered to be dry.

Before any wells are drilled (that is, before any cards are turned face up) what's the likelihood of finding oil? It is 13/52, right? There are 52 elements in the sample space each having an equal chance of occurrence, and of these 52 elements 13 are defined as diamonds. So if we selected one of the 52 cards and turned it face up the likelihood of it being a diamond is exactly 13/52.

Suppose we did turn up a card and it was a diamond. Now what's the likelihood of the second card we turn up also being a diamond? As long as we've left the 51 face-down cards undisturbed and the original card is still face up (and hence cannot be selected on any subsequent trial) the answer is 12/51. Of the remaining 51 face-down cards there are $13 - 1 = 12$ diamonds left, so the chance of a diamond on the second card turned face up, given that one diamond has already been "discovered," is 12/51.

This probability is a conditional probability – it's conditional on already having turned up one diamond. The sequence of turning up the two cards in the manner just described is said to be a series of dependent events. The odds for finding a diamond on the next trial change after each trial, depending on what has been turned face up thus far.

Suppose we had selected and turned up five cards and none of the five cards was a diamond. What's the probability that the sixth card turned face up will be a diamond? The answer is 13/47. There are 47 cards still face down, and none of the diamonds had been turned up so all 13 of the original diamonds remain in the group of 47 cards lying face down on the table.

The analogy of the 52 cards lying face down on a table is exactly analogous to the real-world of petroleum exploration. Nature has distributed N prospects in a basin or exploration area, some of which have oil (hopefully!) and some are dry. As we drill each prospect we are, in essence, turning one of the cards face-up to see if it's oil or dry. But this card is then left face up and there is one less in the sample space of outcomes on the next trial. In the real-world once a prospect has been drilled it is removed from the total of those remaining to be drilled. Each system is one of sampling *without* replacement.

What would the equivalent analog be if the card experiment represented a series of independent events? The experiment would be the same, and on the first card selected the odds of a diamond would be exactly the same, 13/52. But then for the sequence of events to be independent we'd have to gather up all 52 cards, reshuffle them thoroughly, and lay them face down before selecting another card.

The odds of this second card being a diamond are 13/52. Whatever the first card was it was placed back in the deck before the reshuffle and became available to be selected again on the second trial of the experiment. The second 13/52 is a probability which is independent of what occurred on the first trial. To continue this sequence, after the second trial had been performed all 52 cards would again be reshuffled, laid face down on the table and another card selected and turned face up.

The odds for this being a diamond? Exactly the same — 13/52. The odds for a diamond by this process will always be 13/52. Even if we repeated the trials 20 times, or 100 times, or 1 million times! This is called sampling *with* replacement, and the repetitive trials are said to be a series of independent events.

Now clearly this last analog does *not* apply to the real-world because of the fact that once we drill a prospect and observe whether it has oil we don't "un-drill" it and put it back in the sample space and make it available to be selected and drilled on a subsequent trial. Once a prospect is drilled that's it. It is taken out of the sample space of prospects which can be drilled on subsequent trials.

As a result of all this I think we must all clearly agree that the sequence of drilling several exploration wells is a sampling without replacement process—a series of dependent events in which the odds change after each prospect has been drilled. The probabilities of what is left to find at any point are dependent upon what has been found thus far. And by recognition of these conditions we can, by the reverse reasoning, conclude that drilling a series of exploration wells within a basin or exploration area is *not* a series of independent events.[1]

With this background let's again look at a simple example problem given back in Chapter 6 on page 286. In this example the exploration area of interest contained 10 seismic anomalies of which 30% were estimated to have oil. The question of interest: "What is the probability of finding two discoveries in a 5 well exploration program?" The answer was given as 0.417 and was computed by using the two-outcome hypergeometric probability equation 6.14. But let's see if we could determine this by logic instead of the equation.

Suppose the first two anomalies tested were oil productive and the next three were dry. Using the symbols P = productive and D = dry, the sequence would be P, P, D, D, D. What is the probability of this sequence occurring? The likelihood of the first anomaly selected being oil productive, all other factors equal, is 3/10. The conditional probability of the second anomaly being oil, given the first was oil is 2/9. (Remember our original analogy of the 52 cards!)

The conditional probability of the third being dry is 7/8. (The eight undrilled anomalies include all seven of the dry anomalies originally hypothesized to have been the state of nature.) The conditional probability of the fourth anomaly being dry, given that two oil anomalies had been found plus one of the dry anomalies is 6/7. Finally, the conditional probability of the fifth anomaly being dry, given that two of the four already tested had been found to be dry is 5/6. The likelihood that this entire sequence of P, P, D, D, D could occur is the product of these five probability terms. That is, the probability of two discoveries followed by three failures is $3/10 \times 2/9 \times 7/8 \times 6/7 \times 5/6 = 3/72$.

Is this the answer to our original question? Definitely not! This is

---

1. We stress this point because in most references on multiwell drilling programs the probability calculations are based on the binomial probability equation—a model which describes independent events. It's much easier to use for one thing, and perhaps some of the authors are unaware of it's assumptions. For whatever the reasons these analyses are not realistic because they fail to recognize the "sampling without replacement" nature of the real-world.

the probability of one particular way to obtain 2 discoveries in 5 wells, but there are other sequences which also could occur. For example PDPDD, or PDDPD, etc. So now what we must do is enumerate the mutually exclusive ways in which the sequence of five trials can include two discoveries, and compute the probability of each particular sequence occurring. The final step is to sum the probabilities of the mutually exclusive ways we could get two discoveries in five trials, and this sum will be the answer we are searching for.

For this example there are 10 mutually exclusive ways of achieving two discoveries in five trials, as listed in Column 1 below. The probabilities of each sequence, based on the condition of sampling without replacement — dependent events — are shown in Column 2.

Mutually Exclusive Sequences (P = anomaly productive D = anomaly dry)	Probability of Sequence Occurring, Given That of the Initial N = 10 Anomalies, 3 Are Hypothesized to Have Oil and 7 Are Dry*
1. PPDDD	$3/10 \times 2/9 \times 7/8 \times 6/7 \times 5/6 = 3/72$
2. PDPDD	$3/10 \times 7/9 \times 2/8 \times 6/7 \times 5/6 = 3/72$
3. PDDPD	$3/10 \times 7/9 \times 6/8 \times 2/7 \times 5/6 = 3/72$
4. PDDDP	$3/10 \times 7/9 \times 6/8 \times 5/7 \times 2/6 = 3/72$
5. DPPDD	$7/10 \times 3/9 \times 2/8 \times 6/7 \times 5/6 = 3/72$
6. DPDPD	$7/10 \times 3/9 \times 6/8 \times 2/7 \times 5/6 = 3/72$
7. DPDDP	$7/10 \times 3/9 \times 6/8 \times 5/7 \times 2/6 = 3/72$
8. DDPPD	$7/10 \times 6/9 \times 3/8 \times 2/7 \times 5/6 = 3/72$
9. DDPDP	$7/10 \times 6/9 \times 3/8 \times 5/7 \times 2/6 = 3/72$
10. DDDPP	$7/10 \times 6/9 \times 5/8 \times 3/7 \times 2/6 = \underline{3/72}$
	$\Sigma = 30/72 = \underline{\underline{0.417}}$

* While the probabilities are changed after each trial to reflect the dependency relationship on the next trial, each of the remaining undrilled prospects is assumed equally likely to be selected. This scheme is called "equiprobable sampling without replacement."

We see by this series of computations the probability for two discoveries in five wells drilled is $30/72 = 0.417$, exactly the same value we obtained back in Chapter 6. In general we don't need to make such a detailed listing because to answer questions of the type being asked we can solve the hypergeometric probability equation (Eq. 6.15) directly. The above listing is given to show what the equation is doing for us.

You should note two other things about this example question. The exact analog with cards (and continuing our assumption that a diamond equals oil, any other suit equals a dry hole) is to take 10 cards (there were 10 seismic anomalies) of which 3 are diamonds and the other 7 cards are of any other suit. The 10 cards are shuffled thoroughly and laid face down on a table. We then select five of the cards and turn them face up. We're asking then, the likelihood of two of these five overturned cards being diamonds.

And you should note also that the probabilities of the next anomaly being productive or dry in the 10 sequences is always changing. The chances of oil may go up or go down — but they are always changing as a result of the sampling without replacement mechanism. If this were a series of independent events the probabilities for sequence 4, for example, (PDDDP) would be $3/10 \times 7/10 \times 7/10 \times 7/10 \times 3/10$. This is clearly a different value than the (correct) value of $3/10 \times 7/9 \times 6/8 \times 5/7 \times 2/6$.

All of this is to say that for a sequence of dependent events — sampling without replacement — we must use the hypergeometric probability model, equation 6.15, to compute probabilities of outcomes of multiple trials. For convenience the equation is listed again:

$$\begin{array}{l}\text{Probability of } x_1, x_2, \\ x_3, \ldots, x_r \text{ outcomes} \\ \text{in a sample of n trials}\end{array} = \frac{\left(C_{x_1}^{d_1}\right)\left(C_{x_2}^{d_2}\right)\left(C_{x_3}^{d_3}\right) \ldots \left(C_{x_r}^{d_r}\right)}{\left(C_n^N\right)} \tag{6.15}$$

where

n = number of trials (wells) in the multiple-well program.

N = Total number of undrilled prospects in the basin or exploration area before the n wells are drilled.

r = number of possible outcomes on any trial.

$d_1, d_2, d_3, \ldots, d_r$ = number of elements in the sample space designated as each outcome before the n wells are drilled ($d_1 + d_2 + d_3 + \ldots + d_r = N$).

$x_1, x_2, x_3, \ldots, x_r$ = number of outcomes which occur in each of the r categories, such that ($x_1 + x_2 + x_3 + \ldots + x_r = n$).

The numerical equivalents for our example problem are as follows:

$n = 5$, the number of wells drilled (one on each of five seismic anomalies).

$N = 10$, the number of undrilled prospects initially in the exploration area.

$d_1 = 3$ ⎫ The number of elements in the original sample space, where
$d_2 = 7$ ⎬ subscript 1 corresponds to the outcome oil and subscript 2 corresponds to the category of dry.

$x_1 = 2$ ⎫ The actual number of outcomes in the category of oil and dry
$x_2 = 3$ ⎬ anomalies (Again the subscript 1 = oil, 2 = dry).

Substituting directly we obtain

Probability of $x_1 = 2$ discoveries
and $x_2 = 3$ failures in an $n = 5$ =
well program

$$\frac{\left(C_2^3\right)\left(C_3^7\right)}{\left(C_5^{10}\right)} = \frac{\left(\frac{3!}{2!\,1!}\right)\left(\frac{7!}{3!\,4!}\right)}{\left(\frac{10!}{5!\,5!}\right)} = \underline{0.417}$$

Let's look at another example. We have conducted some reconnaissance seismic surveys in a new offshore area and have identified 20 anomalies of sufficient size to be considered viable exploration prospects. By comparison to a nearby, geologically similar area we estimate that 70% of the anomalies will be dry, 15% will have reserves less than 100 million barrels, 10% will have reserves from 100 million barrels to 500 million barrels, and only 5% of the anomalies will have reserves of 500 million to 1000 million barrels.

If we drill six of the anomalies what is the probability that a.) all six will be dry, b.) we will only find one large field and two small fields, c.) we find 3 fields of reserves less than 100 million barrels, and one field with reserves in the range of 100–500 million barrels, and one field with reserves in the 500–1000 million barrel range?

First we can define N as being 20 and $n = 6$, the number of wells being drilled in the multiwell exploration program. We can also assign the subscript categories and compute the "d" terms as follows:

Subscript	Corresponding Outcome Category	"d" Terms, the Number of Elements Originally in Each Category
1	Anomaly is dry	$20 \times 0.70 = 14 \leftarrow d_1$
2	Anomaly has reserves of < 100 MM bbls	$20 \times 0.15 = 3 \leftarrow d_2$
3	Anomaly has reserves in the range of 100–500 MM bbls	$20 \times 0.10 = 2 \leftarrow d_3$
4	Anomaly has reserves in the range of 500–1000 MM bbls	$20 \times 0.05 = 1 \leftarrow d_4$

Note that in Column 3 to obtain the "d" terms we multiplied the total number of prospects in the area (N = 20) by the proportion of the total estimated to be of each category. We obtained these proportions by analogy to the nearby area. From the nearby area we estimated there was a 5% chance of reserves in the range of 500–1000 MM barrels (as being, for example the right tail of a reserve distribution), but this relative fraction or percentage is not enough. We must also be able to say 5% of how many total prospects will have reserves of this amount. Hence the N = 20 is the vital proportionality constant we mentioned earlier on page 323 with regard to ultimate reserve distributions.

     Finally, to answer the specific questions:

     a. *The probability all six will be dry.*
        For this case $x_1 = 6$, $x_2 = 0$, $x_3 = 0$, $x_4 = 0$
        Solving Equation 6.15 gives

        Probability of

$$\left.\begin{array}{l} x_1 = 6 \text{ dry} \\ x_2 = 0 \text{ discoveries in} \\ \qquad < 100 \text{ MM} \\ \qquad \text{range} \\ x_3 = 0 \text{ discoveries in} \\ \qquad 100\text{–}500 \\ \qquad \text{MM range} \\ x_4 = 0 \text{ discoveries in} \\ \qquad 500\text{–}1000 \\ \qquad \text{MM range} \end{array}\right\} = \frac{\left(C_6^{14}\right)\left(C_0^3\right)\left(C_0^2\right)\left(C_0^1\right)}{\left(C_6^{20}\right)} = \underline{\underline{0.0775}}$$

     in n = 6 wells drilled   (rounded to 4 digits)

     [Note: Recall that $\left(C_x^n\right)$ is defined as $\left[\dfrac{n!}{x!(n-x)!}\right]$, where n! is the product of all the integers up to and including n, and 0! is defined as 1.]

     b. *This question asked the probability of 1 large field ($x_4 = 1$), 2 small fields ($x_2 = 2$) and the rest dry (i.e. $x_1 = 3$, $x_3 = 0$). So we solve the same equation but with different values of the "x" terms.*

Probability of

$$
\left.
\begin{array}{l}
x_1 = 3 \text{ dry} \\
x_2 = 2 \text{ discoveries in} \\
\qquad < 100 \text{ MM} \\
\qquad \text{range} \\
x_3 = 0 \text{ discoveries in} \\
\qquad 100{-}500 \\
\qquad \text{MM range} \\
x_4 = 1 \text{ discovery in} \\
\qquad \text{the } 500{-} \\
\qquad 1000 \text{ MM} \\
\qquad \text{range}
\end{array}
\right\}
= \frac{\left(C_3^{14}\right)\left(C_2^3\right)\left(C_0^2\right)\left(C_1^1\right)}{\left(C_6^{20}\right)} = \underline{0.0282}
$$

in $n = 6$ wells drilled

Finally, question c.) concerned the probability that the six wells drilled would find 3 fields with reserves < 100 MM barrels (i.e. $x_2 = 3$), one field with reserves in the 100–500 MM barrel range ($x_3 = 1$), and one field with reserves in the 500–1000 MM barrel range ($x_4 = 1$). This implies of course, that one of the six anomalies would have had to have been dry ($x_1 = 1$). Again solving Equation 6.15 gives

Probability of

$$
\left.
\begin{array}{l}
x_1 = 1 \text{ dry} \\
x_2 = 3 \text{ discoveries in} \\
\qquad \text{the } < 100 \\
\qquad \text{MM range} \\
x_3 = 1 \text{ discovery in} \\
\qquad \text{the } 100{-} \\
\qquad 500 \text{ MM} \\
\qquad \text{range} \\
x_4 = 1 \text{ discovery in} \\
\qquad \text{the } 500{-} \\
\qquad 1000 \text{ MM} \\
\qquad \text{range}
\end{array}
\right\}
= \frac{\left(C_1^{14}\right)\left(C_3^3\right)\left(C_1^2\right)\left(C_1^1\right)}{\left(C_6^{20}\right)} = \underline{0.0007}
$$

in 6 wells drilled

These two examples should give you an idea of how the analysis method based on the condition of dependent events can be applied to multiwell drilling programs. Equation 6.15 is tedious to solve, but it is really the only alternative if we want the probability analysis to be consistent with the "sampling without replacement" nature of the drilling of a series of exploration wells.

As a comment here I would mention that to make this analog work we must know how many of the elements of the sample space existed in each category before the n wells were drilled. That is to say, we have to be able to estimate or compute the d_1, d_2, d_3, d_4, . . . , etc. terms. We were able to do this in the example because we could multiply the proportions in each reserve range (as observed from the nearby area) by N, the total number of prospects in the area.

In general you will find when applying this to the real-world the estimation of N may be a rather sticky problem. In an offshore, structural environment we would probably run seismic to determine the number of structural prospects (if any) that exist in our license or operating area. In this case N will probably be defined (by default) as being the number of anomalies that are large enough to be worth testing. In such cases N would be more appropriately defined as the number of undrilled prospects of interest and not the total number of prospects in the basin.

These same comments would also apply to onshore structural areas with the exception that since our exploration and development costs are lower we could probably afford to look for smaller structures. This probably would result in N being larger. In a basin or area where the trapping mechanism is purely stratigraphic we probably won't be able to enumerate N from seismic and/or geologic control. Rather, we may have to estimate or extrapolate a value from an analogous drilled area. Such an extrapolation might be in the form of a number of prospects per 100 square miles, or a number of prospects per cubic mile of sedimentary strata.

These ratios could then be multiplied by the corresponding area or volume of the new basin to get at least an order of magnitude estimate of N. In these types of plays, however, N is a difficult number to quantify, so the analyst needs to make a sensitivity analysis using varying values of N to see its effect on our overall investment strategies.

At the beginning of this section we posed three questions about the hypothesis that the drilling of a sequence of wells is a series of dependent events. Hopefully the card experiment we discussed, together with the two numerical examples illustrate the notion of dependent events. How can we tell if a real-world series of wells are dependent or independent?

The answer is they will always be dependent except for the possible instances listed in the right portion of Table 7.2.

The general rule is that a sequence of wildcats is a series of dependent events. The choice we have is not whether they are dependent or independent, but whether we choose (or are able) to use the more complicated hypergeometric probability model of Equation 6.15. The third question asked whether it matters which system we use.

You can draw your own conclusions on this. The numerical probabilities computed by each system will be different, particularly if N is of the same order of magnitude size as n, the number of wells being drilled. As N becomes very large (relative to n) the differences become minimal, and the results of each method will be about equal. If we wish to base our decisions on outcome probabilities that are representative of the real-world sequence of events we must choose the dependent event/sampling without replacement method.

In some of my short-course discussions on this issue I have been asked on several occasions the following question.

"Paul, I think what you have just tried to explain is a bunch of poppy-cock. You've been trying to tell us that probabilities change after each wildcat is drilled, and that the odds for discovery are dependent on the results of having previously drilled other structures in the area. So here I am with a structure—you mean to tell me the probability of finding oil in my structure is dependent on whether the next nearest structure 50 miles away had oil? That's nonsense.

"There is either oil or no oil in my structure anomaly, but whether there's oil in the nearby structures is immaterial. The odds for oil in my structure have no relationship to any of those other structures that have been drilled. So everything you've been trying to tell us about dependent events is nonsense!"

How do you react to that viewpoint? Do you think his point is well taken?

My reply to this question is that he has failed to catch the important point as to the definition of sample space. His viewpoint implies that the sample space of interest to him is the set of outcomes which could occur from the drilling of his particular, single structure. But the sample space we've been concerning ourselves with here is the set of outcomes which could occur from the drilling of n wells.

We're talking about two entirely different probabilistic phenomena. It's similar to trying to compare the probability of a win on the next spin of a roulette wheel with the probability of at least one win in n spins of the wheel. They aren't even the same questions, as we pointed out earlier in Chapter 6 on page 222.

I mention this point because I have a feeling that many explorationists, when first exposed to the issue of dependent events versus independent events, try to understand the concepts by a line of thought similar to the above question. And the point here is that such a line of thought can lead one astray because the logic relates to a different sample space of outcomes. The best way to try to grasp the issue, in my judgment, is to go back to the playing card analog we mentioned earlier in this section.

B. *Analyses Based on the Condition of Independent Events*

While we have just tried to demonstrate that the realities of the drilling of a sequence of exploration wells is one of dependent events, there are several special instances where the parallel system of independent events applies. These were listed in Table 7.2 and include the case where each exploration well in the n well program is in a different basin, and the case where the stratigraphy varies so rapidly (on an areal basis) that each well drilled is nearly a separate reservoir. These are admittedly, rather special cases.

Perhaps a more useful situation for considering the series of wells as being independent events is when N is much larger than n — specifically, when N is at least ten times larger than n. In these instances the results obtained from an independent event analog will be approximately equal to those we would have obtained using the hypergeometric distribution. The motive for choosing the approximation is that, in general, the independent event equations are much easier to work with from a computational standpoint.

The mathematical analogs or models of interest here are those based on the condition that a series of events are statistically independent of one another. The probability models are the multinomial distribution and a special case of the multinomial for two outcomes called the binomial distribution. The binomial probability model is widely used in statistics and quality control, and of the probability methods for analyzing multiple-well drilling programs (hypergeometric, multinomial, binomial) it is by far the simplest and easiest to use. So let's take the easy one first and discuss the binomial probability model as it relates to multiwell drilling programs.

Explanation of the binomial probability equation (6.11) was given in detail back in Chapter 6 so we will just repeat the equation here for convenience.

$$\begin{array}{l}\text{Probability of x} \\ \text{successes in n} \\ \text{independent trials}\end{array} = \left(C_x^n\right)(p)^x(1-p)^{n-x} \qquad (6.11)$$

where

n = number of independent trials.

x = number of successes in the n trials ($o \le x \le n$).

p = probability of success on any given trial ($0 \le p \le 1.0$).

$\left(C_x^n\right)$ = combination of n things taken x at a time = $\dfrac{n!}{x!(n-x)!}$.

Inasmuch as the binomial equation is a two-outcome model we can only use the model to distinguish between success or failure, discovery or dry hole. We cannot include in the model various levels of reserves (or levels of profitability) given that a discovery is made. The other specific qualifications of the model are that each trial is an independent event (i.e. based on a sampling with replacement process), and the probability of a success or failure remain constant over time, no matter how many times the process is repeated. Of course the really severe restriction to the oil business is the notion of independent events. But none the less it can be useful to gain insights in certain types of scenarios in petroleum exploration. One such application might be as follows.

EXAMPLE: Our exploration staff is evaluating a multiwell drilling program in a relatively new play in a basin in which stratigraphic traps are the primary exploration objective. The basin is large and it is thought that the number of possible traps (prospects) in the basin, N, is large—perhaps on the order of 200 or more. By some very preliminary correlations with an analogous area the staff estimates the chance of finding reserves at 0.12.

Our engineers have advised that if we plan to drill 10 exploration wells we would have to make at least two discoveries to yield a minimum acceptable return on the 10 well drilling program. What is the probability of achieving this minimum return? What is the probability of no discoveries in 10 wells? Of only one discovery in 10 wells?

SOLUTION: Since N was thought to exceed 200 prospects in the basin and we are considering an n = 10 well drilling program it is clear that N ≥ 10n; hence we can use the independent event model as an approximation to the actual sampling without replacement process. The other parameters required to solve this problem are the probability of success (discovery), $p = 0.12$, and x, the number of successes of interest. For the first question we are interested in knowing the probability of at least $x = 2$ successes in the 10 trials, given that $p = 0.12$.

We can read the answer to this question from the tables of cumulative binomial probabilities in Appendix F (page 619).

Binomial Probability of $x \geq 2$ successes in $n = 10$ trials, $p = 0.12 = \underline{0.3417}$

The equation we would have to solve for this question, in lieu of using Appendix F is:

Binomial Probability of $x \geq 2$ successes in $n = 10$ trials, $p = 0.12$

$$= \sum_{x=2}^{x=10} \left(C_x^n \right)(p)^x(1 - p)^{n-x}$$

There is a lot of arithmetic involved in solving this equation, however, and most find the tables of Appendix F much more convenient for working with binomial probabilities.

For the second question we are interested in determining the probability of no discoveries, $x = 0$, in the ten wells. We can read this also from Appendix F (on page 619) as:

Binomial Probability of $x = 0$ successes in $n = 10$ trials, $p = 0.12 = \underline{0.2785}$

And again, if you prefer you can solve equation 6.11 directly rather than use Appendix F. The parameters you would use in the equation for this second question are $n = 10$, $x = 0$, and $p = 0.12$.

Finally, the probability of only one discovery in 10 wells would be the solution of equation 6.11 for the parameters $n = 10$, $x = 1$, and $p = 0.12$. Or, we could read this value from the convenient tables of Appendix F as:

Binomial Probability of $x = 1$ success in $n = 10$ trials, $p = 0.12 = \underline{0.3798}$

To summarize this example problem we have about a 28% chance of no discoveries in the 10 well program, a 38% chance of only one discovery, and a 34% chance of two or more discoveries required to meet the minimum profit objectives.

As an alternative question suppose the manager, after observing there was only a 34% chance of at least two discoveries in 10 wells, asks the following question. How many wells would he have to drill to have at least a 70% chance of at least two discoveries?

In this question the unknown is n, the number of wells required to have a 70% chance or more of at least two discoveries, where $p = 0.12$. We can use the binomial model for questions of this type, also. All that is required is to determine binomial probabilities for successively higher values of n until the chance of at least two discoveries exceeds 70%. We merely enter the tables of Appendix F a number of times as shown in the following summary tabulation:

n, the Total Number of Wells Drilled	Binomial Probability of at Least Two Successes ($x \geq 2$) in the n Trials, Given That $p = 0.12$ (From Appendix F)
n = 10	0.3417 ← Calculated previously
n = 14	0.5141
n = 16	0.5885
n = 18	0.6540
n = 19	0.6835
n = 20	0.7109

From this analysis we can conclude that he would have to drill n = 20 wells or more to have at least a 70% chance of two or more discoveries. Analyses of this type are useful to get an order of magnitude estimate of the number of exploratory wells that would be required to yield a specified probability of achieving an objective of a given number of discoveries.

This basic logic can be extended to another possible real-world application. Suppose we are drilling in a basin in which N, the total number of prospects in the basin, is thought to be very large, perhaps greater than 200 or so. The objective strata are gas productive, and we have been negotiating a drilling program in which we must find enough gas reserves to produce a specified daily rate during the first two years of the contract.

Our engineers have estimated we'd need at least three gas discoveries to meet the minimum contract producing rates. We do not know what the chance of finding gas is but we estimate it to be in the range of $p = 0.20$ to $p = 0.40$. Our concern is to try to determine an order of magnitude estimate of the number of wells we'll have to drill (and hence the amount of exploration capital required) to have a reasonably good chance (say 80 or 90%) of at least three discoveries.

We can gain useful insights to this sort of problem with the binomial probability model in much the same manner as our previous example. The steps involved include estimating a value of p (say for example, $p = 0.20$) and entering the cumulative binomial tables of Appendix F to read the probability of at least three successes ($x \geq 3$) for various values of n, the total number of wells. These data are recorded and used in the plotting of a useful graph to summarize the data. Then the value of p is changed and the repetitive entries in Appendix F are again made. This is repeated for various values of p. Results of such a series of steps are listed in the following table. These data are plotted in Figure 7.5 to give a useful summary of the data obtained from the binomial probability model.

CASE 1		CASE 2		CASE 3	
n, the Number of Trials	Binomial Probability of x ≥ 3, Given That p = 0.20 (Appendix F)	n, the Number of Trials	Binomial Probability of x ≥ 3, Given That p = 0.30 (Appendix F)	n, the Number of Trials	Binomial Probability of x ≥ 3, Given That p = 0.40 (Appendix F)
5	0.0579	5	0.1631	4	0.1792
8	0.2031	7	0.3529	6	0.4557
10	0.3222	9	0.5372	8	0.6846
14	0.5519	11	0.6873	10	0.8327
16	0.6482	13	0.7975	12	0.9166
20	0.7939	15	0.8732	14	0.9602
25	0.9018	17	0.9226	16	0.9817
		19	0.9538		

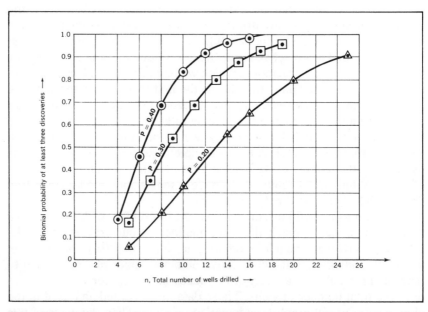

Figure 7.5 *Graph of the binomial probability of at least three discoveries (x ≥ 3) as a function of n, the total number of wells drilled for various values of p, the probability of success on any trial.*

The interpretation of Figure 7.5 would be as follows. Suppose the decision maker wishes to have at least an 85% chance of three discoveries. If the probability of finding gas is as high as 0.40 he would have to drill at least 10 exploration wells. If the probability of gas is as low as 0.20 the minimum number of wells required would be 22. These values are read from Figure 7.5 by entering on the vertical scale at a value of 0.85, proceeding horizontally to the right to the appropriate curves of p = 0.40 and 0.20, then vertically downward to the n scale. So if he wants at least an 85% chance of 3 or more gas discoveries a minimum of 10 wells would be required under the best condition (highest value of p) and a minimum of 22 wells under the worst condition (lowest value of p). Or, alternately, he could use the graph in the reverse manner. Suppose he can arrange financing for the drilling of only 15 wells. What are the chances of achieving his objective of at least three discoveries?

To determine this we enter the graph on the horizontal axis at a value of n = 15, then proceed upward to the various values of p, and read the corresponding probabilities of at least 3 discoveries from the vertical axis. For this case, if he drills 15 wells he has a 60% chance of at least 3 discoveries if p = 0.20, an 87% chance of at least 3 discoveries if p = 0.30, and a 98% chance of at least 3 discoveries if p is as high as 0.40.

Does a graph such as Figure 7.5 tell us what the true value of p is? Certainly not. Again the statistical concepts of decision analysis can't tell us geological information such as the chance of finding gas in this particular basin. We must specify what p is, or the range of possible values of p. Having specified this we can then use decision analysis, statistics, and graphs such as Figure 7.5 to help us devise feasible investment strategies.

These examples give ideas about the use of binomial probabilities in the analysis of multiwell drilling programs. Be sure and remember, however, that the binomial model is based on the notion of independent events, and is thus *not* an exact analog of the drilling of a sequence of exploration wells. The reason we were able to use it in these examples is that in both cases it was felt that the total number of undrilled prospects, N, was large relative to n, the number of wells being considered. And for values of N at least ten times greater than n the independent event model (binomial) will yield probabilities approximately equivalent to those we would have obtained using the sampling without replacement model (hypergeometric).

The other obvious shortcoming of the binomial model is that it can only account for two possible outcomes. This shortcoming can be sidestepped by considering the more general case of the independent event model — the multinomial probability distribution.

For convenience, the multinomial probability equation is given again at this point.

Probability of
$x_1, x_2, x_3, \ldots, x_r$
outcomes in n $\qquad = \dfrac{n!}{x_1! \, x_2! \, x_3! \, \ldots \, x_r!}(p_1)^{x_1}(p_2)^{x_2}(p_3)^{x_3} \ldots (p_r)^{x_r}$
independent trials

$$(6.13)$$

where

n = total number of independent trials.
r = the number of possible outcomes.
$x_1, x_2, x_3, \ldots x_r$ = the number of times the outcomes occur in n trials, where $(x_1 + x_2 + x_3 + \ldots + x_r = n)$.
$p_1, p_2, p_3, \ldots p_r$ = probability of occurrence of the outcomes on any given independent trial, with the condition that $(p_1 + p_2 + p_3 + \ldots + p_r = 1.0)$.

The multinomial probability model is based on the condition that each of the n trials is an independent event and the probabilities of occurrence of each of the r possible outcomes remain constant with time. It is based on a sampling *with* replacement scheme, so this model is *not* an exact analog of the sampling without replacement process characteristic of the drilling of a sequence of exploration wells.

But again if N is very large relative to the number of wells being considered the results of the multinomial model will be approximately equivalent to the probabilities obtained from the hypergeometric probability model of Equation 6.15. The advantage of the multinomial (over the hypergeometric) in these instances is that it may be slightly easier to handle from a computational standpoint.

To cite an example of how this distribution could be used suppose we are considering an exploration program in an extension area of a basin whose trapping mechanism is stratigraphic as defined by extensive drilling. The new extension area is very large (in terms of area) but has had little exploration because it is substantially deeper (and more costly to explore) than the older drilled area. But recent increased crude prices and shortages have stimulated a new interest in the deeper extension area.

The number of prospects in the new area, N, is not known, but if we extrapolate an equivalent density of prospects per unit of area to the new area from the older area N would be on the order of 150–170. We are considering an 8 well exploration program on some recently purchased acreage. Since N = 150–170 is substantially larger than n = 8 wells we can use the multinomial distribution as an approximation to the actual sampling without replacement process. The likelihood of discovery in the new area is estimated to be about 12%.

To get an idea of the various levels of possible reserves per field our staff has tabulated reserves from the old, densely explored portion of the basin, with the following results:

Reserve Range,* Bbls/Field	Number of Fields with Reserves in the Range	Proportion of the Fields with Reserves in the Range	Probability of Encountering Reserve Range in New Extension Area, Given That the Chance of Finding Oil Is 0.12
Small	35	35/50 = 0.70	0.70 × 0.12 = 0.084
Medium	10	10/50 = 0.20	0.20 × 0.12 = 0.024
Large	5	5/50 = 0.10	0.10 × 0.12 = 0.012
	50	1.00	

* To simplify the arithmetic, the reserve range was divided into just three ranges. We could have subdivided the range of reserves per field into a larger number of intervals if we so desired. The ranges could be defined numerically also, rather than with adjectives small, medium, etc.

This tabulation is, in essence, the result of dividing a distribution of reserves per field into three intervals, or ranges and determining the conditional probabilities of each range by reading areas under the distribution curve. These conditional probabilities are the data of Column 3, and are conditional upon finding oil in the first place. In Column 4 we then multiply the conditional probabilities by the chance of finding oil, $p = 0.12$, so as to obtain an unconditional probability of encountering reserves of various amounts in the new extension area. With this data we wish to determine answers to the following questions:

Assuming we believe the information from the older area is representative of what we could expect to find in the new extension area, what is the probability that

a. all eight of our wells will be dry?
b. the only reserves we find are two small fields?
c. drilling of the eight wells results in the discovery of one large field, one medium-sized field, and two small fields?

We can answer these questions using the multinomial probability equation (6.13). If we assign the subscripts on the x and p terms as $1 = $ dry hole, $2 = $ small reserves, $3 = $ field of medium-sized reserves, and $4 = $

field of large reserves, we can tabulate the parameters we need to solve the equation for question a:

$$n = 8$$
$$p_1 = \text{Probability of dry hole} = 0.880$$
$$p_2 = \text{Probability of small reserves} = 0.084$$
$$p_3 = \text{Probability of medium reserves} = 0.024$$
$$p_4 = \text{Probability of large reserves} = 0.012$$
$$x_1 = 8 \text{ dry prospects}$$
$$x_2 = x_3 = x_4 = 0 \text{ prospects having any oil}$$

We now substitute directly into equation 6.13 to compute the probability that none of the eight wells encounter oil.

a. Probability of

$$\left.\begin{cases} x_1 = 8 \\ x_2 = 0 \\ x_3 = 0 \\ x_4 = 0 \end{cases}\right\} \text{outcomes} = \frac{8!}{8!\ 0!\ 0!\ 0!}(0.88)^8(0.084)^0(0.024)^0(0.012)^0$$

in n = 8 trials   = 0.3596

For question b. the only terms that change are the x terms. In this case $x_2 = 2$, $x_3 = 0$, $x_4 = 0$, and x_1, the number of dry holes = 6. Again we substitute directly into Equation 6.13:

b. Probability of

$$\left.\begin{cases} x_1 = 6 \\ x_2 = 2 \\ x_3 = 0 \\ x_4 = 0 \end{cases}\right\} \text{outcomes} = \frac{8!}{6!\ 2!\ 0!\ 0!}(0.88)^6(0.084)^2(0.024)^0(0.012)^0$$

in n = 8 trials   = 0.0918

For question c. the x terms change to $x_4 = 1$, $x_3 = 1$, $x_2 = 2$, and $x_1 = 4$.

c. Probability of

$$\left.\begin{cases} x_1 = 4 \\ x_2 = 2 \\ x_3 = 1 \\ x_4 = 1 \end{cases}\right\} \text{outcomes} = \frac{8!}{4!\ 2!\ 1!\ 1!}(0.88)^4(0.084)^2(0.024)^1(0.012)^1$$

in n = 8 trials   = 0.0010

Remember that the only reason we used the multinomial distribution in this example (rather than the more realistic hypergeometric) is that our staff estimated N was large as compared to the value of n = 8. For this condition the multinomial yields about the same results as if we had used the hypergeometric equation, and is slightly easier to compute. We must also remember that the multinomial distribution is an independent event model and the probabilities of each outcome (the p terms) remain constant with time.

C. Multiwell Drilling Program Analyses Which Include Various Levels of Reserves

In 1961 an early paper on multiwell exploration risk analysis was presented (Reference 7.5) which suggested a way of combining the multiwell analysis methods we've just discussed with the probabilities that any discoveries we make will yield at least a minimum specified level of profitability.

To use this method the analyst specifies the number of exploratory wells being considered, their costs, any initial seismic costs or lease bonuses which must be recouped, the minimum level of profitability desired, and the distribution of reserves describing the possible range of reserves per field that could occur, given a discovery.

Having specified these parameters the method can be used to determine the probability that results of the multiwell drilling program will yield at least the minimum level of profitability. As such it can be a useful analysis technique to evaluate and compare alternative multiple-well exploration programs.

The analysis method itself has a series of preliminary calculations that are required to yield the numerical data for the final computation. It has been my experience in presenting this method to explorationists that many tend to get lost in these preliminary calculations and lose sight of how the steps all fit together. To try to reduce the chances of this happening I've prepared a pictorial diagram, or map, of the entire analysis method in Figure 7.6. So before we discuss a numerical application to show how this method is used we should probably study the diagram of Figure 7.6 in a bit more detail.

To use this method we specify the number of exploration wells being considered in the multiple-well program (n), the drilling costs per well C, the amount of net present value dollars (NPV_1) we wish to recover in addition to earning a rate of return equal to i_0, the overall, discounted

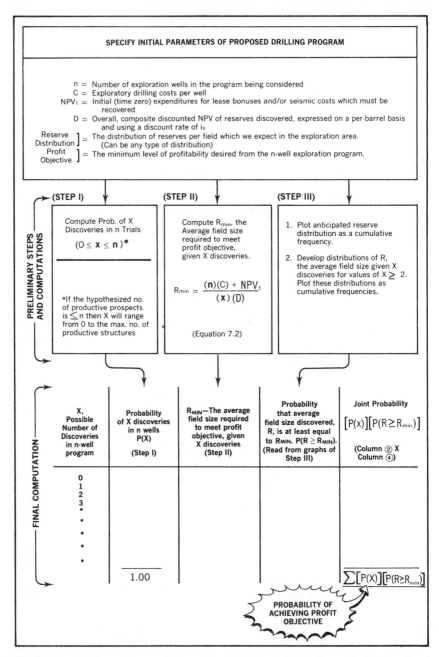

Figure 7.6 *Outline of the analysis method which combines the probabilities of outcomes of multiwell exploration program with the probabilities that reserves discovered will be of sufficient magnitude to meet a stated profit objective.*

value of reserves expressed on a per-barrel basis, and the distribution of possible values of reserves per field that we anticipate in the exploration area being considered.

The profit objective is expressed as a positive net present value sufficient to recover the exploratory drilling costs, initial expenditures for leases and seismic (NPV$_1$), plus earning a rate of return equal to i$_0$, the discount rate used in computing D. The method will then yield the probability of achieving the stated profit objective with the n-well program. This probability is the sum of the joint probabilities of Column 5 of the final computation shown in Figure 7.6. That is, sum of Column 5 in the lower right hand corner of Figure 7.6 is the probability term we are seeking to determine by this method.

Regarding the initial parameters, the n and C terms are obvious and should require no further explanation. NPV$_1$ is the amount of net present value dollars *in excess* of a rate of return equal to i$_0$ that we wish to recover from the anticipated discoveries of the n-well exploration program. It can be set at any level and can include initial exploration expenses of lease bonuses, and seismic programs. If these expenses do not bear on the program being considered, NPV$_1$ can be set equal to zero.

The D term, the overall discounted net present value of reserves is computed by a separate engineering calculation of the type explained in Appendix I. The parameter is expressed on a net present value per barrel basis using a discounting rate of i$_0$. It is, essentially, a conversion factor to get from gross reserves discovered to the net present value of the reserves after consideration of development well costs, operating expenses, taxes, and royalties.

It is computed by projecting an average or typical production decline schedule over the life of the field and finding the net present value of the entire series of cash flows using anticipated crude prices. This total net present value is then divided by the total oil produced by the projected schedule to yield the net present value of the revenues on an overall, composite per barrel basis.

The aspect of this method which makes it unique with respect to the previous probability models we've discussed in this section on multiple-well programs is specification of the anticipated reserve distribution. The explorationist can use any type of distribution he feels applicable (log-normal or otherwise!), and the distribution can have any range of values.

The random variable of this distribution is reserves per field, or, if you prefer, reserves per prospect. The method can be used in an oil province or a gas province. And if there exists considerable uncertainty as to the shape and range (limits) of the reserve distribution the analyst will certainly want to make the analysis using several possible distributions to ascertain the sensitivity of various feasible reserve distributions on the overall exploration strategy.

The preliminary, or intermediate parts of the method consist of three separate "data preparation" steps, as shown in the middle of Figure 7.6. Step I is simply a computation of the probabilities of various numbers of discoveries in an n-well drilling program using the techniques previously discussed in this section on multiple-well programs.

These methods were summarized in Table 7.2, and consist of the hypergeometric, binomial, and multinomial models depending on whether the series of wells is considered to be one of dependent events or independent events. For reasons outlined previously the real-world analog will usually be the "dependent-event" model (hypergeometric). There may be some occasions, however, when the independent event model may be an adequate approximation, in which case the computations may be somewhat simpler.

Step II introduces into the analysis the concept of how large the reserves must be to meet the profit objective if we make $x = 1, 2, 3, \ldots$ discoveries in the n-well program. The relationship which ties these two dimensions of the program together is Equation 7.2:

$$R_{min} = \frac{(n)(C) + NPV_1}{(x)(D)} \qquad (7.2)$$

where R_{min} is defined as the minimum average field size required to meet the profit objective, given x discoveries. The other terms are as defined in Figure 7.6.

The numerator term of Equation 7.2 represents the (time zero) exploratory drilling costs of the multiple-well program (the number of wells multiplied by the cost per well) plus the additional net present value dollars desired in addition to a rate of return of i_0. The numerator sum thus has units of dollars. Dividing this sum by D, the net present value of reserves per barrel, yields the number of barrels which must be found to meet the stated profit objective. For example, suppose we were planning to drill ten wells at a cost of $400,000 each, we had spent $1 million for a signature bonus, and the estimated value of discovered oil is $1.75 per barrel, discounted at $i_0 = 15\%$.

The numerator terms would thus be $[(10)(\$400,000) + \$1,000,000] = \$5,000,000$. Dividing by $1.75 per barrel gives 2.86 million barrels. This means, then, that if we make $x = 1$ discovery it must be at least 2.86 million barrels for us to recover our $4,000,000 drilling costs, our $1,000,-000 bonus, and still make a minimum rate of return of $i_0 = 15\%$, the discount rate used to compute D.

Notice that Equation 7.2 also includes an x term in the denominator. This is the term which makes R_{min} conditional on the number of discoveries actually realized from the n-well program. If we make two discoveries we still need to have found a minimum of 2.86 million barrels (for our illustration of the previous paragraph), but the individual fields themselves do not both have to be greater than 2.86 million barrels.

The only thing required is that the *average* of the two fields must be greater than $2.86/2 = 1.43$ million barrels. Hence the parameter R_{min} — the average field size required to at least achieve the profit objective. It is an average *per field,* and becomes progressively smaller as the number of discoveries increases. This of course, is logical. As the number of discoveries increases the average field size can be smaller to still yield a total from all fields of at least $[(n)(c) + NPV_1]/D$ barrels. Solution of Equation 7.2 for the various possible values of x generates the numerical data for Column 3 of the final computation.

Step III of the method requires some plotting of distributions on a cumulative frequency basis and a separate set of computations to generate a series of distributions of the average field sizes, given x discoveries. These cumulative frequency graphs are then used to determine the probabilities that the average field size, given x discoveries, exceeds R_{min}, the average field size required to achieve the profit objective.

Let's explain Step III in some semblance of logic using the example numbers of the previous paragraphs. If we make just one discovery $(x = 1)$ we reasoned that it would have to be at least 2.86 million barrels to achieve our stated objective. What is the probability this will happen? Well, we would have previously specified an anticipated reserve distribution and the probability of the reserves of a single discovery being greater than 2.86 million barrels would simply be the area in the right tail of the reserve distribution corresponding to reserves of at least 2.86 million barrels.

We could, of course, read this value directly if we have plotted the anticipated reserve distribution on a cumulative frequency basis. Hence, the first part of Step III is to plot the originally hypothesized reserve distribution as a cumulative frequency.

The case of 2 or more discoveries provides a problem, however. If we made 2 discoveries we reasoned the *average* of the two field sizes must exceed 1.43 million barrels. What is the likelihood that if we make two discoveries the (arithmetic) average of the two will exceed 1.43 million barrels? Do you have any idea?

Certainly we can say one obvious thing which is *not* right. That is, we could enter the original reserve distribution and read the area to the right of 1.43 million barrels. But this would give the probability that a field is at least 1.43 million barrels. And that is not what we are asking! We need the likelihood that the average field size of two discoveries is greater than 1.43 million barrels.

The trick at this point is that we must first develop distributions of the *average* field size of $x = 2$, $x = 3$, $x = 4$, etc. discoveries, given that the possible values of reserves per field are as defined in the original reserve distribution. To do this we need to pursue a series of steps in which we sample possible values of reserves for groups of two fields, three fields, etc. Consider for example that our initial reserve distribution had a range of reserves per field from a low of about 50 M barrels to a high limit of 15,000 M barrels. And suppose further that we had made two discoveries whose actual reserves were 175 M barrels and 4,200 M barrels.

The total reserves of these two discoveries is 175 M + 4200 M = 4375 M barrels. The *average* field size of these two discoveries is 4375 M barrels/2 = 2187.5 M barrels. This value represents one sample, or one combination of the possible values of reserves if we had made two discoveries. By a careful sampling procedure we could make many "samples" of possible average field sizes for two discoveries.

Each time the reserves of each field are from the originally hy- . pothesized reserve distribution, and the numerical value of importance, the average field size of the two discoveries is computed as a simple arithmetic average. After many such samples (say, 100 or 200) we would have enough values of the average field size, given two discoveries to define a full distribution. The random variable of the distribution, R, would be defined as the (arithmetic) average field size, given $x = 2$ discoveries. The distribution is then plotted as a cumulative frequency so that areas under the distribution can be read directly.

This same sampling process is then repeated for the case of $x = 3$ discoveries. In this instance three reserve values are sampled from the original reserve distribution. These three reserve values are added together and the sum divided by three to yield one possible value of the average field size, given three discoveries. After many such samples the distribution of these average values is determined and plotted as a cumulative frequency. The process then is repeated for $x = 4$, $x = 5$, x = 6, etc. until x has reached a numerical value of n or the maximum number of prospects considered to be productive, whichever comes first.

This special sampling technique may sound rather mysterious at the moment but the concept will be explained in much further detail in Chapter 8 when we discuss the notion of sampling from a distribution. Although it may sound complicated it can be done extremely efficiently and quickly on a computer at a very nominal cost. We will see examples of the special distributions of the average field sizes for x = 2, 3, 4, . . . etc. in a numerical example which will follow shortly.

Having completed Step III of the "data preparation" phase of this technique we have merely to combine all the numerical data in a five column calculation shown in the lower portion of Figure 7.6. The column 5 joint probability terms represent the likelihood of making x discoveries *and at the same time* having the average field size of the x discoveries exceed R_{min}, the minimum average field size required to at least reach the profit objective, given x discoveries.

These mutually exclusive joint probabilities are then summed over all possible values of x to yield the desired probability that the entire n-well exploration program will at least achieve the desired profit objective.

This all sounds very lengthy and complicated, but it really isn't as difficult as it sounds. The following numerical example should help clarify the mechanics of the method. We should remember that the two important dimensions of the decision strategy considered by this analysis method are (1) it provides a way of uniting a reserve distribution with the various number of discoveries that are possible in an n-well program, and (2) it provides a means of determining the probability of at least reaching a stated profit objective from the total exploration program.

These considerations make it a useful analysis method to "put a dollar sign" to the hypergeometric or binomial probability models. The size of the field found obviously affects the value of a discovery. The hypergeometric model can tell us the likelihood of x discoveries in n trials, but the added dimension we must also consider is the likelihood these x discoveries will be of sufficient size to meet a profit objective. Hence the value of this particular analysis technique.

An example: Suppose we are considering a five-well exploratory drilling program in a basin in which the drilling costs are estimated to be $500,000 per well. We recently acquired the drilling rights in the area for a signature bonus of $200,000. We then ran extensive seismic costing

$400,000. So in addition to the $2,500,000 to be spent for the five exploratory wells we have an additional $600,000 which we wish to recoup from the revenues of any discoveries we make.

We have identified 20 prospects, and by some comparisons to a correlative basin nearby we estimate that 30% of the prospects will have oil. (This implies that $N \times 0.30 = 20 \times 0.30 = 6$ of the prospects are hypothesized to have oil and $N \times (1 - 0.30) = 20 \times (1 - 0.30) = 14$ of the prospects are hypothesized to contain no hydrocarbons.) The overall, composite discounted NPV of any reserves found, expressed on a per-barrel basis, is $1.29/bbl.

This figure was computed from an estimated production schedule of the discoveries and discounted at 18%. The net revenues used in the discounting included consideration of crude prices, development well costs, operating expenses, taxes, and royalties. If we make a discovery the distribution of reserves per field which we feel will describe the magnitudes of reserves is estimated to be an approximately lognormal distribution in which there is a 60% chance reserves will exceed 150 M barrels per field and a 10% chance they will exceed 7500 M barrels per field.

Our profit objective is to recover the exploration drilling costs ($2,500,000), our initial bonus and seismic expenditures ($600,000) and make *at least* a rate of return of 18% per year, the discounting rate used in our NPV computations. What is the probability the five-well exploration program will result in at least this level of profitability?

The solution: The first step in the solution of this example is that we must compute the likelihoods of even finding oil in the five prospects in the first place. That is, we must compute the probabilities that all five prospects will be dry, that one of the five will have oil, that two of the five will have oil, . . . etc. . . . , or that all five will have oil. To obtain these probabilities we need to solve the hypergeometric probability Equation (6.15) using the following parameters:

$N = 20$, the total number of undrilled prospects.

$n = 5$, the number of prospects we will test in the multiwell exploration program being considered.

$r = 2$, the number of possible outcomes on any given trial—here taken to be just dry hole or oil productive. (The various levels of reserves, given we make a discovery, and their probabilities of occurrence will be accounted for later). We will assign the subscript 1 to be the structure is dry and subscript 2 to be the case where the prospect contained oil.

$d_1 = 14$, the number of undrilled prospects hypothesized to contain no oil (computed by multiplying N by the proportion of the prospects considered to be dry, $N \times (1 - 0.30) = 20 \times (1 - 0.30) = 14$).

$d_2 = 6$, the number of undrilled prospects hypothesized to contain oil, ($N \times 0.30 = 20 \times 0.30 = 6$).

Now we must compute the probabilities of the various possible number of discoveries, x_2, in the five wells drilled.

a. The probability of no discoveries:

$x_2 = 0$
$x_1 = n - x_2 = 5 - 0 = 5$

Solving Equation 6.15:

$$\left.\begin{array}{l} \text{Probability of} \\ x_1 = 5, x_2 = 0 \\ \text{in } n = 5 \text{ trials} \end{array}\right\} = \frac{\left(C_5^{14}\right)\left(C_0^6\right)}{\left(C_5^{20}\right)} = \underline{0.1291}$$

b. The probability of one discovery:

$x_2 = 1, \therefore x_1 = n - x_2 = 5 - 1 = 4$. Solving Equation 6.15:

$$\left.\begin{array}{l} \text{Probability of} \\ x_1 = 4, x_2 = 1 \\ \text{in } n = 5 \text{ trials} \end{array}\right\} = \frac{\left(C_4^{14}\right)\left(C_1^6\right)}{\left(C_5^{20}\right)} = \underline{0.3874}$$

c. The probability of two discoveries:

$x_2 = 2, \therefore x_1 = n - x_2 = 5 - 2 = 3$. Solving Equation 6.15:

$$\left.\begin{array}{l} \text{Probability of} \\ x_1 = 3, x_2 = 2 \\ \text{in } n = 5 \text{ trials} \end{array}\right\} = \frac{\left(C_3^{14}\right)\left(C_2^6\right)}{\left(C_5^{20}\right)} = \underline{0.3522}$$

d. The probability of three discoveries:

$x_2 = 3, \therefore x_1 = n - x_2 = 5 - 3 = 2$. Solving Equation 6.15:

$$\left.\begin{array}{l} \text{Probability of} \\ x_1 = 2, x_2 = 3 \\ \text{in } n = 5 \text{ trials} \end{array}\right\} = \frac{\left(C_2^{14}\right)\left(C_3^6\right)}{\left(C_5^{20}\right)} = \underline{0.1174}$$

e. The probability of four discoveries:

$x_2 = 4$, $\therefore$ $x_1 = n - x_2 = 5 - 4 = 1$. Solving Equation 6.15:

$$\left.\begin{array}{l}\text{Probability of} \\ x_1 = 1, x_2 = 4 \\ \text{in } n = 5 \text{ trials}\end{array}\right\} = \frac{\left(C_1^{14}\right)\left(C_4^{6}\right)}{\left(C_5^{20}\right)} = \underline{0.0135}$$

f. Finally, the probability of five discoveries:

$x_2 = 5$, $\therefore$ $x_1 = n - x_2 = 5 - 5 = 0$. Solving Equation 6.15:

$$\left.\begin{array}{l}\text{Probability of} \\ x_1 = 0, x_2 = 5 \\ \text{in } n = 5 \text{ trials}\end{array}\right\} = \frac{\left(C_0^{14}\right)\left(C_5^{6}\right)}{\left(C_5^{20}\right)} = \underline{0.0004}$$

The next step of the "data preparation" phase (Step II), is to compute the minimum average field size, R_{min}, that is required to achieve our stated profit objective. To do this we solve Equation 7.2 using the following parameters:

$n = 5$; the number of exploratory wells being drilled

$C = \$500,000$; the exploratory well costs

$NPV_1 = \$600,000$: the desired net present value profit in addition to making a rate of return of i_0 and recovering our drilling costs so as to recoup our lease bonus and seismic costs

$D = \$1.29$ per barrel; the net present value of the reserves discounted at 18% and expressed on a per-barrel basis.

Solving Equation 7.2 for values of $x = 1, 2, 3, 4$, and 5 gives the following values of R_{min}:

x, Possible Number of Discoveries in the $n = 5$ Well Program	R_{min}, the Minimum Average Field Size Required to Achieve the Profit Objective (from Equation 7.2) $$R_{min} = \frac{(5)(\$500,000) + \$600,000}{(x)(\$1.29/bbl)}$$
1	2,400,000 bbls = 2,400 M bbls
2	1,200,000 bbls = 1,200 M bbls
3	800,000 bbls = 800 M bbls
4	600,000 bbls = 600 M bbls
5	480,000 bbls = 480 M bbls

These results mean, in words, that if we make one discovery it must be at least 2,400 M bbls to yield our profit objective; if we make 3 discoveries the (arithmetic) average of the three field sizes must exceed 800 M barrels, etc. These computed values of R_{min} will occur in Column 3 of our final computation.

The final step in our "data preparation" is to plot cumulative frequencies of the initial reserve distribution and the distributions of average field size for x = 2, 3, 4, and 5 discoveries. From these graphs we will then be able to read the probabilities that the average field sizes, R, will exceed R_{min} for the various values of x. These are the Column 4 numbers of the final computation.

In the introductory description of this numerical example it was stated that the reserve distribution which the analyst anticipated was approximately lognormal with a 60% chance reserves will exceed 150 M barrels per field and a 10% chance they will exceed 7500 M barrels per field.

Plotting these two cumulative percentages and reserve values on log probability graph paper provides the points needed to construct the lognormal reserve distribution on a cumulative frequency basis, as shown in Figure 7.7. (Recall from Chapter 6 that the cumulative frequency of a lognormal distribution plots as a straight line on log probability graph paper, and that only two values of the random variable and their corresponding cumulative percentiles are required to construct this straight line.)

By performing a sampling of the reserve distribution for groups of x = 2, 3, 4, and 5 discoveries we were able to determine the corresponding distributions of the average reserves per field, given x discoveries. These distributions have also been plotted on a cumulative frequency basis on Figure 7.7 and identified along the lower portion of the curves as 5, 4, 3, and 2 — representing the distributions of average field size for the case of x = 5, 4, 3, or 2 discoveries respectively.

The sampling procedure was briefly described before the example and used a sampling technique of the type described in Chapter 8. Note that while the original reserve distribution in this example was lognormal the distributions of the *average* reserves per field are not lognormal, even though the reserve values used in the computation of the averages were sampled from a lognormal distribution.

We have now completed the three "data preparation" steps of this analysis method, and we can proceed to the final five column computation as shown in Table 7.3.

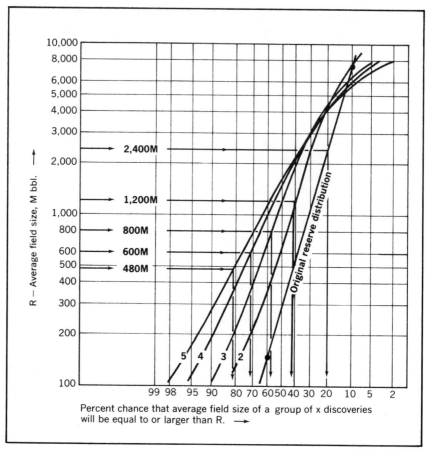

Figure 7.7 *Cumulative frequency graphs of the reserve distribution used for example and the distributions of average field size, given x = 2, 3, 4, and 5 discoveries.*

The data for Columns 2 and 3 were computed by solving the hypergeometric model and Equation 7.2 respectively. The probabilities in Column 4 are read from the cumulative frequency graphs of Figure 7.7. For the case of $x = 1$ discovery we read the value 0.20 as being the probability that the reserves per field will exceed R_{min} of 2400 M bbls. (Enter vertical axis at $R = 2400$ M bbls as shown on Figure 7.7, read across to the original reserve distribution frequency line, and down to the horizontal axis).

If we make two discoveries the average field size must exceed 1200

Table 7.3

Computation of the Probability of Achieving the Profit Objective of Returning a NPV Profit of $[(n)(C) + NPV_1]$ Plus a Rate of Return of $i_0\%$ for Example Problem. (5 Well Exploration Program, $C = \$500,000$ Per Well, $NPV_1 = \$600,000$, and $i_0 = 18\%$)

x, Possible Number of Discoveries in n-Well Program	Probability of x Discoveries in n Wells P(x) (Step I)	R_{min} — the Average Field Size Required to Meet Profit Objective, Given x Discoveries (Step II)	Probability That Average Field Size Discovered, R, Is at Least Equal to R_{min}. $P(R \geqslant R_{min})$ (from Figure 7.7)	Joint Probability $[P(x)][P(R \geqslant R_{min})]$ Column ② × Column ④
0	0.1291	—	—	—
1	0.3874	2,400 M bbls	0.20	0.0775
2	0.3522	1,200 M bbls	0.41	0.1444
3	0.1174	800 M bbls	0.57	0.0669
4	0.0135	600 M bbls	0.71	0.0096
5	0.0004	480 M bbls	0.81	0.0003
	1.0000			0.2987

M bbls. We enter the vertical axis at 1200 M bbls, read across to the cumulative frequency curve corresponding to $x = 2$ discoveries, and down to the vertical axis to read 41%. The remaining values of Column 4 are read from Figure 7.7 in a similar manner.

The final step in Table 7.3 is to multiply the probabilities of Column 2 and 4 together and sum these product terms. The 0.1444 term in Column 5 represents the joint likelihood of making two discoveries in the five well exploration program and at the same time having the average field size of these two to be greater than $R_{min} = 1200$ M bbls, the minimum needed to make our profit objective. Each of the entries in Column 5 represents the probability of a mutually exclusive way of meeting our profit objective, and the sum of these mutually exclusive probabilities is the total probability of meeting the objective.

Hence, we can conclude from this example that we have roughly a 30% chance (29.87% more precisely) of recovering our drilling costs of $2,500,000, our initial seismic and lease bonus of $600,000 and still make a rate of return from the five well exploration program of $i_0 = 18\%$. QED. This implies, of course, that we have a 70% chance of *not* achieving the stated profit objective.

This example should be useful to illustrate how the concept can be applied to multiple-well exploration program analysis. As a final comment we should observe that there are several modifications that can be made to the basic analysis method to fit certain types of exploration options.

The probability terms of Column 2 (the probabilities of x discoveries in n trials) can be computed using the binomial probability model instead of the hypergeometric if the analyst is satisfied that the drilling of the sequence of n wells is approximately a series of independent events. And the D term, the discounted net present value of the reserves expressed on a per barrel basis, can be considered to vary with x rather than be treated as a constant as in the example.

The reason one might wish to make D a function of x would be to account for the possibility that operating many small fields may not be as profitable as operating one or two larger fields. With this scheme different values of D would be used in solving Equation 7.2 for R_{min}, depending on the value of x. Normally it would be expected that D would decrease as x increases. This will result in R_{min} being higher (as x increases) than the R_{min} values that would be computed using a constant value of D.

Three Level Estimates of Risk

Over the years the usual approach to prospect analysis has been to compute the "average," or "most probable" value of profitability from a discovery. This single value of profitability is computed using a single "average" value of pay thickness, an "average" recovery factor, "average" drilling costs, most probable crude sale price, etc. All these single values of each parameter are combined to yield a so-called "average," or "most probable" value of profitability.

This value is then weighed against the costs of a dry hole in some arbitrary and/or subjective manner to establish the relative desirability of the drilling prospect. Such a single-point estimate was essentially what Henry Oilfinder presented in the first part of the Blackduck Prospect Case Study given at the end of Chapter 3.

The limitations of such an analysis are fairly obvious. First of all it only presents to management two discrete levels of value—a dry hole and an average, or most probable value, given a discovery. And we all can readily agree there are many, many possible levels of profitability besides these two discrete values. The oil exploration game is simply not an "either/or" situation.

Another critical limitation is that single values of net pay thickness, porosity, recovery factor, drilling costs, crude prices, etc., when combined to yield a single value of profitability may or may not represent a profitability value which is "most probable." In general it will not, for reasons which we'll discuss in more detail in the next chapter.

Consequently, in an attempt to be able to define more discrete possible levels of profitability explorationists proposed describing three discrete values of each variable (rather than an average, or most probable value) and then computing all the combination of ways the three values of each variable could occur with the other variables.

Typically the three values specified were classified as a minimum, most likely, and maximum value. Such an approach is definitely a step in the right direction in that it provides to management more information about the range of possible levels of profitability he could expect if the prospect is drilled. This approach is generally called a three-level estimate of risk, and examples of early papers on the method include References 7.6 and 7.7.

To illustrate how the method works suppose we have an investment opportunity in which profit is a function of three independent variables A, B, and C. Suppose further that the algebraic relationship which ties the variables together is $Profit = (A)(B)(C)$. Now suppose that we were able to specify three discrete values of each random variable—a minimum, or low value, a most likely value, and a maximum, or high value. These might be as shown in the summary below:

Variable	Minimum Value	Most Probable Value	Maximum Value
A	$A_1 = 2$	$A_2 = 5$	$A_3 = 10$
B	$B_1 = 6$	$B_2 = 7$	$B_3 = 8$
C	$C_1 = 2$	$C_2 = 8$	$C_3 = 20$

We are implying by such a statement that uncertainty still exists in that we don't know which value of each variable will occur of the three possible values. This leads to the possibility of many different combinations of A, B, and C. Twenty seven, in fact, for this example. These twenty seven values of profit include:

$$Profit_1 = (A_1)(B_1)(C_1) = (2)(6)(2) = 24$$
$$Profit_2 = (A_1)(B_1)(C_2) = (2)(6)(8) = 96$$
$$Profit_3 = (A_1)(B_1)(C_3) = (2)(6)(20) = 240$$
$$Profit_4 = (A_1)(B_2)(C_1) = (2)(7)(2) = 28$$
$$Profit_5 = (A_1)(B_2)(C_2) = (2)(7)(8) = 112$$

.
.
.

etc.

If we computed all 27 values of profit representing the possible combinations of A, B, and C and tabulated them we would end up with the following results:

Range of Values of Profit	Number of Times Profit Fell in the Range	Percent of Times Profit Fell in the Range
0–200	11	40.7%
201–400	7	26.0%
401–600	3	11.1%
601–800	3	11.1%
801–1000	0	0
1001–1200	1	3.7%
1201–1400	1	3.7%
1401–1600	1	3.7%
	27	100.0%

This is clearly more information about the possible levels of profitability than if we had just computed an average, or most probable value of profit of $(A_2)(B_2)(C_2) = 280$. Now we are able to see the range, or limits, of possible values of profit in addition to information about the likelihoods of various ranges of profit. And all we had to do to gain this added insight is specify three discrete possible values of each variable (rather than just the average) and compute values of profit which could occur from all the possible combinations of ways the variables could occur.

For an investment prospect having r variables there are 3^r possible combinations of values. In this simple example there were only three variables so the total number of possible values of profit was $3^3 = 27$. If we had considered three values of five different random variables there would be $3^5 = 243$ different combinations.

For those of you who like to think in terms of decision trees the 27 combinations of values of profit represent the 27 mutually exclusive terminal points of a decision tree having a series of chance nodes in sequence. The first chance node would have three branches corresponding to the three possible values of variable A (i.e., A_1, A_2, and A_3). At the ends of each of these three branches would be a chance node having three branches corresponding to the three possible values of variable B. There would be a total of nine terminal points at this point. (Sketch it out on a sheet of scratch paper if you don't believe it!) Finally on the ends of each of these nine branches would be another chance node having three branches corresponding to the three possible values of variable C. This then, results in $9 \times 3 = 27$ terminal points representing the number of possible combinations of ways that three discrete values of each of three variables can occur.

The simple illustration just discussed demonstrates the general logic of the approach. There is, however, a further consideration that is usually included in the analysis. That is, the relative likelihoods of occurrence of each of the three values of each variable. In the example above each value was assumed to be equally likely to occur. That is, $P(A_1) = P(A_2) = P(A_3) = 1/3$; $P(B_1) = P(B_2) = P(B_3) = 1/3$; and $P(C_1) = P(C_2) = P(C_3) = 1/3$. As a result the 27 computed values of profitability all had exactly the same probability of occurrence $- 1/27$.

But suppose the analyst wished to "weight" the three values of each variable in the following manner:

Variable A		Variable B		Variable C	
Possible Value	Probability of Occurrence	Possible Value	Probability of Occurrence	Possible Value	Probability of Occurrence
$A_1 = 2$	0.20	$B_1 = 6$	0.10	$C_1 = 2$	0.30
$A_2 = 5$	0.50	$B_2 = 7$	0.80	$C_2 = 8$	0.60
$A_3 = 10$	0.30	$B_3 = 8$	0.10	$C_3 = 20$	0.10
	1.00		1.00		1.00

How would the method work now?

The answer is exactly the same, except that for each of the 3^r values of profit we would also be able to compute a specific probability of occurrence. For example:

$$\text{Profit}_1 = (A_1)(B_1)(C_1) = (2)(6)(2) = \underline{24} \quad \text{(as before)}$$

Probability of Profit 1; $P(\text{Profit}_1) = P(A_1)P(B_1)P(C_1)$
$$= (0.20)(0.10)(0.30) = \underline{0.006}$$

$$\text{Profit}_2 = (A_1)(B_1)(C_2) = (2)(6)(8) = \underline{96} \quad \text{(as before)}$$

Probability of Profit 2; $P(\text{Profit}_2) = P(A_1)P(B_1)P(C_2)$
$$= (0.20)(0.10)(0.60) = \underline{0.012}$$

.
.
.

etc.

The sum of all the 3^r probability terms thus computed will be 1.00. With this modification, to find the probability of profit being less than or equal to, say 100 we'd merely sum all the probability terms for profits of 100 or less. And we could calculate a mean value profit (EMV) by simply multiplying the 3^r values of profit by their corresponding probabilities of occurrence and summing all 3^r product terms.

Three level estimates of risk, the method just described, clearly give management more information about the various possible levels of profitability than a single average value. To make it work the analyst need only describe two additional numerical values of each variable (besides the "average" value he had been using), plus the weighting factors or probabilities of occurrence of the three values of each variable. Describing the three numerical values is no problem.

Indeed, the analyst probably goes through a thought process of defining a minimum, or low value, and maximum, or high value, when he is trying to define the average or most probable value. The weighting factors cause a little more trouble to the analyst. He will have to compare in his mind whether the minimum value is more likely to occur than the maximum, whether the intermediate value is more likely to occur than either of the other two, etc. But since it is probably the exceptional case when the three values are all equally likely to occur this is a necessary part of the analysis.

While three level estimates of risk clearly tell us more about the prospect than a single, most likely value of profit plus the dry hole cost, the method has largely been superseded by another new risk analysis method called simulation. We'll talk more about this in the next chapter.

Briefly, the reason why simulation is usually chosen over three level estimates of risk is that simulation allows us the opportunity to consider the *entire distribution* of possible values of each variable. Not just a single, average value, or three discrete values – but the *entire distribution* of possible values! So while three-level estimates of risk give us much more information than a single average value, simulation can provide even more insight than three level estimates of risk.

From a historical point of view the method of three level estimates of risk was a precursor of simulation. And now that simulation is gaining acceptance and widespread use the utility of three level estimates of risk is probably declining somewhat.

SUMMARY AND CONCLUSIONS

Those who have taken the time to dig into the details of the analysis methods described in this chapter have no doubt by now begun to sense the complexity and frustrations of petroleum exploration risk analysis.

It seems as though nearly all the methods have certain restrictive assumptions and/or limitations, and none of the methods represent an all-inclusive "fit" of the real-world system we are trying to study and analyze.

And those who have actually tried using some of these methods can appreciate the difficulties of applying them. The available data isn't sufficient, you bog down in the computations, the decision maker doesn't understand the concepts, and any number of similar hurdles seem to appear. Sometimes I have the feeling we make these analyses with the thought that if they confirm our (preconceived?) ideas and biases we'll use them — otherwise we'll rely on older ways of analyzing prospects.

The frustrations of it all are akin to the story in Reference 6.3 (page 32) on how they weigh hogs in Arkansas.

> "What you do is to get a long straight board. You find a convenient place — a fence or a log on the ground — on which to balance the board at its middle. After placing the board so that it can be balanced, you tie the hog to one end. You very carefully weight the other end with different sized rocks until you get one so that the hog and the rock balance perfectly. *Then you guess the weight of the rock!"*

All these comments are to stress the point we made at the beginning of the chapter — when overwhelmed by the futility of probabilities and risk analysis just ask yourself "What's the alternative?" Sure it's tough. But if we ignore probabilities, risk, and uncertainty we haven't accomplished anything, and our investment strategies without some consideration of these factors are going to be pretty grim, unless we are exceptionally lucky! So my feeling is that despite the complexities and frustrations the concepts provide the only real, rational basis upon which to cope with risk and uncertainty.

On a positive note, while there are some voids in determining certain probability terms using the methods discussed in this chapter, there is a new concept that in many cases can fill in the voids where no present methods are adequate. It is the method of risk analysis called simulation which we will discuss in the following chapter. The method can greatly expand our capabilities for assessing probabilities. So there's more to the story on exploration risk analysis, and perhaps the concept of simulation can mitigate some of the frustrations we no doubt feel at this point regarding prospects for dealing with the subject of petroleum exploration risk analysis.

And as long as we are momentarily philosophizing on numerical methods of quantifying risk and uncertainty we need to remind ourselves that numbers are not necessarily ends in themselves. There are no absolutes in risk analysis. Hayward, an early writer on exploration risk analysis, said it quite well back in 1934:

". . . we must not fall into the error of believing that because we have attached a number to a chance that we have thereby made a successful issue more sure, or have in any way altered its probability. Further, we must be ever on the watch for that most insidious and widespread superstition that assumes that mathematical manipulation, if sufficiently accurate, involved and prolonged can transmute doubtful data into positive scientific fact." — Hayward, 1934 (7.1).

His elegant statement stands on its own, with no further comment needed.

8 Risk Analysis Using Simulation Techniques

Intuition and the scratch pad must give way to the computer and the whole machinery of modeling. For it is only through modeling and simulation that the critical variables and their probability distributions can be strung together and tested for their interactions and for the outcomes which they will conspire to produce.

—Carter, 1972

In this chapter we will discuss a new approach to petroleum exploration risk analysis called simulation. Briefly stated, the concept of simulation allows the analyst the option of describing risk and uncertainty in the form of distributions of possible values which the uncertain parameters such as pay thickness, recoveries, drilling costs, etc. could have.

These distributions are then combined to yield a distribution of the possible levels of profitability which could be expected from the prospect. From such a distribution it is only a small, final step to compute an expected value parameter for use in the decision making process. The method is a continuous outcome model of risk and uncertainty, as opposed to the discrete outcome models we have discussed thus far in the book.

Simulation as a means of risk analysis in decisions under uncertainty is a fairly new approach. It was proposed for general business decisions in 1964, and its use in petroleum exploration investments began to appear in our literature beginning in about 1967. Today it is being used in varying degrees by nearly every major petroleum company or agency involved in the active exploration for oil and gas.

The simulation method has several synonyms, including random simulation, Monte Carlo simulation, and the Monte Carlo method. We should point out, however, that the simulation we will be discussing here has absolutely nothing to do with a reservoir engineering analysis method called reservoir simulation. Nor does it have anything to do with certain computer programming languages which have the word simulation in their name. The simulation method we will be considering is more in the sense of a method to work with and combine probability distributions of random variables.

The subject of using simulation for petroleum exploration risk analysis is fairly broad and encompassing. So we probably won't be able to discuss in full detail every possible facet of the method. However, hopefully our discussion here will be of sufficient detail to present the major aspects of the method and how it can be applied to prospect analysis. The outline of our discussions will include the following principal topics:

I. The Logic of Simulation—An Overview
II. The Mechanics of Simulation Analyses
III. Applications and Examples
IV. Using Computers to Make Simulation Analyses
V. On the Philosophical Defense of Simulation
VI. Henry Oilfinder Revisited—Part III of the Blackduck Case Study

I. THE LOGIC OF SIMULATION—AN OVERVIEW

Before we submerge ourselves in the myriad of details I think it is important to gain an overview of the method as it relates to petroleum exploration. The primary reason we will be making a simulation analysis in the context of decision making under uncertainty is to define the distribution of profit which could be anticipated. Once we know this distribution we can easily find its mean value using the type of calculations described in Table 6.9. And since the mean value of a distribution is, by definition, the same as its expected value it follows then that the mean value of a distribution of net present value profit discounted at i_0 is also the expected net present value profit, or EMV.

This is the parameter management can use to determine the feasibility of the drilling prospect. In addition to being able to compute the EMV directly, having the distribution of profit offers numerous graphical options for presenting information to the decision maker about the range and likelihoods of occurrence of possible levels of profit and loss. A picture such as a probability distribution is sometimes worth a thousand words.

So the objective of simulation is to determine the distribution of profit from a proposed drilling prospect. Now we take a step backward to see how the profit distribution comes into the picture in the first place. Suppose we were considering a hypothetical investment in which ultimate net profit is a function of three separate, independent variables X, Y, and Z. And let's suppose the relationship (equation) which tied these parameters together was as follows:

$$\text{Profit} = 3.4X + \left[\frac{1}{\sqrt{14Y}} + 17.9\right]\left[0.4Z^2\right] \tag{8.1}$$

This of course is just an abstract situation in which X, Y, and Z are said to be the independent random variables and Profit is the dependent variable. In the oil business the equivalent relationship would probably have NPV profit on the left side of the equal sign and on the right side would be the equation(s) by which we compute recoverable reserves, convert reserves to net revenues, subtract operating expenses, drilling costs, taxes, royalties, etc. and discount the net revenues at i_0. The independent variables would include all the factors used in the calculations to compute NPV profit: net pay thickness, recovery factors, productive area, drilling costs, crude prices, operating expenses, etc., etc. But for this overview let's just think in terms of a simpler, abstract model such as Equation 8.1.

Now suppose we know the exact values which the random variables will have in an investment. Do we know the resulting profit? Certainly! We'd merely substitute the (known) values of X, Y, and Z into Equation 8.1 and solve it directly for profit. This would be what the mathematicians call a deterministic computation. There is no uncertainty with regard to the variables or the resulting value of profit.

But such an ideal case never exists in prospect analysis. At the time of decision making we usually do not know exact values for most of the random variables affecting NPV profit. Perhaps all we may know about them are the ranges of possible values. In this case which value from the range do we use in the profit calculation?

Well, the answer is that using the concept of simulation we can consider *all* the values within the range for each random variable. And these are all combined to yield the range and distribution of profit. The usual situation in exploration prospect analysis is that we do not know exact values of each parameter—but if we can describe a range and distribution of possible values of each random variable we can use simulation to derive the resulting distribution of values of profit. The scenario, then, is as in Figure 8.1. The model we are trying to analyze is the equation of profit for the hypothetical investment of Equation 8.1.

The quantitative expressions of uncertainty are given as distributions

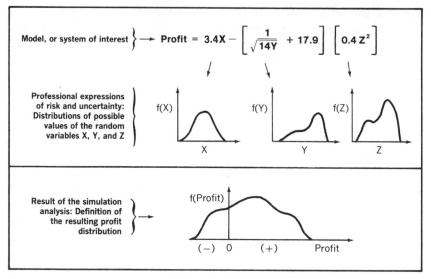

Figure 8.1 *Schematic of simulation analysis of the hypothetical investment of Equation 8.1.*

of possible values of the random variables X, Y, and Z. We should note here that we are speaking of the full range and distribution of possible values of the random variable, and not just an average or most likely value, or three discrete values such as discussed in the last chapter under the heading of three-level estimates of risk. Rather, we are considering the entire, continuous range of possible values of X, Y, and Z.

We must observe one additional point. The actual state of nature will be that variable X will have one specific value, variable Y will have only one specific value, and variable Z will only have one value. But the trouble is that before the investment is accepted we do not know what these values will be. All we can say about the variables is the range and relative likelihoods of possible values of each variable.

To define the scenario to this point is nothing new. We are all aware of the model or system we are trying to analyze in prospect analysis. And we can agree that we usually do not know, in advance, the actual net pay thickness (despite a geologist's pride in his isopach map!), but we may be able to estimate a range of possible values from, say, 0 to 100 feet.

Where the problem comes in is solving the equation once we've defined distributions for each random variable. It's not immediately obvious what each of the three distributions is in Figure 8.1, and it probably isn't too obvious how the equation can even be solved if the random variables are expressed as distributions, rather than single, deterministic values.

This is precisely where simulation analysis enters the picture. It is a simple, easy way of analyzing an investment prospect where some (or all) of the independent random variables are expressed as probability distributions. We use simulation in this situation probably because the two obvious alternative ways of solving the problem won't work. Let's mention these two ways first.

In Chapter 6 we mentioned that every probability distribution has what's called a probability density function. It's the f(X), f(Y), and f(Z) functions on the ordinates which, when integrated over the range of possible values of each variable yields dimensionless 1.0, the area under all probability distributions. So one thing we could do if we had a problem such as shown in Figure 8.1 would be to consult a mathematician and ask him to determine the probability density functions, insert these analytic functions into the equation and solve for the corresponding density function of the dependent variable profit.

A fine idea; but it simply won't work because the complexities of an analytic solution such as this are totally intractable. While there are a few special situations where the analytic approach might be possible (See Appendix H), the analytic approach is, in general, not feasible in petroleum exploration prospect analysis.

The second approach that some are tempted to try so as to simplify the problem is to determine the mean values of the X, Y, and Z distributions and insert these discrete values into the equation and solve for profit. The idea would be that this single value of profit would correspond to the mean value of the profit distribution. And since the mean of a profit distribution and EMV are synonymous, we'd then have the parameter desired for decision analysis. But this logic doesn't work either!

In general, you *cannot* compute the mean value of the dependent variable by substituting the mean values of each random variable distribution in the equation.[1] It simply won't work for exploration prospect analysis. And another thing which won't work is to take the most likely values of each distribution (the modes), substitute these values into the

1. There are a few exceptions to this statement. See Appendix H for further details. These exceptions are not applicable to prospect analysis, however, so the statement stands as written with respect to prospect analysis.

equation and solving the equation in hopes of computing the most likely value of profit. Either of these "short cut" methods will not be of use to us.

All of this is to say that we use the mechanics and logic of simulation by default – it's the only way we can solve the problem. If we have defined an investment opportunity as shown above the horizontal dividing line in Figure 8.1 the only way we can convert all this into a description of the resulting profit distribution is by simulation.

In a nutshell here is how the dependent (profit) variable distribution is defined. A series of repetitive calculations of possible values of profit are made. Each value is computed using a value of X, Y, and Z selected from within their respective distribution ranges. Each value of profit computed in this manner represents one possible state of nature, or possible combination of X, Y, and Z. The specific values of X, Y, and Z for each computation of profit are chosen so that over a series of repetitive calculations the values selected will be in the same frequency distribution as the original distribution specified by the analyst. Each of these repetitive computations of values of the dependent variable is called a *simulation pass,* or more simply, a *pass.* The value of the independent variable used for each pass is obtained by *sampling from its original distribution* in a manner which honors the shape and range of the distribution.

These repetitive computations, or passes, are continued until a sufficient number of values of the dependent variable are available to define its distribution. This usually requires at least 100 passes and perhaps as many as 1000 or 1500 passes. Obviously this many arithmetic solutions of an equation imply a great amount of work – and indeed the only practical way of accomplishing it is with a computer. But fortunately it can be done on a computer very quickly, efficiently, and for a nominal cost of computer time.

And that's all there is to it. A series of repetitive solutions of the equation or model of interest – each time with values of the independent variables which are sampled (selected) from their respective distributions. After a sufficient number of passes have been made the computed values of the dependent variable are tabulated as a distribution – the objective of the analysis.

Of course, there are a lot of mechanical details we'll have to dig into later, such as how to sample values of the independent variable distributions for each pass, how many passes are required, how are the distributions of the independent variables defined in the first place, what if variable X and Y are related in some manner, etc. But the general logic of simulation is simply to define the objective distribution by a series of repetitive samplings.

This approach for exploration risk analysis has several very important advantages:

■ It allows the explorationist to describe risk and uncertainty as a range and distribution of possible values for each unknown factor, rather than a single, discrete average or most likely value. The resulting profit distribution will reflect all the possible values of the variable. The method accounts for the degree of risk and uncertainty by considering the variability of each of the variables.

■ It can be applied to any type of calculation involving random variables. We'll be talking primarily about NPV profit of drilling prospects, but the logic can be applied to describe a distribution of recoverable reserves, a distribution of water saturation from well log data, the bottomhole position of a directionally drilled well, etc. The logic is the same — only the model or equation of interest and its corresponding variables change.

■ There is no limit to the number of variables which can be considered. If profit is a function of 45 different variables, and we can define distributions for each variable simulation can still be used effectively. The bookkeeping in the computer will be a bit longer, and the program may require a few more seconds of computer time to run — but that's the only handicap.

■ The distributions used to define the possible values for each random variable do not have to be of a specific form such as lognormal, normal, etc. Rather, the distributions can be of any type whatsoever. If the analyst can draw the picture of the distribution on paper that's all that is required. This is important because it means the analyst does not have to be able to describe the distribution other than with a picture.

■ The expertise of the firm can be used more effectively because the judgments about the distribution of possible values of each variable can be made by the person most knowledgeable about the parameter. The geologist can define the net thickness and productive area distributions, the engineer can define the distribution of recovery factors and production schedules, and the drilling engineer can specify the probable drilling cost distribution, etc. The method does not require that one person provide all the input to the analysis.

■ The computer programming of simulation analyses is relatively straight forward and running a simulation analysis requires very little computer time. In fact, reading in the initial distributions and printing out the final dependent variable frequency data will usually require more computer time than the actual computations.

■ The method lends itself to sensitivity analyses. Indeed one of the important aspects of making a simulation analysis is to be able to define the one or two or three factors which have the most significant effect on the resulting values of profit. Such analyses are made by using different distributions (either in terms of the shape and/or the ranges) for each variable to observe the extent to which the dependent variable distribution is changed.

Finally, we should observe that the critical part for the explorationist is defining the distributions of each random variable. We'll talk about this in detail later in the chapter, but for the moment we should realize that the distributions can be based on statistical data if available, by analogy with other similar producing areas, or perhaps even by a subjective judgment. If we are forced to rely on subjective opinions the opinions enter the analysis in the form of the distributions. One variable at a time — one step at a time.

This is the distinctly new approach we mentioned early in Chapter 7 in our discussion of subjective probability estimates. If we must rely on subjective judgments it seems reasonable to expect that we will do a better job of it if we consider one parameter at a time, rather than trying to assess how all the uncertain parameters will occur collectively in a drilling prospect.

Having discussed an overview of the method, let's now turn our attention to some of the mechanical details involved in a simulation analysis.

II. THE MECHANICS OF SIMULATION ANALYSES

In this section of the chapter we concern ourselves with the "nuts and bolts" of the simulation technique of risk analysis. There are lots of miscellaneous details that we must consider in organizing the analysis and preparing the data. And it is important that the analyst have a good, firm understanding and grasp of the basics. Without this understanding, simulation can be a dangerous tool.

For those of you who have a general understanding and/or have used simulation before it should be no problem to continue on into these details at this time. If the concept is completely new to you and the previous section seemed a bit mysterious I would suggest you skim over this section and proceed to the next section on applications and examples. Then having followed through some examples you may have a better understanding of how all the pieces fit together. At that time you can return to this section to pursue the details of each of the individual pieces of the analysis.

If we are analyzing a drilling prospect and wish to quantify the degree of risk and uncertainty using a simulation analysis we need to follow six general steps:

Step 1: Define all the variables. We must specify the measure of value of interest to us (e.g. net present value profit after taxes), and all the variables which affect value.

Step 2: Define the relationship which ties all the variables together. The relationship of interest here is the equation(s) or series of numerical computations by which we compute the value (profitability) of a drilling prospect. This relationship, or series of stepwise computations, is the model or system we are trying to analyze.

Step 3: Sort the variables affecting value into two groups — the variables whose values are known with certainty and the unknown, or random variables for which exact values cannot be specified at the time of decision making.

Step 4: Define distributions for all the unknown, or random variables.

Step 5: Perform the repetitive simulation passes so as to describe the resulting distribution of value.

Step 6: Compute Expected Value of the profitability distribution and prepare graphical displays of analysis procedure and results.

Having enumerated the steps or pieces of a simulation analysis let's look at the details involved in each of these steps.

Defining the Variables (Step 1)

This initial step is the obvious starting point of any quantitative analysis of a drilling prospect: define the measure of value of interest to us and the factors which affect it. Typically, the measure of value is chosen as an after-tax NPV profit discounted at i_0, however, we are free to choose any measure of value we like (rate of return, discounted profit/ investment ratios, etc.). The variables which affect value are the obvious parameters we deal with daily: net pay thickness, porosity, water saturation, recovery factors, number of wells to be drilled, productivity of the wells or of the field as a whole, drilling costs, operating expenses, platform and pipeline facility costs if offshore, crude and associated gas sales prices, taxes, royalties, etc.

While our principal concern here is the analysis of the prospect from a profitability viewpoint there are many instances where we may be interested in determining the distribution of some of the intermediate dimensions such as reserves. In these instances the dependent variable of interest may be recoverable reserves from the structure (rather than profitability), and the variables affecting reserves would include only the parameters relating to reserves. At the end of the next section we give brief reference to some of these intermediate, or auxiliary simulation schemes.

Defining the Analysis Model (Step 2)

This step is the logical followup to the listing of the variables—defining the relationship or equation which ties all the variables together. It may take the form of a single equation, several equations, or even a series of step-wise computations. For the hypothetical investment of the initial section of this chapter the relationship was Equation 8.1. In prospect analysis it is the series of computations by which we compute recoverable reserves, convert reserves to discounted revenues, and subtract all the development and operating costs and expenses.

Inasmuch as the relationship will change from prospect to prospect (depending on which country of the world we are in, whether the prospect is onshore or offshore, etc.) it probably wouldn't serve a very useful purpose to list specific examples at this point. Most explorationists are intimately familiar with these steps, and the specific relationship to use for the simulation will pretty much follow from logic for each prospect.

As an observation in passing these first two steps are fundamental to any analysis, simulation or otherwise. In completing these two steps we are in essence, defining what it is we are trying to analyze, what factors affect it, and how they affect it. It is only as we proceed to Step 3 that we begin to take a new direction for analyzing risk and uncertainty.

Sorting the Variables into Groups (Step 3)

With this step we must divide all the variables which we listed in Step 1 as affecting value (profit) into two groups. One group consists of all the variables or parameters for which their exact values are known. The second group includes all the parameters or variables for which we do not know exact values at the time of analysis. In exploration prospect analysis most of the variables affecting profitability will, unfortunately, end up in the second group!

Examples of some of the parameters which might be included in the "known" group are royalty and tax rates. If we have already signed a lease the royalty we pay to the owner of the mineral rights is known with certainty — it's specified in the lease. And assuming we are operating in a politically stable country we would probably know the exact rates for computing taxes due.

The group of "unknown" parameters will usually include all the variables relating to the size of the structure or prospect, the amounts of recoverable gas and oil, productivity schedules, factors relating to costs and future operating expenses, and the future crude and gas sales prices. These are the random variables which will be considered in the analysis in the form of their respective distributions.

When we make this sorting of variables we need to remember that when we speak of a variable as being "unknown" we mean that we do not know its exact value at the time of the analysis. For instance — net pay thickness. It is clearly a variable which affects the magnitudes of recoverable reserves, and hence profit, but it is probably a variable which we have to classify in the "unknown" group. This is because we do not know the exact pay thickness of the objective formation before we drill the well.

We can't say for certain it will be 45 meters thick, or 71 meters. We may be able, however, to specify a range and distribution of possible values which net pay could assume in the prospect, in which case we would classify the variable net pay thickness as an unknown, or random variable. The point I raise here is that "unknown"/"known" in this sense relate to whether or not we can specify the variable's exact numerical value with certainty at the time the analysis is being made.

Another subtle, but important point in classifying the variables is the question of whether or not we know enough about a variable to ever be able to describe a distribution of possible values. Some analysts will say, for example, that a parameter such as recovery factor is an unknown factor, but since we are in a rank wildcat area we have no basis for defining a range or distribution of possible values, so let's just use an average recovery factor of, say, 15% of the oil in place.

As a consequence they would put recovery factor into the group of "known" variables for which specific single values are used in the profit computations. In the strict sense of what we are trying to accomplish with a simulation analysis such a viewpoint (although quite commonly used) is not valid. The statement implies the following logic:

    a. If we had perfect information about recovery factor we would know its precise numerical value, say RF_1.

    b. If we had imperfect information, but good control and experience from a nearby, correlative area we might be able to estimate a range of possible values of recovery factor from RF_{MIN} to RF_{MAX}.

    c. As our control and correlative data become poorer and/or fewer in number the range might have to be expanded to include a wider range of possible values of recovery factors. The wider range could be representative of greater uncertainty.

    d. If we are in a rank wildcat area with no way to estimate RF we have no basis upon which to determine a range and distribution; therefore, use a single average or most probable value in the computations.

And this logic is clearly circular! It implies that the less we know about something the more we know about it!

The whole purpose of a simulation analysis is to be able to account for variability in our profitability analysis. Uncertainty implies variability, and this variability is expressed as a range and distribution of possible values of the variable. To do otherwise is to ignore the very dimension of risk and uncertainty we are trying to analyze.

The reason I stress this point is that I have repeatedly observed explorationists who first begin working with simulation make this mistake. When they define a variable such as recovery factor as a single value I ask if they mean, then, that they know (in advance) what the precise value of recovery will be. Their reply is: "No, I don't know exactly what it will be, but I have no available data or information upon which to establish a range so I just used the value of 15%." I have heard such statements so frequently that I am convinced that it is one of the points most people fail to recognize when using simulation.

So the moral here is that if we don't know an exact value of a variable it should be classified in the group of "unknowns," the random variables. *Period!* If it is your preference to say that you know so little about a variable that you can't assign or define a range—and choose, instead, to use a single "average" value then you don't need simulation. We'll speak on this important, but subtle, point again in Section V of this chapter.

Defining Distributions for the Unknown, Random Variables (Step 4)

This is the step where professional expertise and judgment really enter. It's one thing to say recoverable reserves is an unknown, random variable (Step 3), but quite another thing to say "OK, what is the range and distribution of reserves that we can anticipate?." We can all agree unequivocally that we don't know at the time of decision making the exact amount of reserves (if any) that the prospect will yield. But we quickly get on thin ice when we try to specify what the distribution of reserves will be.

Nonetheless, it is the variability which results from not knowing reserves that we are trying to take recognition of with simulation. So this step is really the key, critical point in the entire analysis.

To define or estimate distributions for each of the "unknown" parameters we have these general guidelines:

- The distributions can be of any shape, range, or form. We do not have to use a few "standard" statistical distributions such as the normal, lognormal, etc. The distributions can be discrete or continuous. Some of the random variables may be related to one another, in which case the dependency relationship must be defined.

- The judgments about the distribution for each variable do not have to be made by a single person. Use the expertise of your staff, and let those most familiar with each parameter make the judgment.

- The distributions can be based on histograms or frequency distributions of data of nearby fields; the distributions can be based on knowledge that certain variables characteristically follow a common distribution; or they might be based on a purely subjective judgment.

- We do not have to be "locked in" to a specific distribution. If opinions vary about the range and distribution of a variable try several possible distributions. As the result of such a sensitivity analysis you may find that the variable isn't even that critical on profitability in the first place.

It is, of course, pointless for us to prepare a list of the usual random variables in prospect analysis and then define a specific range and distribution which should be used for each variable. But we can make some general observations about some of the variables relating to reserves and productivity of a field.

As we mentioned in Chapter 6, a parameter which can be fairly accurately described with a normal distribution is core porosity. The lognormal distribution is frequently a fairly good representation for distributions of core permeability, thicknesses of sedimentary strata in a basin, oil recovery (barrels per acre foot) in a given formation producing by a common reservoir mechanism, and in some instances, reserves per field. When analyzing a prospect we may choose to use these functional distribution forms if we have no other data or basis upon which to define some other type of distribution.

Knowing the shape of the distribution (as being normal or lognormal for the corresponding variables) the only other thing we would have to determine is the position of the distribution on the horizontal axis – that is, the range. If we have available statistical data from a nearby correlative area the range will take care of itself. Or, we may be able to specify (by analogy) a mean and standard deviation, in which case we can define the entire distribution (see discussion of normal and lognormal distributions in Chapter 6).

A third alternative which is sometimes useful is to compute or estimate a low and high value of the random variable which are thought to be near the limiting points of the range of possible values of the random variable. Knowing two discrete values and estimating their corresponding cumulative percentiles one can "force-fit" a lognormal or normal distribution through the points.

For example, suppose we wish to describe oil recovery in barrels per acre-foot (BAF) as an unknown random variable having the shape of a lognormal distribution. But our problem is that we have no statistical data to use to establish its range.

We could possibly resolve this problem by computing a near-limiting low value and a high value near the upper end of the range using the equation for oil recovery:

$$\frac{\text{Oil Recovery (BAF)}}{\text{(barrels per acre-foot)}} = \frac{(7758) \, (\phi)(1 - Sw)(R.F.)}{\beta o_i}$$

ϕ is the porosity expressed as a decimal fraction; Sw is the connate water saturation, as a decimal fraction; R.F. is the (decimal) fraction of the oil in place which is recovered; and βo_i is the (dimensionless) initial oil formation volume factor.

Suppose we estimate the probable lower limits of porosity and R.F. for the prospect and the probable upper limiting values of Sw and βo_i. Substituting these numerical limits into the equation will give a low value of BAF which is near the lower end of oil recoveries. Similarly we could estimate a high value of ϕ and R.F. and low values of Sw and βo_i to compute a high value of BAF. To be more specific, suppose the values of oil recovery computed in this manner were 50 BAF and 450 BAF.

Next, we must estimate the likelihood that oil recovery will be less than 50 BAF or greater than 450 BAF. Suppose we estimate these to be 2% and 3% respectively. We can now combine all these bits of information and speak of a specific, unique distribution for oil recovery: a lognormal distribution in which there is a 2% chance recovery will be less than or equal to 50 BAF and a 97% chance recovery will be less than 450 BAF. The distribution can be obtained graphically, as in Figure 8.2, by plotting the 50 BAF and 450 BAF on logprobability graph paper opposite the 2% and 97% percentiles respectively and connecting the points with a straight line.

Thus we have, in essence, "force-fit" a lognormal distribution through two near-limit values of the random variable that were computed using minimal and maximal values of the factors influencing recovery. So in summary, there are a few random variables in prospect analysis which frequently can be represented by a normal or lognormal distribution. If we have no evidence to suggest that a different type of distribution might apply we at least have a basis upon which to estimate the shape of the distribution.

The range, or position on the horizontal scale of the random variable can be established by statistical data, if available; by knowing the distribution's mean and standard deviation; or by "force-fitting" the lognormal or normal distribution using two discrete near-limit values and their corresponding cumulative percentiles.

Some variables may be defined as discrete distributions. Generally, these will consist of random variables which can have only certain discrete values or are countable. An example of a discrete probability distribution is Figure 6.9. When defining discrete distributions we must remember that the sum of the probabilities of occurrence of all the possible discrete values which could occur must add up to 1.0. With simulation we can even use a discrete distribution in which some event either occurs or it does not occur.

For instance, suppose we were analyzing an offshore exploration venture in an area of severe storms. One of the uncertainties which could affect ultimate profitability is the likelihood of a storm destroying our production platform in any given year. If the likelihood of this occurring is, say, 3%, we could describe a discrete probability distribution having

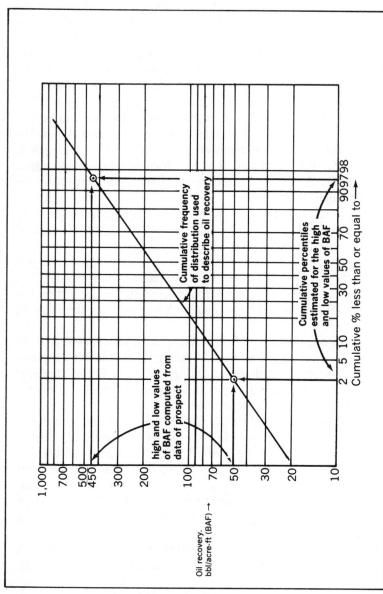

Figure 8.2 Graphical method of "force-fitting" a lognormal distribution through the near-limit values of 50 BAF and 450 BAF. These values are computed using data from the prospect. They are then plotted opposite the estimated percentages that BAF will be less than or equal to. Finally, a straight line is drawn through the points to yield the cumulative frequency of the lognormal distribution. This cumulative frequency graph would then be used in the simulation analysis to represent the distribution of possible values of oil recovery from the prospect.

two values: "Yes, there was a storm," and "No, there was no storm during the year." The heights of the probability lines above these two points would then be 3% and 97% respectively.

In the simulation analysis the distribution would be sampled on each pass to determine if a storm occurred. And 3% of the time the "value" of the variable would be yes, in which case the program would be directed to the series of computations which consider the losses of a platform. The other 97% of the time the "value" would be that no storm damage occurred, and profitability would be computed accordingly. This illustration should suggest to you that we have considerable flexibility in describing discrete distributions in a simulation analysis.

For some variables we may have some statistical data from nearby, correlative areas upon which to base the distributions for the prospect being considered. An example of how the data can be used to describe a distribution is given in Tables 6.5, 6.6, and Figure 6.12.

The usual way to represent statistical data is with a histogram or relative frequency distribution of the type shown in Figures 6.12 and 6.13. Either form is acceptable. If there is a large amount of data and the bars of the histogram are narrow we may wish to smooth a distribution curve through the tops of the bars. The decision whether to leave the distribution as a histogram or whether to fit a smooth curve is relatively unimportant if there is a fairly large number of bars to the histogram.

It is basically a question of whether to represent the distribution as a smoothed curve the shape of which is approximately described by the observed data. My feeling is that this point is generally of minor significance, and since it is easier to work with a histogram I prefer to use it, rather than a smooth curve fitted through the tops of the bars.

We should remember that when using statistical data to describe a distribution we do not have to be concerned about its shape. The histogram can be of any shape and range imagineable. This should be obvious, but nonetheless I've frequently been asked the following question:

"Paul, I've got some statistical data from a nearby area on pay thickness. I arranged the data in terms of cumulative frequencies (as in Table 6.7 for example) and plotted them on normal probability paper. But it wasn't a straight-line so I plotted the data on logprobability paper. But it wasn't a straight-line on it either. So the data aren't normally distributed or lognormally distributed. It doesn't fit anything, so what do I do?"

My answer—don't worry about it! If you have statistical data which you think is representative of what you expect for the prospect being considered just use it as is. The distributions we use in simulation do not have to be just a normal or lognormal – they can have whatever shape we feel is representative. It seems to be such an obvious point that it doesn't really require much explanation. But it's a human foible I guess to try to get everything to be a straight line, and some explorationists seem to get upset if the data don't fit some nice, well-behaved straight line.

A bit stickier situation is when we wish to define a distribution for a random variable but have no data available and have no idea what the shape of the distribution is, or should be. What then? In these cases we need to first try to at least establish the range of values – a minimum value and a maximum value. Next we would need to determine if any value, or range of values within the limits might be more likely to occur than other values. That is, does the distribution have a mode, or most likely value?

If the answer is yes we may then wish to represent the variable as a triangular distribution such as Figure 6.27. If we have no reason to assume one value is any more likely than any other value perhaps the best we can do is describe the variable as a uniform, or rectangular distribution as in Figure 6.26.

And indeed these two distributions are frequently used to represent the variability (uncertainty) of a random variable when the only things we can say about the parameter are a minimum, a maximum, and whether or not the distribution has a mode. In essence, the distributions are used as expressions of ignorance about the random variable. But if we wish to account for variability we may have no other choice than to use the uniform or triangular distributions.

Some writers on simulation have taken the position it is rare that we ever know much more than a minimum, most likely, and maximum value. They have even carried their view to the point where they intimate that we are "gilding the lily" to even consider using any distributions more explicit or detailed than a triangular distribution. My feeling is that triangular distributions should be used if we have nothing else to go on, but if we have data or information to suggest a different type of distribution we should certainly use it.

More specifically, if the only thing we can estimate is a minimum value and a maximum value we should probably use the uniform or rectangular distribution. These two limiting values completely define the distribution. The analysis will then be made on the premise that any value of the random variable between the two limits is just as likely to occur as any other value. As a passing note there are some variables which may actually be distributed in this manner, in which case the uniform distribution is a precise fit, rather than an approximation.

The triangular distribution requires that we specify a minimum value, a most likely value, and a maximum value. The mode value can be located at any point between the limits, or it can be coincident with either the minimum or maximum value. These three parameters completely define the triangular distribution. Pretty simple, right?

But I should warn you of some dangers of using the triangular distribution in a somewhat "indiscriminate" manner. First of all, explorationists frequently error in their definition of the minimum and maximum values. To illustrate the problem we can get into, suppose we have the following set of data of a random variable x:

10, 11, 12, 12, 12, 12, 16, 17, 19, and 24.

If you had to define the set of data as a triangular distribution what values would you use for the three parameters? I guess we could all agree that 10 is the minimum, 12 is the most likely, and 24 is the maximum. Right? The resulting triangular distribution would be:

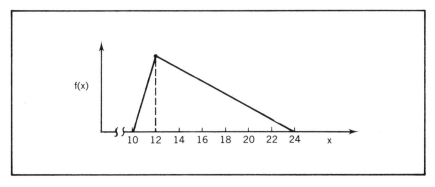

But now suppose, instead, that the available data of the random variable x is:

10, 10, 10, 11, 11, 12, 12, 12, 12, 12, 14, 17, 18, 20, 23, 24

How would you fit a triangular distribution through this data?

I find that many would observe that 10 is the minimum value, 24 is the maximum value, and the value $x = 12$ occurs most frequently. So they end up describing another triangular distribution having exactly the same parameters (minimum, most likely, and maximum values) as the picture above representing the first set of data. But are they the same?

I think the answer should be no. The reason is that minimum *numeri-cal* value for the second set of data is, indeed, 10, but there appears to be a fairly significant likelihood of the value 10 occurring. It occurred three times in just the sixteen observations. But if we assign 10 as the minimum value of a triangular distribution the probability of x having a value within the range of (10 + α) shrinks to zero as α approaches zero. So perhaps a more meaningful representation of the second set of numerical data would be a distribution such as this:

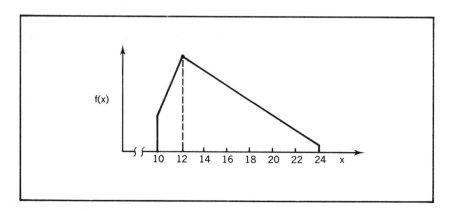

The whole point to remember here is that when we speak of minimum and maximum values of a triangular distribution we imply values for which the probability of occurrence vanishes to zero as we consider ranges closer and closer to the limits. This is distinctly different than speaking about a minimum or maximum *numerical* value of the random variable whose likelihood of occurrence may not be zero. This is a very common mistake, and we should all be alert to the meaning of the mini-mum and maximum values of a triangular distribution.

Another point to remember when describing distributions is that if you ask a person to define a minimum, most probable, and maximum value that a variable could assume he will undoubtedly specify the most probable value at the midpoint of the range. The logic for this, according to some psychologists, is that he is interpreting most probable as "best estimate" and is striving to be an unbiased forecaster.

If he selects his best estimate at the midrange (median) he reasons there is just as good a chance the variable will be greater than his best estimate as lesser than his estimate. The idea being if he always selected the most probable value at mid-range he minimizes his chances of being a consistently high or consistently low estimator. The point we should remember here is to stay with the term most probable value, rather than best estimate, and to recognize again that a triangular distribution does not have to be symmetrical. The most probable value can be at either end of the range or any place in between.

A final note to remember about using triangular distributions is that they are, in general, very poor representations of highly skewed data. Suppose, for example that a random variable y is distributed as in part (a) of Figure 8.3. The minimum, most probable, and maximum values of the distribution are y_1, y_2, and y_7 respectively.

If we chose to represent this distribution as a triangular distribution, the parameters would be y_1, y_2, and y_7, as in Figure 8.3(b). But this is a poor approximation because it gives much more weight (probability) to higher values of y than the original distribution.

If we wished to represent the highly skewed data of (a) with an approximation we would be much better advised to use a four-sided polygon shown in part (c). So the warning here is to be very cautious in using

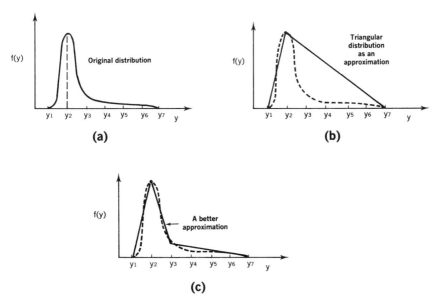

Figure 8.3 *An illustration to show that a triangular distribution is a poor fit (approximation) to a highly skewed random variable. A much better approximation is a four sided polygon as in (c).*

triangular distributions to represent a random variable having highly skewed values.

A final comment about defining distributions of the unknown, random variables regards the issue of dependency. In our simple hypothetical investment of the previous section we more or less assumed that the random variables X, Y, and Z were independent of one another. This implies that in the repetitive simulation pass calculations the values used for X had no affect on the value of Y or Z.

On a particular pass the value of X could be towards the high end of the X distribution and the value of Y could be low. And whatever the values X and Y were, they had no affect on the numerical value of Z chosen for the pass. But what if, in fact, there existed a relationship, or dependency, between variables X and Y such that if one had a high value the other one would also tend to have a high value?

When we sort out the variables in Step 3 the group of (unknown) random variables in prospect analysis may very well have some variables which are dependent upon one another. For example, net pay thickness is an unknown, and so is well (or field) producing rate. So both of these parameters would be on our list of variables for which we need to describe distributions.

But thickness and well productivity are related by virtue of Darcy's equation (flow rate is directly proportional to reservoir thickness). So this would mean that when we describe distributions for these two parameters we will have to somehow account for the dependency. Other examples of possible dependencies: connate water saturation (Sw) generally increases as porosity (ϕ) decreases, field productivity is related to thickness, to the number of wells in the field, to the physical producing capacity of the well bore, and to storage capacity and tanker arrival schedules, pipeline costs are related to field productivities, etc.

All this suggests that one of the concerns we will have to take recognition of when defining distributions are possible relationships or dependencies among some of the random variables. This is an extremely important issue, and the results obtained from a simulation analysis can be very misleading if the analysis does not honor the dependent relationships.

We need to be able to first recognize whether a dependency exists between several variables and then have some way to account for the dependencies in the simulation computations. It's all fairly easy to do, but before we get into the details we need to talk about a few other aspects of simulation. So we will come back to the important issue of dependency in a later portion of our discussion of the mechanical details of how the actual simulation computations are performed.

Performing the Simulation Passes (Step 5)

Up to this point of the simulation analysis we have: defined the value parameter of interest and the variables which affect it (Step 1), defined the relationship which ties all the variables together (Step 2), we have classified the variables affecting profit as known values or variables for which exact values are not known (Step 3), and we have specified distributions for all the unknown random variables. At this point the model, or system of interest together with judgments about the variability (uncertainty) have been defined.

The work of the explorationist is completed for the moment and what follows are the mechanics of solving the problem. It is mechanical in the sense that the computations and procedures of this step follow an exact set of rules which require no further interactions on the part of the analyst. These mechanical computations and procedures can be performed by a technician, but it is many times more efficient to complete this step of the analysis with the aid of the computer. Consequently our comments on this step will be in the general sense of what the computer will be doing to generate the data needed to define the final profit distribution.

As outlined previously, a simulation analysis consists of a series of repetitive computations of value, or profit. Each value of profit which is computed represents one combination of values of each of the random variables affecting profit. Although there are, in most cases, an infinitely large number of possible combinations the dependent variable distribution will be fairly well defined by a number of repetitive computations in the 100–1500 range. For each of these repetitive passes a value for each random variable is obtained by sampling its originally specified distribution.

This sampling process is the key part of the analysis, and must be done in a manner such that the sequence of values sampled will be distributed in exactly the same manner as its original distribution. That is, if we specify oil recovery to be a lognormal distribution passing through the points of 50 BAF and 450 BAF at the 2% and 97% percentiles respectively we need to have a sampling scheme whereby the actual values of oil recovery used in the repetitive simulation passes will be distributed in exactly the same manner.

One sampling scheme which accomplishes this objective is to use random numbers as entry points on the cumulative percentile scale (or cumulative probability) of a cumulative frequency graph of the random variable distribution. The random number is set equal to the numerically equivalent percentile and the corresponding value of the random variable is read from the graph.

This is the value of the random variable for a particular pass. On the next pass another random number is obtained, it is set equal to its numerically equivalent percentile, and another value of the random variable is read from the graph. Over a series of such samplings the sampled values of the random variable will be distributed in exactly the same form as the originally specified distribution. This sampling scheme, although just one of several possible ways, is by far the most efficient method on a computer and it is used almost exclusively in performing the simulation passes.

Now let's go back and talk about the details involved in actually doing what we just described. The first thing we must do is convert all the distributions into their corresponding cumulative frequency distribu-

Type of Probability Distribution	Characteristics of the Corresponding Cumulative Frequency Distribution, and/or Procedures for Computing Cumulative Frequency
Normal	Cumulative frequency plots as straight line on normal probability graph paper (Examples: Figure 6.21(b), Figure 6.22)
Lognormal	Cumulative frequency plots as straight line on lognormal probability graph paper (Examples: Figure 6.25, Figure 8.2)
Uniform (Rectangular)	Cumulative frequency plots as straight line on coordinate graph paper. The 0% percentile corresponds to the minimum value of the random variable, and the 100% percentile corresponds to its maximum value (Example: Figure 6.26(b))
Triangular	Must use Equations 6.9 and 6.10 to convert to cumulative frequency. Equation 6.9 is used to compute areas to the left of the most likely value. Equation 6.10 is used to compute cumulative areas for values of the random variable to the right of the most likely value. (See example computations on pages 275 to 278.)
Histogram, or Relative Frequency Distribution	The plotting points for the cumulative frequency are determined by accumulating the relative frequencies (or percentages) from the minimum to the maximum values of the random variable. (See Tables 6.5, 6.6, and 6.7, and Figures 6.12, 6.13, and 6.14.)
Continuous Distribution (of any shape)	Use technique described on pages 236 to 243. The method is used to convert the distribution of Figure 6.16 to its equivalent cumulative frequency, Figure 6.18. This technique can be used for any continuous distribution, regardless of its shape (or lack thereof).

tions. Most of the details of converting distributions to their equivalent cumulative frequencies were given in Chapter 6.

For the case of continuous probability distributions the characteristics of and/or the procedures required are summarized here for the principal types of distributions. The cumulative frequency distributions are usually plotted on coordinate graph paper. However, normal and lognormal probability paper is sometimes used when working with normal and lognormal distributions to make use of the straight line characteristics of these distributions when plotted on probability graph paper.

If the probability distribution is discrete the corresponding cumulative frequency graph has a slightly different form. To illustrate how a set of data representing a discrete probability distribution is converted to its equivalent cumulative frequency form suppose our analysis includes a random variable, D, which can have the discrete numerical values given in the first column of Table 8.1. The likelihoods of occurrence of Column

Table 8.1
Example of a Random Variable Having
Discrete Values

Possible Values of Random Variable D	Estimated Likelihood of Occurrence of the Value of D (Relative Frequency)
D = 1	0.20
D = 2	0.40
D = 3	0.25
D = 4	0.10
D = 5	0.05
	1.00

2 were estimated by the explorationist as being representative for the prospect being considered.

If we were to construct a picture of this discrete probability distribution it would appear as in Figure 8.4. The cumulative frequency graph of this distribution would have the form of Figure 8.5. The heights of the vertical segments of the cumulative frequency correspond to the likelihoods of occurrence of each of the possible values of the random variable. Each successive vertical segment begins at the cumulative relative frequency at which the previous vertical segment ends. The vertical seg-

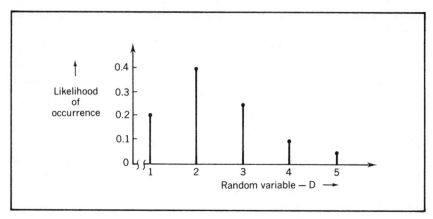

Figure 8.4 *Discrete probability distribution of random variable D. (From data of Table 8.1)*

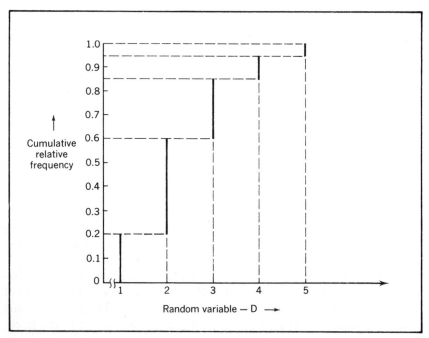

Figure 8.5 *Cumulative frequency distribution of the discrete probability distribution of Table 8.1 and Fig. 8.4. The dashed lines are merely to help in the interpretation of the graph.*

ment corresponding to the maximum possible value of the random variable terminates at a cumulative relative frequency of 1.0. The vertical axis can, alternately, be expressed as cumulative percentage by multiplying the relative frequency values by 100%.

There are several shapes for these types of graphs which are not valid cumulative frequency graphs, and we need to avoid the mistake of using these types of graphs in simulation analyses. These graphs are shown in Figure 8.6. The graph in Figure 8.6(a) is not a valid cumulative frequency distribution in that it implies negative probabilities. Recall in Chapter 6 we pointed out that a cumulative frequency distribution is a graph of the cumulative area under the probability distribution to the left of the value of the random variable as the value ranges from the minimum to the maximum.

As we move from the minimum value to the maximum value the cumulative area continues to increase. We observe in Figure 8.6(a) the cumulative area less than or equal to the random variable increases from RV_{MIN} to RV_1 (as it should), but then begins to decrease from RV_1 to RV_2. This could only occur if we began to *subtract* area under the curve. But this violates the fact that probabilities are always positive — there are

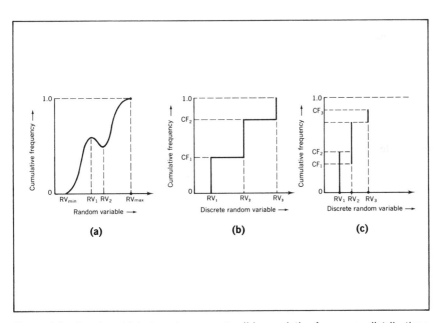

Figure 8.6 *Graphs which do not represent valid cumulative frequency distributions.*

no negative valued probabilities. Hence, a cumulative frequency distribution curve must always be a monotonically increasing function as the random variable ranges from the minimum to the maximum. Figure 8.6(a) violates this monotonicity requirement.

Figure 8.6(b) is invalid as a representation of a discrete random variable because of the horizontal portions of the graph at cumulative frequency values CF_1 and CF_2. The vertical segments should have discontinuities over the portions of the random variable scale which cannot occur. The sampling technique used in simulation will become inoperable when a cumulative frequency graph has a horizontal segment. The reason is that if the random number selected corresponds to CF_1 or CF_2 we are not able to determine a unique value of the random variable. That is, if the random number corresponded to CF_1 we would be unable to determine whether the corresponding value of the random variable is RV_1, RV_2, or some value in between. The rule is no horizontal lines on the cumulative frequency graphs!

The graph in Figure 8.6(c) is not valid for two reasons. First of all, it is not complete in that the sum of the relative frequencies for the three permissible values add to only CF_3, rather than 1.0. The sum of the likelihoods of occurrence of all possible numerical values of a discrete random variable must add to 1.0 by definition. So either the three probabilities shown in (c) must be normalized so that they add to 1.0, rather than CF_3, or we must specify what will occur the proportion of the time corresponding to $(1.0-CF_3)$. The second reason the graph of Figure 8.6(c) is invalid is that the vertical segments representing the relative frequencies of RV_1 and RV_2 overlap in the range CF_1 to CF_2. This again violates the monotonicity constraint.

Cumulative frequency graphs can be expressed in terms of cumulative fractions (probabilities) or cumulative percentages. In the latter case, the vertical scale is merely the cumulative fractions or probabilities multiplied by 100%. Most computer simulation analyses will operate with cumulative fractions, but it is a bit simpler to explain simulation using cumulative percentages. So our subsequent discussions may use these two equally valid scales interchangeably. Also, it is perfectly satisfactory to plot the cumulative frequency graphs on a "cumulative percent greater than or equal to" basis (such as Figure 6.15) rather than a "cumulative percent less than or equal to" basis. Either type of graph will be satisfactory for the sampling schemes of simulation.

The next detail we need to be concerned with is the definition and meaning of random numbers. We slipped random numbers into some of the previous descriptions of how the sampling of random variable distributions is accomplished without really defining what they are and why they are used. The use of random numbers in simulation is strictly for the purpose of having an easy, unbiased method of sampling values from the random variable distributions for each simulation pass such that the values sampled will be distributed in exactly the same form as the random variable distribution originally specified.

Using random numbers to perform this sampling process is just one of several ways to sample distributions. But it is by far the most efficient method and used almost universally in simulation analyses, so we won't even bother to explore some of the alternative sampling schemes.

Random numbers are dimensionless, positive numbers. A sequence of say, 2-digit random numbers (the digits 00, 01, 02, . . . , 98, 99) is said to be a sequence of random numbers if: a) there is no pattern in the order in which the numbers appear, and b) each of the 100 2-digit numbers is equally likely to occur in the sequence. Random numbers all have equal probabilities of occurrence.

At any point in a sequence of random numbers one number is just as likely to occur as any other numbers. We can speak of 2-digit random numbers, 3-digit randoms (000 up to and including 999), or random numbers expressed as decimal fractions over the range 0.0 to 1.0. It will be a bit more convenient to talk in terms of 2-digit random numbers in our discussions here. However, on computers random numbers are normally expressed as 8 or 16 digit decimal fractions.

Random numbers can be obtained from several sources. If we were interested in obtaining 2-digit random numbers we could simply toss a fair die which had one of the 100 2-digit numbers printed on each of 100 faces of the die. Repetitive tosses of such a 100-sided fair die would generate a sequence of 2-digit random numbers. Another source of random numbers is the sequence of digits to the right of the decimal point of irrational numbers (such as π, e, $\sqrt{2}$, etc.).

Still another source is the use of what are called random number generators. These are a series of simple (but very carefully designed) equations which when solved many times will result in a sequence of numbers which are approximately random — or so-called "pseudo-random numbers." Indeed this is the way random number tables, such as Appendix G, are obtained, and it is the way that random numbers are generated by a computer.

The random number generator algorithms are normally programmed in most present-day scientific computers as a standard library subroutine function. The programmer does not even have to be aware of how the numbers are generated, just how to access the subroutine to obtain a random number. The issue of whether or not a sequence of numbers satisfy various tests of randomness involves some highly mathematical and statistical concepts which we will not get into here. Suffice it to say at this point that although random number tables and random number generators produce sequences of pseudo-random numbers, for the purposes we will be using random numbers the issue of "pseudo" versus "truly" random will be of no concern to us.

For the sampling of probability distributions we can safely consider the random numbers obtained from a table or from a computer algorithm as being essentially random. (Those wishing to pursue the mathematical theory of random numbers should consult Reference 8.7 and/or pages 470–484 of Reference 1.8.)

We have included a few pages of a random number table in Appendix G. The details of how to enter and use the table are given at the beginning of the appendix section.

Now that we have defined random numbers the next obvious question is how are they used to sample distributions in a simulation analysis? The answer is that the random number is set equal to the numerically equivalent percentile (or fractile if using random numbers expressed as decimal fractions) to serve as the entry point on the cumulative frequency

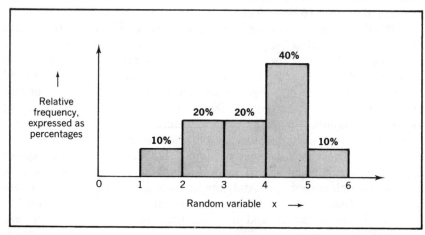

Figure 8.7 *Histogram of the probable distribution of values which the random variable x could assume in the exploration prospect being considered. The distribution was defined in step 4 of the analysis by using data from nearby, correlative fields.*

graph of the random variable being sampled. The sampled value of the random variable is the value corresponding to the cumulative percentile entry point.

Let's illustrate this with an example. Suppose one of the random variables affecting profitability, x, is distributed in the form of the relative frequency, or histogram of Figure 8.7. This histogram could represent, for example, a set of statistical data that might have been obtained by looking at x-data from a nearby, correlative area. The histogram is then converted to its equivalent cumulative frequency form as shown in Figure 8.8. Now suppose we are making the actual simulation passes in which we need to sample a value of x for each of the passes.

On a given pass suppose the random number obtained from the tables is 55. We then enter the vertical ordinate at the 55 percentile (the percentile numerically equivalent to the random number), read across to the

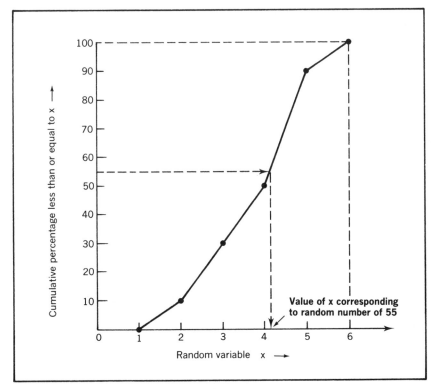

Figure 8.8 *Cumulative frequency graph of the histogram of Fig. 8.7.*

cumulative frequency curve, then down to the abscissa to read the corresponding value of the random variable x as about 4.1.

This is the value of x which is then used in the equation to compute profit on the simulation pass. On the next pass we would sample another value of x by obtaining another random number, entering the ordinate at the numerically equivalent percentile, and reading the corresponding value of x from the abscissa.

If the procedure for sampling values of x were repeated for, say, 200 passes we would have obtained 200 values of x. And if we had taken the time to tabulate and plot the 200 sampled values the resulting histogram of sampled values would have the exact form as the histogram of Figure 8.7. The mechanical scheme we've just described using random numbers will insure for us that the values we sample from a random variable distribution over a series of repetitive passes will be distributed exactly as we specified the variables to have been distributed in Step 4 of the analysis.

This is a very significant point, and you should be sure you completely understand the mechanics of this sampling scheme. Some people, for example, will get to this point and then dismiss simulation because they feel that when we introduced random numbers into the analysis we suddenly reduced the system of uncertainty to one of complete randomness.

But we need to remember that the random numbers have no effect on the system or resulting profitability — but rather are used for just the mechanical step of sampling values from each random variable distribution for each of the passes. It's a method which insures that we, in fact, honor the specific uncertainty of the system that we have described via the distributions of the random variables. We could have accomplished the same result without even using random numbers — but the alternative procedures are much more tedious and do not lend themselves readily for computer solution.

To demonstrate in perhaps another way why the random number sampling scheme works we can consider the distribution in Figure 8.9(a). By observing the distribution we can probably all agree that most of the values of W we sample should be within the range of W_1 and W_2. Only seldom should the sampled values of W be less than W_1 or greater than W_2. The approximate form of the distribution's cumulative frequency is given in part (b) of Figure 8.9.

If we enter the abscissa at the ranges W_1 and W_2 and proceed upward to the curve, then across to the ordinate we see that we have encom-

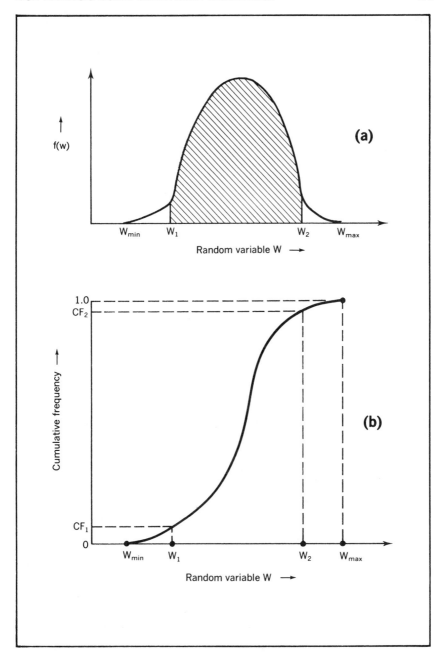

Figure 8.9 *A distribution of random variable W and its corresponding cumulative frequency graph.*

passed nearly the entire vertical scale between the corresponding cumulative frequencies CF_1 and CF_2. Since our entry point on the vertical scale is random, this implies then that most of the time our entry point will be within the range CF_1–CF_2. And this in turn implies that most of the time the corresponding values of W will be within the range W_1–W_2 as we originally hypothesized. The only chance we will sample a value of W less than W_1 is when the random number obtained is numerically less than CF_1.

We can generalize these comments to say that even though our entry point is random on the ordinate the corresponding value of the random variable is not random. As the slope of the cumulative frequency curve increases it exposes a larger proportion of the vertical axis for a given range on the horizontal axis. As the slope becomes less there is a smaller proportion of the vertical axis associated with a given range on the abscissa.

The slope of the cumulative frequency curve is the rate at which we are adding cumulative area under the probability distribution curve as we range from the minimum to the maximum value of the random variable. As we get near the center of the W distribution of Figure 8.9(a) we are adding area under the curve at a rapid rate — hence the slope of the cumulative is highest, meaning there should be greater chance of sampling a value of W in that range.

As we near W_{max} the rate at which we are adding cumulative area under the distribution diminishes — and the corresponding slope of the cumulative frequency diminishes, exposing less distance on the vertical scale for a given range of W. Even though our entry point on the ordinate is random the slope of the cumulative frequency controls the relative position of the value sampled from the abscissa. And this scheme will insure, therefore, that the values sampled will be in exactly the same form as the original distribution.

The only time the values of the random variable will also be randomly distributed is when the cumulative frequency curve has a constant slope from the minimum to the maximum value, as in Figure 6.26(b). But this corresponds to the cumulative frequency curve of a uniform, or random distribution in which we would have expected the values sampled to have been randomly distributed in the first place.

The sequence of repeated simulation passes is performed by sampling a value of each of the random variables, substituting these values into the profit equation(s) and solving for profit. Each pass requires that we sample from each of the random variable distributions. And we must use a different random number in the sequence to sample each distribution. Thus, if there are five random variables affecting profit we need five random numbers for each pass.

Note that it is *not* correct to use the same (single) random number to sample all the distributions on a pass. The reason for this is that using the same random number would automatically imply fixed values for all

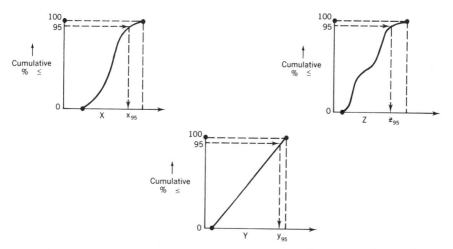

Figure 8.10   *Illustration to show why it is not correct to use the same random number to sample more than one distribution. Such a scheme implies that if the random number is high (say, 95) all of the values of X, Y, and Z will be high. If the number is low the values sampled would all be low. We must use a separate random number to sample each distribution.*

variables. For example, if the random number used to sample the X, Y, and Z distributions was, say, 95, as in Figure 8.10, we would obtain values of X, Y, and Z which were all near their upper limit, X_{95}, Y_{95}, and Z_{95}.

If the random number happened to be low the three values would all be low. Using the same random number to sample more than one distribution would not allow the possible combination of say a high value of X, a low value of Y, and a mid-range value of Z. So the rule is that we must use a separate random number to sample each distribution.[1]

After each pass is completed the computed value of the dependent variable profitability is written down, or stored in the computer. Each value of profit computed in this manner represents one state of nature, or possible combination of ways in which the random variables interact. The passes are continued until a sufficient number of values of profit are available to be able to define the profitability distribution.

Unfortunately, there are no prescribed rules to tell us exactly how many passes are required. The reason is that the number of passes required to define the dependent variable distribution is a function of: a) the number of distributions being sampled on each pass, b) the shapes and magnitudes of the individual distributions, c) the relationship which ties all the variables together, etc. As a general rule, however, we can state that as the number of distributions being sampled increases the number of passes required increases. In any case, the minimum number of passes is no less than about 100, and may range up to as many as 1000 or 1500.

In the absence of a rule about the minimum number of passes it is the usual practice to make, say, n passes and compute relative frequencies of various ranges of profit and the corresponding mean value. Then another increment of, say, 100 or 200 passes is made and the relative frequencies and mean of the (n + 100) passes are compared to the relative frequencies previously computed after just n passes. If they are essentially the same the profit distribution has stabilized (in terms of its shape and range) and you've made enough passes. If the relative frequencies and mean values after n and (n + 100) passes are not close together more passes are required.

On a computer such a "compute-test-compute-test" sequence may take more time in terms of input and output than if we simply made a very large number of passes in the first place, say 1000, or 1500. The computing time for making the actual passes is very small, and it may require

---

1. It is quite possible in obtaining three random numbers to sample the X, Y, and Z distributions of Figure 8.10 that two or three of the numbers might be the same. For instance, the three numbers obtained in sequence from the tables might be 95, 95, and 14. This is OK, and we need not be alarmed. The occurrence of the second 95 was just as likely in the sequence as any other 2-digit number. The important point is to use a random number to sample only one distribution. To sample the second distribution go to the next random number in the sequence.

less computer time in the long run to "overkill" and make an excess of passes and eliminate the intermediate input/output steps.

Generally it has been my experience that those who work with simulation on a regular basis begin to get a feel for the number of passes required. Typically they will start high and gradually work the number downward over a series of similar simulation analyses.

We have now essentially covered all the mechanics of simulation analysis except for the important issue of dependency. We mentioned near the end of our discussion of Step 4 that some of the random variables affecting profitability may, in fact, be related to one another. If this is the case we need to be able to sample from distributions of the random variables in a manner which will duplicate and honor the dependency.

So the questions before us are (a) how do we recognize that a dependency may exist between two or more random variables; and (b) given that we can recognize a dependency relationship how do we modify our sampling technique so as to be able to account for the dependency in the sampling scheme?

At this point I can relate the manner in which I first became aware of the need to account for dependency. One of the early analyses which I did back in 1966 related to an exploration play in which my colleagues had specified, among other things, distributions of net pay thickness and initial production rate per well.

In my ignorance I put each of these distributions into the analysis separately, and on each pass the two distributions were sampled separately as if they were independent of one another. To make a long story short the results from the analysis were wild, so I began to back track to see where I had gone astray. It only took a short while to find the mistake. The program sampled net pay thickness separately from IP (initial potential, bbls/day/well) and went through a decline curve analysis to compute the life of the field for each pass so as to be able to determine a NPV of the revenue.

If the value of thickness happened to be high (which implied large reserves) and at the same time the value of IP sampled happened to be low the resulting field lifes were as much as 243 years! And at the other extreme if the random numbers were such that thickness was low (implying low reserves) and IP was high the field was depleted in as little as 17 days! Needless to say, I had learned an important lesson on the importance of treating dependencies in simulation analyses of risk and uncertainty.

There are various statistical measures of correlation (or the lack thereof) which can be used to ascertain whether two or more variables are dependent upon one another. But I think an easier way that the practicing explorationist can use without having to learn a lot of statistical theory is to make a cross-plot of values of two variables of concern.

For instance, suppose we were concerned about whether variables A and B were related to one another, and we had some numerical data from nearby wells or fields which we considered representative. We could make a cross-plot of variable A versus variable B by plotting the numerical values of A and B for each data point. And suppose if we did this the cross-plot looked like the graph of Figure 8.11.

What would we conclude about any dependencies between the variables? Well, I think we could probably all agree there appears to be no relationship. We can have high values of A and low values of B, high values of B and low values of A, and about everything between. We can probably safely conclude in this instance there is no relationship or dependency, so we can sample from each distribution separately in the analysis. This is the one extreme where the random variables are independent.

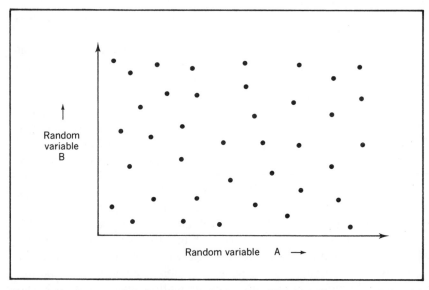

Figure 8.11 *A cross-plot of random variables A and B which shows no relationship between the variables. In this case we would describe distributions for each variable and sample each distribution separately in the simulation passes.*

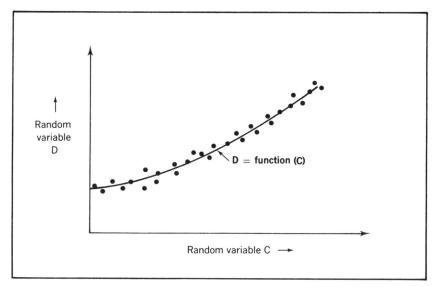

Figure 8.12 *A cross-plot of random variables C and D in which they are completely dependent. In this case we need specify only a distribution of C. Knowing a value of C we can enter the graph (or solve the equation D = f(C) with the value of C) to determine the corresponding value of D to use in the pass.*

Now suppose we make a similar cross-plot between two other random variables, C and D and the results are as shown in Figure 8.12. What now? This represents the other extreme of complete dependency. If we know C, we automatically know D (or vice versa). So in terms of the simulation analysis we only need to specify a distribution of one of the variables (whichever is easiest to define). For each pass we would sample the distribution to determine its numerical value.

We would then enter the cross-plot with that value to read the corresponding value of the other variable to use in the pass. Or, alternately if this were being done on a computer we would determine the function (equation) of the curve and use the equation to establish the value of D given C (or vice versa).

These two instances represent the either/or case of dependency, and are simple to handle. But now let's look at the case where there is sort of a "half-way" dependency. Again suppose we had numerical data of two (unknown) random variables X and Y from a nearby area, and that the cross-plot of X–Y data appeared as in Figure 8.13. Now we're really in trouble! The X–Y data are not clearly independent (as in Figure 8.11),

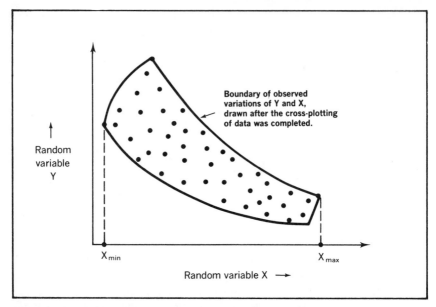

Figure 8.13 *Cross-plot of random variables X and Y in which there is a partial dependency between X and Y, but the dependency is not a specific function as in Fig. 8.12. After the data points were plotted a boundary, or envelope was drawn around the observed data. Based on this information any combination of X-Y values outside of the boundary would not be included in the simulation analysis.*

and there does not appear to be an explicit dependency relationship such as Figure 8.12.

The variable Y has a substantial range of values, given X which suggests the need to sample from a Y distribution for each pass. But it also should be rather obvious that the range of variation of Y changes as X increases. This means that the distribution of Y that we need to sample from must somehow be related to the specific value of X that we sample on each pass.

The situation shown in Figure 8.13 represents the "gray" area between the extreme black and white instances of complete dependency or complete independence. So the obvious question then is how do we modify our sampling scheme so that we can sample from both the X and Y distributions on each pass but in a way which will honor the observed partial dependency of Figure 8.13?

Well, we have our work cut out for us at this point. The way to do this is very simple, but it is a bit tricky and involved to explain. So before we plunge into the "swamp" let's at least know where we are going.

First, I'll try to explain Table 8.2 which lists the initial data preparation steps which we must follow to handle the situation of partial dependency. Then we will walk through the conceptual steps of actually sampling from two variables which exhibit a partial dependency. This will all be given in Table 8.3. Next we'll go through the same steps as if it were being done on a computer (Table 8.4).

This will then be followed by a numerical example. Finally, we shall mention some general points about recognizing and treating partial dependencies in simulation. By that time you'll either be an expert on partial dependency or you'll be thoroughly confused (or both!). It's really very simple to handle dependencies once you get the hang of it. So . . . right on!

To account for partial dependencies in simulation analysis we must first follow a few "data preparation" steps. These steps are given in Table 8.2. The first two steps are self explanatory, and at the completion

Table 8.2

Data Preparation to Account for Partial Dependencies Between Random Variables in Simulation Analyses

1. Prepare a cross-plot of available numerical X–Y data as in Figure 8.13.
2. Draw a boundary, or envelope around the observed set of X–Y data points. This boundary defines the limits within which X and Y can vary. Any combination of values of X and Y outside the boundary will be excluded and considered as having zero probability of occurrence.
3. Determine the variation of Y within the boundary *as a function of X,* and define a normalized distribution of Y which represents the observed variation. The limits of this normalized distribution are 0 and 1.0, where 0 corresponds to the minimum possible value of Y for a given value of X and 1.0 corresponds to the maximum possible value of Y for the given value of X. The dimensionless, normalized distribution is given the symbol YNORM on its random variable axis. The distribution can be of any form.
4. The input to the simulation model are:
   a. The X distribution cumulative frequency
   b. The cumulative frequency of the normalized Y distribution, YNORM
   c. The following equation used to compute the value of Y for each pass:

$$Y = YMIN_x + (YMAX_x - YMIN_x)(YNORM) \qquad (8.2)$$

   where: $YMIN_x$ = Minimum possible value of Y, given a sampled value of X
   $YMAX_x$ = Maximum possible value of Y, given a sampled value of X
   $YNORM$ = Value of the dimensionless normalized Y distribution that is sampled each pass
   Y = Computed value of random variable Y to use for the pass.

of these two steps we would have a bounded, cross-plot such as Figure 8.13. It should be noted that the boundary around the observed X–Y plotting points does not have to be drawn as straight lines. It should be drawn, however, so as to include all possible (observed) X–Y values. The reason is that the subsequent sampling scheme will exclude all X–Y combinations outside the boundary.

In the third step of Table 8.2 we must specify the type of variation of Y within the boundary *as a function of X*. Conceptually this is equivalent to cutting away the graph paper outside of the boundary of Figure 8.13 and looking only within the boundary to see how Y is distributed as a function of X. That is, within the boundary are the Y values scattered vertically at random? Are most of the values near the upper or lower boundary? Is there a clustering of data midway between the boundaries?

Be sure to note here that we are trying to ascertain the variation of Y *as a function of X*, which means for a given value of X we are concerned about the distribution of Y values along a vertical ordinate erected above the value of X.

For the data of Figure 8.13 we could probably agree that the variation of Y as a function of X is essentially random. This means that the distribution which describes the variability of Y as a function of X is a uniform, rectangular distribution.

We have the choice of defining the variability of one variable as a function of the other with any type of distribution which approximates the data. In Figure 8.14 are given sketches of some possibilities in this regard. In Part (a) the plotted X–Y data suggest that the variability of Y within the boundary is uniform. The particular uniform Y distribution corresponding to the specific value X_1 is shown in the inset above. The intersection of the X_1 ordinate with the upper and lower bounds define the limits of this specific, unique Y distribution, and are given on the graph as YMIN and YMAX.

For Figure 8.14(b), the variability of Z as a function of W is not random. Rather, there appears to be a concentration of data points along a loci of points approximately midrange between the upper and lower bounds. In this case we might choose to represent the variability of Z within the bounded region as a series of symmetrical triangular distributions. The particular distribution corresponding to a value W_1 appears in the upper inset of Figure 8.14(b). We could have also possibly represented the variability of Z as a series of normal distributions.

Figure 8.14(c) shows an instance where the variability of T as a functions of S could be represented by triangular distributions skewed

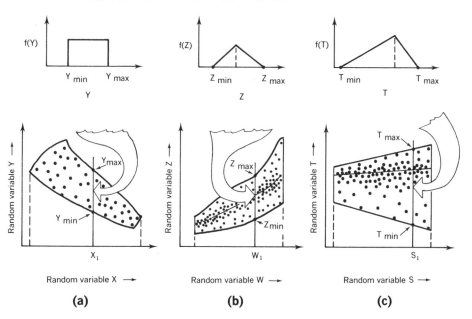

Figure 8.14 *Examples of various types of variability within the bounded regions of partially dependent random variables. In graph (a.) the variability of Y as a function of X is random. In graph (b.) the variability of Z as a function of W could be approximated by a symmetrical triangular distribution (or perhaps, alternately, a normal distribution). Graph (c.) shows a variability of T as a function of S which could be represented by a skewed triangular distribution with its most likely value near the maximum value to reflect the relative density of points in the corresponding area of the cross-plot. The inset distributions represent conceptually the Y, Z, and T distributions corresponding to the vertical ordinates drawn as X_1, W_1, and S_1 respectively.*

toward the maximal values of T to reflect the concentration, or clustering of data in the upper portion of the bounded area. The skewed triangular distribution corresponding to S_1 is shown in the inset. In lieu of such a skewed triangular distribution we could perhaps have chosen to represent the variability with a lognormal distribution having its mode value near the maximal values of T.

At this point, we presumably recognize what is meant by describing the variation of Y within the bounded region as a function of X. It is simply a matter of trying to characterize the scatter of plotted X–Y data points by a type of distribution such as uniform, triangular, etc. But having described the form, or shape of the distribution of variability of Y is only half the problem.

The other dimension we must be concerned with is the fact that there exists a specific, unique distribution of Y (in Figure 8.14(a) for instance) *for each value of X!* And since X is drawn as a continuous random variable there are an infinite number of possible values of X between its minimal and maximal values, X_{MIN} and X_{MAX}. For each value of X there is a separate unique uniform distribution of Y, caused by the fact that the numerical limits of Y, YMIN and YMAX, vary as X varies. For those of you who like to (and are able to) think in 3 dimensions, the 3 dimensional surface we get when we add a frequency scale to the X–Y cross-plot of, say, Figure 8.13 is as shown in Figure 8.15.

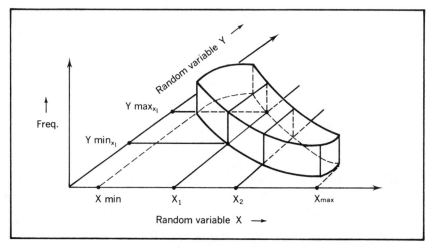

Figure 8.15 *Three dimensional representation of the partial dependency of Fig. 8.13. The third axis is the frequency scale. The projection of the top surface of the figure down to the X-Y plane is the boundary of Fig. 8.13. This figure represents the case where the variability of Y as a function and X is uniform, or random.*

The way such a figure evolves can be thought as follows. Suppose we erect a vertical wall around the bounded area of X–Y data in Figure 8.13. Then since the variability of Y within the bounded area was considered to be random we lay a flat "roof" on the vertical boundary. The result is a 3-dimensional surface of Figure 8.15.

Our problem of sampling X and Y values for each pass, as related to Figure 8.15, is that once we sample (in the usual manner) a value of X, say X_1, from the X distribution we have fixed our position on the X axis of the 3-dimensional graph. To then sample a value of Y we would, in essence, cut the surface with a knife to look at the distribution profile along the face of the cut. It would be a rectangular distribution having its "walls" at $YMIN_1$ and $YMAX_1$ and a flat "roof."

This would be the exact distribution from which we would sample a value of Y. On the next pass the value of X we sample may be X_2. If we cut the surface of Figure 8.15 at this point (X_2) we see yet another uniform distribution. Its shape is the same but the numerical values of its limits are different. And indeed for every possible "cut" of the surface we would expose a different Y distribution. The shapes would be the same (two vertical "walls" and a flat "roof"!) but the numerical values of the limits would all be different.

So far so good. But the catch is that what we have just described implies that we need to store, or write down the specific Y distribution for each possible value of X. In this manner once we establish a value of X each pass we could go to the Y cumulative frequency graph *corresponding to that value of X* to sample the value of Y for the pass.

If there are many (infinite?) possible values of X this means that we have to graph many (infinite?) cumulative distributions of Y. And that may be an extremely tedious job! So what we need is some way to represent all the many, distinct Y distributions in some form so that we don't have to spend the rest of our working careers plotting Y distributions.

Fortunately, the technique to do that is relatively straight-forward, and this method is really the linchpin which holds this simple(?) procedure for treating partial dependencies together. It involves the normalized distribution mentioned in Step 3 of Table 8.2. To normalize a probability distribution means to convert the random variable scale to an equivalent dimensionless scale which ranges from zero to one.

The equation, in general form, which accomplishes this normalization is:

$$X' = \frac{X - X_{MIN}}{X_{MAX} - X_{MIN}} \qquad (8.3)$$

where X is the random variable, X_{MIN} is the minimum value of the distribution, X_{MAX} is its maximum value, and X' is the transformed random variable of the normalized distribution. Notice that when $X = X_{MIN}$ solution of equation 8.3 leads to a value of X' of zero. When $X = X_{MAX}$ the equation yields a value of X' of 1.0. Since I've been trying to keep this discussion in terms of the general partially dependent variables X and Y, and since we are interested in normalizing the variability of Y we'll rewrite the equation in the symbols of Table 8.2 as:

$$YNORM = \frac{Y - YMIN}{YMAX - YMIN}$$

where

YMIN = Minimum value of Y
YMAX = Maximum value of Y
YNORM = Normalized, dimensionless random variable

The reason it will be extremely useful to deal with a normalized distribution is that it is a representative distribution which describes all the infinite possible Y distributions as functions of X. For each pass we could sample a value of YNORM from the single, representative distribution, and this would tell us the fractional distance along the line from YMIN to YMAX to locate the desired sampled value.

The only final step then would be to associate this fractional distance to the specific numerical values of YMIN and YMAX associated with the value of X previously sampled. This final step is accomplished using the equation (8.2) of Table 8.2.

With reference to Figure 8.14, all the infinite number of Y distributions would have as their equivalent normalized distribution a uniform distribution of the parameter YNORM, having as its minimum value zero and its maximum value 1.0. For the case of the symmetrical triangular distributions of Figure 8.14(b) the equivalent normalized distribution would be triangular having as its parameters a minimum value of zero, a most likely value of 0.5 (by symmetry the mode is midway from the minimum value to the maximum), and a maximum value of 1.0.

For the case of Figure 8.14(c) in which the most likely values appear to be about three-fourths of the distance toward the upper bound the equivalent triangular distribution would have a minimum value of zero, a most likely value of 0.75, and a maximum value of 1.0.

In all cases the minimum numerical value of the Y variable will correspond to zero on the normalized distribution and the maximum numerical value of Y will correspond to 1.0 on the normalized distribution. And having determined the distribution of the dimensionless, normalized variable YNORM we need to express the distribution in its equivalent cumulative frequency form so as to be able to sample a value of YNORM for each simulation pass using a random number.

All the things we have just described regarding the treating of partial dependencies will insure that over a series of repetitive samplings of X and Y all the X–Y values will fall within the originally defined bounded area, and the distribution of Y as a function of X will be in the exact form as we originally specified in Step 3 of Table 8.2. Or stated more simply, the scheme we've just described will insure that our sampled values of X and Y over a series of simulation passes will have precisely the same partial dependency that we would have observed when we first made a cross-plot such as Figures 8.13 and 8.14. *Amazing!*

Let's see how this would work in the actual simulation analysis. In Table 8.3 I have prepared a schematic drawing of the conceptual steps to sample two variables exhibiting partial dependency. The adjective conceptual here means this is the way it would be done if we were sampling values by hand with the actual cumulative frequency graphs before us. As we'll see in a moment the actual sequence of steps on a computer has a few modifications.

Table 8.3 shows how one pass would proceed. Since we are sampling two random variables we would need two random numbers, even though in this case the variables are related to one another. We can use the first random number, RN_1, to sample the X distribution and the second, RN_2, to sample the normalized YNORM distribution (Step 1 of Table 8.3). Having determined a value of X for the pass we use the X value as our entry point into the X–Y cross-plot.

By erecting an ordinate above the value of X we can determine YMIN and YMAX from the intersection points of the ordinate and upper and lower boundaries. These limiting values of Y, together with the sampled value of YNORM are placed in equation (8.2) to solve for the value of Y to use for the pass. It's as simple as that. For successive passes the steps of Table 8.3 are merely repeated so as to sample new values of X and Y.

If the simulation analysis is being done by computer (as it normally is) the steps of Table 8.3 are modified somewhat. The principal reason we need to modify the procedure slightly is that the actual cross-plot used in Step 2 of Table 8.3 will not be available in the computer. So instead, we will have to input to the computer the equations of the lower boundary and the upper boundary as functions of X.

In this manner once the value of X is sampled it is substituted into the equation of the lower boundary to compute the value of $YMIN_x$ (as opposed to reading the value directly from the cross-plot). Similarly, the value of $YMAX_x$ is determined for the pass by solving the equation of the upper boundary for the sampled value of X. These values and the sampled value of YNORM are then used in equation (8.2) to solve for Y in the same manner as before.

If the equations of the upper and lower boundary lines are not known the analyst can read into the computer a few X–Y points on each boundary and let the computer derive equations using a polynomial approximat-

TABLE 8.3

CONCEPTUAL STEPS TO SAMPLE VALUES FOR EACH PASS OF TWO PARTIALLY DEPENDENT RANDOM VARIABLES X AND Y

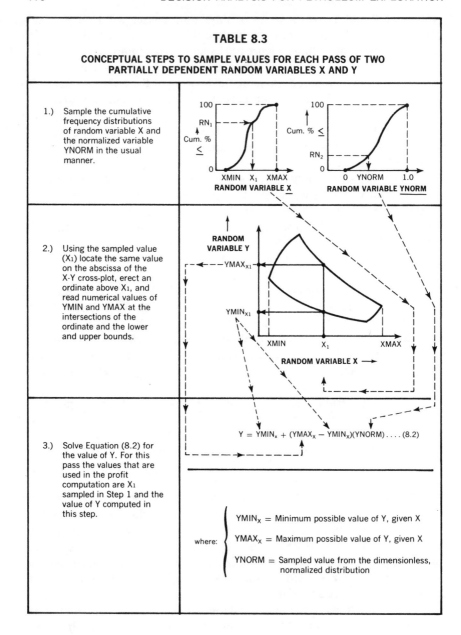

1.)	Sample the cumulative frequency distributions of random variable X and the normalized variable YNORM in the usual manner.
2.)	Using the sampled value (X_1) locate the same value on the abscissa of the X-Y cross-plot, erect an ordinate above X_1, and read numerical values of YMIN and YMAX at the intersections of the ordinate and the lower and upper bounds.
3.)	Solve Equation (8.2) for the value of Y. For this pass the values that are used in the profit computation are X_1 sampled in Step 1 and the value of Y computed in this step.

$$Y = YMIN_x + (YMAX_x - YMIN_x)(YNORM) \dots \dots (8.2)$$

where:
$YMIN_x$ = Minimum possible value of Y, given X

$YMAX_x$ = Maximum possible value of Y, given X

$YNORM$ = Sampled value from the dimensionless, normalized distribution

TABLE 8.4

STEPS TO SAMPLE VALUES OF TWO PARTIALLY DEPENDENT RANDOM VARIABLES X AND Y ON A COMPUTER

METHOD #1 (Equations of the boundaries on the X-Y plot are known)	METHOD #2 (Equations of the boundaries on the X-Y plot are not known)
Step 1.) Input to computer: a.) Cumulative frequency distribution of random variable X b.) Cumulative frequency distribution of normalized Y distribution, YNORM c.) Equation 8.2 d.) Equations of upper and lower boundaries as functions of X	**Step 1.)** Input to computer: a.) Cumulative frequency distribution of random variable X b.) Cumulative frequency distribution of normalized Y distribution, YNORM c.) Equation 8.2 d.) Four to eight X-Y coordinate points from the upper and lower bounds.

Method #1 diagram: RANDOM VARIABLE Y (vertical axis) vs RANDOM VARIABLE X (horizontal axis). Curves labeled $Y_{MAX} = f(X)$ and $Y_{MIN} = f(X)$.

Method #2 diagram: RANDOM VARIABLE Y (vertical axis) vs RANDOM VARIABLE X (horizontal axis). "Upper boundary Coordinate Points read as X-Y Coordinates" and "Lower bound coordinates".

METHOD #1	METHOD #2
Step 2.) Sample values of X and YNORM in usual manner. **Step 3.)** Substitute sampled value of X into equations for YMIN and YMAX to compute YMINx and YMAXx. **Step 4.)** Substitute computed values of YMINx and YMAXx, and sampled value of YNORM into Equation 8.2 and solve for Y.	**Step 2.)** Using coordinate X-Y values from lower bound have computer fit a polynomial equation through the points of the form YMIN = f(X). Repeat for the upper boundary points to derive polynomial equation of upper bound, YMAX = f(X). **Step 3.)** Sample values of X and YNORM in usual manner. **Step 4.)** Substitute sampled value of X into equations for YMIN and YMAX to compute YMINx and YMAXx. **Step 5.)** Substitute computed values of YMINx and YMAXx, and sampled value of YNORM into Equation 8.2 and solve for Y.
Note: Step 1.) is done only once. Steps 2.), 3.) and 4.) are repeated for each pass.	**Note:** Steps 1.) and 2.) are done only once. Steps 3.), 4.), and 5.) are repeated for each pass.

ing subroutine (a standard library subroutine algorithm on most scientific computers). All these steps for computer applications are summarized in Table 8.4 on page 417.

At this point let's look at a practical, real-world situation involving partial dependency between two random variables.

Numerical Example

Bob Petersen, a staff geologist for "Success" Oil Company, has prepared a new exploratory well recommendation in ABC County. Although he had some data from a nearby field, he considered the prospect to be in the "rank wildcat" category. To evaluate some of the uncertainty he decided to use the method of simulation.

The first step in his analysis was to formulate an economic model for the prospect. After carefully studying the technique of simulation he devised a model having three basic sub-sections:

1. A routine to generate a distribution of total recoverable field reserves. Random variables: productive acres, net pay thickness, and primary recovery factor in units of bbls./acre-foot.
2. An accounting of total development expenditures. Random variables: number of producing wells, number of development dry holes, drilling costs.
3. A projection of future revenues and operating expenses so as to be able to compute a discounted net present worth of future revenues. Random variables: initial well potentials, operating costs.

Each simulation pass would include these three sub-sections, together with related parameters such as probability of finding oil, crude prices, discount rates, etc. The final dependent variable would be a net present worth of the prospect. After running sufficient passes he thought he would be able to estimate the distribution of total profitability, as well as the mean or expected value profit.

The next step in his analysis was to ascertain distributions of possible values which each of the random variables could assume in the prospect area. From a review of the nearby field he concluded that net pay thickness and initial potential data from the 31 wells in the field would probably be representative of expected pay thicknesses and potentials in the prospect area. His first step was to tabulate data from the 31 wells in the field, as shown in Table 8.5.

Bob studied the initial potential (IP) data and noted wide variations in net pay and potential. For example, Well C-5 with 20 feet of pay had an IP of only 75 BOPD, while Well B-1 with 10 feet of pay was completed for 197 BOPD. He recognized that when initial potential values are sampled in part 3

Table 8.5

Table of Net Pay Thickness and Initial Potential Data from 31 Wells in a Nearby Field

Well No.	Net Pay Thickness, Feet	Initial Well Potential, BOPD
B-3	5	25
C-1	5	102
D-4	6	160
C-7	9	72
B-1	10	197
D-6	10	175
C-8	10	112
E-1	11	134
A-5	12	68
D-7	15	98
C-6	17	195
D-1	19	172
E-7	20	150
C-5	20	75
E-4	23	110
D-8	26	172
A-3	28	200
B-2	30	140
D-2	31	93
C-4	37	190
E-3	38	139
E-6	40	225
A-2	45	245
A-4	47	182
C-3	54	230
E-2	59	170
A-1	65	225
E-5	66	185
D-3	70	275
D-5	78	237
C-2	80	300

of the model the sampling should include a dependency on net pay if such a dependency existed. But from looking at the data he was not certain if IP and net pay were related to one another in some manner.

Question: Is there a dependency between net pay thickness and initial potential? If so how can the partial dependency be described and used in his simulation analysis?

Solution: Now that we are experts on handling partial dependencies we can advise Bob Petersen on how to proceed. First, we need to make a cross-plot of the net pay and IP data to see if a partial dependency even exists in the first place. So that we stay with consistent nomenclature let's let net pay thickness be random variable X, and IP be random variable Y. The X–Y cross-plot of the data of Table 8.5 is given in the graph of Figure 8.16.

We can observe from the cross-plot that there is, indeed, a partial dependency between the two variables. As thickness increases the per well productivity increases (as it should by virtue of Darcy's equation), but there is, nonetheless a considerable "band" of variation of productivity for given values of pay thickness.

So it would probably be advisable to sample values of thickness and potential using the techniques for partial dependency that we've been discussing. To accomplish this for Bob we need to follow the data preparation steps of Table 8.2.

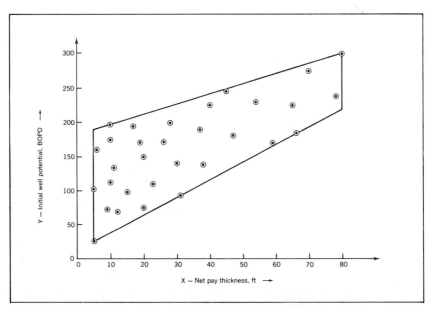

Figure 8.16 *Cross-plot of net pay thickness and initial potential (IP) data of Table 8.5. Each plotted point represents one of the 31 wells listed. A partial dependency appears evident, and the area of observed X-Y data has been bounded by the four-sided polygon shown.*

The first step, plotting of the cross-plot, is finished. The boundary around the area of possible X–Y combinations has been drawn on Figure 8.16 as a four sided polygon. The choice of straight lines for the upper and lower bounds is, of course, discretionary. But it seems to be a fairly good "fit," given the limited number of observed data points. In Step 3 we need to describe the variability of IP within the bounded region as a function of net pay.

I would judge the variation to be essentially random. (Do you agree?) Assuming this, the normalized distribution of IP as a function of thickness, YNORM, will be a uniform, or rectangular distribution having its minimum value at zero and its maximum value at 1.0. And since the cumulative frequency of a uniform distribution is a straight line (as in Figure 6.26b) we can graph the cumulative frequency of the normalized variable YNORM directly, Figure 8.17.

The next bit of input information we need to give Bob is the distribution of X, net pay thickness, expressed on a cumulative frequency basis. The thickness data of Table 8.5 define the distribution of the X variable, but since we only require a distribution's equivalent cumulative frequency graph in simulation we can go directly to a cumulative and not even bother with the distribution itself.

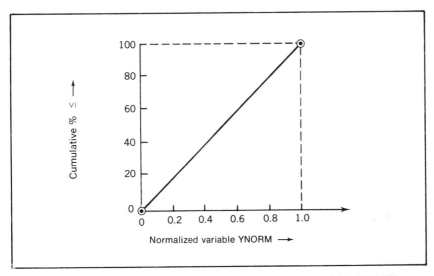

Figure 8.17 *Cumulative frequency graph of the normalized variable YNORM representing the variability of initial potential as a function of pay thickness in Fig. 8.16.*

Since the thickness data are arranged in numerically ascending order we can obtain the cumulative frequency plotting points with only a minimum of side calculations. More specifically:

Value of Random Variable, X, Net Pay Thickness (Feet)	Number of Wells Having Thickness Equal to or Less Than X	Percent of Wells Having Thickness Equal to or Less Than X
5	2	6.5%
10	7	22.6%
15	10	32.3%
20	14	45.0%
30	18	58.0%
40	22	71.0%
54	25	80.7%
70	29	93.5%
80	31	100.0%

We can plot these computed percentages versus the value of X on co-ordinate graph paper, as shown in Figure 8.18.

Now that all our graphs are plotted let's go through a pass following the steps of Table 8.3. We'll need two random numbers from the random number table, and for this example let's assume the random numbers obtained were 65 and 31. The steps to determine a value of X (thickness) and Y (IP) for this pass are:

1. Sample values of X and YNORM:

   Entering Figure 8.18 with the random number 65 gives a value of X of *35 feet*

   Entering Figure 8.17 with the random number 31 gives a value of YNORM of *0.31*

   (You should check these values yourself!)

2. Using the sampled value of X = 35 feet we enter the X–Y cross-plot of Figure 8.16 on the abscissa at 35 feet and erect an (imaginary) ordinate above a value of X = 35 feet. The intersection of the ordinate with the minimum and maximum values of Y give values of:

$$\text{YMIN}_{35} = 102 \text{ BOPD} \left.\right\} \leftarrow \begin{array}{l}\text{Do you check}\\ \text{these values?}\end{array}$$
$$\text{YMAX}_{35} = 234 \text{ BOPD}$$

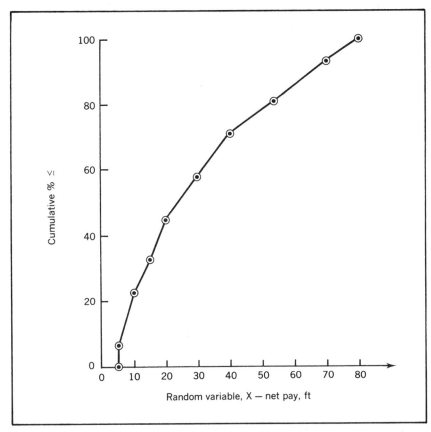

Figure 8.18 *Cumulative frequency graph of random variable X, net pay thickness. Data are from Table 8.5.*

3. Solving equation 8.2 for Y:

$$Y = YMIN_x + (YMAX_x - YMIN_x)(YNORM)$$
$$Y = 102 \text{ BOPD} + (234 \text{ BOPD} - 102 \text{ BOPD})(0.31)$$
$$Y = \underline{143 \text{ BOPD}}$$

The values we would use for the pass for X and Y would be X = 35 feet and Y = 143 BOPD. If you would like to practice with another pass use the random numbers of *80* and *55*. You should end up with values of X = 52 feet and Y = *208* BOPD. Did it check?

If Bob Petersen planned to run this analysis on a computer we would need to follow the steps of Table 8.4. First we'd have to input to the computer the cumulative frequency graphs of X and YNORM (that is, Figures 8.18 and 8.17). We'll explain how to do this in a later section of this chapter dealing with computers, so for now just assume we can do it reasonably efficiently. Step 1c of Table 8.4 requires that we also give the computer Equation 8.2.

No problem there, it just requires one card with the equation punched in it. Step 1d requires that we specify the equations of the upper and lower bounds of the X–Y cross-plot of Figure 8.16. Since the bounds were given as straight lines all we need to do is find the equations of each boundary as functions of X.

Here's how to do it. From our algebra classes of some years back(!) we probably recall that the equation of a straight line is given as:

$$y = mx + b \qquad (8.4)$$

where m is the slope of the line and b is the intercept at $x = 0$. The equation of a straight line can also be computed using the x-y coordinates of two points along the line with the following equation:

$$\frac{y - y_1}{x - x_1} = \frac{y_2 - y_1}{x_2 - x_1} \qquad (8.5)$$

Let's use Equation 8.5 to find the equation of the upper boundary line of Figure 8.16. To do this we need to read two points off the line:

$$X_1 = 10 \rightarrow Y_1 = 197$$
$$X_2 = 80 \rightarrow Y_2 = 300$$

We could, of course, read any other points along the upper boundary line and we'd still end up with the same equation. Substituting these values into Equation 8.5 (and changing to X and Y, rather than x and y) gives:

$$\frac{(Y - 197)}{(X - 10)} = \frac{(300 - 197)}{(80 - 10)}$$

After a bit of algebra and simplification this reduces to:

$$Y = 1.472X + 182$$

But since the upper bound is YMAX in the nomenclature of Table 8.4 we can rewrite this as:

$$YMAX = 1.472X + 182$$

This is the equation we would enter into the computer as the functional representation of the upper boundary of the cross-plot. By following the same method of reading two points off the lower boundary line of Figure 8.16 (e.g., $X_1 = 31$, $Y_1 = 93$; $X_2 = 66$, $Y_2 = 185$) we can obtain a correlative lower boundary equation of:

$$YMIN = 2.63X + 12$$

This explanation of handling partial dependency has been, of necessity, a bit long and involved, but I think we have covered the main aspects of the method. There are, however, a few points which require further comment.

1. Our description has been in the terms of random variables X and Y. We can define these variables as we choose, including the possibility that they may be clusters of data. For example, we may wish to analyze partial dependency between initial potential per well, X, and oil recovery in barrels per acre foot, Y. In this case Y is, itself, a grouping of several random variables: porosity, water saturation, recovery factor, and oil formation volume factor.

2. We have predicated the approach to handle partial dependency on having a cross-plot of X–Y data, such as Figures 8.11–8.14 and Figure 8.16. But if we do not have the statistical data available in the first place how do we obtain the all-important cross-plot? Certainly an important question, and one which has particular relevance in new exploration areas for which little or no data are available.

   Without the data we obviously won't be able to "model" the partial dependency in quite the degree of accuracy, but it is nonetheless just as important to be concerned with dependency. That is, we should not fall into the trap of deciding to sample each variable separately (independently) because we have no information to do otherwise. Even though we have no data initial potential and thickness, for example, are still related to one another!

   One approach is to make various preliminary engineering calculations to try to define the limits (boundaries) of probable variation between two partially dependent random variables. By running a series of "what if" cases it may be possible to at least define the boundaries of the dependency variation. We'll be more

precise on this point and discuss an actual example in the following section on applications and example.

As an alternative, we may be able to judge a probable trend
between two variables and then assign a range, or band on either
side of the trend line so as to allow variability of Y as a function
of X in the analysis. This is illustrated in Figure 8.19. With such
an approach we can at least define the bounded region within
which values of X and Y can occur. As far as describing the distribution of Y within the bounds as a function of X we would
probably have to use a uniform distribution if any Y value within
the range is considered equally likely to occur. If the center trend
line is thought to be more probable we could represent the variation of Y within the boundaries with a triangular distribution.

3. In Figure 8.14(b) and (c) we suggested that a symmetrical and a
skewed triangular distribution could be used in the respective
instances to represent the observed variation of Z and T. These
specific distributions could be used because the loci of most
probable values were always the same distance from the lower

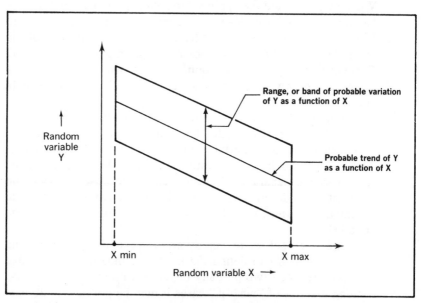

Figure 8.19 *Example of estimating a trend line and assigning a range on either side
of the trend line to approximate probable variation of Y as a function of X.*

boundary to the upper boundary for all values of W and S. But what if the position of the loci of most probable values changes (with respect to its distance between the bounds) as X ranges from X_{MIN} to X_{MAX}, as in Figure 8.20?

We can no longer represent all the infinite Y distributions (one for each value of X) by a single, representative distribution. In fact, each of the Y distributions has a separate, specific shape. Towards the low end of the X range the mode is skewed toward YMIN. As X increases the most probable value moves toward YMAX, as shown by the upper insets in Figure 8.20.

The solution is that an extra input is given to the computer: the equation of the line representing the loci of most probable values as a function of X. Thus, we need to find the equation of the curve as a function of X, or give the computer a set of co-

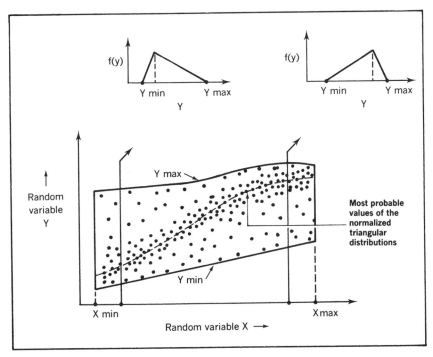

Figure 8.20 *Cross-plot of X-Y data in which the position of the most probable values changes between XMIN and XMAX. For this case the position of the most probable values must be treated as a function of X.*

ordinate points from the line and let the computer fit a polynomial equation through the points.

Knowing the equation of the loci of most probable points doesn't mean we've reduced the infinite number of YNORM triangular distributions to just one, but it does solve our problem because the computer can sample from the specific YNORM distribution given a value of X. We'll explain the mechanics of how the computer does this in Section IV of this chapter.

The point I make here is that if the variability of Y as a function of X is as given in Figure 8.20 we can handle the observed partial dependency by specifying the equation of the loci of most probable values as a function of X.

4. In our discussions on partial dependency we've always stressed the fact that the variability we are concerned with on the X–Y cross-plot is Y *as a function of X*. But what about the variability of X? How do we account for the apparent fact that in Figure 8.16 most of the data (when viewed as a variation along the X axis) occur in the range of low values of X? If you study the cross-plot you can observe this clustering of data in the range of X = 5 feet to about X = 40 feet.

    This will all be taken care of by the fact that we first sample the X distribution so as to determine the entry point into Figure 8.16. And the observation we have made merely implies that the net pay distribution is skewed with most of its area near the low end of the range. But by the mechanics we described earlier of why the sampling procedure using a cumulative frequency graph works we will exactly duplicate this observed variation.

    The cumulative frequency distribution will control the value of X for each pass. And it will control the variability of X. The scheme for handling partial dependency takes full account of this, and over a series of passes the X–Y values sampled will exhibit the exact relationships of partial dependency of Figure 8.16.

5. The procedures for treating partial dependencies in simulation analyses involved a bit of extra work on the part of the analyst. Is it worth it? For instance, if there existed a partial dependency such as shown in Figure 8.21 why couldn't we just fit a least-squares function through the observed data points and treat Y as a direct function of X using the equation of the function?

    The answer is yes, we could do that. And indeed, it would simplify the analysis. For each pass we would simply sample a

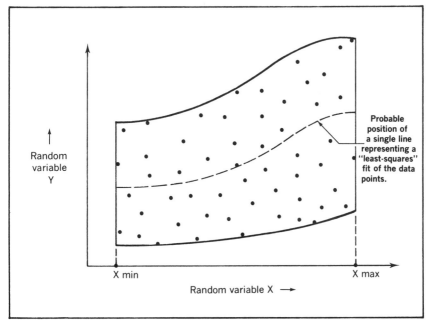

Figure 8.21 *Cross-plot of two partially dependent random variables in which the dashed line represents a least-squares fit of the data points.*

value of X, substitute it into the least-squares function and solve for Y (in lieu of sampling a value of **YNORM** and solving equation 8.2).

To follow such a course, however, is to ignore the very thing we are trying to account for by using simulation—*variability*. The resulting profitability values would not reflect the substantial variation of values Y could have on either side of the function.

It's almost as if we had decided to build a new garage to protect our shiny new automobile and then when the garage was completed we decided to leave the auto parked outside on the driveway because it's too much extra work to get out of the car to open the garage door each time! Enough said on this point. You can draw your own conclusions as to the feasibility of such a short-cut approach.

6. Finally, we mentioned earlier in this section on the mechanics

of simulation that it is not valid to use the same random number to sample two or more distributions (see Figure 8.10 and related discussion). Some authors on simulation have proposed that using the same random number to sample several distributions is, in fact, a way to treat partial dependencies, in lieu of procedures such as we've been discussing. But proposals to use the same random number twice to account for dependency are not valid! Here's why.

In Figure 8.10 suppose variables X and Y are partially dependent and the bounded region within which the dependencies exist is as shown in Figure 8.22. If we used the same random number to sample the X and Y distributions separately (as in Figure 8.10) it will turn out that over a series of passes the sampled values of X and Y will all lie along a line represented by the dashed curve. If the random number happened to be low

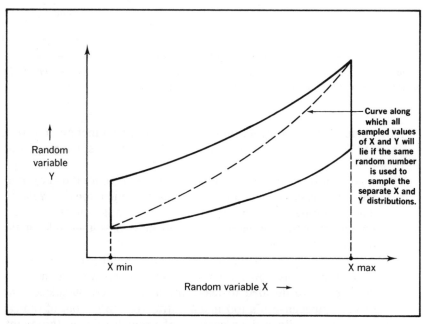

Figure 8.22 *Cross-plot of two partially dependent random variables X and Y. If the same random number is used to sample the separate distributions (as in Fig. 8.10) the loci of sampled points will lie along the dashed curve. All values would lie on the curve, and no variability of Y as a function of X would be possible.*

both X and Y would have low values. If the random number was large X and Y would be large.

The curve itself may have a slightly different shape (depending on the shapes of the X and Y distributions), but all sampled points would lie exactly along the curve. This of course again defeats the objective of trying to account for the variability of Y as a function of X. If we used the same random number twice (in hopes of treating partial dependency) we would not be able to account for the partial dependency observed in Figure 8.22. So the rule still stands—do not use the same random number to sample two or more distributions.

Computing EMV and Preparing Graphical Displays (Step 6)

After a sufficient number of repetitive simulation passes have been completed the analysis is essentially completed. The only remaining task is to use the computed profitability values to determine the expected value profit (EMV) and make various graphical displays useful to management. To illustrate how this is accomplished, suppose we were making a simulation analysis of an exploratory drilling prospect in which the profitability was defined in terms of a NPV profit using a discounting rate of 15 percent. And suppose further, that after making 150 simulation passes the actual computed values of NPV profit were as given in Table 8.6, column 2.

Such a tabulation results from dividing the entire range of computed profits into a series of sub-ranges (nine in this case) and then counting the number of computed values which fell in each of the nine ranges. For example, three of the passes resulted in NPV losses of between −$50 M and −$75 M, six passes resulted in losses of between −$25 M and −$50 M, etc. And of course, the sum of all the frequencies of column two is the total number of passes that were made.

Column 3 simply expresses the frequencies of Column 2 as a fraction of the total number of passes. At this point all we have done is arrange the 150 numerical values of profit into a table of relative frequencies.

The next step—and indeed an important one in the whole scheme of simulation—involves an assumption. The assumption is that the actual relative frequencies of column 3 are representative of the *probability* of having a profit in each range. We are saying here that the relative frequencies observed by a carefully designed "sampling" of 150 combinations of ways the various variables could occur in nature are representa-

Table 8.6

Results of 150 Simulation Passes of an Exploratory Drilling Prospect Simulation Analysis. Profit Expressed As NPV Profit (Thousands of Dollars), Discounted at $i_0 = 15\%$.

Range* of NPV Profit ($M)	Number of Computed Values of NPV Profit Which Fell in Each Range	Relative Frequency in Each Range
−$ 75 to −$ 50	3	3/150 = 0.020
−$ 50 to −$ 25	6	6/150 = 0.040
−$ 25 to 0	12	12/150 = 0.080
0 to +$ 25	15	15/150 = 0.100
+$ 25 to +$ 50	60	60/150 = 0.400
+$ 50 to +$ 75	30	30/150 = 0.200
+$ 75 to +$100	12	12/150 = 0.080
+$100 to +$125	10	10/150 = 0.067
+$125 to +$150	2	2/150 = 0.013
	150	1.000

* Minimum value: −$72M.
Maximum Value: +$148M.

tive of the proportions we would have obtained if we had sampled all the possible combinations (which for continuous random variables is an infinitely large number).

It's analogous to sampling the reading habits of, say, 200,000,000 adults. We could speak to every person if we had the time and resources, but a simpler procedure would be to interview a few persons selected in a manner to eliminate any bias in the sampling procedure. If 12% of this carefully selected sample regularly read the Paris edition of the Herald-Tribune we would probably be on safe ground to project that 12% of the entire 200 million people read the Herald-Tribune. Conceptually we do the same thing with simulation.

The computed values of profit for each of the repetitive represents a sample of profit from the theoretical sample space (or population) of possible values of profit. But the computed values are just not any value — they are carefully sampled values such that the distributions of values used to compute profit are in the exact same form as they were originally specified (in Step 4).

So when we complete a simulation analysis and compute relative frequencies of the actual computed values of profit we assume these are representative of the probabilities of occurrence of the various levels of profit. This assumption is perfectly valid if we have made a sufficient number of passes. As long as we make the checks to insure that a sufficient number of passes have been made we have nothing to fear regarding this subtle, but important assumption.

With that bit of explanation, and the assurance that for this analysis 150 passes was a sufficient amount, we can translate, or interpret the relative frequencies of column 3 of Table 8.6 as being probabilities. The final EMV computation is then straight-forward (Table 8.7), and the expected NPV profit of +$41.33 M is the decision parameter which management would use to determine whether or not to drill the prospect. Since the expectation is positive the prospect would be an acceptable investment. Of course, this investment must be compared to other available investment alternatives if any constraints on capital exist.

Table 8.7
Expected Value Computation for Example Results of Table 8.6

Range of NPV Profit ($M)	Probability of Occurrence (Column 3 of Table 8.6)	Midpoint Value of NPV Profit ($M)	Expected NPV Profit ($M) (Column 2 × Column 3)
−$ 75 to −$50	0.020	−$ 62.5	−$ 1.25
−$ 50 to −$25	0.040	−$ 37.5	−$ 1.50
−$ 25 to 0	0.080	−$ 12.5	−$ 1.00
0 to +$25	0.100	+$ 12.5	+$ 1.25
+$ 25 to +$50	0.400	+$ 37.5	+$15.00
+$ 50 to +$75	0.200	+$ 62.5	+$12.50
+$ 75 to +$100	0.080	+$ 87.5	+$ 7.00
+$100 to +$125	0.067	+$112.5	+$ 7.54
+$125 to +$150	0.013	+$137.5	+$ 1.79
	1.000		+$41.33

Expected NPV Profit at $i_0 = 15\%$ (EMV) →

As we pointed out in Chapter 3, the expected value of a decision alternative is the fundamental basis upon which to accept or reject decisions involving uncertainty. Consequently, having computed an EMV from the output of a simulation analysis (such as Tables 8.6 and 8.7) fulfills our objective. It is useful, however, to also present to the decision maker the results of the simulation analysis in various graphical

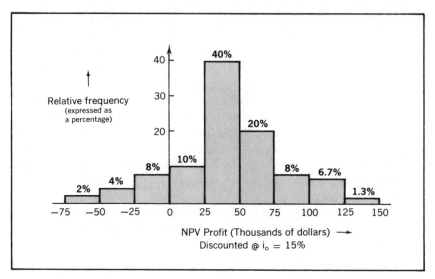

Figure 8.23 *Relative frequency histogram of simulation results of prospect analysis, Table 8.6.*

displays. One obvious display is a "picture" showing the entire profit distribution obtained from the simulation analysis.

For the output data of Table 8.6 the distribution would appear as in Figure 8.23. A picture such as this can give the decision maker a feeling for the ranges of possible profits, the relative likelihoods of losses versus gains, an idea of the range of profits which are most probable, etc.

Another useful display is to convert the profit distribution to its equivalent cumulative frequency. The advantage of having the cumulative frequency graph available is that we can read the probabilities of given levels of profit directly, without having to stop and add all the areas of each bar of the histogram.

These graphs can be plotted in either of two ways: the cumulative chance of a given profit or less, or the cumulative chance of a given level of profit or more. The former is constructed by accumulating the areas under the distribution as profit increases. The latter is constructed by accumulating the areas under the distribution as profit decreases. Usually the decision maker is concerned with the likelihood of *at least* a given level of profit, so the latter of the two options is usually the more useful. For the distribution of Figure 8.23 such a cumulative frequency would appear as in Figure 8.24.

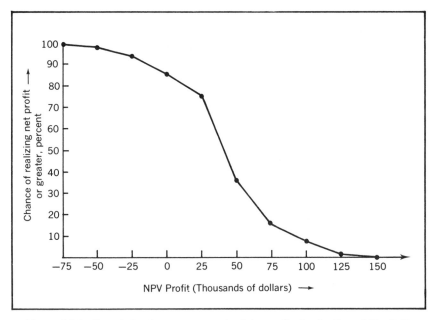

Figure 8.24 *Cumulative frequency graph of the profit distribution of Fig. 8.23.*

The decision maker can use Figure 8.24 to answer all sorts of questions. He may be concerned, for example, about the likelihood of making at least a 15% return. Since the discount rate used in the NPV computations was 15% he is, in essence, concerned about the likelihood that the NPV profit will be at least 0. To determine this he can enter the abscissa at 0 NPV profit, read up and across to the ordinate and read the chance of at least 0 NPV profit (i.e. at least a 15% rate of return) of 86%.

Suppose in order to secure the rights to drill we have to pay a $50,000 signature bonus. If we burden the prospect with this time zero expenditure what is the likelihood of at least a 15% rate of return? The answer is *36%*. If we pay an extra $50 thousand for a bonus the prospect must ultimately yield a NPV profit of at least $50 thousand. The likelihood of this occurring is found from reading the cumulative frequency graph directly.

Finally, we should note that once we obtained the results of the 150 simulation passes (Table 8.6) the analyst was *not* required to make any final subjective risk assessments. All the judgment and professional expertise went into the analysis in Steps 1–4 in the form of probability distributions, dependency relationships, etc. Everything past that point (steps 5 and 6) was simply arithmetic and bookkeeping.

I think this is one of the plus factors in favor of simulation. If we must rely on subjective judgments it seems reasonable to expect that the final results will be less shaded by biases and emotions if these judgments are made early in the analysis on one variable at a time.

It is pretty difficult to predict ahead of time how all the probability distributions will interact and affect ultimate profit, hence it is less likely that one individual can influence or force an answer. Conversely, using subjective judgment at the end of a discrete outcome analysis, such as expressed in Chapter 7, can be biased very strongly by the analyst. He can make or break the deal simply by his (arbitrary?) assessment of a risk factor.

With these comments we should also note that in general it is not possible to predict ahead of time the range and shape of the dependent profit distribution. I sometimes hear people casually say that the profit distribution will be normal, or lognormal, or some other type. We should be cautious about such statements, because it is a very treacherous bit of estimating indeed.

The shape of the final profit distribution is influenced by the shapes of the random variable distributions, the magnitude of the numerical values of the distribution, and the relationship (equation) which relates all the variables to profit. And I for one will readily admit that I lack the intuitive perception to know how all these considerations will interact to yield the profit distribution.

III. APPLICATIONS AND EXAMPLES

In analyzing exploratory drilling prospects there are many, many different variations, or ways to organize and structure the analysis. This is because the variables being described as distributions may vary from prospect to prospect, partial dependencies between several variables may apply on one prospect but not others, etc. So each prospect will usually require minor changes in the structure of the simulation logic scheme. But despite the fact that there are many variations the analyses all include major similarities.

These common attributes include the provision to account for exploratory success and failure probabilities, the need to convert reserves discovered to an equivalent NPV cash flow stream, procedures to ascertain the number of development wells to be drilled, and an accounting for all investment costs, expenses, and taxes.

These considerations will apply whether the prospect we are considering is onshore or offshore, a gas prospect or an oil prospect, or whether the prospect is in North America or Southeast Asia. The specific tax laws and government regulations may change, and certainly the magnitudes of scale change from prospect to prospect, but analyzing a drilling prospect involves basically the same series of steps everywhere.

In this section of the chapter we will consider the general framework of a simulation analysis of a drilling prospect. Several schematic drawings will be discussed to provide a visual overview of how to organize the simulation analysis. We will then look at an actual numerical example of a prospect analysis. This example should be useful to show how the logic scheme of the analysis is formulated. Next we will make some general observations about incorporating the parameter of initial productivity into the analysis. This is a parameter which we must deal with very carefully because of its usual dependencies on other dimensions of the analysis.

Finally, we will briefly cover some applications of simulation other than profitability analysis. One of these is a simulation model which can be used to estimate the distribution of total basin reserves in a new exploration area. The other applications will be mentioned in the form of a brief literature review.

The General Organization of a Simulation Analysis

The analyses of most drilling prospects can be organized in the form of a partial decision tree, as shown in Figure 8.25. The tree is partial in the sense that the other investment alternatives are not included. The objective of analyzing the prospect is to determine the expected NPV profit (or equivalent measure of value) at chance node A. This expected value would be the decision parameter used to accept or reject the project and/or compare its relative desirability to other available drilling prospects.

For the outcome "Dry Hole" the terminal point losses include the dry hole well cost plus any writeoffs to be applied to the prospect for lease bonuses, rentals, seismic surveys, etc. If we make a gas or oil discovery the terminal point values are not single values, but rather distributions of profitability. The reason it is appropriate to consider distributions of profit is that at the time we are trying to decide whether to accept the prospect we do not know the exact size and value that a gas or oil discovery will be.

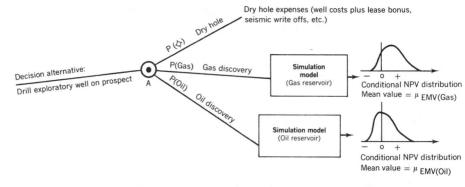

$$\left[P(\Diamond) + P(Gas) + P(Oil) = 1.0 \right]$$

Figure 8.25 *Schematic of a partial decision tree showing basic components of a simulation analysis of an exploratory drilling prospect.*

That is, if we find oil will it be 100 million barrels or 10 million barrels? Will the discovery require two (offshore) platforms or just one? Etc. To determine these profit distributions we can use a simulation model for the gas and oil case. The results of the simulation will be a conditional profit distribution, given a gas discovery or an oil discovery.

The mean value of these respective distributions represents the expected value given that it was a certainty that gas or oil would be found. But since this is obviously not the case we must make a final calculation in which the conditional mean values, EMV_{gas} and EMV_{oil}, are weighted by their respective likelihoods of occurrence, P(Gas) and P(Oil), in a calculation of the form of Equation 8.6:

$$EMV_A = \left[P(\Diamond) \times \binom{\text{Dry Hole}}{\text{Expenses Plus}}{\text{Writeoffs}} + P(Gas) \times \binom{\text{Conditional}}{\text{Mean Value,}}_{\mu_{EMV(Gas)}} \right.$$

$$\left. + P(Oil) \times \binom{\text{Conditional}}{\text{Mean Value,}}_{\mu_{EMV(Oil)}} \right] \qquad (8.6)$$

In this equation EMV_A is the expected value at chance node A (our objective), and the other terms are as defined in Figure 8.25.

The net effect of formulating the analysis as in Figure 8.25 is to divide the prospect analysis into two parts: one part consisting of a simulation analysis to determine a conditional profit distribution given a discovery, with the second part being to combine the conditional expected values with the likelihoods of discovery and the dry hole costs via Equation 8.6. This two-part approach has certain computational advantages.

For one thing it reduces by one the number of parameters which must be sampled with a random number each pass. This is because the simulation analysis only begins once a discovery has been made and does not require the analyst to include a random sampling scheme to determine each pass whether or not the exploratory well found oil or gas.

Another advantage is that the two-part approach is more amenable to a sensitivity analysis of the discovery probabilities. The conditional mean values given a discovery are independent of the likelihood of discovery. Thus, if we wish to evaluate the effect on EMV_A of different chances of discovery we merely change the $P(\diamond)$, $P(Gas)$, and $P(Oil)$ terms and re-solve Equation 8.6. If we included the discovery probabilities within the simulation we would probably have to rerun the simulation analysis to make a sensitivity. With this approach we need run the simulation only once, with the discovery probabilities treated separately in Equation 8.6.

It should be mentioned that the partial tree of Figure 8.25 is clearly not a complete representation of how we might wish to analyze some situations. For example, we may wish to include an option to consider drilling a second exploratory well if the first well on the prospect is dry. This would mean that another decision and chance node would be attached to what is now the dry hole terminal point of Figure 8.25.

Also, we should note that if we are in a predominantly oil play the term $P(Gas)$ can be set equal to zero, making the outcomes either an oil discovery or a dry hole. The gas reservoir of Figure 8.25 is taken to mean a non-associated gas discovery. The associated gas produced in conjunction with an oilfield would be accounted for within the oil reservoir simulation model.

With this brief description of how the pieces all fit together let's talk a bit more about the general characteristics of the "simulation model" shown in Figure 8.25. The model consists of all the variables and computation steps that are used to determine the conditional distribution of value of an oil or gas discovery. These computational steps essentially

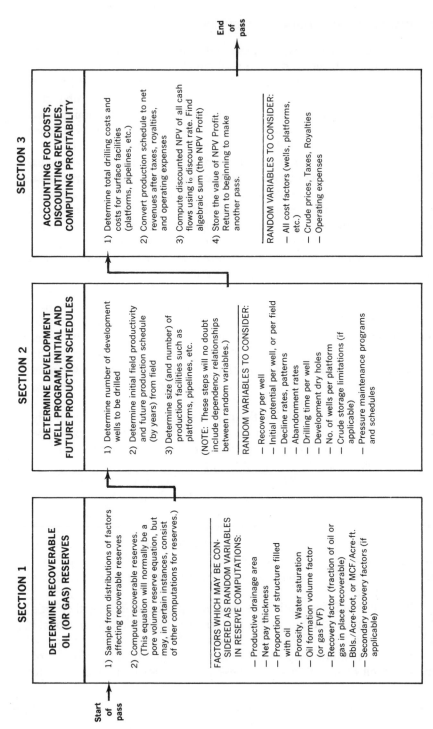

Figure 8.26 *Schematic drawing of the principal parts of the simulation model shown in Fig. 8.25. The drawing shows the factors considered in one pass of the simulation process.*

SECTION 1

DETERMINE RECOVERABLE OIL (OR GAS) RESERVES

Start of pass

1) Sample from distributions of factors affecting recoverable reserves

2) Compute recoverable reserves. (This equation will normally be a pore volume reserve equation, but may, in certain instances, consist of other computations for reserves.)

FACTORS WHICH MAY BE CON-SIDERED AS RANDOM VARIABLES IN RESERVE COMPUTATIONS:

— Productive drainage area
— Net pay thickness
— Proportion of structure filled with oil
— Porosity, Water saturation
— Oil formation volume factor (or gas FVF)
— Recovery factor (fraction of oil or gas in place recoverable)
— Bbls./Acre-foot, or MCF/Acre-ft.
— Secondary recovery factors (if applicable)

SECTION 2

DETERMINE DEVELOPMENT WELL PROGRAM, INITIAL AND FUTURE PRODUCTION SCHEDULES

1) Determine number of development wells to be drilled

2) Determine initial field productivity and future production schedule (by years) from field

3) Determine size (and number) of production facilities such as platforms, pipelines, etc.

(NOTE: These steps will no doubt include dependency relationships between random variables.)

RANDOM VARIABLES TO CONSIDER:

— Recovery per well
— Initial potential per well, or per field
— Decline rates, patterns
— Abandonment rates
— Drilling time per well
— Development dry holes
— No. of wells per platform
— Crude storage limitations (if applicable)
— Pressure maintenance programs and schedules

SECTION 3

ACCOUNTING FOR COSTS, DISCOUNTING REVENUES, COMPUTING PROFITABILITY

1) Determine total drilling costs and costs for surface facilities (platforms, pipelines, etc.)

2) Convert production schedule to net revenues after taxes, royalties, and operating expenses

3) Compute discounted NPV of all cash flows using i_o discount rate. Find algebraic sum (the NPV Profit)

4) Store the value of NPV Profit. Return to beginning to make another pass.

RANDOM VARIABLES TO CONSIDER:

— All cost factors (wells, platforms, etc.)
— Crude prices, Taxes, Royalties
— Operating expenses

End of pass

consist of the three main sections shown in Figure 8.26. The terms and steps given in Figure 8.26 should be reasonably clear to anyone who has ever computed the value of a discovery.

In the lower portions of each section are listed some of the factors which will probably have to be considered in the computations. The lists are not necessarily complete, and it is not necessarily implied that all the factors must be used as distributions in the analysis. There will also probably be a number of factors which must be considered in more than one section. For example, we may sample a value of pay thickness in the first section as one of the parameters for computing reserves. This same value of net pay thickness may then be retained as an entry point into a partial dependency cross-plot to sample a value of IP per well in the second section.

As we mentioned at the beginning of this section there are many variations possible in analyzing exploration prospects, but they will usually have the general characteristics outlined in Figure 8.26. A diagram such as Figure 8.26 is obviously simplified for the sake of generality, but it should provide at least the framework upon which to build a specific analysis.

To interject a little "real-world" meaning to Figures 8.25 and 8.26 let's now look at a numerical example of the analysis of an exploratory drilling prospect using simulation.

Example

In this section we will look at a simulation analysis that was made to determine whether to offer a $1 million bonus on a 20,000 acre block of open acreage. From some earlier reconnaissance seismic and geological correlations the exploration staff had located a large structural anomaly under the block. The anticipated productive interval was the A sand. Nearest A sand production was about 50 miles to the north in the same basin. The staff estimated the likelihood that this prospect, called the Bryarwood prospect, would have oil to be in the range of 0.10 to 0.20. With the analysis they were attempting to determine if they could justify a $1 million bonus, and if so, the minimum probability of discovery that would be required to yield an expected return of at least 16% (i.e. expected NPV @ $i_0 = 0.16 \geq + \$1$ MM).

First, a few general comments about the analysis and how it was organized:

1. They considered two exploratory dry holes to be the maximum they would drill on the anomaly. These wells had an expected cost of \$200 M each.

2. They defined the following parameters as random variables described as probability distributions:

   a. Productive area — acres
   b. "A" sand net pay thickness, h — feet
   c. Primary recovery — barrels/acre-foot
   d. Initial potential, IP, per well — BOPD/well
   e. Development well drilling costs — \$/well
   f. Development dry hole costs — \$/well
   g. Margin — \$/barrel
   (Margin was defined as gross sales price per barrel for oil and associated gas minus taxes, royalties, and operating expenses.)

   From analysis of data from the field to the north, the staff determined that a definite partial dependency existed between well potential (IP) and thickness (h). From a cross-plot of h versus IP per well they were able to establish a boundary around the region of possible values. They further observed that the variability of IP as a function of h was approximately represented by a symmetrical triangular distribution. These partial dependency relationships, together with the other distributions used in the analysis are shown in Figure 8.27.

3. Their approach to establish the number of development wells required was to sample a value of IP per well, compute the recovery per well over a 20 year exponential decline, and divide recovery per well into total field reserves. This had the effect of accounting for the variability of productivity with an IP/well then drilling enough wells to deplete the field in 20 years. They assumed the wells would be on (capacity) decline from the start of production, and would be abandoned in 20 years at a rate of 10 BOPD per well.

   In a moment we will discuss the actual computations of the simulation pass, and the equations of Steps 5, 6, and 13 relate to exponential decline curve analysis. For a more detailed explanation of these equations you should consult Appendix I. The equations in Steps 5, 6, and 13 are equations (1), (2), and (5) respectively in Appendix I.

4. Having determined the number of development wells to be

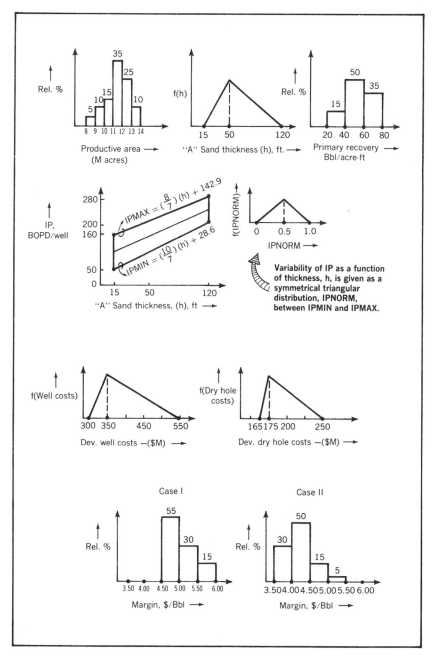

Figure 8.27 *The distributions used in analysis of the example drilling prospect, and the partial dependency cross-plot between sand thickness, h, and IP per well.*

drilled, the number of development dry holes likely to be drilled around the productive limits of the field was determined from the following correlation:

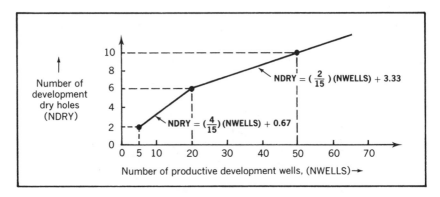

5. They estimated that 16 wells per year could be drilled if they discovered oil in the prospect. The time required to complete drilling was determined by dividing the total number of field wells (completions and development dry holes) by 16 wells per year. For purposes of simplifying the discounting of drilling costs they assumed that the total drilling expenditure occurred at a time when one-half of the wells had been drilled.

This point in time was also assumed to be the start of production, for purposes of finding NPV revenues. (*Comment:* This simplifying assumption was made to make this example slightly easier to follow. In actual practice if the drilling extended over a substantial period of time the discounting should be based on the actual times when wells are completed and come on to production.)

6. NPV computations were based on continuous discounting using a rate of 16% per year; that is, $i_0 = 0.16$.

7. The simulation analysis method was used to develop and describe a conditional distribution of NPV profit, given a discovery. The mean value of this conditional profit distribution, $\mu_{NPV_{oil}}$, was then weighted by the likelihood of finding oil in the first place, P(OIL), to yield the expected value of the outcome of a discovery. The expected value of the outcome of the structure being dry was merely the loss of two dry holes, −$0.4 MM, multiplied by P(DRY HOLE). The partial decision tree (of the type of Figure 8.25) describing all this is:

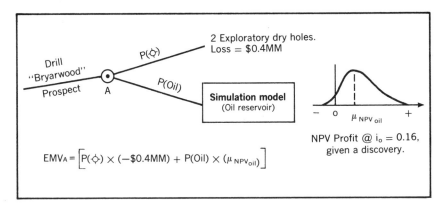

8. The staff considered the estimates of margin to be quite speculative, due to the uncertainties of future trends of world crude oil
prices and economic conditions. As a result they ran two cases—
one using the Case I distribution of margin given in the lower left
corner of Figure 8.27 and one using the Case II distribution in
the lower right corner of Figure 8.27. Case I was considered the
optimistic situation and Case II was considered to be the more
pessimistic case. When Case II was run the distributions of all
other variables were held the same as Case I. Consequently a
comparison of results of the two simulation runs would show the
effects of different assessments of the margin per barrel.

Logic Scheme of the Simulation Analysis

The series of steps representing one simulation pass were as follows:

1. Sample the productive area, "A" sand thickness, and primary
recovery (barrels per acre-foot) distributions and compute total
field reserves:

   TOTAL RESERVES = (Area)(thickness)(Bbls/acre-foot)

2. Using the sampled value of net sand thickness, (h), compute
the minimum and maximum values of the dependent initial potential distribution:

$$\text{IPMIN} = \left(\frac{10}{7}\right)(h) + 28.6$$

$$\text{IPMAX} = \left(\frac{8}{7}\right)(h) + 142.9$$

3. Sample a value of IPNORM from the normalized IP distribution having as its shape a symmetrical triangular distribution. That is, sample from:

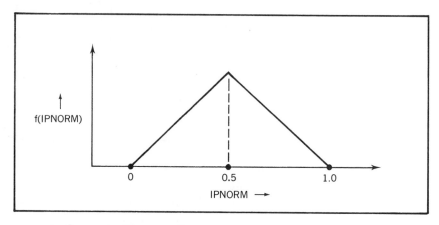

4. Compute IP per well:

$$IP(BOPD/well) = IPMIN + (IPMAX - IPMIN)(IPNORM)$$

(*Note:* Steps 2, 3, and 4 involve the procedures to sample a value of IP/well given a partial dependency with the random variable thickness. They are the steps described in general in Table 8.3 and specifically in Table 8.4 for the case where variables X and Y are thickness and IP/well respectively.)

5. Compute the exponential decline rate, a:

$$a = \frac{\ln\left[\dfrac{q_i}{q_f}\right]}{t}$$

where

q_i = initial decline rate, bbls/year
    = IP × 365
q_f = abandonment rate, bbls/year
    = (10 BOPD × 365)
t = time on decline = 20 years

6. Compute recoverable reserves per well:

$$DELN = \Delta N = \frac{1}{a}\left[q_i - q_f\right]$$

$$DELN = \Delta N = \frac{1}{a}\left[(IP \times 365) - (10 \times 365)\right]$$

7. Compute number of field development wells, NWELLS:

$$\text{NWELLS} = \frac{\text{TOTAL RESERVES}}{\text{DELN}} \quad \begin{array}{l}\text{(Round off to nearest}\\ \text{whole integer)}\end{array}$$

8. Determine corresponding number of development dry holes, NDRY, as:

IF NWELLS > 20 then NDRY $= \left(\frac{2}{15}\right)$ (NWELLS) + 3.33

IF NWELLS $\le$ 20 then NDRY $= \left(\frac{4}{15}\right)$ (NWELLS) + 0.67

(Round off to nearest whole integer)

9. Sample drilling cost distributions.
10. Compute total undiscounted drilling costs, DCOST:

$$\text{DCOST} = (\text{NWELLS}) \begin{pmatrix}\text{Sampled Value}\\ \text{of Dev. Well}\\ \text{Drlg. Costs}\end{pmatrix}$$

$$+ (\text{NDRY.}) \begin{pmatrix}\text{Sampled Value}\\ \text{of Dev. Dry Hole}\\ \text{Costs}\end{pmatrix}$$

11. Compute time required to complete drilling of all field wells, based on 16 completions per year:

$$\text{TIME} = \frac{\text{NWELLS} + \text{NDRY}}{16} \quad \text{(Time will have units of years)}$$

12. Discount drilling costs to time zero at $i_0 = 0.16$ assuming the total expenditure occurs when one-half of the field wells have been drilled:

$$\text{NPVDCOST} = (\text{DCOST}) \times (e^{-0.16 \times \text{TIME}/2})$$

13. Sample value of margin and compute discounted value of revenue as of the start of production, REV:

$$\text{REV} = (\text{TOTAL RESERVES})(\text{MARGIN}) \left[\frac{a}{a+j}\left\{\frac{1 - e^{-t(a+j)}}{1 - e^{-ta}}\right\}\right]$$

where

a = exponential decline rate (Step 5)
j = discount rate, 0.16
t = years on exponential decline, 20 years

14. Discount REV back to time zero, assuming production starts at TIME/2:

$$\text{NPVREV} = (\text{REV}) \times (e^{-0.16 \times \text{TIME}/2})$$

15. Compute NPV Profit:

$$\text{NPV PROFIT} = \text{NPVREV} - \text{NPVDCOST}$$

16. End of Pass. *Store value of NPVPROFIT* and return to Step 1.

To further illustrate this logic scheme let's make the numerical calculations for one hypothetical pass. Remember again that the simulation analysis is only being used to describe a profit distribution given a discovery. As a result exploratory dry hole costs and probabilities do not enter into the simulation analysis.

1. Sampled values of area, thickness, and primary recovery are:

$$\text{area} = 11,840 \text{ acres}$$
$$\text{h, thickness} = 48 \text{ feet}$$
$$\text{recovery} = 50 \text{ barrels/acre-foot}$$

TOTAL RESERVES = (11,840 acres)(48 feet)
$\qquad$ (50 barrels/acre-foot) = 28,416,000 Bbls

2. For a value of thickness, h, of 48 feet compute IPMIN and IPMAX:

$$\text{IPMIN} = \left(\frac{10}{7}\right)(48) + 28.6 \cong 97 \text{ BOPD/well}$$

$$\text{IPMAX} = \left(\frac{8}{7}\right)(48) + 142.9 \cong 198 \text{ BOPD/well}$$

3. Sampled value of IPNORM: 0.51
4. Compute IP per well:

$$\text{IP} = \text{IPMIN} + (\text{IPMAX} - \text{IPMIN})(\text{IPNORM})$$
$$= 97 + (198 - 97)(0.51) = 148.5 \text{ BOPD/well}$$

(*Comment:* For this pass we have discovered a field with 28.4 million barrels having wells with average initial productivities of 148 BOPD. Steps 2, 3, and 4 were to obtain a value of IP from the conditional distribution of IP, given that thickness was 48 feet. The specific conditional distribution of IP that we sampled was a symmetrical triangular distribution having as its minimum and maximum values 97 BOPD and 198 BOPD respectively.)

5. Compute exponential decline rate for IP = 148.5 BOPD:

$$a = \frac{\ln\left[\dfrac{148.5 \times 365}{10 \times 365}\right]}{20} = 0.135$$

6. Compute recoverable reserves per well, DELN:

$$DELN = \frac{1}{0.135}\left[(148.5 \times 365) - (10 \times 365)\right]$$

$$\cong 375{,}000 \text{ Bbls/well}$$

7. Compute number of wells, NWELLS:

NWELLS

$$= \frac{TOTAL\ RESERVES}{DELN} = \frac{28{,}416{,}000 \text{ Bbls}}{375{,}000 \text{ Bbls/well}} \cong \underline{\underline{76 \text{ wells}}}$$

(*Comment:* Once we determined the average productivity per well in Step 4 the program next computes the average recovery per well, given IP, a 20 year exponential decline, and a 10 BOPD abandonment rate. For this pass it was 375,000 Bbls. per well. Knowing total field reserves and average recovery per well Step 7 computes the number of completions needed to deplete the field in 20 years.)

8. Determine number of development dry holes, given that NWELLS = 76:

$$NDRY = \left(\frac{2}{15}\right)(76) + 3.33 \cong \underline{\underline{13}} \text{ dev. dry holes}$$

9. Sampled value of well costs (for this pass):

Completed Well Costs: $355,000 each
Dev. Dry Hole Costs: $181,000 each

10. Compute total drilling costs:

$$DCOST = (76)(\$355{,}000) + (13)(\$181{,}000) = \underline{\$29{,}333{,}000}$$

11. Compute time required to complete field drilling:

$$TIME = \frac{76 + 13}{16} = 5.56 \text{ years}$$

12. Compute NPV drilling costs:

$$\text{NPVDCOST} = (\$29{,}333{,}000)(e^{-0.16 \times 5.56/2}) = \underline{\$18{,}797{,}000}$$

13. Sampled value of margin (Case I): \$4.80/bbl
Compute discounted value of revenue as of start of production:

REV

$$= (28{,}416{,}000 \text{ Bbls})(\$4.80/\text{bbl}) \left[\frac{0.135}{0.135 + 0.16} \left\{ \frac{1 - e^{-20(0.135 + 0.16)}}{1 - e^{-20(0.135)}} \right\} \right]$$

$$= \underline{\$66{,}713{,}000}$$

14. Discount REV back to time zero from TIME/2:

$$\text{NPVREV} = (\$66{,}713{,}000)(e^{-0.16 \times 5.56/2}) = \underline{\$42{,}751{,}000}$$

15. Compute NPVPROFIT:

$$\text{NPVPROFIT} = \$42{,}751{,}000 - \$18{,}797{,}000 = \underline{\underline{\$23{,}954{,}000}}$$

16. End of pass.

On this hypothetical pass we discovered a field having 28.4 million barrels of primary reserves, it took 76 wells to produce the field in 20 years, drilling required about $5\frac{1}{2}$ years, and the discovery was worth a net of \$23.9 million as of time zero using a discount rate of 16% per year.

Results of the Simulation Analyses

The simulation model as outlined was run on a CDC 6600 computer using 500 passes. Central processing unit time required for each run was less than 8 seconds. The resulting conditional NPV profit distributions were as follows:

	CASE I (Margin Ranges from \$4.50 to \$6.00 Per Barrel)	CASE II (Margin Ranges from \$3.50 to \$5.50 Per Barrel)
MINIMUM NPV PROFIT	+\$2.414 MM	−\$2.084 MM (loss)
MAXIMUM NPV PROFIT	+\$74.880 MM	+\$63.562 MM
MODE VALUES	+\$24.154 MM to +\$27.777 MM	+\$24.174 MM to +\$27.456 MM
MEAN VALUE ($\mu_{\text{NPV}_{\text{oil}}}$)	+\$30.063 MM	+\$21.541 MM

The cumulative frequency graphs of each NPV profit distribution are given in Figure 8.28. From this graph the decision maker can associate probabilities of achieving various levels of profit, given a discovery. The final EMV calculations, assuming a probability of discovery of 0.10, are as follows:

CASE I $P(\diamondsuit) = 0.9$ $P(OIL) = 0.1$ $\mu_{NPV_{oil}} = +\$30.06$ million
Exploratory dry hole losses $= -\$0.4$ million

$EMV_A = [(0.9)(-\$0.4 \text{ MM}) + (0.1)(+\$30.06 \text{ MM})] = \underline{+\$2.646 \text{ MM}} \leftarrow$

CASE II $P(\diamondsuit) = 0.9$ $P(OIL) = 0.1$ $\mu_{NPV_{oil}} = +\$21.54$ MM
Exploratory dry hole losses $= -\$0.4$ MM

$EMV_A = [(0.9)(-\$0.4 \text{ MM}) + (0.1)(+\$21.54 \text{ MM})] = \underline{+\$1.794 \text{ MM}} \leftarrow$

So in answer to their first question it is feasible to spend $1 million for a lease bonus and still have an expected return of at least 16%. This is because the EMV for both cases (at a discovery probability of 0.10) is greater than an expected NPV of zero (corresponding to a rate of return

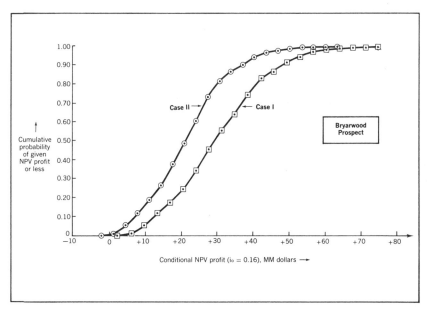

Figure 8.28 *Cumulative frequency graphs for the example simulation analysis. Case I is for the optimistic margin distribution and Case II is based on the lower, more pessimistic distribution of margin.*

of 16%) plus $1 million for the bonus. That is, EMV is > NPV = +$1.00 MM.

To answer the second question about the minimum probability of discovery needed to have an expected return of at least 16% they constructed a graph of EMV versus the probability of finding hydrocarbons, Figure 8.29. The plotting points used to construct the graph were the EMV values computed above for a discovery probability of 0.10 and the −$0.4 million loss when the probability of discovery is zero.

The ordinate EMV values are exclusive of lease bonus. From this graph we can determine that the minimum discovery probability required to yield an expected return of 16% if we pay a $1 million bonus is between 0.06 and 0.07 for Case II, and between 0.04 and 0.05 for Case I. (Note: These probabilities are the abscissa values corresponding to an EMV of +$1.0 million.) Since the staff estimate of the likelihood of discovery was in the range of 0.10 to 0.20 the conclusion was they could, in fact, justify the expenditure of $1 million to secure the drilling rights and proceed to drill an exploratory test on the Bryarwood Prospect.

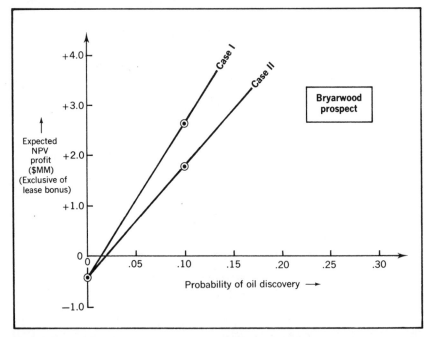

Figure 8.29 *Graph of expected NPV profit (exclusive of lease bonus) versus the probability of oil discovery for each case of the Bryarwood Prospect analysis.*

This numerical example should be helpful to show what an actual simulation analysis looks like, and how the results of the simulation-derived profit distribution are used to compute the expected value for various decision strategies. And I would mention again the importance of a graphical presentation such as Figure 8.29 to display the sensitivity of the discovery probability on EMV.

Estimating the likelihood of discovery is a pretty nebulous process to say the least, and from graphs such as Figure 8.29 we may be able to formulate positive decision strategies, even though we may not know the exact value of the discovery probability at the time we are evaluating the decision options.

Treating Initial Productivity and Determining Development Programs in the Simulation Model

The initial productivity of the wells or field is a random variable which has an obvious effect on the NPV of revenues, and may be an important factor in establishing the number of development wells to be drilled in the field. It also is a variable which we can rarely predict with certainty before the wells are drilled. As a result we will normally wish to describe the variability (uncertainty) associated with productivity as a distribution of possible values.

But this is where the problem gets a bit sticky. As we have mentioned on several previous occasions in this chapter productivity is usually dependent on several other parameters or dimensions in the analysis, such as net pay thickness (by Darcy's law), physical limits on producing rate due to wellbore and/or surface lift equipment, the number of wells in the field, and possibly even storage capacity limitations due to delayed tanker arrivals.

All this is to say that the first two steps of Section 2 of Figure 8.26 are sometimes not as simple to handle as it may seem at first glance. The problem is to be able to develop a production schedule and determine the number of wells required in a manner which recognizes all the interrelated dependencies and dimensions of the prospect *on each pass*.

There are, of course, a number of ways to accomplish this. We will list several possible approaches to give you an idea of some of the ways to handle IP and development well programs. But the choice will vary from prospect to prospect, so you need to be very careful and explicit in structuring this part of the analysis.

1. If data are available to prepare a cross-plot of IP/well versus net pay thickness we can follow the steps used in the example just described.

The value of net pay sampled for the reserve computation is used as the entry point of the partial dependency graph to sample IP/well. Knowing IP/well and the parameters of abandonment rate, producing life, and decline pattern we can compute recoverable reserves per well. This parameter is then divided into reserves to give the number of producing wells to be drilled for the pass.

2. Knowing total field recoverable reserves (from Section 1) we can back-calculate the initial field production rate that would be required to deplete the field in, say, 15 or 20 years to a specified economic limit. Then we could separately sample a value of IP/well from an IP/well versus thickness cross-plot. Finally, dividing total field IP by the sampled value of IP/well gives the number of wells to be drilled.

3. If a statutory spacing pattern applies (such as 80 acres per well, 160 acres per well, etc.) we could divide the spacing per well into the sampled value of total field productive area used in the reserve calculation to set the number of wells. We can then divide total field reserves (of Section 1) by the number of wells to yield an average recoverable reserves per well.

The next step would be to use a correlation of IP/well versus reserves per well to sample an IP from within a feasible range, given the reserves per well, so as to be able to estimate the production schedule. This correlation may require that we make some preliminary calculations to establish the partial dependencies, and in our earlier discussion on partial dependencies we mentioned that we would later talk about how to establish the bounds of an X–Y cross-plot without having access to a set of statistical data.

So let's digress here and show one way this could be done. Suppose we can specify that the range of possible initial potentials per well is from, say, 50 BOPD/well to 400 BOPD/well. And suppose further that the range of producing life of the wells is from 8 to 24 years and that the wells will produce at capacity on an exponential decline pattern to an economic limit of, say 5 BOPD. What we have specified with these statements is the variability of initial rate and producing life, and the expected pattern of declining rates and an economic limit.

These factors are sufficient to compute total recoverable reserves per well, and by a series of engineering calculations we could establish the correlation between IP/well and reserves per well. The calculations are straight-forward and consist of repetitive solution of the equations which relate reserves to IP, producing life, and abandonment rate. For example, we could first compute the reserves produced by a well which started at a maximum rate of 400 BOPD and depleted itself in a minimum time of 8 years, the schedule given as the (1) curve in Figure 8.30. Then the life could be increased to 10 years and the resulting reserves computed.

Each successive computation would be for a longer producing life until the maximum point of 24 years is reached. For each possible production schedule we would have computed a value of recoverable reserves. Then we could decrease the starting rate to, say, 350 BOPD, set time back to 8 years and repeat the calculations for producing life ranging from 8–24 years. Then the rate would be lowered to 300 BOPD and the process repeated, then 250 BOPD, then 200 BOPD, . . . , and finally 50 BOPD. This sounds like a great deal of work but in fact is very simple to do on a computer using nested DO LOOPS.

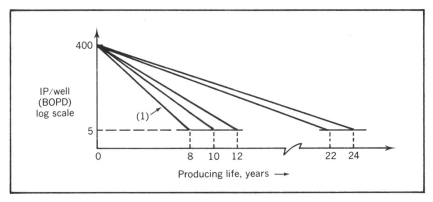

Figure 8.30 *Exponential decline curves for preliminary computations to determine correlation of IP/well versus reserves per well.*

But the results are what we want, so the final step is to make a cross-plot of the IP/well values used in the repetitive computations versus the recoveries per well that were computed for the various values of producing life. The cross-plot would appear as in Figure 8.31.

This correlation establishes the bounds within which IP/well can vary for given values of reserves per well and still honor the producing life limits, the total reserves for the pass computed in Section 1, and the average recoveries per well computed previously by dividing the sampled value of area by the spacing pattern. The method of preliminary calculations described doesn't help us to define the variability of IP/well within

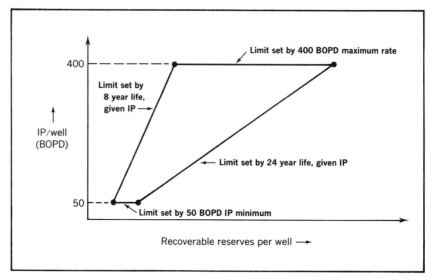

Figure 8.31 *Correlation of IP/well versus reserves per well obtained from the preliminary calculations of possible decline schedules.*

the limits, but it does provide a means of at least establishing the boundaries of the partial dependency. So this suggests an alternative way to define the partial dependency cross-plot if we do not have a set of X–Y data points available.

To recap, this alternative for handling the number of wells and initial potential is as follows:

a. Sample area, thickness, and Bbls/Acre-foot distributions and compute total field reserves.
b. Divide the sampled value of area by spacing pattern (acres by acres per well) to determine required number of development wells.
c. Divide total field reserves by number of development wells to get reserves per well.
d. Using the computed value of reserves per well as the entry point in the cross-plot of IP/well versus reserves per well sample a value of IP/well from within the bounded region. This IP is then used to define the entire production schedule by years.

This sequence of steps would be repeated on each simulation pass, and would provide a way for treating IP/well as a random variable—but in such a way that it honors all the inter-related dimensions on each pass.

4. Knowing total field recoverable reserves we could determine the number of wells to be drilled by an iterative test in which additional development wells would be drilled (to recover the same amount of reserves) until such time that the incremental increase in NPV profit from another well is less than the cost of drilling the additional well.

This criterion for establishing the number of wells says, in essence, that we continue to add "straws in the container" so as to accelerate recovery of the reserves until the incremental gain from an additional "straw" is exceeded by the cost of the "straw." This scheme is sort of an optimization approach to maximize NPV profit without regard to spacing patterns.

These four approaches should provide some ideas on how to handle initial potential and the development well program on each pass of the simulation analysis. Again we stress that this is a very important part of the model, and careful consideration should be given to these or alternative approaches for handling IP and the number of wells to be drilled on each pass.

Other Applications of Simulation

So far in this chapter we have presumed that the dependent variable we are trying to describe by simulation is some measure of value such as NPV profit discounted at i_0 per year. But there are other useful types of analyses for which the concept of simulation can be effectively used. In these alternative applications the methodology of simulation remains the same (i.e. describing random variables as distributions, sampling from the distributions using random numbers, etc.) and only the types of variables change.

One example is use of a simulation analysis to try to assess the range and probable distribution of total recoverable reserves in a sedimentary basin after only a limited amount of exploratory drilling. Consider the following scenario:

Several structures in a new exploration area have been drilled with the result of several shut-in gas discoveries and two dry holes. The discoveries confirm that hydrocarbons are present in the basin, and subsequent reconnaissance seismic has indicated the basin contains many structural anomalies similar to the few that have been tested thus far. It appears the conditions for a worthwhile exploration program exist, but the problem is that the area is remote and no pipeline facilities exist for selling the gas.

Management is reluctant to commit large amounts of capital until it seems more certain that a market will become available in the near future. The pipeline companies, on the other hand, are reluctant to commit to the huge expenditures to lay a line until they have a more firm indication of whether the basin will contain enough reserves to justify the line. How can the impasse be resolved? How could we get some idea of the range of possible reserves in the basin at this time?

Examples of areas where such a scenario exists at this time (early 1975) include the Northwest Shelf area of Australia and the basins in the Arctic Regions of Canada and Alaska. The questions posed in the scenario are indeed important and timely issues that will occur in most areas where the reserves are in a remote area relative to markets for the oil and gas.

What follows is a simulation model which might be used to gain insights into the questions. The model will be loosely called a basin reserve model, although we'll see later that it has the flexibility to be used for assessing many other dimensions of a new exploration play. Probably the unique feature of the model is the use of a random number sampling scheme to simulate the results of (hypothetically) drilling all the observed and/or projected prospects in the basin.

The model itself is later displayed as a flow diagram of the computer program, so to follow the logic involved we need to first define the coding symbols used, as shown in Table 8.8. Now suppose further the sedimentary basin consists of a large number (100 for example) of structural prospects as in Figure 8.32. These anomalies may have been defined by reconnaissance seismic or it may be that the existence of some of the

Table 8.8
Coding Symbols Used in Basin Reserve Model of Figure 8.33.

RESERVE = Reserves computed for a given productive structure (oil or gas)
TOTRESERVE = Cumulative (total) reserves from all tested structures having gas or oil
N = Number of undrilled structures
NZERO = Total number of structures in basin
P = Fraction of undrilled structures having hydrocarbons
PZERO = Ultimate success ratio (i.e., the ratio of productive structures to total structures after all of basin had been drilled up)
PROD = Number of undrilled structures having hydrocarbons
I = Index to keep track of number of simulation passes
K = Index to keep track of separate structures
(K = 1, 2, 3, . . . , NZERO)

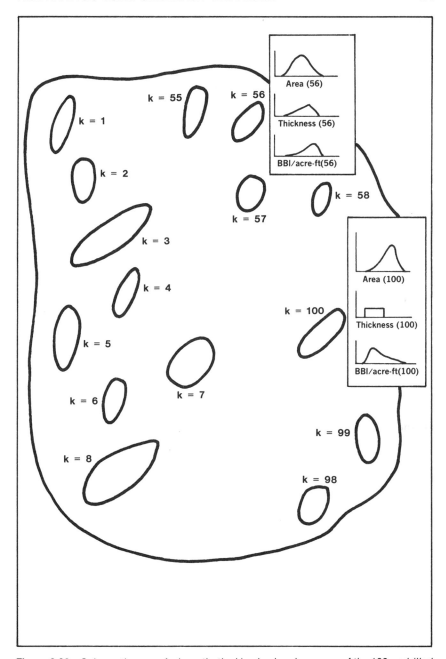

Figure 8.32 *Schematic map of a hypothetical basin showing some of the 100 undrilled prospects. For each anomaly we can describe distributions of productive area, thickness, and oil or gas recoveries per acre/ft. The K values are merely index numbers to keep track of each prospect in the model.*

anomalies has been hypothesized by extrapolating the locations and density of anomalies identified by seismic to the portions of the basin for which no seismic has yet been run.

For each anomaly we describe a distribution of area and net pay thickness to reflect the variability (uncertainty) of the physical size of the prospects and a distribution of barrels per acre-foot (or MCF per acre-foot if in a gas province) to reflect the ranges of possible hydrocarbon recoveries. Each anomaly is identified by a K number, and the three distributions for each anomaly are indexed by the appropriate K subscript.

The input data which the analyst must specify to use this model are:

a. NZERO   (Total number of structures in basin, say 100, for example. This number would be an estimate or determined from regional seismic data.)

b. PZERO   (Ultimate ratio of productive structures to total structures, say 0.15 for example. This means 15% of the 100 structures will have gas or oil; i.e., $100 \times 0.15 = 15$ structures. So PROD at time zero = NZERO × PZERO)

c. Area Distribution
   Net Pay Thickness Distribution
   MCF/Acre-ft Distribution } For each of the NZERO structures
   (or Bbls/Acre-ft if it is an oil province)

The model itself is described by a computer program flow diagram in Figure 8.33. For those not accustomed to thinking in terms of flow diagrams, Figure 8.33 probably looks very complex and imposing. But it really isn't very complicated if we just stop to walk through the steps. To do this let's suppose we estimate that there are 100 structures in the basin and that 15% are hypothesized to have oil. This means that NZERO is set equal to 100 and PZERO is 0.15.

The first step for each pass is to set some of the parameters to initial conditions. Initial here is in the sense of before any of the NZERO prospects have been drilled.

a. TOTRESERVE is set to 0 — to reflect that cumulative reserves found thus far are zero.

b. N is set to NZERO — N is the number of undrilled structures, which initially is equal to the total number in the basin, NZERO.

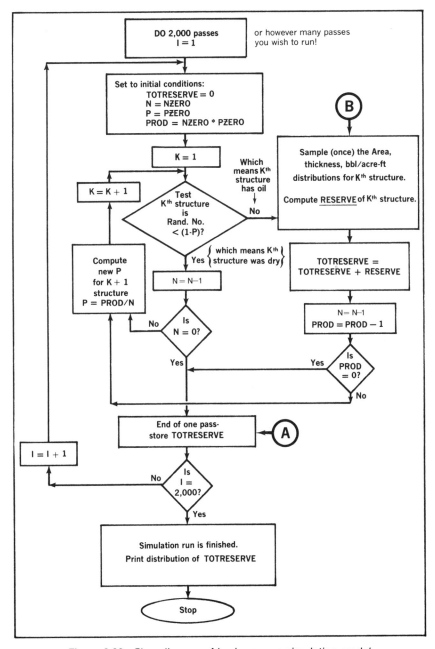

Figure 8.33 *Flow diagram of basin reserve simulation model.*

    c. P is set to PZERO — P is the probability of discovery on the next structure tested, and at time zero corresponds to 0.15, the value specified as PZERO in this example.

    d. PROD is set equal to ⎤ PROD is the number of remaining
        the product of NZERO ⎬ — structures which have oil, 15 in this
        multiplied by PZERO ⎦ case at time zero.

Next the program sets K equal to 1, meaning we are now going to "drill" the K = 1 structure. A random number expressed as a decimal is next obtained and is tested to see if it is less than (1 − P). P is the probability of success, so (1 − P) is the dry hole probability. For our numerical illustration P is 0.15 for the K = 1 structure, so (1 − P) is 0.85. Consequently if the random number is less than 0.85 the analog is we've drilled a dry hole.

Following the arrows out of the bottom of the diamond shaped box the program next computes a new value of N as being the previous N minus 1 (for our example this new N would now be 100 − 1 = 99: There are 99 undrilled structures left). After testing to see that N is not zero it computes a new probability of success for the next structure as P = PROD/N (for our example this would be 15/99). It then sets K = K + 1 = 1 + 1 = 2 and repeats the sequence by "drilling" the K = 2 structure.

If the random number for our initial structure was equal to or greater than 0.85 the analog would be that we had made a discovery and we would exit the diamond shaped box to the right to the rectangle identified as B. To determine the magnitude of reserves discovered the computer would sample (one time) values of area, thickness, and recovery per acre-foot from the K = 1 distributions.

These three parameters would be multiplied together to yield RESERVE, the recoverable reserves of the structure. Following the arrows downward the program next adds this amount of reserves to the previous total (TOTRESERVE) to yield the new cumulative total reserves found thus far in the pass. Next, the parameters N and PROD are reduced by one to reflect the fact there is one less undrilled structure and the remaining number of productive structures is reduced by one because of the discovery we have just made.

The next step is to test if PROD equals zero. If it does it means all the productive structures have been found and the pass ends. If PROD is not zero the drilling sequence continues by drilling the K + 1 structure using a new probability of discovery corresponding to PROD/N.

The pass ends when PROD is zero and the program prints out the value of TOTRESERVE, the total reserves found. Then I is set equal to I + 1 for the next pass and the whole sequence is cycled back to the top of the flow diagram to initial conditions for the next pass. After doing, say, 2000 passes the values of TOTRESERVE are tabulated as output to describe the distribution of total reserves in the basin. The whole 2000 passes can be run on present-day computers in just a matter of seconds, so the computer time and expense is negligible.

We should notice several important characteristics of this model:

■ Probabilities of discovery (P) are revised after each structure has been tested to reflect a sampling without replacement process.

■ On each pass the structures which have oil will vary. On the first pass oil may be found in the K = 3, 10, 11, 14, 21, . . . structures, on the next pass the discoveries may occur at the K = 1, 9, 12, 14, 36, 41, 56, . . . structures, etc. Since we are only concerned in this analysis with trying to define the distribution of total reserves it is of no concern to us which specific structures were productive.

By virtue of the random number scheme being used to test whether the Kth structure is productive we are, in essence, randomly redistributing the reserves in the various structures after each pass.

■ The drilling sequence on each pass will continue until all NZERO structures have been tested or until all the (NZERO × PZERO) structures hypothesized to be productive have been found, whichever comes first. In our example (NZERO × PZERO) was 15, so the drilling sequence would continue until 15 discoveries had been made. When that occurs the pass ends.

Having described the basic logic of the model let's look at some of the options we have.

Option 1. The program as described by Figure 8.33 will print out (or tabulate) the total reserves for each pass, at box A. But we could also have printed out the value of K at the time the computation sequence reached the end of the pass. This value represents the number of structures that had been drilled on each pass before discovering all the (NZERO × PZERO) productive structures. After 2000 passes we'd have 2000 values of K which would describe a distribution of the number of structures tested before discovery of all the hydrocarbon-bearing structures. The minimum value of this distribution would be K = NZERO × PZERO, the case where no dry holes were drilled. The maximum value of the distribution would be K = NZERO, the case where the last productive structure was not found until the very last structure was drilled. With a distribution of K values we could gain insight as to the amount of exploratory drilling required to find all the reserves in the basin. That is, we could answer questions such as:

a. "How many exploratory wells would we have to drill to have a 95% certainty that we would have found all the Basin reserves?"
b. "If we drilled 30 exploratory wells what is the probability that we would have found all the basin reserves?"

Option 2. Suppose we were interested in assessing the distribution of reserves found by, say, a 20 well exploration program (where 20 < NZERO). No problem. We merely need to add another exit point out of the pass loop after K = 20 wells had been drilled, regardless of whether all (or any) of the originally hypothesized productive structures had been found. The output reserve distribution would furnish insights as to the results we could expect from drilling 20 wells. And a further modification would be that if we could drill, say, 4 wells per year this would give us some idea of the timing involved before enough reserves had been found to justify starting a pipeline.

Option 3. What if some of the structures are much larger than the others and would probably be tested first — how can the model be used to reflect this? Simple! The program tests structures in the sequential, or serial order of K = 1, 2, 3, 4, . . . etc. But we are free to index the structures in any way we choose. So if a given structure is likely to be drilled first (due to real-world considerations such as — it's the largest structure, or shallowest, or closest to a pipeline, . . .) simply designate it as the K = 1 structure! If a specific order of drilling is anticipated subscript the structures in the same order.

Option 4. We have many options regarding the addition of economic and invested capital considerations to the analysis. For example, at the end of each pass (box A) we would know total reserves discovered and K, the number of structures tested. Of these K wells drilled the number of shut-in discovery wells would be equal to (NZERO × PZERO), and the number of exploratory dry holes drilled in the pass would be [K − (NZERO × PZERO)]. These numbers of wells could be multiplied by the appropriate shut-in discovery well costs and dry hole costs to obtain the total exploration investment to find the total reserves of the pass. At the end of 2000 passes we would have a distribution of exploration investment required to find the reserves of the basin.

Also, the total basin reserves computed at end of each pass could be multiplied by a NPV/Bbl factor (computed after operating costs, taxes, transportation costs, development drilling costs, etc.) to give a NPV of revenue from the total basin reserve. After repeated passes this would result in a distribution of NPV Revenue. With this distribution and the distribution of exploratory drilling capital the manager could get an idea of the values expected from a total drilling capital outlay of x dollars.

Option 5. To be even more precise, a value or distribution of drilling costs and NPV/Bbl could be assigned to each structure. These would consider the specific drilling depth, distances from a probable pipeline, etc. of each structure. With this scheme, if the Kth structure tests oil the costs and revenues would be computed at the same time as reserves (B, Figure 8.33). These costs, revenues, and reserves would be accumulated as the pass progressed, and at the end of each pass values of total reserves, total exploratory drilling costs, and total NPV revenues would be stored in the computer. After the 2000 passes output distributions of total basin reserves, total exploratory drilling costs, and total NPV of reserves found would be printed out.

With this option the costs and revenues would be specific for each structure, given its geographic location, depth of burial of sediments, etc. The NPV/Bbl factor used in this option will undoubtedly be dependent on the amount of reserves per structure. To construct this correlation may require some side, hypothetical case studies of various possible development schemes for given levels of reserves in the structure.

It is no doubt apparent at this point that the simulation model of Figure 8.33 has many possible applications for assessing results of multiple well drilling programs, as well as trying to assess the probable range and distribution of reserves in a new basin.

Since the concept of simulation first began to appear in the petroleum literature in about 1967 there have been numerous suggested applications other than profitability analysis. Some of these other applications include:

■ The analysis of the variability of S_w, connate water saturation, as calculated from well log measurements. (Reference 8.2)

■ The evaluation of whether to spend more money on continued bench testing in the lab of a new refining process or take the gamble of going full scale without knowing for sure whether the process will work. (Reference 8.9)

■ The length of time required to lay offshore pipelines. The unique part of this application is the modeling of weather conditions in the simulation logic scheme. (Reference 8.12)

■ The determination of the range of variation which might occur in computing the bottom-hole position of a directionally drilled well. (Reference 8.14)

In the final analysis the concept of simulation to describe the analytic combination and inter-relationships of random variables expressed as probability distributions is so general that the number of possible applications is only limited by our ability to define and describe the systems of uncertainty. As time goes on we'll probably notice many other applications of simulation. It's pretty much just a matter of getting the analyst and the problem together!

IV. USING COMPUTERS TO MAKE SIMULATION ANALYSES

In this section I would like to offer a few comments about computer programming of simulation analyses. It clearly is the only way to make all the repetitive calculations required to define the profit distributions. Although some explorationists tend to feel they lose control of the problem once it goes into the "black box," we really have no feasible alternative. The comments here will be general in nature.

I don't plan to get into specific programming because everyone seems to have slightly different job control procedures and/or programming languages. These general comments should be helpful, however, in guiding your company's programmers as they set up procedures to handle simulation.

First, a few words about general purpose risk programs. With the advent of increased use and acceptance of simulation many companies set out to develop a general purpose, off-the-shelf risk analysis program (using simulation) that could be used throughout their organization. A number of consulting firms also got in the act and offered to the industry general purpose risk analyses for some nominal cost such as $1.50 per run.

The goal is admirable, but the practical result in many instances has been a sacrifice in the flexibility or options so as to make it general. Often the geologist in Casper, Wyoming, finds he may have to "shoe-horn" his prospect into the general format of the computer program at his company's home office in Houston.

The opposite viewpoint is usually taken by those who have had experience programming and have access to computer facilities. As we will see in a moment the computer program remains essentially intact from prospect to prospect except for the all-important description of the equations which relate all the variables to profit—the analytic function. Because of this they usually prefer to change the few cards necessary to make the simulation analysis specific for each prospect. The obvious advantage is that the program is tailored to exactly fit each particular exploration prospect.

As with most things there is probably some sort of mid-position which maximizes the advantages of both views. My feeling is that since not everyone on our exploration staffs are programmers a basic general program enlarges the accessibility of the technique to more explorationists. But I also think that the general program should be designed to permit easy changes to the equations used to compute profit. In this regard it is most useful to have at least one explorationist in each company location who can make the needed programming changes on the spot.

But whatever your position about general purpose programs versus individual programming I think we can all agree that it needs to be designed so that the input of data (from the user viewpoint) is as simple and efficient as possible. This is particularly true regarding the mechanics and options for reading into the computer the various random variable distributions the user wishes to use. More about this in a moment.

A computer program for making a simulation analysis consists of about five parts, or facets:

1. Procedures for reading into the computer the random variable probability distributions.
2. Procedures for reading into the computer the equations (or series of steps) used to compute profitability, numerical values for the known parameters, and any equations relating to partial dependencies.
3. A random number generator.
4. The programming required to make the actual simulation passes.
5. Procedures for output of the computed data relating to profit distributions, the computed mean value of the profit distribution, and any (optional) graphical displays generated by the computer.

Of these five parts only the second part varies from analysis to analysis. Parts 1, 3, 4, and 5 remain essentially the same for each analysis.

Procedures for Reading Distributions into the Computer (Part 1)

This is probably the single most critical part of the program in terms of trying to simplify the input procedures for the user. In fact I think the consideration and diplomacy used here can sometimes mean the difference between success or failure. What is desired, of course, is to pre-program the capabilities to handle as many different types of probability distributions as the user might ever wish to use.

These might include the uniform (rectangular), triangular, normal, lognormal distributions, and relative frequency histograms. But what if the user had a distribution such as Figure 6.16 which doesn't fit any of the standard options? Does he have to "shoe-horn" Figure 6.16 into a normal distribution? To read in normal and lognormal distributions we can give the computer the μ and σ and it could internally handle the distributions. But what if some of the staff members don't know how to compute σ?

A number of years back, I was asked to review a general purpose simulation risk analysis program developed for a company's exploration department by their research center mathematicians. In this program the option for using any type of continuous distributions other than the uniform and triangular was the Weibull distribution.

My immediate response was probably the same as yours—What in the world is a Weibull distribution? I had never heard of it at the time, and had to search through about six textbooks before I even found mention of it! Well, as it turns out it's a distribution useful in representing the distributions of the life of a piece of machinery. It's used widely in systems reliability analysis, and by careful selection of the numerical parameters of the density function the Weibull distribution can have about any shape imaginable.

It was a brilliant choice as a general purpose distribution—from a mathematician's point of view. But what is poor Henry Oilfinder located in Monahans, Texas, supposed to do if he has never heard of a Weibull distribution?

The point of all this is that we must strive to make the input procedures as simple and non-technical as possible while still having the flexibility of being able to accomodate as many different types of distributions as possible. On final balance, when the value of simplicity is weighed against the value of flexibility I am convinced that the most versatile set of input options are:

A. The *uniform* distribution—The user merely specifies the minimum and maximum values of the random variable.
B. The *triangular* distribution—The user only needs to specify the minimum, most probable, and maximum values of the random variable.
C. The set of X–Y coordinates from the cumulative frequency graph of any distribution (other than uniform or triangular).

By allowing the user to read in a set of X–Y coordinates of the cumulative frequency distribution the program could use virtually any type of distribution he chooses. The only requirement from the user standpoint is that he would first have to convert the distribution he wishes to use to its equivalent cumulative frequency so as to be able to read X–Y coordinates.

The input data sheet for uniform distributions would only need to include blanks for the following information:

Variable name __NPAY__ (Net pay thickness, feet)

Minimum value __20__ (feet)

Maximum value __50__ (feet)

To read in a triangular distribution such as the drilling cost distribution on page 275 all we would need is:

Variable name __DRLCOST__ (Drilling costs)

Minimum value __100__ (thousands of dollars)

Most likely value __130__ (thousands of dollars)

Maximum value __200__ (thousands of dollars)

To enter a distribution such as Figure 6.16 we provide as input data values from the cumulative frequency graph (Figure 6.18):

Variable name __DRLCOST__ Drilling costs, thousands of dollars

Value of the variable	100	108.5	115	123	131	140
Corresponding cum. freq.	0	.0178	.0696	.1719	.3127	.5126

Value of the variable	**146**	**154**	**163**	**174**	**181**	**190**
Corresponding cum. freq.	**.6459**	**.7959**	**.9006**	**.9674**	**.9916**	**1.0000**

Simple! And it offers complete flexibility to handle *any* distribution. The user does not have to know how to compute a μ or σ, or know what a Weibull distribution is. Whatever the distribution he only needs to convert it to a cumulative frequency (using the techniques we described in Chapter 6) and read in a series of X–Y coordinates from the cumulative frequency. If he has plotted a variable on lognormal probability graph paper all he has to do is read in a series of X–Y points from the straight line plot. These three options are so simple and versatile I find it difficult to recommend going much beyond these types of distributions.

At this point, it might be of interest to you to understand how the computer actually operates on these data so as to sample values with random numbers. That is, what does the programmer have to do in terms of pre-programming these options? It merely consists of having the computer programmed with a few equations and keeping the coding straight. For this brief explanation let's use the symbol X to be the random variable, CF to be a cumulative frequency, and RN to be a random number expressed as a decimal fraction.

Uniform Distribution: For this distribution the cumulative frequency is merely a straight line as shown. The equation of a straight line, given the coordinates of $X_1 - CF_1$, $X_2 - CF_2$ is:

$$\frac{CF - CF_1}{X - X_1} = \frac{CF_2 - CF_1}{X_2 - X_1}$$

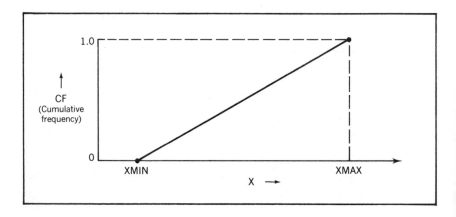

Rearranging to solve for X gives:

$$X = X_1 + \left[\frac{(CF - CF_1)(X_2 - X_1)}{(CF_2 - CF_1)}\right]$$

But for the uniform distribution $X_1 = $ XMIN, the minimum value specified on the input sheet, $X_2 = $ XMAX, the maximum value specified by the user, $CF_1 = 0$, and $CF_2 = 1$. So we can substitute these values into the equation as:

$$X = \text{XMIN} + \left[\frac{(CF - 0)(\text{XMAX} - \text{XMIN})}{(1.0 - 0)}\right]$$

or

$$X = \text{XMIN} + [(CF)(\text{XMAX} - \text{XMIN})] \qquad (8.7)$$

This is the equation which is programmed in advance. XMIN and XMAX are specified as input data. To sample from the X distribution the computer obtains a random number, RN, sets it equal to CF and solves for X on each pass.

Triangular Distribution: The logic is essentially the same in this case. The equations which the computer uses are rearranged forms of Equations 6.9 and 6.10. In these equations we solve for the cumulative probability less than or equal to a value of X. To sample values we do the reverse. We set the random number equal to cumulative frequency and solve for X. Consequently Equations 6.9 and 6.10 in rearranged form to solve for X are:

For values of $X \leq X_2$

$$X = X_1 + (X_3 - X_1)\left[\sqrt{(CF)\left(\frac{X_2 - X_1}{X_3 - X_1}\right)}\right]$$

For values of $X \geq X_2$

$$X = X_1 + (X_3 - X_1)\left[1 - \sqrt{(1 - CF)\left(1 - \left\{\frac{X_2 - X_1}{X_3 - X_1}\right\}\right)}\right]$$

where $X_1 = $ minimum value of triangular distribution, $X_2 = $ most likely value, and $X_3 = $ maximum value.

But if we code the minimum, most probable, and maximum values specified by the user as XMIN, XMODE, and XMAX we can substitute these coded parameters for X_1, X_2, X_3 respectively:

For values of $X \leq $ XMODE

$$X = \text{XMIN} + (\text{XMAX} - \text{XMIN})\left[\sqrt{(CF)\left(\frac{\text{XMODE} - \text{XMIN}}{\text{XMAX} - \text{XMIN}}\right)}\right] \qquad (8.8)$$

For values of X ≥ XMODE

$$X = XMIN + (XMAX - XMIN)\left[1 - \sqrt{(1 - CF)\left(1 - \left\{\frac{XMODE - XMIN}{XMAX - XMIN}\right\}\right)}\right] \quad (8.9)$$

The only problem now is that when we get a random number and set it equal to CF we'll have to know which equation to solve. We won't know X so we won't know if X is less than XMODE or greater than XMODE. But the trick is that the cumulative frequency less than or equal to a value of $X = X_2$ will always be equal to $(X_2 - X_1)/(X_3 - X_1)$. So we can rewrite the test as being CF ≤ (XMODE − XMIN)/(XMAX − XMIN) or greater. As a result of this bit of logic, the procedure used by the computer to sample a value from a triangular distribution having as its parameters XMIN, XMODE, and XMAX is:

Obtain a Random Number, RN
Set CF = RN

If CF is $\leq \dfrac{(XMODE - XMIN)}{(XMAX - XMIN)}$ → Solve Equation 8.8 for X

If CF is $\geq \dfrac{(XMODE - XMIN)}{(XMAX - XMIN)}$ → Solve Equation 8.9 for X

For the General Distribution: The input data specified by the user are a set of X–CF coordinates. These coordinates would have been read as a series of points from the cumulative frequency graph of the random variable as in Figure 8.34. We can observe now that instead of just one or two equations to solve for X there are in this case six specific straight line segments to the curve. While each segment is a straight line the slopes are different. Consequently when we obtain a random number and set it equal to CF we are going to have to determine which straight line segment to use.

To accomplish this we need to pre-program the following general equation of a straight line, applicable for up to r sets of X–CF coordinates:

$$X = X_N + \left[\frac{(CF - CF_N)(X_{N+1} - X_N)}{(CF_{N+1} - CF_N)}\right] \quad (8.10)$$

for N = 1, 2, 3, 4, . . . , (r − 1)
and where CF = RN, the random number, (0 ≤ RN ≤ 1.0)

The numerical values of X_1, CF_1; X_2, CF_2; . . . , X_r, CF_r are read in as input and stored by the computer. When it comes time to sample a value of X the computer will follow this sequence:

Obtain a Random Number, RN
Set CF = RN
Test: Is CF ≤ CF_2?   → If yes, set N = 1 and solve Equation 8.10
                            for X
                         If no, go to test ②.

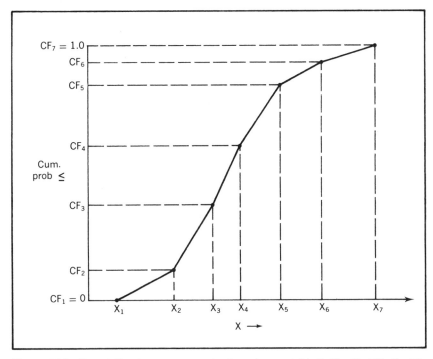

Figure 8.34 *Cumulative frequency graph of random variable X. The X_1, CF_1; X_2; CF_2; X_3, CF_3; . . . . . .; X_7, CF_7 coordinates are read into the computer as input data.*

②   Test: Is CF $\leq$ CF_3?   →   If yes, set N = 2 and solve Equation 8.10
                                                for X
                                                If no, go to test ③.

③   Test: Is CF $\leq$ CF_4?   →   If yes, set N = 3 and solve Equation 8.10
                                                for X
                                                If no, go to test ④.

     etc.

     .

     .

     .

(r − 2)   Test: Is CF $\leq$ CF_{r-1}?   →   If yes, set N = r − 2 and solve Equation
                                                8.10 for X
                                              If no, set N = r − 1 and solve Equation
                                                8.10 for X

     End of sampling of X.

This series of steps, together with Equation 8.10 will provide a means to sample any distribution expressed in its cumulative frequency X–CF coordinates.

If you are familiar with programming these comments should be of sufficient detail to show you how to program the capabilities to receive the uniform, triangular, and cumulative frequency coordinates as input data. For those of you that aren't programmers the comments probably are a bit confusing. But the important point to observe is that when a computer samples a distribution it normally is solving the pre-programmed equations such as 8.7, 8.8, 8.9, and 8.10.

In some instances it may be computationally more efficient to read in (or generate internally) 100 values from the cumulative and store them as subscripted values of the random variable. Then to sample a value of the variable the random number is set equal to the subscript and the corresponding numerical value of the random variable is recalled from storage. Whether this approach is more efficient or less efficient will depend on the storage capacity and the configuration of the computer process unit and its peripheral equipment.

We should mention a few final points about reading into the computer continuous probability distributions.

1. If the program has the capability of receiving X–CF coordinate data of the cumulative frequency the extra options of a uniform and triangular distribution are, in fact, redundant. The uniform distribution could be read in as X–CF data as:

Variable Name __X__

Value of variable **XMIN XMAX** _____ _____ _____ _____

CF, Corresponding __0__ __1.0__ _____ _____ _____ _____
cum. freq.

The cumulative frequency of the triangular distribution (such as the figure on page 277) could be used to read about 6–8 sets of X–CF coordinates and entered as the same type of data. But the reason for leaving the uniform and triangular distributions as separate options is that they both are used fairly frequently in simulation analyses, and it is probably a little easier for the user to merely list the parameters as described previously.

2. Back in the discussion on treating partial dependencies we men-

tioned that the procedure for treating a dependency such as Figure 8.14(b) was to specify a normalized distribution ZNORM which describes all the possible Z distributions as functions of W. The parameters of the normalized triangular distribution would be 0, 0.5, and 1.0 – the 0.5 to reflect the fact that the mode for each of the Z distributions was always positioned mid-way from ZMIN to ZMAX.

Then later, on page 427, we discussed the situation where the position of the mode changes as the random variable of Figure 8.20 increases. In this case we can not input one specific normalized triangular distribution. The procedure, instead, is to input to the computer the equation of the line through the modes of the Y distributions in Figure 8.20 as a function of X (in addition to the equations of the upper and lower bounds of Y as a function of X).

When the computer samples a value of X and Y for each pass the steps followed in Table 8.4 have one additional part to Step 4. That is to solve for the mode value as a function of the sampled value of X. This computed mode value is then inserted into Equations 8.8 or 8.9 to sample the corresponding value from the appropriate triangular distribution of the partially dependent variable.

Procedures for Reading in the Computation Steps, Numerical Values for Known Parameters, and Partial Dependency Relationships (Part 2 of the Program)

This is the part of the simulation program which may change from prospect to prospect. It includes the equations or series of steps involved in making the passes, the input of numerical values for all the known parameters (constants), and any special equations relating to treating partial dependencies. Because these aspects of the program change it is pointless to try to suggest specific input formats or procedures.

For those companies choosing to set up a general purpose simulation risk analysis model this part of the program will be specific for the types of drilling prospects being considered (i.e., a model for onshore prospects, offshore prospects, etc.) For those preferring a more flexible approach in which the explorationist can modify the analysis to suit his needs this is the section of the overall program which he would be changing.

Random Number Generator (Part 3)

As we have mentioned on several previous occasions most present-day computers have a random number generator subroutine already programmed as a library function. Thus the only concern here is to insure that the coding is kept straight and the proper procedure for calling the subroutine.

Programming of the Actual Simulation Passes (Part 4)

This part of the program consists of all the required programming instructions to actually perform the repetitive passes, making comparison checks to see if enough passes have been made, and the accessing of input data and storing of computed values of profitability of each pass. There are no useful tricks or shortcuts I can offer here, but this is usually a straight-forward bit of programming that should cause no particular problem for a competent programmer.

Procedures for Output of Data (Part 5)

There are many different ways to have the output of the analysis displayed. The analyst will normally not wish to have all 2000 values of profit printed out (he'll drown in paper, for one reason!), so one of the usual output procedures is to arrange the computed values of profitability in tabular form, such as Table 8.6. Since the analyst will normally not know ahead of time what the range of values of profit will be, it is necessary to program a scheme for automatically establishing the ranges of each interval of Column 1, Table 8.6. This can be done by having the computer search out the largest negative value of profit and the largest positive value. Then all that is required is to divide this range into, say, 15–20 equal intervals. Such a scheme will automatically fix the interval ranges for each run.

Once the intervals are established the computer is then programmed to count how many values fall in each range, compute relative frequencies, (as in Table 8.6), multiply relative frequencies by the midpoint of each interval (as in Table 8.7), and sum the product terms to yield the expected value. All this tabular data is printed out automatically.

Depending on computer capabilities and personal preferences some people have the profit distribution printed out as a graphical display, as well as printing out a cumulative frequency graph of profit.

All these output procedures and formats are pre-programmed and will not change from analysis to analysis.

When writing a new program to do all this there are several checks that can be made along the way to check programming integrity. You can program Part 1 and 3 and then check to see if the input distributions are being handled correctly by sampling 300 or so sampled values of the variable and printing them out. You can then plot the sampled values to see if they duplicate the distributions you read in to start with. Following this check you can program the output procedure (Part 5).

This is then checked by reading in specific distributions, sampling 300 or so values of each variable, then having the sampled values given as tabular relative frequencies and a mean value. If the output distribution matches the input distribution you have confirmed that your programming of Part 5 is correct. Finally, parts 2 and 4 are added.

To check these about the best you can do is read in a prospect analysis (such as the numerical example of the preceding section) and have the computer print out, say, on every 20th pass all the internal computations. This printout should include the random numbers used for the pass and the corresponding sampled values of the random variables.

In this way you can go through the calculations by hand to see if the computer was doing what you had expected (or hoped) it would be doing. My experience has been that in making spot checks such as this I usually found more errors in the logic scheme than I did in the programming! Either way, the periodic dumping of all internal calculations of a pass serve as a useful way to check the black box!

V. ON THE PHILOSOPHICAL DEFENSE OF SIMULATION

Now that we've discussed all the myriad of details relating to simulation I believe it important to step back and look at the concept from a distance to keep a perspective on things. Certainly the primary reason for using simulation is to gain the insights which result from expressing uncertainty as a range and distribution of possible values of the unknowns, rather than using a single most likely profitability and a final subjective risk assessment (the first approach mentioned in Chapter 7).

We as explorationists have always recognized uncertainty in statements such as, "I don't know exactly how thick the reservoir will be, it could be as thin as 20 feet or as much as 300 feet." And historically we usually resolved the dilemma by using a single value we considered to be a typical, or average, or mid-range, or most probable value. But simulation offers the opportunity to include our entire initial statement (that thickness could range from 20 to 300 feet) in the analysis.

We no longer are forced to make a (subjective? arbitrary?) judgment about a representative "average" value, and the resulting information we present to the decision maker is no longer restricted to just one value of profitability.

Other important advantages include the following:

■ Simulation allows professional judgment to be applied on one variable at a time, rather than how all the unknown parameters are going to occur together.

■ The professional judgments about each parameter can be made by whomever is the most knowledgeable on the subject. It's a risk analysis method which effectively blends the expertise of the firm in analyzing exploration prospects.

■ The method is conducive to sensitivity analysis. By varying distributions the analyst can isolate the one or two most critical variables. He can then show the decision maker his best judgments (in the form of distributions) of each variable. This helps guide the decision maker through the mass of data and information so he can focus his attention on the consideration of the key variables.

■ The graphical display of profitability distributions can provide much more information to the decision maker on the levels of profit which are likely and their corresponding likelihood of occurrence.

■ The concept is completely general and can be used for analyzing any situation involving uncertainty. There are no restrictions on the number of distributions which can be considered or the shapes of the distributions. Only a minimum amount of statistical expertise is required to use simulation.

While its advantages are apparent, the application of the concept to petroleum exploration will sooner or later raise two important questions.

1. Can simulation be used where there is no statistical data or information available upon which to base a distribution?
2. Is a professional judgment expressed as a range and distribution of possible values a better or poorer judgment than a single value estimate?

What are your feelings about these points? I think it is very important that you think about these questions before you embrace simulation. If you prepare a simulation analysis but are not really committed to the notion the manager will no doubt detect your apprehension at once. And if you can't defend the merit of the method you probably shouldn't be using it in the first place.

In the course I've been teaching throughout the world on risk analysis I have, on many occasions, given the students a simulation problem in which the following data from a nearby field is given to them:

> Area: 6,000 acres
> Thickness: Average = 50 feet. Range is from 15 to 90 feet
> Porosity: Varies from 10% to 30%. Average value: 18%
> Water Saturation: Varies from 20% to 30%
> Primary recovery factor: 16% of the original oil in place
> Oil Formation Volume Factor: 1.4
> etc.

They will almost invariably formulate the simulation model relating to reserves as follows:

"We first will generate a reserve distribution using the pore volume equation parameters of drainage area, thickness, porosity, water saturation, recovery factor, and oil formation volume factor. To do this we'll sample from a triangular distribution of thickness (minimum 15 feet, mode 50 feet, maximum 90 feet), a triangular distribution of porosity, and a uniform distribution of S_w. We will consider area, recovery factor, and formation volume factors as constants.

Next we . . ."

At this point I interrupt, and the following exchange usually takes place:

Paul: Why did you consider area, recovery factor, etc. as constants? Do you know for certain that the recovery factor for this undrilled prospect is going to be exactly 16%?

Student: Well, no we don't, but we had no other information about recovery factor, so we could not describe a distribution to use. That's why the analysis is based on just a single value of 16% of the original oil in place.

This logic implies that simulation is OK when you have lots of statistical data, but in the absence of the data it can't be used. And this, in turn, leads to the often expressed conclusion that simulation is a good analysis tool in heavily drilled areas such as West Texas but can not be used in virgin new exploration areas such as offshore Greenland.

But this conclusion clearly misses the point of simulation. It's analogous to the circular reasoning we mentioned back on page 380 in which the less we knew about a variable the more precise our single value estimate becomes! Single point estimates of random variables imply certainty when no certainty may exist. In reality recovery factor in his prospect is *not* known with certainty to be 16%. In fact, he can't even say for sure the structure has oil in it!

It is a natural (but incorrect) tendency to let the presence or absence of data dictate whether we treat a variable as a known parameter or as a random variable. But the decision should be based on whether we can predict or define a parameter's exact value before the well is drilled. Period! No qualifications! If the answer is no we can't, then we need to try and describe the variable as a range and distribution — whether there is available data or not.

So I think the answer to the first question must clearly be yes. It is in those areas where there is very little data that the uncertainty is the greatest. And we can't get much of an impression of uncertainty (variability) if we base our profit computation on single average values of each parameter.

I have observed that the second question is often raised by explorationists and/or managers who are skeptical of the whole idea. They view an expression of uncertainty in the form of a distribution to be weaker than a single value.

EXAMPLE: "Mr. Oilfinder, you tell me thickness could be anywhere from 20 to 200 feet, drilling costs could be anywhere from $500,000 to $800,000, etc. . . . It almost seems like you couldn't make up your mind about anything — all you could say are ranges."

The manager making such a statement is probably implying that if Mr. Oilfinder had said thickness will be 85 feet, well costs will be $627,495, etc. it would have been a much more precise, confident appraisal of professional expertise.

What do you think? This point is clearly a personal choice, and if you subscribe to the view just expressed by the manager there isn't much benefit to be gained from simulation. The feeling of most professionals, however, is just the opposite. That is, most will take the view that expressing a range of values about a parameter you can't measure or evaluate before the well is drilled is a more honest, realistic appraisal of uncertainty.

One of the students attending my course (a geologist, by the way) said it quite eloquently. He stated that for a geologist to ignore or refuse to consider variability is to be completely naive about the world of geology. He went on to challenge the skeptical geologists to go to a geologic outcrop and take measurements of dip angle, strike, thickness, etc., then move 100 feet away and measure the same parameters. His point was that the whole system of geology is one of variability and to fail to express variability is to fail at making a realistic appraisal of the geological system. This question is one you must resolve.

I firmly believe expressing uncertainty as a range is much more realistic than to say thickness will be exactly 85 feet. But my opinion is not the one which should dominate. You must formulate your own position on the issue. And you would be well advised to consider this issue before you begin using simulation. If you aren't psychologically committed to the notion that distributions are better expressions of professional judgment the tedious work involved will have little benefit to you.

On final balance simulation is not a cure-all. Those who take the position 'When in doubt simulate' are probably just as guilty of extreme views as those who despair of the advent of simulation by warning that the exploration game is now 'Strangulation by simulation.' It forces us to make careful, explicit considerations of all the variables and facets of the prospect. It can provide management with much more insight about various levels of profitability and their likelihoods of occurrence than most other analysis concepts. But we can't expect simulation to tell us what the reserve distribution in a basin is, or what the probability of finding oil is, or whether we should explore for oil on the north or south end of the basin. It isn't a replacement for professional expertise. Rather, it is a powerful new analysis method which can greatly expand an explorationist's capabilities for communicating the value of the prospect to the decision maker.

To make certain we all keep our feet on the ground I think it appropriate to close with the story given on the following page about how to win a no-count lease.

VI. HENRY OILFINDER REVISITED

At the end of Chapters 3 and 4 we listened in on a presentation by Henry Oilfinder regarding a drilling prospect on the Blackduck acreage. Let's pick up the story again after Henry had analyzed the prospect using simulation. The conversation is between Henry and Marvin Lee, a simulation expert working for Profit Oil and Gas Company.

Notice particularly the type of questions Marvin used to obtain

How to win a no-count lease

WE HAVE ALWAYS been impressed by the scrupulously scientific approach which oil companies use in deciding how many millions to bid on offshore leases.

The computer—that Mr. Know-All of the modern age—tells them right down to the cent-per-acre how much it will cost to locate, drill, produce, market, and collect a profit on a given quantity of oil or gas.

We've had a nagging suspicion, however, that these calculations are something less than precise, despite such meticulous care. How can you explain a $33 million bid by one company on a tract that another company, which had the control wells, valued at only $5.3 million?

Well, we found out a short while back. An executive of a large company was sharing the experience.

The exploration department, he relates, asked for an evaluation per acre for a tract in 80 ft of water, with 1,000 proved acres, four gas sands each 20 ft thick, and two oil sands 15 ft thick, one of which had a 412-acre gas cap averaging 11 ft thick. Exact depths were given.

The company's figure fiends then went to work. They figured the recoverable reserves. They computed the exploratory drilling costs, using the industry's success ratio for dry holes. They worked out the cost of development drilling and erecting platforms. They calculated depletion on a 20-year life at current crude prices. They applied present worth factors.

Then they fed all these data into the company computer and found that if they bid a certain price per acre, they would, if successful, get a fair return on investment.

"We passed the results back to exploration," recalls our storyteller, "and you would have thought we were their worst enemy. Three days later, back they came with a new set of parameters resulting from a restudy of their seismic, which I later find out was a 'stacked' survey. This I can believe.

"Twelve computer runs later we had increased the size of the field from 1,000 to 2,000 acres, the number of sands from six to eight—of which six were now oil—and, in some mysterious fashion, the thickness of the sands also had increased.

"The recovery factors jumped 50%. The price escalated. And the field depleted in 5 years, not 20. These reservoirs were all water drive, and therefore compressors and pumping units would never be needed. We, of course, could beat the industry average and would drill only two dry holes for both exploratory and development.

"We finally made them happy and sent in the sealed bid."

You want to know the end of the story? Did they have the high bid? Did they find all those oil sands? Did they get rich?

"You guessed it," our grinning tale-bearer concludes. "Six other companies bid higher than we did. And the subsequent exploratory well was a dry hole."

Howard M. Wilson
Gulf Coast Editor

(Reprinted with permission from the February 9, 1970 issue of the Oil and Gas Journal.)

judgments from Henry in the form of distributions. A similar line of questioning may be required to obtain distributions from members of your staff who might be unfamiliar with simulation.

A Case Study

The Blackduck Prospect (Part III)

The technique that Bill Davis had read about which considers all (infinite) possible outcomes is called simulation. We have just discussed how simulation works, its logic, and the method of using the results to compute expected profitability parameters. Perhaps it would be of interest to see how the Blackduck Prospect would have looked if Henry Oilfinder had applied a simulation analysis to his recommendation.

Most of the data and judgment required for the simulation study were given in Henry's two committee presentations. A few points have to be clarified, however, so let's send Marvin Lee, a simulation expert, down to call on Henry to answer a few questions.

MARVIN LEE: Henry, you have already provided most of the data for our special simulation analysis. But we need to check out a few additional points about the prospect.

HENRY OILFINDER: Sure, fire away.

MARVIN LEE: Well, let's talk about the variable net pay thickness. You indicated that average net pay would be 50 feet, with the minimum and maximum values being 30 feet and 70 feet respectively. Was that right?

HENRY OILFINDER: Yes sir.

MARVIN LEE: Do you think net pay could be greater than 70 feet or less than 30 feet, or do these values represent the extreme values which you consider will occur?

HENRY OILFINDER: Well, of course that's hard to say, because we really have no control in the Blackduck area. We have seen wells with 35 feet, 65 feet, and 52 feet in the nearby Pinetree field, and we think Blackduck should be similar to Pinetree. I guess I picked the range of 30 feet to 70 feet as covering the range of what we have seen in Pinetree so far.

MARVIN LEE: I see. Well Henry what do you think is the probability of encountering, say, 40 feet?

HENRY OILFINDER: Boy I don't know. You see you are asking me for precise numbers about things I really have very little knowledge of because we've done no drilling in the area.

MARVIN LEE: About all you can say about it then, is that the range of possible values of net pay is from 30 to 70 feet and that you feel the most likely, or most probable value of net pay is about 50 feet. Is that correct?

HENRY OILFINDER: Yes sir, and I'm stretching my imagination a bit to even be able to be that definitive.

MARVIN LEE: But don't you agree there are many possible values of net pay which *could* occur within the range of 30 to 70 feet?

HENRY OILFINDER: Oh yes, I certainly agree to that. Trouble is I just don't know how to assign a probability to each of those many possible values, and I think you would need to know them.

MARVIN LEE: Yes that's right, but we can represent what you've said about your knowledge of net pay values in what is called a triangular distribution. The frequency is proportional to probability. So the higher the distribution curve is above the horizontal axis the more likely it is that the value of net pay will occur. The distribution is highest about 50 feet and you said that was the most likely value. Values of net pay approaching the end values of 30 feet and 70 feet become increasingly less likely to occur. This distribution also implies that no values of net pay less than 30 feet or greater than 70 feet can occur. Does that seem like a logical representation of your professional judgment?

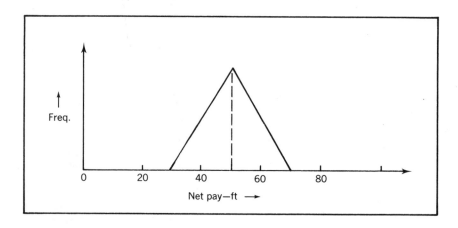

HENRY OILFINDER: Well yes, I guess so. You really filled in a lot of gaps in my judgment with that simple graph, but it sounds pretty reasonable.

MARVIN LEE: Well of course I realize that it is, at best, only an approximation of the distribution of values of net pay, but if all a person can say about a variable is a minimum, maximum, and most likely value then it is a pretty good approximation. Another parameter which you said could vary was the areal size of the field. I believe you said 400 acres for the minimum case, 1600 acres for the average, or most likely case, and 2400 for the maximum case.

HENRY OILFINDER: Yes sir, I believe that was right. These numbers were based largely on our interpretation of the "seis" data and the geological control we had.

MARVIN LEE: Then you probably don't have sufficient data to form opinions as to the likelihood of intermediate values of areal size either. Right?

HENRY OILFINDER: That's right. We're not in much better shape than estimating net pay. Could we use a triangle to describe probabilities of various values of area, just as we did for net pay?

MARVIN LEE: Yes, and it would look like this:

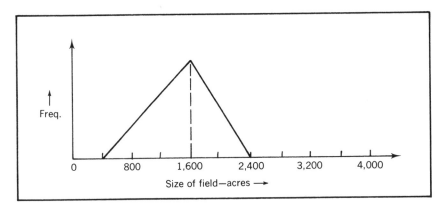

HENRY OILFINDER: But that triangle isn't symmetrical like the net pay distribution.

MARVIN LEE: Correct, but that's no problem. It can be symmetrical, or skewed in either direction. Now let's talk about recovery, barrels per acre-foot. How would you describe possible values of recovery and the associated probabilities?

HENRY OILFINDER: Well here again we didn't have much to base it on. It was sort of a compromise with the engineers, using data from the completed wells in Pinetree. I feel 200 barrels per acre-foot is certainly the most probable value, but they thought it could be as low as 150 barrels per acre-foot because we don't know for sure what our recovery on primary, as a fraction of original oil in place, will be. The upper value of 225 was based on the unusually high porosity we saw in the Perkins Unit, the dryhole 18 miles from Blackduck.

MARVIN LEE: What about intermediate values? Do you have any feeling as to how likely they would occur?

HENRY OILFINDER: No sir, not for this prospect.

MARVIN LEE: Apparently then, about the best we can do is to use another triangular distribution.

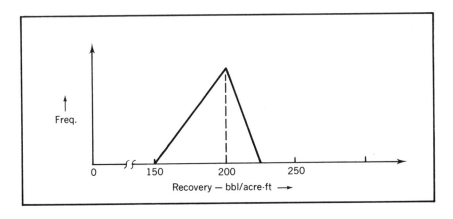

Now the advantage we will have by doing this is that we can consider all intermediate values of the three parameters rather than just the three discrete sets of parameters which you used in your analysis last Friday. By the way Henry, do you feel that these three variables are independent of one another?

HENRY OILFINDER: What does that mean?

MARVIN LEE: Well, as an example, do you think size of the field is related to how many feet of pay you get? If the field was large do you feel that you could still expect the same range of possible values of net pay?

HENRY OILFINDER: Well there might be a correlation, but I'm not sure I could describe it.

MARVIN LEE: Is recovery factor independent of the other two parameters?

HENRY OILFINDER: Probably not. As recovery factor increases it means that porosity is increasing and/or water saturation is decreasing and/or the primary recovery fraction is increasing. It seems logical that recovery would be better from a thick sand, but I don't know how to describe it with numbers. We'd have to ask the engineers for some help on that one. But to be frank about it, we have so little control I don't know whether we'd be able to define or defend such a correlation even if it existed.

MARVIN LEE: I see, well I guess if we make the assumption for now that they are independent it is about the best we can do. The final thing to consider is the well cost variations. The way you handled it in your analysis, that is, by using an expected average well cost of $1.18 million each is OK. I'll just use the same number in my study. This means then that in any of the possible field sizes that 10% of the wells drilled will cost about $.8 million each, 70% will be about $1.0 million each, and 20% of the wells would cost about $2.0 million each.

HENRY OILFINDER: That's my understanding of what engineering concluded from their well cost study.

MARVIN LEE: Well Henry, thanks a lot. That's all I need. Everything from here on is just mechanical.

HENRY OILFINDER: You mean that's all you need from me?

MARVIN LEE: That's right. You just about had it all done when you went through the decision tree calculation.

HENRY OILFINDER: Well I'm sure surprised by this. I had a terrible time trying to get three possible field sizes built into the analysis. When Mr. Miller started talking about thousands or millions of possible intermediate outcomes I figured it would be so complex, if even possible, that it would take a Ph.D. mathematician.

MARVIN LEE: No, simulation is really very simple, and requires no formal mathematics on the part of the geologist and engineer.

HENRY OILFINDER: What will the result of your study be?

MARVIN LEE: We will use your judgment (the distributions) and generate or compute a distribution of profit. This will tell us the range and probabilities of values of profit if we find oil. The other outcome, of course, would be a dry hole. It is called a conditional distribution, because it applies only for the case of finding oil in the first place. It will be what replaces the three branches you had in your decision tree. All of the infinite number of possible levels of reserves (or profit) would be included.

HENRY OILFINDER: But how would we associate probabilities of occurrence to all of these many outcomes?

MARVIN LEE: You wouldn't have to do anything. We could get these probabilities by reading areas under segments of the distribution. All of the probability numbers we need to make an expected value calculation would be there. That's why I said I had all the information I needed from you.

HENRY OILFINDER: Well I'm sure anxious to see your results. If it works it should really give us a good handle on trying to get these blasted probability numbers. Say by the way, what if I knew enough about a variable to describe a distribution other than those triangles?

MARVIN LEE: It would be no problem. Simulation can handle any type of distribution and any number of them. And you don't have to use any math to describe the equations of the distributions. Say, why don't you let me buy you a cup of coffee and I'll explain to you how it all works?

HENRY OILFINDER: Sounds great—let's go!

Following Marvin's discussion with Henry he wrote a simple computer program to simulate the conditional values of profit, given that the wildcat hit oil. The model sampled from the three independent distributions and computed a value of reserves and total net revenues. From the size of field it computed the number of wells by dividing area by 80 acres per well. Then the program sampled from the well cost distribution for each well and accumulated total development costs. Net profit was then computed by subtracting total well costs from total net revenues. The resulting conditional profit distribution is shown in the following table and distribution.

Range of Profit ($MM)	Relative Frequency (fraction)
$-13 \leq P < -9$	.005
$-9 \leq P < -5$	.028
$-5 \leq P < -2$	.078
$-2 \leq P < 0$	.091
$0 \leq P < +2$	.139
$+2 \leq P < +4$	.163
$+4 \leq P < +6$	.131
$+6 \leq P < +8$	.111
$+8 \leq P < +10$	.069
$+10 \leq P < +12$	.052
$+12 \leq P < +14$	.046
$+14 \leq P < +16$	.033
$+16 \leq P < +18$	.030
$+18 \leq P < +20$	.013
$+20 \leq P < +24$	.011
	1.000

From these data it is a simple step to compute the expected value profit for the decision alternative of drilling. Using Henry's estimate of a 10% chance of finding oil we have to multiply all of the relative frequencies of the conditional profit distribution by 0.1. (This is because there is only a 10% chance of finding oil in the first place.) The final expected value profit computation is given below—for the case of a 10% chance of finding oil.

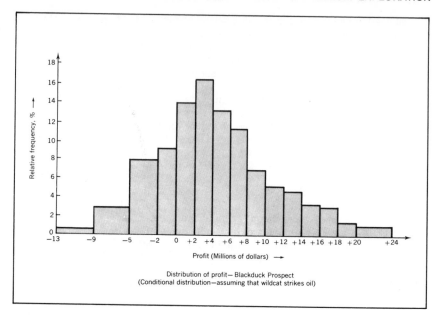

Distribution of profit— Blackduck Prospect
(Conditional distribution—assuming that wildcat strikes oil)

BLACKDUCK PROSPECT

Expected Value Computation using 10% chance of finding oil, 90% chance of dry
hole costing $1.18 million dollars.

Programmer: Marvin Lee, Economic Analysis Group

Outcome ($MM)	Probability of Occurrence	Midpoint of Profit Range ($MM)	Expected Value of Outcome ($MM)
−$1.18 (Dryhole)	0.9000	−$1.18	−$1.0620
−13 to − 9	.0005	−11.00	− .0055
− 9 to − 5	.0028	− 7.00	− .0196
− 5 to − 2	.0078	− 3.50	− .0273
− 2 to 0	.0091	− 1.00	− .0091
0 to + 2	.0139	+ 1.00	+ .0139
+ 2 to + 4	.0163	+ 3.00	+ .0489
+ 4 to + 6	.0131	+ 5.00	+ .0655
+ 6 to + 8	.0111	+ 7.00	+ .0777
+ 8 to +10	.0069	+ 9.00	+ .0621
+ 10 to +12	.0052	+11.00	+ .0572
+ 12 to +14	.0046	+13.00	+ .0598
+ 14 to +16	.0033	+15.00	+ .0495
+ 16 to +18	.0030	+17.00	+ .0510
+ 18 to +20	.0013	+19.00	+ .0247
+ 20 to +24	.0011	+22.00	+ .0242
	1.0000		−$0.5890

EXPECTED VALUE PROFIT
FOR ALTERNATIVE "DRILL"

This expectation compares to the +$0.148 million which Henry computed using just three values of possible reserves.

If Henry wanted to consider different values of the chance of discovering oil it would be a simple calculation and would not require running the simulation program again. If he wanted to see if the expectation was positive for a chance of success of 15%, the first probability number in the above table (corresponding to a dry hole) would then be .85 rather than .90. All of the remaining probabilities would be obtained by multiplying the relative frequencies of the conditional profit distribution (output of the simulation program) by 0.15. This is a convenient feature of formulating the simulation model on a conditional basis.

But Marvin even had an additional short-cut up his sleeve! The functional relationship between expected value profit and chance of finding oil is a straight line. Marvin already computed expectation at a chance factor of 10% (−$0.589 million),

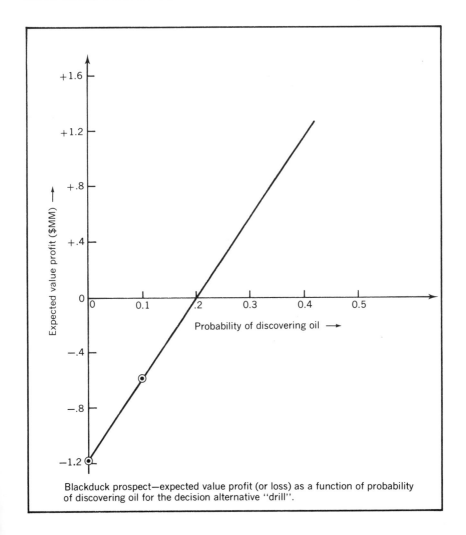

Blackduck prospect—expected value profit (or loss) as a function of probability of discovering oil for the decision alternative "drill".

so he only needs one additional point to define the relationship. The second point can be at zero chance of success. The expectation is simply the dry hole loss of −$1.18 million. From the resulting graph the decision maker can find the expectation for any value of success. In the Blackduck example we can determine, for example, that Profit Oil and Gas would have to expect a chance of success over 35% to get an expected profit of +$1.0 million. Using similar calculations any other decision alternative can be plotted on the same graph for comparison purposes.

9 Implementing Risk Analysis Methods

There is no doubt that courage to act boldly in the face of apparent uncertainty can be greatly bolstered by the clear portrayal of the risks and possible rewards. To achieve this security requires only a slight effort beyond what most companies exert in studying capital investments.

— Hertz

WE have essentially completed our discussion of the basic logic and techniques of exploration risk analysis with the exception of a few remaining specialized types of analyses (to be discussed in the final two chapters). Thus, it is timely to talk about implementing these risk analysis procedures into the overall analysis of a prospect. As you no doubt suspect the implementation of these ideas is not a "bed of roses." In fact, it sometimes raises very strong emotions and opinions on the part of management decision makers, geologists, engineers, management scientists, etc.! Some examples:

■ Management science has grown so remote from and unmindful of the conditions of "live" management that it has abdicated its usability. (Ref. 9.1)

■ The manager doesn't understand the power of the (decision analysis) tools. He isn't sympathetic to systematic decision making and would rather fly by the seat of his pants because this is safer for his ego. (Ref. 9.1)

■ In the oil business we're fooling ourselves to try to be quantitative because risk is just too difficult to quantify.

■ The analysis methods which now exist are well in advance of the ability of most decision makers to use them.

■ Within 10 years, decision theory, conversational computers, and library programs should occupy the same role for the manager as calculus, slide rules, and mathematical tables do for the engineer today. The engineer of Roman times had none of these, but he could make perfectly good bridges. However, he could not compete today, even in bridge building, let alone astro-engineering. Management is today at the stage of Roman engineering. (Ref. 9.2)

■ Application of probabilities will often yield entirely different and better decisions.

■ We are trying to use decision analysis but it's no use — our managers don't understand it and won't listen to what we have to say. All they want to know is the rate of return.

■ Management in recent years has spent much time in developing economic formulas for prospect evaluation, which in the last analysis have planted only more trees in the already cluttered forest of the explorationist. (Ref. 9.3)

■ . . . if capital budgeting standards were applied to risk analysis it would never pay its way. The organizational and computer costs, given the low impact of a single project on a multibillion dollar corporation, and given our limited knowledge about how to use the results, make it more expensive than it's worth. (Ref. 9.4)

■ "There is absolutely no way to quantify geology. and I resent you even suggesting or implying that I need risk analysis. If you engineers and economists want to run out a bunch of numbers that's your business, but don't clutter up my analysis of the prospect with any of that stuff."

■ My company feels these ideas are too new and too radical to use today on really big decisions. But these ideas are too promising to ignore altogether. We are encouraging our management to experiment with them and to actually carry out decisions based on these analyses for selected medium-sized problems in some departments where middle management feels that it makes sense to do so. . . . As yet, we don't allow the formal procedures to contaminate our intuitive analyses of major problems, but after a decision has been made we sometimes like to compare notes with those fellows who have been formally analyzing the problem on the side. When they're way off they complain they are not privy to the counsel of the top people, but when they're right they sometimes can raise some embarassing questions. I think we'll adopt these techniques for some big problems in the future. (Ref. 9.5)

■ If we don't use risk analysis what other viable alternative is there?

My feeling is that most oil companies are definitely committed to a goal of making better use of the quantitative decision analysis methods. But the transition from our historic ways of evaluating prospects to some of these new approaches is indeed painful and frustrating. The current state of transition is akin to the meshing of the two gears, or cogwheels of Figure 9.1. The larger gear represents all of us in the oil business who

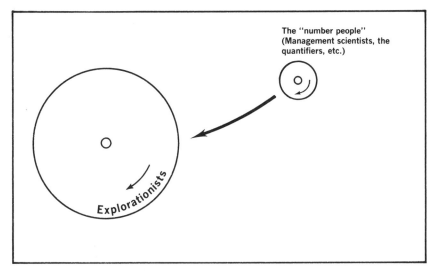

Figure 9.1 *An analogy of the current state of affairs with regard to using risk analysis in petroleum exploration. The objective is to mesh together the explorationists and the management scientists to provide better insights about how best to make decisions under uncertainty. But rather than meshing together often times the two groups clash, as when two gears rotating in the same direction are brought together.*

have been going about our tasks in seemingly good fashion, turning to the right as we go.

Now along comes a group of management scientists, statisticians, computer experts, etc., the so-called "number people," who say to the explorationist: *"Say, Mr. Oilman, you have a classic situation of decision making under uncertainty and we've got some neat new tools that will permit you to do a better job of investing your money. Let's get together and we'll show you all about it."* Well, it seems the number people are usually turning to the right also. And when the two gears are brought together for the purpose of meshing their capabilities you can bet what happens. Sparks fly!

The gears are stripped and both cease to turn! The oilmen leave the meeting muttering that the number people don't even comprehend the problem. The "quantifiers" leave wondering how the oil business has survived even this long using some of their "seat-of-the-pants" evaluations.

These all too frequent encounters are unfortunate. Obviously both parties need to communicate more effectively. But perhaps the really unfortunate result is a hardening of views on either side of the issue and endless arguments between the believers and the non-believers. The effect is to shift the focus of the arguments away from the key issue. And that is—can these new risk analysis methods give our management a better insight about how best to invest exploration capital? Can the feasibility (viability, profitability, desirability) of a prospect be established without engaging in quantitative analyses?

If the answer to the last question is yes then we as oilmen are justified in telling the number people to leave us alone. If the answer is no then perhaps we had all better sit down together and try to see how the ideas can be integrated into our prospect analyses. We must remember the decision theory advocates aren't attempting to run our business, or tell us where, or when to drill. They are clearly an auxiliary gear in the machinery. But the presumption is that if we mesh together we may have a smoother running and more powerful machine.

Perhaps I need to be more positive about all this, but as I work with different companies throughout the world I observe over and over again the sharp divisions within the organization of those who accept risk analysis and those who reject it. Certainly one of the probable causes is lack of understanding by decision makers.

They do not understand all the new terms and feel their role as a decision maker is being challenged—or indeed threatened. What is needed are operating executives who think and communicate comfortably in the formal language of probabilities. But also the "quantifiers" need a much larger supply of patience, tact, diplomacy, and understanding than I sometimes see displayed.

It serves no useful purpose, however, to continue describing the points of conflict. Rather, let's emphasize the positive and mention a few ways in which the transition can perhaps be smoothed somewhat. I think there are several areas where we can take important steps to speed the implementation and acceptance of these new approaches:

- Motivation for using risk analysis
- Organization: Where should the specialists be located?
- Communication
- Education
- The need for teamwork in developing risk analysis procedures

Motivation for Using Risk Analysis

As I see it, two things are required here—high level executive support and lower management level, grass roots acceptance of the need to try to do a better job of evaluating and accounting for risk and uncertainty. Both are essential. If top level management decrees that the ideas will be used beginning next Monday (come hell or high water!) many of the lower level management may take only minimal interest in trying to understand and use it. A formal edict is likely to be counter productive. The problem is just as difficult if lower level staffs make an honest effort to improve their analyses of risk and uncertainty, only to find unreceptive or unresponsive upper level management.

I believe a strong "champion" of the cause at very top levels within the company or organization is absolutely essential. He is needed to urge, foster, and encourage his subordinate managers to experiment with and learn more about decision making under uncertainty. He must nurture a spirit of willingness to listen to new and better ideas. To the extent possible he needs to provide his lower level decision makers assurance that he is open to new ideas, and that their job security will not be in jeopardy for trying new analysis techniques.

But a top level manager can't implement successful risk analysis procedures by himself. The various lower level, operating decision makers must be involved also. And if their active involvement is by choice (rather than by edict) the likelihood of success is much higher. They must be willing to learn, to study, and to become involved with the analysis of risk and uncertainty. I think sometimes our middle and lower level managers tend to become too inflexible at times. Perhaps this is caused by their concern for job security, or a desire to not "rock the ship." But these positions in the organization are important — most of the day-to-day decisions are made at these levels — and these decision makers must become actively involved. They need to continually assess the way they make decisions and to compare their present methods with alternatives to see if there is a better way.

All of this is to counsel an open mind on the part of all management. The top level executives need to encourage their subordinate managers to try new ways. And the open mind of the lower level managers is essential to foster a desire on the part of the explorationist to use the new concepts. A closed mind can be terribly deadening. I am aware of one organization which had a capable, alert staff of geologists and engineers who were making some rather sophisticated analyses of risk and uncertainty — only to find their work totally rejected by an implacable manager who simply did not want to give the new ideas a try. I have also observed another organization who had a senior level executive who was alive to the insights that could be gained from risk analysis and expected values.

He had an open mind, he encouraged his staff, he asked the right kind of questions, and in general fostered an open spirit of trying to do a better job of decision making. Within a few years the result was a superb management-analyst team that is probably better equipped to analyze exploration prospects than any I've seen.

Organization: Where Should the Specialists Be Located?

When a company decides to improve its capabilities for incorporating risk and uncertainty in the decision making process an obvious first step is to secure (or train) a number of explorationists who are skilled with the risk analysis concepts. Where should these specialists be located within the organization?

There seems to be two schools of thought on this. One is to build a small staff of specialists at the corporate headquarters who would serve as internal consultants to the various operating divisions.

The other approach is to integrate these specialists into the line organization of each of the operating divisions or regions.

The special management science teams at headquarters have the advantage of concentrating the specialists into a highly capable team who can focus their collective skills on major problems. Their location at headquarters has the advantage of their availability for immediate top-level evaluations, and of coordinating the risk analysis procedures throughout the company. It is usually this group which prepares and guides the company analysis procedures, such as developing general purpose simulation programs.

The advantages of the second approach result from having a specialist closer to the day-to-day problems. He works as a team with division exploration staff and his allegiance and job satisfaction are in concert with theirs. It eliminates the need for the division staffs to deal with an outsider (the corporate level management scientists), thus eliminating a possible breakdown of trust and communication.

Having a specialist in each operating division also greatly facilitates the on-the-spot changes that are sometimes required to modify a general purpose simulation program to fit a special condition.

While the choice is somewhat dictated by the size and organizational structure of the company, and by the number of risk analysis specialists available, I generally favor the second approach. The principal weakness of clustering the few specialists in a headquarters group is their remoteness from the problem. I think it is a much better choice to initially place these few specialists out in the divisions as part of the day-to-day prospect evaluation team.

The changes may not be quite as dramatic initially, but in the long term I think the ideas can be implemented with much greater chance of success. As people work together ideas are exchanged more readily. Other staff members will have greater opportunity to be exposed to risk analysis — and, on the other side of the coin, the management scientist specialist will certainly gain a much better understanding of the strengths and limitations of various methods when he is an integral part of the action.

Communication

Communication between the analyst and decision maker, and between the various levels of management is clearly important. A good number of the comments and feelings about risk analysis that were mentioned at the beginning of this chapter are simply caused by the inability of the management scientist and the decision maker to communicate with one another.

It seems to be a rather common weakness of some specialists to view the decision process too narrowly. They are comfortable with their numerical methods and tend to view the decision making process as an exact, black and white issue, when in fact it's usually inexact and consists of many shades of gray. The specialist needs to continually receive input from the perspective of the man who finally assumes responsibilities for the decisions.

And on the other hand, the decision maker needs to adopt a more active stance in learning about the new quantitative methods. He needs to find out from the specialist the strengths and weaknesses of the methods, what methods are even available, how they compare with present company procedures, etc.

All this suggests the need for open and continuing communication between the manager, the explorationist, and the management science specialist. All too frequently we tend to build castles, or empires, within our organizations. Sometimes it is almost as if one "empire" will refuse to help or to ask for help from the other "empire." Unfortunate, really, because we are all working within the company for a common goal. We must communicate! Risk analysis requires the talents and input of all professionals, and to implement a new approach of this complexity requires continuous communication at all times.

Education

Education is certainly an important prerequisite to implementing risk analysis. The education is needed early and at all levels. The explorationists who will be actually using the methods need to learn all the tedious details involved, for obvious reasons. This may include short courses on probability and statistics for those who have never had any academic training on the subject, short courses on the logic and techniques of the methods, and continuing self-study by each person.

The decision makers also must become involved in the education process. It is important that they, too, learn as much as possible about risk analysis. As a very minimum they should be exposed to the overall philosophy of quantitative risk analysis methods, and to the types of information that can be obtained from things such as decision trees, simulation, etc. The managers must also become thoroughly conversant in the meaning and implication of the expected value concept.

The form of this education process can vary — company short courses, outside courses offered by universities and/or various training organizations, one and two day management seminars, etc. Whatever form it may take is probably not as important as recognition that the education process cannot be short-circuited. The learning process is slow and tedious and I know of no way we can speed it along or collapse it into just an afternoon seminar.

Some companies take great pains to smooth the transition to quantitative risk analysis but then fail to provide sufficient (or any) education opportunities for those who will be involved in using the new methods. As with most scientific endeavors a little knowledge is sometimes more dangerous than none at all.

As an educator I am often requested to come to a company's office for a day or two to teach their exploration staff risk analysis. I usually advise them to forget the idea — you simply can't learn it all in just two days, and to try to do so is a waste of money and time. I also frequently hear executives lament that they would like to attend a course on risk analysis but they can't take the time away from the office. Admittedly I find it difficult to offer much sympathy in these cases!

There can really be no alternative to a complete, on-going education process. It's hard to study and learn for some, and education courses cost money and require management and explorationists to be away from their normal duties. But to eliminate education, or to cut corners to save a little money here or there is, in the end, false economics. Without educating the explorationists it's going to be pretty difficult for them to wade through the complexities of hypergeometric distributions, Bayesian analysis, etc. on their own.

The Need for Teamwork in Developing Risk Analysis Procedures

As you start out developing quantitative risk analysis procedures for your company I strongly recommend a team effort. Set up a task force, or team, to develop the analysis procedures and supporting computer programs, rather than relying on one or two specialists to do it all. The team should consist of representatives from geology, engineering, management, geophysics, management science, and computer programmers. A team effort results in much more input into the development, better continuity, and better documentation.

The broader input is obvious. If each of the professional disciplines is involved in formulating the procedures and programs the result will certainly be more realistic than if one person had tried to anticipate the needs and types of input required of a general risk analysis procedure. Such an approach also has the intrinsic benefit that the various professional disciplines will all feel involved in the general procedures.

When the procedures and computer programs are developed by a team effort another benefit is better documentation and continuity. If one person writes an entire simulation program and he is the person who will be running the program for the various divisions he will probably not make much of an effort to document the program. Why should he—he wrote it so he knows exactly what it does! But what happens when he goes on vacation, or leaves the company? With a team effort this problem will likely not arise. Since many persons are involved in its preparation there will be many aware of how the programs work. So when one person leaves or is on another assignment there will be others available who are knowledgeable about the whole set of procedures and programs.

*   *   *

And a final comment, in case all else fails!

10   Decisions to Purchase Imperfect Information

There is no such thing as a sure thing!

IN the petroleum industry we sometimes have the opportunity to purchase additional information so as to better define (or reduce) the uncertainty associated with important decisions. For example, we may decide to defer our decision about drilling a geologic prospect until we run a seismic survey to better define the structure and its physical dimensions. Other examples of information purchased to better define uncertainty include logging surveys, drill stem tests, and the drilling of additional delineation wells on a newly discovered structure before building production platform facilities.

If the additional information is perfect (i.e., there is no error in the interpretation and it will tell us precisely the true state of nature) a relatively straight-forward analysis will suggest whether it is feasible to purchase the information. If the information is imperfect the analysis of whether to purchase the information is considerably more complex. Fortunately, however, the analyses of decisions to purchase imperfect information all follow a single, consistent format, and it is the purpose of this chapter to explain the logic and mechanics of the analysis.

Imperfect information is probably the rule, rather than the exception in the oil business, and the analysis technique we will be discussing has many potential applications for the explorationist.

First we will discuss the general logic and approach for analyzing decisions to purchase imperfect information, including a generalized decision tree. Then we will illustrate the analysis technique using two practical, real-world problems:

EXAMPLE NO. 1: Analyzing whether to run a logging survey to detect whether we are approaching an over-pressured formation which may cause a potential blowout.

EXAMPLE NO. 2: Analyzing whether to drill an additional downdip delineation well on a recently discovered offshore oil structure so as to be able to determine the areal extent of the field before sizing the production platforms.

501

THE GENERAL APPROACH TO THE ANALYSIS

In considering the logic to support the analysis of decisions to purchase imperfect information we must first recognize that the reason we are buying the information is to defer our point of decision until after we have the information. Thus, our time-zero decision as to whether to purchase the information is going to be influenced by later, sequential decisions that will be made after receipt of the information. This means that we will have to use a decision tree in the analysis.

The second point we need to remember is that the later, sequential decisions required after the information is available are made without knowledge of whether the new evidence or interpretations of the information are correct or incorrect. That is, because the information we are purchasing is imperfect we still have decisions under uncertainty. We will hopefully have better perceptions about the likelihoods of the states of nature, but because the information is imperfect we will not have complete, perfect information about the actual state of nature.

The decision tree which describes the sequence of choices and outcomes follows a definite series of steps.

1. First, we purchase the imperfect information, whatever it may be (run logs, drill additional delineation well, etc.).
2. Next, we determine what the information tells us. These are the new interpretations, or evidences which become available.
3. At this point the decision maker must decide how to proceed, that is, which choice should he now make? These will be the same choices he had at time-zero in lieu of purchasing the information. But his decision choice will now be based on different expected values because he has new perceptions about the states of nature from the evidence received.
4. Finally, having decided on a strategy the actual state of nature (outcome) will emerge from those that were possible.

All of this is summarized in a generalized decision tree for the option "purchase additional information" in Figure 10.1. This is a schematic diagram and the analysis of decisions to purchase imperfect information will always follow this general pattern. The tree in Figure 10.1 is drawn for only two possible interpretations of the information (B and B'), only two sequential decision choices, and only three possible states of nature (E_1, E_2, and E_3).

If there were more than two possible interpretations you would

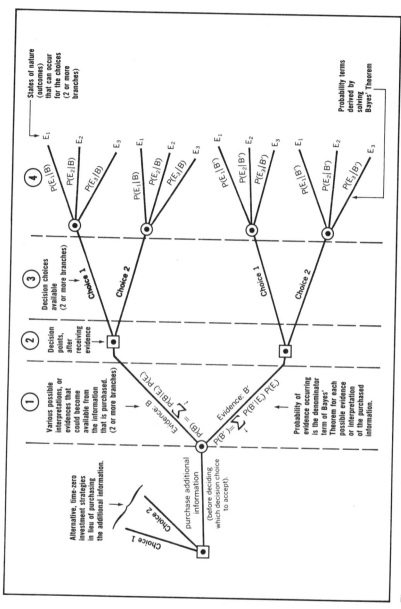

Figure 10.1 *Generalized decision tree used to determine the feasibility of purchasing additional information.*

503

simply have additional branches in Section I. Same for Sections III and IV—more choices and/or more possible states of nature would merely imply additional branches from the decision and chance nodes.

The probabilities on the chance node branches in Section IV are obtained by solving Bayes' Theorem (Equation 6.16 and Table 6.17). The probability terms represent the revised perceptions of the likelihoods of the various states of nature, given the new evidence or interpretation. The probability terms in Section I represent the denominator terms of Bayes' Theorem. These terms are the sums of Column 4 if the Bayesian analysis is performed using a 5-column table such as Table 6.17. Because of this fact it is suggested that when solving the tree you determine the Section IV probabilities first. In the process of determining these terms the Section I probabilities will become available immediately as the respective sums of Column 4.

Well, that's all there is to it! However, assuming you are somewhat bewildered by it all let's look at two real-world problems and then we'll come back to these general steps and look at them again. By that time you will probably be in a better position to recognize the general pattern or scheme of things as displayed in Figure 10.1.

Example No. 1

The Problem

Our company is planning to drill a deep exploratory well in an area where a high pressure formation, located about 500 feet above the objective oil formation, has been encountered in some wells. The normal drilling practice is to drill to TD, run open-hole logs, and set a production string of casing if a discovery is indicated. The objective formation is expected to have normal pressures, and relatively low mud weights are required to maintain control of the well. If the over-pressured zone is present above the objective formation substantially higher mud weights are required to maintain control of the well.

This usually results in breaking down the incompetent strata near the top of the long open-hole section and losing circulation. To protect against the lost circulation problem (and resulting loss of control) we could plan to drill to a point just above where we anticipate the over-pressured zone to be, run a string of intermediate casing to seal off the incompetent strata, increase mud weight and continue drilling.

With this strategy we'd be protected from any possible blowouts, but at the expense of an extra string of intermediate casing costing $100,000. This expense added to the estimated cost of $300,000 to drill the well in the conventional manner, would result in a total well cost of $400,000.

An alternative drilling program would be to eliminate any plans for running intermediate casing and gamble that the over-pressured zone is not present. This would reduce the drilling costs to $300,000 unless we encountered an over-pressured zone.

If this happened we would have to pay an additional amount to regain control of the well and set an intermediate string of casing. It is estimated that this contingency would cause well costs to climb to $450,000 ($300,-000 base cost, $100,000 for intermediate casing, and $50,000 for extra rig time, mud costs, etc. needed to maintain control until the intermediate casing is set.) Our geologists estimate that the probability of encountering the over-pressured zone is 0.3.

As a third alternative, we could make a series of logging runs above the possible high pressure zone and compute geostatic pressure gradient trends as measured from the resistivity changes of the shale sections above the zone. This interpretation technique can sometimes be used effectively to determine whether the well is approaching an over-pressured zone.

The reliability of this interpretative technique is estimated to be 0.9. (That is, whatever the interpretation we make from the logs there is only a 90% chance the interpretation is correct.) The cost of these logging surveys is $20,000. If we ran the logs we'd of course defer our decision whether to run intermediate casing until we obtained the additional (and imperfect) information from the logs.

These options lead to the following important concerns:

 I. Assuming we're interested in minimizing expected drilling costs which would be the preferred strategy to drill the well?
 II. How would our strategies be affected if the probability of encountering the over-pressured zone was 0.4 instead of 0.3? How high does this probability have to be to make the "safe" strategy of setting intermediate casing the best choice? Would our strategies be changed if the reliability of the logging interpretation was only 0.8?
 III. The "penalty" for encountering the high pressure zone without having previously set intermediate casing is only $50,000 in the problem. (That is, we will only have to pay an extra $50,000 to regain control of the well so that drilling can be stopped to set the intermediate casing.) Conceivably, it is possible that if we lose control of the well

it may cost a great deal more to regain control. How would we formulate the analysis to include the possibilities that if we encounter high pressure without intermediate casing there is a 20% chance that we can regain control for $50,000, but there is a 50% chance that $400,-000 would be required to control the well, and a 30% chance that $700,000 would be necessary to regain control?

The Solution

First, before reading on to see the solution to this problem I would challenge you to take a sheet of scratch paper and see if you can draw the decision tree for this problem. (Watch out though—it's not as obvious as you may think!)

To outline the solution we would proceed as follows. We have three strategies available to us and we are trying to determine the one having the lowest expected costs.

1. "Safe" Strategy: Run casing—well costs are $400,000.
2. "Gamble" Strategy: Drill with no plans of running intermediate casing unless high pressure is encountered. If high pressure is encountered total well costs will be $450,000; if no high pressure is encountered costs will only be $300,000.
3. "Purchase additional information" Strategy: This will permit deferring the decision whether to run casing until we have analyzed the geostatic pressure gradient trends obtained from the logs. The information may tell us whether or not we are approaching an overpressured zone. Then our decision choices are whether or not to run intermediate casing. The decision is made *after* we've analyzed the logs but *before* we know whether or not our interpretation of the logs was correct—hence it's still a decision under uncertainty.

The decision tree which describes these options is shown in Figure 10.2. (How does the tree you drew initially compare?) The significant part of the analysis relates to what follows if we initially elect to purchase the additional information, and you should take the time to compare this portion of the tree to the generalized scheme shown in Figure 10.1. So as to focus our attention on just the structure of the decision tree the probability terms and actual solution have been omitted from Figure 10.2.

The monetary values given at the terminal points are all costs, expressed in thousands of dollars. Our choice of strategies will, therefore, be the one which minimizes expected value costs. All the terminal points (d) through (k) include an extra $20 M for logging costs. Note that if we

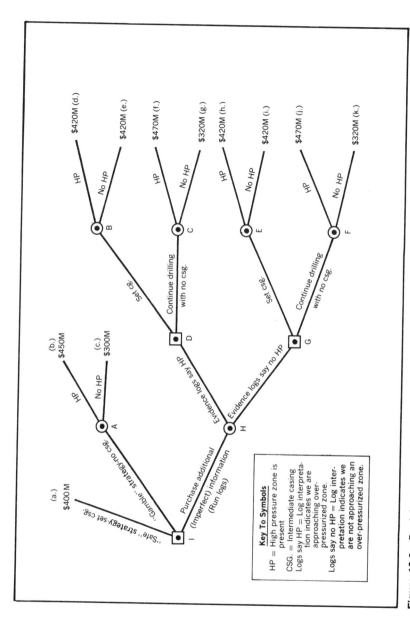

Figure 10.2 *Decision tree required to answer question I of Example 1. All terminal point monetary values are costs expressed in thousands (M) of dollars. Probability terms and the resulting solution of the tree are given in Fig. 10.3.*

Key To Symbols

HP = High pressure zone is present
CSG. = Intermediate casing
Logs say HP = Log interpretation indicates we are approaching over-pressurized zone.
Logs say no HP = Log interpretation indicates we are not approaching an over-pressurized zone.

"Safe" strategy-set csg.
(a.) $400 M

"Gamble" strategy-no csg.
A
HP
(b.) $450M
No HP
(c.) $300M

Purchase additional (imperfect) information (Run logs)
I

Evidence logs say HP

Evidence logs say no HP
H

Set csg.
D

Continue drilling with no csg.

B
HP
$420M (d.)
No HP
$420M (e.)

C
HP
$470M (f.)
No HP
$320M (g.)

Set csg.

Continue drilling with no csg.
G

E
HP
$420M (h.)
No HP
$420M (i.)

F
HP
$470M (j.)
No HP
$320M (k.)

507

had decided not to run casing at decision node D and later encountered high pressure we would have the same potential blowout problems and resulting costs, encountered at terminal point (b). The same applies at terminal point (j).

Before we proceed to determine the necessary probabilities we must observe one very important point. The tree shown in Figure 10.2 implies that the decision about running casing is *not* automatic, depending on what the logs say. That is to say, if we interpret the logs as saying we are approaching high pressure it is not an automatic decision that we must run casing. Rather, the decision will be based on which choice at decision node D results in the lowest expected costs. The same applies if the logs say we are not approaching high pressure — we make the choice at decision node G based on minimizing costs.

I mention this because the usual reaction is to do what the logs suggest. If the logs say we're approaching high pressure we run casing; if they say we're not approaching high pressure then we don't run casing. People presumably defend this "automatic" decision rule on the basis that if we don't believe the logs (and do what they imply) why even run them in the first place? But the very important point they are missing is that the reliability of the log interpretation is not 100%!

Their rule would apply if the log interpretation was perfect, but by definition of the problem the logs are not perfect. We may, in fact, decide to be safe and set casing at decision node G even though the log interpretation would imply that we do not need casing.

This choice would follow, for example, if the potential disaster at terminal point (j) is very large and we're not willing to expose ourselves to a chance of this happening via an error in our interpretation of the logs (i.e. we interpret the logs to say no high pressure, but in fact there is HP). So before proceeding with the solution to the tree you should make sure you understand the management choices at decision nodes D and G. These choices are the central issue in the analysis of decisions to purchase imperfect information, and we cannot simplify the analysis based on some implied, automatic decision rules.

To proceed we must next determine probabilities for all the branches radiating from chance nodes A, B, C, E, F, and H. The probabilities around chance node A represent our initial (time zero) perceptions as to the likelihoods of encountering high pressure and no high pressure. Our geologist estimated the likelihood of encountering the over-pressured zone to be 0.3, so these probabilities are 0.3 and $(1 - 0.3) = 0.7$ respectively.

Next we come to chance node B. The probabilities on the branches leading to terminal points (d) and (e) are the *revised* perceptions as to the likelihoods of high pressure or no high pressure *given the new information that the "logs say high pressure."* We must compute these probabilities with Bayes' Theorem (Equation 6.16). Using the solution of Bayes' Theorem in the form of the five column computation of Table 6.17 we define the following terms:

E_1 = state of nature is high pressure

E_2 = state of nature is that there is no over-pressured zone

$\left.\begin{array}{l} P(E_1) = 0.3 \\ P(E_2) = 0.7 \end{array}\right\}$ Our time zero perceptions as to the likelihoods of the states of nature

B = Evidence: "Logs say HP"

This information gives us the first two columns of the solution to Bayes' Theorem:

Possible States of Nature	Original Risk Estimates
E_1 = HP	0.3
E_2 = No HP	0.7
	1.0

Column 3 of the Bayesian computation lists the conditional probabilities that the evidence, B, could have occurred, given the states of nature. That is, the column 3 values we need are $P(B \mid E_1)$ and $P(B \mid E_2)$. We've defined B as the evidence "Logs say HP," so now we must ask ourselves the questions:

$P(B \mid E_1)$: What is the probability of the logs indicating high pressure when in fact, the true state of nature is that there is high pressure?

$P(B \mid E_2)$: What is the probability of the logs indicating high pressure when, in fact, the true state of nature is that there is no high pressure?

The answers are 0.9 and 0.1 respectively! The only way that the logs can say high pressure when there is high pressure is when we interpret the logs correctly. The only way the logs can say high pressure when there is, in fact, no high pressure is for us to have made an incorrect interpretation of the logs – probability of 0.1. So we see that the reliability of the log interpretation is the conditional probability we need for Column 3 of a Bayes' Theorem solution. With this discovery everything else falls into place, and the probabilities required around chance node B are computed as follows:

Table to Compute Probabilities at Chance Node B, Figure 10.2
Evidence B: "Logs say HP"

Possible States of Nature	Original Risk Estimates	Conditional Probability That Evidence B Could Have Occurred, Given the States of Nature	Joint Probabilities	Revised Risk Estimates
$E_1 = HP$	0.3	0.9	0.27	$0.27/0.34 = 0.795 \leftarrow P(E_1 \mid B)$
$E_2 = $ No HP	0.7	0.1	0.07	$0.07/0.34 = 0.205 \leftarrow P(E_2 \mid B)$
	1.0		0.34	1.000

The probability on the branch leading to terminal point (d) is the revised perception of the likelihood of high pressure, given the evidence "Logs say HP," or in symbols, $P(E_1 \mid B) = 0.795$. The probability on the branch leading to terminal point (e) is the revised perception of the likelihood of no high pressure, given the evidence "Logs say HP," or $P(E_2 \mid B) = 0.205$.

The probabilities at chance node C are the same as at chance node B because they are conditional on the same evidence of "Logs say HP." So we can just enter the 0.795 and 0.205 directly on the branches of chance node C. Recall also that the sum of Column 4 of a Bayesian computation is the likelihood of the evidence occurring in the first place, given our initial perceptions about the likelihood of the states of nature occurring. This implies that the likelihood that the logs will indicate high pressure (one of the branches at chance node H) is simply the sum of Column 4 above, or 0.34.

Finally, we have to compute the probabilities at chance nodes E and F (both will be the same because in both cases they represent revised perceptions based on the evidence "Logs say no HP"). Everything is much the same as the logic we used to compute probabilities at chance node B except the evidence is different:

E_1 = state of nature is high pressure
E_2 = state of nature is that there is no overpressured zone
$P(E_1) = 0.3$
$P(E_2) = 0.7$
B' = Evidence: "Logs say no HP"

Table to Compute Probabilities at Chance Node E, Figure 10.2
Evidence B': "Logs say no HP"

Possible States of Nature	Original Risk Estimates	Conditional Probability That Evidence B' Could Have Occurred, Given the States of Nature	Joint Probabilities	Revised Risk Estimates
E_1 = HP	0.3	0.1	0.03	$0.03/0.66 = 0.045 \leftarrow P(E_1 \mid B')$
E_2 = No HP	0.7	0.9	0.63	$0.63/0.66 = \underline{0.955} \leftarrow P(E_2 \mid B')$
	1.0		0.66	1.000

The computation above is exactly parallel to the approach we previously used at chance node B. The only numerical difference is the conditional probabilities of Column 3 (and the resulting numerical changes in Columns 4 and 5). The questions we ask to determine the conditional probabilities of Column 3 are:

$P(B' \mid E_1)$: What is the probability of the logs indicating no high pressure when in fact there is high pressure? (Answer is 0.1 – the result of an incorrect interpretation).

$P(B' \mid E_2)$: What is the probability of the logs indicating no high pressure when in fact there is no high pressure? (Answer is 0.9 – the likelihood of a correct interpretation).

The Column 5 probabilities above are the probabilities required at chance nodes E and F. By the same reasoning mentioned previously the sum of Column 4, 0.66, is the likelihood of the evidence "Logs say no HP" ever occurring in the first so we can place this probability on the appropriate branch radiating from chance node H.

We have now determined all the probabilities that are required to solve the tree. These terms, together with the solution to the tree are given in Figure 10.3. The step-wise numerical solution of the tree is as follows.

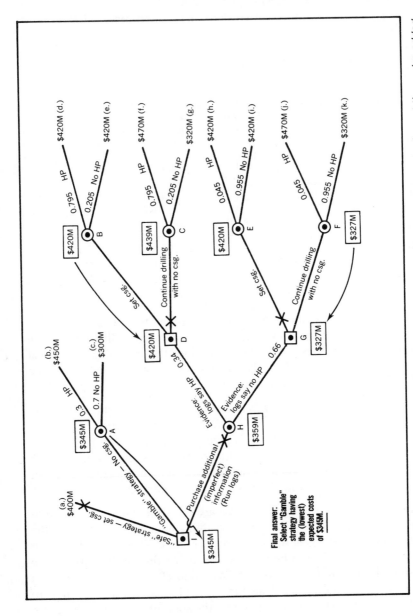

Figure 10.3 *Decision tree of Example 1 and Fig. 10.2 with probabilities and expected values at the nodes added. Final decision which minimizes expected costs (for these cost figures) is to select the "GAMBLE" strategy. Symbols are as defined in Fig. 10.2. Costs are in thousands of dollars.*

512

1. *Chance Node A:* $EV_A = (0.3)(\$450\text{ M}) + (0.7)(\$300\text{ M}) = \underline{\$345\text{ M}}$
2. *Chance Node B:* $EV_B = (0.795)(\$420\text{ M}) + (0.205)(\$420\text{ M}) = \$420\text{ M}$
3. *Chance Node C:* $EV_C = (0.795)(\$470\text{ M}) + (0.205)(\$320\text{ M}) = \$439\text{ M}$
4. *Decision Node D:* Choices are
   Set CSG $\rightarrow$ EV costs are $420 M
   NO CSG $\rightarrow$ EV costs are $439 M
   Decision: Set CSG (it has lower expected costs)
   Thus EV at decision node D is $\underline{\$420\text{ M}}$.
5. *Chance Node E:* $EV_E = (0.045)(\$420\text{ M}) + (0.955)(\$420\text{ M}) = \$420\text{ M}$
6. *Chance Node F:* $EV_F = (0.045)(\$470\text{ M}) + (0.955)(\$320\text{ M}) = \$327\text{ M}$
7. *Decision Node G:* Choices are
   Set CSG $\rightarrow$ EV costs are $420 M
   No CSG $\rightarrow$ EV costs are $327 M
   Decision: Continue drilling with no plans for casing.
   Thus EV at decision node G is $\underline{\$327\text{ M}}$.
8. *Chance Node H:* $EV_H = (0.34)(EV_D) + (0.66)(EV_G)$
   $= (0.34)(\$420) + (0.66)(\$327\text{ M})$
   $= \underline{\$359\text{ M}}$
9. *Decision Node I:* The initial time-zero choices are
   Safe strategy – Set CSG $\rightarrow$ EV costs are $400 M
   "Gamble" strategy – No CSG $\rightarrow$ EV costs are $345 M
   Purchase information – Run logs $\rightarrow$ EV costs are $359 M
   Decision: *Select the "Gamble" strategy of no CSG.* <u>QED</u>

The numerical solution just given suggests the optimal decision is the "Gamble Strategy" having expected costs of $345 M. This compares with the higher expected costs of running casing ($400 M), or the "purchase of additional information" strategy with expected costs of $359 M. So the answer to Question I) of this example problem would be drill with no plans to run logs and no plans to run intermediate casing unless the suspected over-pressured zone is encountered.

Wow! That was a bit complex wasn't it! Remember at the outset of this chapter we indicated that the analysis of decisions to purchase imperfect information can get a bit complex. But there really is no way we can simplify the analysis — it just happens to be a decision problem requiring a little extra work.

Before we proceed to resolve Questions II) and III) of this example problem I would like to draw your attention to two things we did in the steps above. The first point relates to the Bayesian calculations we used to compute probabilities at chance nodes B, C, E, and F. The purpose of the Bayesian computation was to be able to update, or revise our initial perceptions about whether an over-pressured zone may exist in view of the additional evidence we obtained from the logging survey.

At chance node B the new perception as to the likelihood of HP, given that the logs say HP is 0.795 (branch leading to terminal point (d)), as compared to 0.3 as of time zero. It's still not a certainty that high pressure exists if the logs say HP because the reliability of our interpretation is only 0.9. But what if the log interpretation was perfect — and would indicate the true state of nature? Well, intuitively we could probably agree that if the logs say HP and the logs are perfect then our revised perception as to the likelihood of high pressure at terminal point (d) should be 1.0. Right?

Well let's see if the Bayesian computation would confirm our intuition. If the logs were perfect the reliability of the log interpretation would be 1.0 rather than the 0.9 used in solving question I. Thus the Bayesian computation given previously to compute probabilities at chance node B would be changed as follows:

Table to Compute Probabilities at Chance Node B, Figure 10.2

Evidence B: "Logs say HP"

Reliability of Log Interpretation = 1.0

Possible States of Nature	Original Risk Estimates	Conditional Probability That Evidence B Could Have Occurred, Given the States of Nature	Joint Probabilities	Revised Risk Estimates
E_1 = HP	0.3	1.0	0.3	$0.3/0.3 = 1.0 \leftarrow P(E_1 \mid B)$
E_2 = No HP	$\underline{0.7}$	0	$\underline{0}$	$0/0.3 = \underline{0.0} \leftarrow P(E_2 \mid B)$
	1.0		0.3	1.0

We see from this Bayesian computation that if the logs say HP and the interpretation is perfect our revised perceptions as to the likelihood of high pressure, $P(E_1 \mid B)$, would be 1.0. It would be a certainty that high pressure existed, and this is just what we previously agreed should be the case based on our intuition.

On the other end of the "reliability of interpretation" spectrum, what do you suppose our revised perceptions would be if the log interpretation was just as apt to be wrong as right? Hopefully we could agree that for this worst possible case our revised perceptions would be no different (or better) than our initial, time-zero perceptions.

Will the Bayesian computation confirm this? Certainly it will. The only difference now is that the reliability of interpretation has dropped to 0.5. The Bayesian computation for this is as follows:

Table to Compute Probabilities at Chance Node B, Figure 10.2
Evidence B: "Logs say HP"
Reliability of Log Interpretation = 0.5

Possible States of Nature	Original Risk Estimates	Conditional Probability That Evidence B Could Have Occurred, Given the States of Nature	Joint Probabilities	Revised Risk Estimates
E_1 = HP	0.3	0.5	0.15	$0.15/0.50 = 0.30 \leftarrow P(E_1 \mid B)$
E_2 = No HP	$\underline{0.7}$	0.5	$\underline{0.35}$	$0.35/0.50 = \underline{0.70} \leftarrow P(E_2 \mid B)$
	1.0		0.50	1.00

From this computation we see that our assessment as to the likelihood of high pressure after receiving the log information is no different (or better) than our original estimate of high pressure of 0.3. The logs didn't tell us anything. Hopefully these two illustrations should help clarify what the Bayesian computation is doing and why we must use it to determine the probabilities at chance nodes B, C, E, and F. This computation is the point where we introduce into the analysis the imperfect nature of the information we have purchased.

The second point I would mention regarding the solution to question I.) regards the manner in which we constructed the decision tree at chance nodes B and E. You will note that the monetary costs at terminal points (d) and (e) are the same, and the costs are also the same for the branches radiating from chance node E (terminal points (h) and (i)).

Recall that in Chapter 4 (page 127) we proved that when all the terminal point values around a chance node are equal the expected value at the chance node is equal to the terminal point value. As a result of this fact we could have omitted the chance node B and its branches and merely placed the $420 M cost at the end of the branch "Set CSG" radiating from decision node D. The same would be true at chance node E. This would have simplified the tree slightly, and either way would have been a correct solution.

Now let's turn our attention to question II.) of this example problem. The question is seeking to determine the effects of varying numerical values of the key parameters on our overall strategy for drilling the well. We can evaluate these effects by making a sensitivity analysis of the entire decision problem. The procedure is simple—we merely vary the numerical value of one parameter (holding all other parameters at constant values) and re-solve the tree. If our initial estimate of HP was 0.4, rather than 0.3, we merely change the probability terms at chance node A and in Column 2 of the Bayesian analysis accordingly. The 0.3 would become 0.4, and 0.7 would become 0.6. The tree would then be re-solved to see if this results in any changes of strategy at decision node I.

With this general approach we could evaluate many, many possible "what if" situations merely by repetitive solution of the tree with different values of the parameters. The tree does not change, nor does our approach for making the Bayesian computations at chance nodes B, C, E, and F. Only the numbers change. Having performed many of these repetitive computations it is most useful to display the expected values for each initial strategy as a function of the parameter being varied. Such a graph provides a helpful visual display to resolve concerns of the type raised in question II).

The results of such a sensitivity analysis are displayed in Figures 10.4, 10.5, and 10.6. The first of these graphs, Figure 10.4, shows the effect of the costs to regain control of the well on our drilling strategies. The three circled points at the left of the graph represent the three expected value costs computed previously (Figure 10.3) for the case where the extra costs to regain control of the well are only $50 M.

To obtain the plotting points required to graph the function of expected costs the extra penalty cost was merely increased to various higher values and the tree solved again. These increased extra costs only occur at terminal points (b), (f), and (j). For example, for a cost to regain control of the well of $200 M the total drilling cost at these terminal points would be $200 M plus the base drilling cost of $400 M, or a total of $600 M.

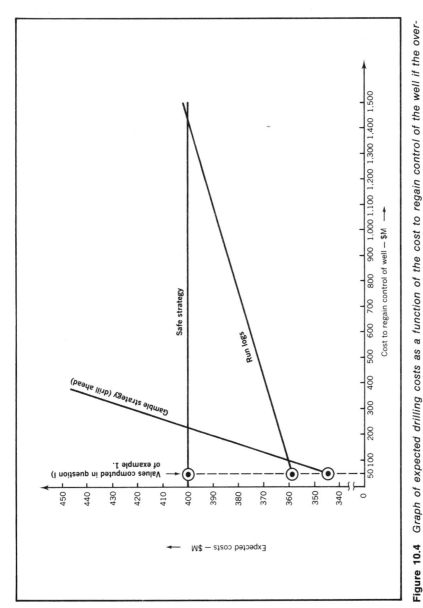

Figure 10.4 Graph of expected drilling costs as a function of the cost to regain control of the well if the over-pressured zone is encountered without having previously set intermediate casing. All other parameters are held constant (i.e., probability of encountering high pressure = 0.3, and the log reliability = 0.9). Circled points are the expected costs computed for Question I in Fig. 10.3.

517

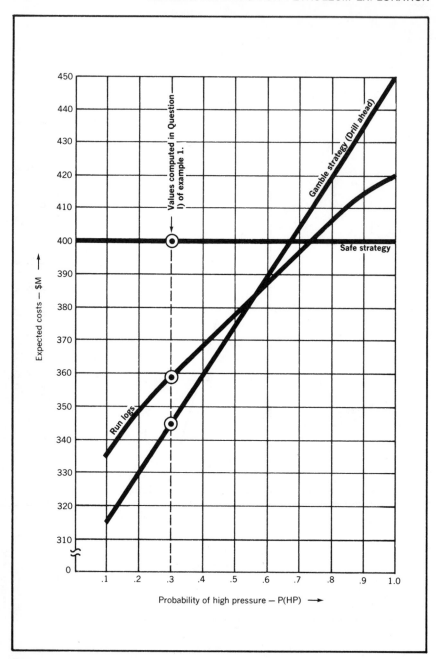

Figure 10.5 *Graph of expected drilling costs as a function of the original estimated probability of encountering the over-pressured zone. All other parameters are held constant (i.e., cost to regain control of the well = $50M, and the log reliability = 0.9). Circled points are the expected costs obtained in the original part of the example, Fig. 10.3.*

From a graph such as Figure 10.4 we can make several general conclusions. If we feel the probable cost to regain control of the well (due to encountering the over-pressured zone without having previously set intermediate casing) will be greater than about $250 M the "gamble strategy" clearly becomes a poor choice. (Remember, we're trying to *minimize* expected costs — not maximize expected values).

For the range of costs to regain control of $250 M up to about $1300 M or $1400 M the "log" strategy would minimize costs. If potential costs of regaining control of the well are thought to probably exceed $1400 M our best strategy is to play it safe and plan to set intermediate casing before drilling into the depth range of the suspected over-pressured zone.

Does such a graph tell us what the cost to regain control of the well will be? Certainly not! We as professional explorationists must evaluate that. We must be able to specify where we expect to be on the horizontal axis. But once we can estimate that (or at least a general range) Figure 10.4 would immediately suggest to us an optimal strategy. Statistics and decision analysis can't run the oil business for us — but if we can specify the general ranges of values of some of these key variables (and uncertainties) decision analysis can provide management with concise values of his available options.

Figure 10.5 shows the effects of various values of the initial estimate of the likelihood of encountering high pressure while holding all other parameters the same as used in solving Figure 10.3. (That is, the penalty cost for regaining control of the well was held at $50 M and the log reliability was held at 0.9.)

The circled values correspond to the expected value costs for the probability of high pressure of 0.3 used in the first part of the example. The general conclusions from this graph would be that if we estimate the likelihood of encountering high pressure to be less than 0.6–0.7 the "gamble" strategy and the "run log" strategy have advantages over the "safe" strategy. If the likelihood of high pressure is greater than this range we're better off to play it safe and run casing.

Incidentally, the reason the "run logs" function is not linear is because of the sequential management options of this strategy at decision nodes D and G. As the probability of high pressure increases the decision choices change at these points, and this destroys any possible direct linear relationships that might exist between expected values and probabilities.

In Figure 10.6 we can assess the effects of the log reliability on our strategies for completing the well. As before everything else was held

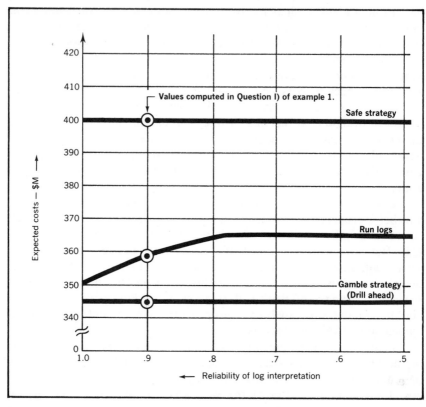

Figure 10.6 *Graph of expected drilling costs as a function of the reliability of the log interpretation. All other parameters are held constant (i.e., cost to regain control of the well = $50M, and the original estimate of encountering high pressure = 0.3). Circled points are the expected costs computed in the original part of the example, Fig. 10.3. A reliability of 1.0 corresponds to perfect information, and a reliability of 0.5 is the worse possible case in which a log interpretation is just as likely to be wrong as it is to be correct.*

constant except the reliability of the log interpretation. The circled values again correspond to the expected costs determined in the first part of the problem (Figure 10.3) for a reliability factor of 0.9. In this graph we can observe that the "safe" strategy and the "gamble" strategy functions are constant for all values of log reliability. Why?

We can also observe that the "gamble" strategy has lower expected costs than running logs, even if the logs were perfect and told us the true state of nature. Note here that we are not implying we should never run logs — we are just saying we can't justify the logs relative to our other options of $345 M expected value costs of the "gamble" strategy. As the log reliability becomes poorer than about 0.8 we notice the "run log" strategy becomes a near-constant value of $365 M, exactly $20 M higher than the alternative "gamble" strategy.

This implies that if the reliability is poorer than about 0.8 the information we get from the logs (if any) is of essentially no value and we are almost back to making the casing decision on our time zero probability estimates. About all we've done is spent an extra $20 M!

We can also observe that if the logs were absolutely perfect (reliability factor of 1.0) the expected value of the "run log" strategy is $350 M. This amount includes costs of the logging survey of $20 M. If we subtracted this cost we would be correct in saying the expected value for the running logs strategy, assuming the logs were free, is $330 M. By default, we can also conclude that if our alternative strategy has an expected value cost of $345 M the *maximum* we could afford to pay for perfect information (i.e. the logs telling us the exact state of nature) would be $345 M − $330 M = *$15 M.*

If we could obtain the perfect information for less than $15 M the logging strategy would have lower expected costs. If we paid more than $15 M for logs the "gamble" strategy would have lower expected costs and be preferred. If we paid exactly $15 M for perfect information the "log" strategy and the "gamble" strategy would be equal in terms of expected costs. By exactly this same reasoning the maximum we could afford to pay for the logs which have a reliability of 0.9 (as in this example) is $345 M − [$359 M − $20 M] = *$6 M.* Right?

Finally, question III) of this example asks how we could modify the analysis to account for several possible levels of costs to regain control of the well, rather than just the single value of $50 M. The modification is really very simple. The only place in the decision tree of Figures 10.2 and 10.3 which would be affected would be terminal points (b), (f), and (j).

These are the points at which we would be encountering the overpressured zone without having previously set the intermediate casing. These are the points where we could have really serious problems — perhaps even a complete blowout. This question regards how to modify the tree so as to reflect the possibility of substantially higher costs than the $50 M used in the first part of the example.

The change that is needed is to add a chance node at the former terminal points (b), (f), and (j) in which the branches radiating from the node reflect the levels of costs required to regain control of the well. For the values given in question III) the new total well costs would be:

Costs to Regain Control	+	Base Drilling Cost	=	Total Drilling Cost	Probability of Total Cost Occurring
$ 50 M	+	$400 M	=	$ 450 M	← 0.20
$400 M	+	$400 M	=	$ 800 M	← 0.50
$700 M	+	$400 M	=	$1100 M	← 0.30

These new terminal point values and the appropriate modifications are shown in Figure 10.7. And again, the only change from the original tree is at the former terminal points (b), (f), and (j). All other parts of the tree, terminal point costs, and probabilities remain unchanged. The revised tree of Figure 10.7 has been solved, and we observe at the initial decision node I the strategy now having the lowest expected costs is to purchase the additional log information before deciding whether to run the intermediate string of casing.

With the conclusion of example 1 you've probably been exposed to everything you ever wanted to know about running intermediate casing — and then some! But the example is a good one to demonstrate the methodology of analyzing decisions to purchase imperfect information. Now let's look at another example — this time a completely different scenario — but still involving whether to spend money to purchase some additional information.

Example No. 2

The Problem

An offshore structure has been tested productive by two exploratory wells drilled near the crest of the structure. Management is now trying to determine whether it is wise to make a definite commitment to begin building a platform at this time or to defer the decision to allow for the drilling of one additional down-dip delineation well. The principal uncertainty at this point regards the areal extent of the field.

If the structure is nearly filled to its spill point the areal extent of the

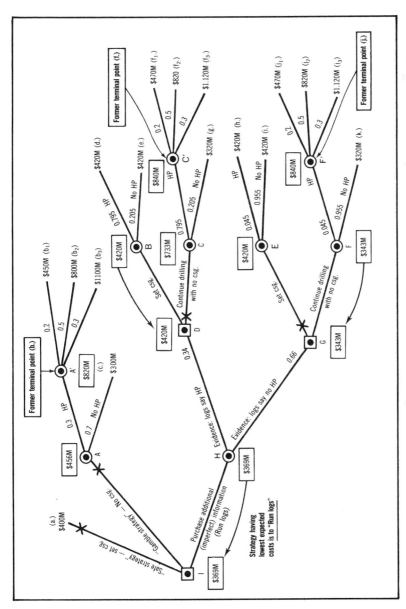

Figure 10.7 Decision tree of Example 1 but with the modification to account for various levels of costs to regain control of the well. Costs are in thousands of dollars. The optimal initial choice is to purchase the additional log information.

523

field will be about 15 square miles and would require installation of a large platform costing $50 MM. If the structure is only partially filled with oil the drainage area will be much smaller and the field could be depleted with fewer wells and a smaller platform costing $30 MM.

Our company does not have a good basis upon which to estimate the likelihood of a large field (areally) or a small field because this is the first structure to be tested productive in the basin. However, it is the subjective judgment of our geological staff that, based on limited evidence available at this time, the likelihood of the structure being completely filled is probably on the order of only 0.4.

So, of course, management has two options at this point: 1) Begin construction of a large $50 MM platform capable of depleting a field of up to 15 square miles, or 2) Begin construction of a small platform for $30 MM that would be capable of depleting the structure only if it were partially filled (say, less than 8 square miles in area), with the realization that if, in fact, the structure is filled to its spillpoint a second platform would have to be added at a later date.

The cost of the second platform is estimated on a net present value basis at $30 MM (construction costs would be higher but the expenditure would be deferred three or four years which would reduce its NPV cost back down to $30 MM). In addition to the cost of the second platform it is estimated that the delay in revenues caused by having to construct another platform at a later date will reduce the NPV of total revenues by $15 MM.

Thus strategy 2 may result in a total overall NPV cost of $75 MM if the structure is filled to the spillpoint, or a total NPV cost of only one small platform ($30 MM) if the productive area of the structure is less than 8 square miles.

Or, as a third strategy, management could defer the decision on platform size until information from one additional delineation well becomes available. The well would be targeted along the flanks of the structure. Data from this well would be used to try to locate the oil/water contact by various means (such as extrapolation of oil-water pressure versus depth graphs for example), but there is the possibility that the data will not provide the conclusive (i.e. perfect) information needed to properly size the platform.

Our geological and engineering staff estimate there is a 10% chance that the data from the down-dip delineation would be insufficient to determine the exact size of the discovery. The well is estimated to cost $2 MM and we have a rig in the area under contract which could begin drilling immediately. The only problem in this regard is that we only have one more well on the contract, at which time the rig must be released to another company.

Our initial plan had been to use the rig to test a new "B" structure located about 50 miles east. Rigs are very scarce in the area and our drilling staff estimate that if we keep the rig on this structure it will probably be another year or so before we can secure another rig to drill the "B" structure. In addition to the delay in testing the "B" structure the rig costs are expected to be at least 25% higher a year from now.

QUESTIONS: I. Which strategy has the lowest expected cost?
            II. If we chose to drill this delineation well rather than move the rig to structure "B" and increased the well cost to include the 25% increase we would pay a year from now, would the preferred strategy be the same? (That is, the present well cost would be burdened with the 25% rig cost increase, so that its effective cost would be $2 MM + [0.25 × $2 MM] = $2.5 MM)
           III. Suppose our estimate for the cost of another platform at a later date is in error and, due to an unforeseen future cost increases, shortages, etc. it will increase to an estimated net present value cost of $40 MM (from the $30 MM used in parts I) and II)). What effect would this contingency have on our choice of strategies today?

The Solution

Here again we are faced with analyzing the expenditure of money to purchase additional imperfect information before making a major investment decision. The additional information in this case is the drilling of a downdip delineation well costing $2 MM (MM equals million dollars). The major decision we have is whether to set a large $50 MM platform initially, or a smaller $30 MM platform. The purpose of spending the extra $2 MM to drill the downdip well is to try to get a little more information about the areal extent of the structure before making this decision.

With this brief introduction the parallel between this problem and the generalized decision tree of Figure 10.1 should begin to emerge. In fact, to firmly establish this similarity let's list the terms and symbols of Figure 10.1 and their equivalent terms of this example.

Terms and Symbols of Generalized Decision Tree of Figure 10.1	Equivalent Terms for This Example
Choice 1	Set large platform, $50 MM
Choice 2	Set small platform, $30 MM
Purchase Additional (Imperfect) Information	Drill Downdip Delineation Well
Evidence from the information	B – Well data indicate large field (in terms of areal extent) B' – Well data indicate small field
States of Nature $\begin{bmatrix} E_1 \\ E_2 \end{bmatrix}$	$\begin{bmatrix} E_1 = \text{Field is large*} \\ E_2 = \text{Field is small*} \end{bmatrix}$
Initial Estimates as to likelihoods of states of nature $P(E_1)$, $P(E_2)$	$P(E_1) = 0.4$ $P(E_2) = 0.6$
Probability that Information will indicate true state of nature	0.9 (Our staff estimated a 10% chance well data would be insufficient to define size of field – thus a 90% chance it will yield correct information)

* *Note:* For this example we are only considering two discrete field sizes – a large field or a small field. In fact there are many different possible field sizes. We've simplified this example to just two so as to reduce the size of the tree. The solution we give here can, of course, be expanded to include a larger number of possible field sizes and platform sizes.

With this listing of equivalent terms and the generalized decision tree of Figure 10.1 we should be able to construct the decision tree for this second example. For practice why don't you try it on a piece of scratch paper and then compare your tree to that shown in Figure 10.8.

The tree itself should be reasonably self-explanatory. The cost of $75 MM at terminal point (b) represents the situation where we initially decided to set a small platform but later (after beginning to drill development wells from the initial platform) determined the areal extent of the field to be large, requiring a second platform. Thus, the total cost would be $75 MM as stated in the initial description of the example.

The costs at terminal points (f) and (j) represent the same contingency except the costs are burdened with an additional $2 MM for the cost of the downdip delineation well. All the terminal points (d) through (k) reflect the additional $2 MM of the well.

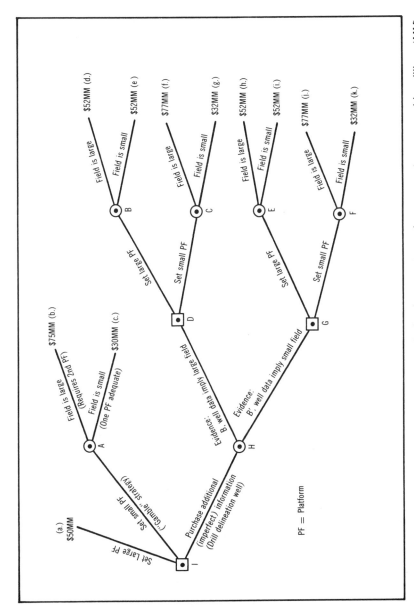

Figure 10.8 *Decision tree for Example 2. All terminal point monetary values are costs expressed in millions (MM) of dollars. Probability terms and the resulting solution of the tree are given in Figure 10.9.*

527

Before we begin to compute the probability terms for chance nodes A, B, C, E, F, and H we must once again remind ourselves that the sizing of the platform is not an "automatic" decision depending on what the delineation well data imply. Just because the data suggest a small field it is *not* an automatic decision that we must set only a small platform. The automatic rules (and probably our intuitive feeling of what to do) would apply if the information was perfect.

But it's imperfect in this example, and the decisions about sizing platforms must consider the evidence, the reliability of the evidence, and the gains and losses associated with making the wrong decisions. So again we see that decision nodes D and G are the vital link in the scheme, and we must be careful about short-circuiting this link with some sort of implied, automatic (real-world?) decision rule.

Our next step is to assess or compute probabilities for all the branches radiating from chance nodes in the decision tree. The probabilities around chance node A represent our initial (time-zero) perceptions as to the likelihoods of the field being large or small. These were estimated as 0.4 and 0.6 respectively by our exploration staff.

Next we come to chance nodes B and C. These probabilities represent our revised perceptions as to the likelihoods of a large or small field given that the delineation well data suggest or imply a large field. The probabilities from both nodes B and C are the same because both nodes are conditional upon the evidence suggesting a large field. As before we must compute these revised perceptions using Bayesian analysis. The computations are as follows.

Table to Compute Probabilities at Chance Nodes B and C, Figure 10.8
Evidence B: Delineation Well Data Imply Large Field
Reliability of Interpretation: 0.9

Possible States of Nature	Original Risk Estimates	Conditional Probability That Evidence B Could Have Occurred, Given the States of Nature	Joint Probabilities	Revised Risk Estimates
E_1 = large field	0.4	0.9	0.36	$0.36/0.42 = 0.857 \leftarrow P(E_1 \mid B)$
E_2 = small field	$\underline{0.6}$ 1.0	0.1	$\underline{0.06}$ 0.42	$0.06/0.42 = \underline{0.143} \leftarrow P(E_2 \mid B)$ 1.000

These revised probabilities of Column 5 are now inserted on the appropriate branches of chance nodes B and C, as shown in Figure 10.9. Also, the sum of Column 4 is the likelihood of the evidence occurring so we can place it on the branch radiating from chance node H for the case of the well data implying a large field.

The exact equivalent series of computations are made to determine the probabilities at chance nodes E and F. The only difference being that the evidence, B′, is different. These revised probabilities are based on the new information that the well data imply the field is small. Specific computations are as follows:

Table to Compute Probabilities at Chance Nodes E and F, Figure 10.8
Evidence B′: Delineation Well Data Imply Small Field
Reliability of Interpretation: 0.9

Possible States of Nature	Original Risk Estimates	Conditional Probability That Evidence B′ Could Have Occurred, Given the States of Nature	Joint Probabilities	Revised Risk Estimates
E_1 = large field	0.4	0.1	0.04	$0.04/0.58 = 0.069 \leftarrow P(E_1 \mid B')$
E_2 = small field	0.6 1.0	0.9	0.54 0.58	$0.54/0.58 = 0.931 \leftarrow P(E_2 \mid B')$ 1.000

As before these revised probabilities of Column 5 are placed on the appropriate branches of chance nodes E and F, as shown in Figure 10.9. The probability of the evidence occurring in the first place is the sum of Column 4, 0.58, and this figure is placed on the appropriate branch at chance node H.

We can now solve the decision tree of Figure 10.9 in the following series of steps.

1. *Chance Node A:* $EV_A = (0.4(\$75 \text{ MM}) + (0.6)(\$30 \text{ MM}) = \$48 \text{ MM}$

2. *Chance Node B:* $EV_B = (0.857)(\$52 \text{ MM}) + (0.143)(\$52 \text{ MM}) = \$52 \text{ MM}$

3. *Chance Node C:* $EV_C = (0.857)(\$77 \text{ MM}) + (0.143)(\$32 \text{ MM}) = \$70.6 \text{ MM}$

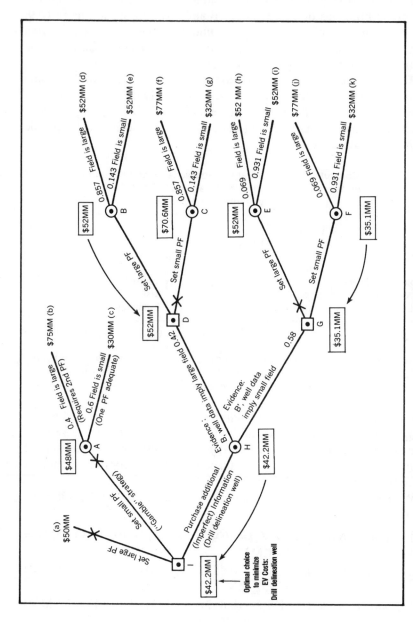

Figure 10.9 *Decision tree of Example 2 and Figure 10.8 with probabilities and expected values at the nodes added. The decision which minimizes expected value costs is to purchase the additional information by drilling the downdip delineation well. Costs are in millions of dollars.*

530

4. *Decision Node D:* Choices are
Set large PF $\rightarrow$ EV costs are $52 MM
Set small PF $\rightarrow$ EV costs are $70.6 MM
Decision: Set large PF (It has lowest EV costs) Thus EV at decision node D is $52 MM.

5. *Chance Node E:* $EV_E = (0.069)(\$52\text{ MM}) + (0.931)(\$52\text{ MM}) = \$52\text{ MM}$

6. *Chance Node F:* $EV_F = (0.069)(\$77\text{ MM}) + (0.931)(\$32\text{ MM}) = \$35.1\text{ MM}$

7. *Decision Node G:* Choices are
Set large PF $\rightarrow$ EV costs are $52 MM
Set small PF $\rightarrow$ EV costs are $35.1 MM
Decision: Set small PF
Thus EV at decision node G is $35.1 MM.

8. *Chance Node H:* $EV_H = (0.42)(\$52\text{ MM}) + (0.58)(\$35.1\text{ MM}) = \$42.2\text{ MM}$

9. *Decision Node I:* The initial time-zero choices are
Set large PF $\rightarrow$ EV costs are $50 MM
Set small PF ("gamble") $\rightarrow$ EV costs are $48 MM
Drill third well $\rightarrow$ EV costs are $42.2 MM
Decision: *Purchase Additional Information by drilling third well along flanks of the structure.* QED

From this we see the answer to the first question of the example is that the strategy offering the lowest expected costs is to purchase the additional (imperfect) information by drilling a downdip delineation well. With this strategy the major investment decision regarding platform size is deferred until the additional information about the probable size of the field becomes available.

Question II is asking, in essence, whether our strategies would change if we burdened the strategy of purchasing additional information with an additional $0.5 MM debit to cover the expected increase in well costs we would have to pay to test Structure B a year from now (rather than at this time in lieu of drilling the delineation well). To resolve this question we simply add an additional $0.5 MM to all the costs at terminal points (d) through (k) of Figure 10.9. These are the terminal points which relate to the strategy of purchasing additional information. The decision tree is then re-solved and the choices at decision node I compared.

We can short-circuit this calculation, however, if we recall from Chapter 3 (page 74) that if we add or subtract a constant amount to all the conditional values received for a given strategy the expected value of the strategy will simply be increased or decreased by a like amount. Using this rule we can simply add $0.5 MM to the expected value at chance node H to compute the new expected value of purchasing additional information. That is:

($42.2 MM)	+	($0.5 MM)	=	($42.7 MM)
↑		↑		↑
The EV at chance node H computed in Question I.), Figure 10.9		The additional debit charge to account for increased drilling costs, Question II.)		The new EV for strategy to purchase additional information, Question II.)

This is exactly the same EV costs we would have obtained at chance node H by adding $0.5 MM to all the terminal points (d) through (k) of Figure 10.9 and re-solving that portion of the tree.

Having accounted for this additional debit charge if we drill the downdip delineation well, rather than move the rig over to structure B the expected value costs for our three time zero options are:

Set large PF: $50 MM ⎫ No change from the initial
Set small PF: $48 MM ⎰ solution of the example.
Purchase Additional   An increase of $0.5 MM
   Information: $42.7 MM   resulting from the
                                 additional well cost debit.

And as before the option to purchase additional information still has the lowest expected costs and would be the preferred choice.

As a possible variation to this question suppose we wish to burden the decision to use the rig on this structure (rather than proceed immediately to test structure B) with a debit which reflects the reduced net present value of Structure B caused by the year's delay in drilling. How could this be accomodated in the context of this example and the decision tree of Figure 10.9?

Well it's quite simple and here's the series of steps we'd have to follow:

1. Compute the expected NPV of the decision to test structure B at this time. (This would have presumably already been computed or we wouldn't have been interested in using the rig to test the B structure. Note also that this is an *expected* NPV in the sense that it is based on the NPV of oil, given that it has oil, the likelihood of it having oil, the likelihood of no oil and the related dry hole costs.)

2. Compute an expected NPV which recognizes the delay of one year that we would experience if we used the rig to drill the delineation well now rather than test structure B. The only things that would probably change would be the discounting of any revenues would account for an additional one year delay, and some of the costs for wells and production facilities may be higher due to inflation. The result will be that the expected NPV for a year's delay will be lower than the value of Step 1.

3. The penalty we would have to charge to the decision to drill the delineation well now rather than to test structure B would be the amount by which the expected NPV of structure B is lowered due to the year's delay. That is,

Penalty = (Expected NPV_B of Step 1 − Expected NPV_B of Step 2.)

4. This penalty term is then added to the EV costs at chance node H to obtain a revised EV cost for the strategy of using the rig to drill the delineation well at this time.

For the data of Figure 10.9 we can observe that the purchase of additional information is more advantageous than the next best strategy of setting a small platform by about $5.8 MM ($48 MM − $42.2 MM = $5.8 MM). Consequently if we went through an analysis of the type just described we would be able to conclude that if the penalty of step 3.) exceeds $5.8 MM we'd be better off to use the rig to test structure B rather than drilling a downdip delineation well on our newly discovered structure.

This follows because if a penalty > $5.8 MM is added to the $42.2 MM EV cost at chance node H the sum will be > $48.0 MM. The total expected costs to drill the delineation well would then be greater than if we took the gamble and set a small platform now (and moved the rig over to structure B). Analyses of these types become very useful in trying to formulate exploration strategies when there are an insufficient number of offshore drilling rigs available in the area.

The final question of Example 2 relates to the effects of increased platform costs in the future on our strategies at the moment. Specifically, we are concerned with the contingency of setting a small platform now and later finding that a second platform is, in fact, required which may cost $40 MM, rather than the $30 MM (net present value costs) used in the initial part of the example.

Recall that if this contingency occurred we computed that total costs would be $75 MM ($30 MM for first platform + $30 MM net present value costs for a second platform three or four years from now + $15 MM to account for the delay in revenues and the resulting reduction of NPV of total field revenues). This contingency occurs at terminal points (b), (f), and (j) in Figure 10.9.

If the second platform costs increase to $40 MM (net present value costs) due to unforeseen inflationary effects, material shortages, labor problems, etc. (rather than the originally estimated $30 MM) the total cost for these contingencies increases to $85 MM. So to answer question III we have to re-solve the decision tree of Figure 10.9 with the following changes:

Terminal Point (Figure 10.9)	Terminal Point Cost, $MM, Used in Question I (Figure 10.9)	Revised Terminal Point Costs for Question III, $MM
(b)	$75 MM	$85 MM
(f)	$77 MM	$87 MM
(j)	$77 MM	$87 MM

All other terminal point costs of Figure 10.9 remain the same, and all probabilities remain the same.

If the decision tree is re-solved using these changed costs at (b), (f), and (j) the resulting expected value costs at decision node I become:

> Set large PF → EV costs: $50 MM
> Set small PF → EV costs: $52 MM
> Purchase Additional
> Information (Drill   → EV costs: $42.6 MM
> delineation well)

The preferred choice would still be to drill the downdip delineation well. This is an important bit of information to the decision maker because we observe that the additional $10 MM cost for a second platform changes the EV of his choice to drill the delineation well only slightly ($42.6 MM versus $42.2 MM). This tells him the correctness of his choice to drill the delineation well is not dependent on a precise estimate of the costs of a second platform. He can thus suppress the second platform cost uncertainty in his overall thought process relating to his decision choices.

Costs of equipment three or four years from now could be quite speculative, but the analysis of this question implies that it will not have a significant effect on overall expected value costs of drilling the third well. This insight reduces by one the many concerns he must consider in decisions of the type illustrated in this example. In these days of rising costs and uncertainties about future events we can be thankful for small favors — like eliminating one of the uncertainties in the decision process!

SUMMARY

Now that we have discussed two real-world examples of decisions to purchase additional (imperfect) information we can perhaps summarize the analysis technique as follows:

■ The analysis of the value of purchasing imperfect information follows a decision tree format of the type shown in Figure 10.1. By virtue of the information being imperfect the investment decisions made after receipt of the information are still decisions under uncertainty.

■ To be able to complete the analysis we must be able to specify or estimate the likelihood that the information will be correct. This factor is usually expressed as a "reliability of interpretation." The term is used in the Bayesian computation of the revised perceptions of the possible states of nature, given the new information or evidence which we have purchased.

■ Sensitivity analyses of the type displayed in Figures 10.4–6 are useful to evaluate the degree to which various investment strategies are altered for various values of the key parameters. Graphs of these types are derived by repetitive solution of the original basic decision tree shown in Figure 10.1.

■ We must be ever mindful of the fact that the decision choices in Section II of Figure 10.1 are decisions based on maximizing or minimizing expected values — and not automatic decisions based on what the purchased information implies. The implied decisions only relate to the case where the purchased information is perfect. If the information is imperfect (the case being studied in this chapter) the automatic decision rules are not valid.

On final balance we see the analysis of decisions to purchase imperfect information is slightly more complex than one probably first imagines. But fortunately the analysis technique follows a distinct and consistent format of the type shown in Figure 10.1 and illustrated in Examples 1 and 2. Hopefully the technique has become reasonably clear and it is hoped that you will now be able to use the approach for the analysis of this important type of decision strategies.

11   Special Problems and Open Issues

When investment choices are made the exact course of future events is unknown.

IN this final chapter we will discuss several special types of analyses related to petroleum exploration and a few important issues which, for the moment, remain unanswered. The three principal topics are generally unrelated, so this chapter represents a collection of ideas given here to round out our overall discussion of petroleum exploration decision analysis.

More specifically, we'll discuss the following topics:

I. Decision Strategies in High Risk Areas and/or Under Conditions of Limited Capital Constraints

a. What is the likelihood of a "gambler's ruin" run of dry holes? Can we afford to drill enough wells to minimize our chances of gambler's ruin?

b. If we are investing with a limited capital constraint how should we play the game — drill a limited number of wells ourselves, or participate with a small interest in a lot of deals so as to spread our limited capital to as many wells as possible? Should we put all our eggs in one basket or put a few eggs in many separate baskets?

II. Profitability Criteria Based on Discounting — A Second Look

a. It is commonly recognized that discount factors begin to approach zero for cash flows occurring after about 20 years. Is rate of return (or NPV at a given discount rate) a valid measure of value for projects whose cash flows extend over very long periods of time, say up to 40 years? If not, what alternative criterion should be used for evaluating long, extended cash flows?

537

b. All discounting criteria implicitly assume that cash flows are re-invested immediately upon receipt in other projects having an annual earnings rate equal to the discounting rate being used. This implies that there must always be other projects available to absorb the monies from the future cash revenues. What if we do not have an unlimited supply of future reinvestment opportunities? Are discounted measures of value such as rate of return, NPV, etc. valid under this condition? If not what other measure of value should we be using?

III. Maximizing Expected Monetary Value—Is It the Only Dimension of Value Management Must Consider?

a. Is expected value, by itself, a sufficient criterion for selecting and comparing investment projects?

b. How should the decision maker compare project A having a high EMV but also a high probability of financial loss with project B having a slightly lower EMV but with virtually no chance of a monetary loss? Is there some way to accomodate a dual objective of maximizing expected gain and at the same time minimizing exposure to financial losses?

As you can see, these issues certainly relate to important considerations in petroleum exploration. The analyses are rather specialized, however, so it seemed appropriate to defer our discussion of these issues until after we had covered the main body of economic and risk analysis techniques. With this editorial apology for not having a common theme for this concluding chapter let's look at each of the three issues in more detail.

I. DECISION STRATEGIES IN HIGH RISK AREAS AND/OR UNDER CONDITIONS OF LIMITED CAPITAL CONSTRAINTS

Whenever we consider a sequence of outcomes from a chance phenomenon we need to be concerned with the possible "runs" of favorable and unfavorable outcomes. If we are playing roulette we would hope (expect?) that periodically we would win a substantial amount of money so as to be able to finance continued play at the game. But the alert gambler must also be aware of the possibility of a long series of consecutive losses such that he loses all his money and must drop out of the game. This is called "gambler's ruin," or a "gambler's ruin run of bad luck."

In the oil business we need to be aware of the possibility of a gambler's ruin run of exploratory dry holes, and shape our investment strategies so as to minimize the chances of this unfortunate turn of events. If we enter a new exploration play and make a significant discovery fairly early (e.g. after two or three wildcats have been drilled) the problem of gambler's ruin becomes moot.

It's when we might encounter a long sequence of consecutive failures that the issue really becomes binding. Generally the characteristics of the play which should cause us to be concerned with gambler's ruin are: areas in which the probability of a large discovery is very low and the probability of failure is high (a so-called high risk area); and/or conditions are such that you only have a limited amount of money to allocate to the area.

The general approach to hedge against the possibility of gambler's ruin is to get out of the play altogether, get a larger supply of money (if possible), or participate with a smaller working interest ownership in as many ventures as possible so as to minimize the likelihood of a total gambler's ruin failure.

To be more specific, the conditions which must exist for us to be concerned about computing likelihoods of gambler's ruin include at least one of the following:

1. Limited capital (or at least not an infinite supply of money).
2. High-risk exploration area—where the probability of large discovery is very low and probability of failure is high.
3. Flexibility in being able to arrange joint participation in many exploratory ventures, if desired. This implies that the exploration area should have a fairly large number of drillable prospects and the leasing and/or operating agreements are such that you can negotiate various schemes for the drilling of many wells with reduced ownership. Certainly a stratigraphic trap play in an area of competitive leasing such as occurs in the U.S. is one example where this flexibility exists. If you have a concession or operating area where you have committed to specific terms and numbers of wells this flexibility may not exist.

I became aware of a good illustration of a recent play having all the conditions necessary for gambler's ruin and an operator who made all the wrong moves. It was a stratigraphic play which became of much interest to everyone following a few spectacular discoveries. The aggressive, alert operators moved in quickly and established a large acreage position while leasing costs were still low. The operator who ultimately made the wrong moves waited for a year or so to make sure the play was going to amount to something before he committed his lease brokers to move.

By this time acreage costs had skyrocketed, resulting in his having to commit a much larger amount of cash just to secure an acreage position. The play happened to be one of incredibly high risks—the chances of a large discovery were very small (perhaps on the order of less than 0.01) and the likelihood of failure was above 0.95. He apparently failed to realize this—and in his intrepid wisdom, he budgeted for the coming year the drilling of five full-interest exploratory wells on his newly acquired acreage.

The results after the five wells were drilled? You guessed it—5 dry holes! And the irony is that he then concluded the area to be a poor one, wrote off the losses and went elsewhere to explore for oil. In the meantime the aggressive, alert operators who understood how to play the game a bit better took a piece of many, many wells and ended up owning a share of a few very large, profitable discoveries and, as a whole, made a good return in the play.

To approach the problem from a quantitative standpoint we would need to specify the following parameters:

1. The amount of capital we can spend in the play.
2. The probability of failure on the successive trials, and
3. The financial losses of each failure (usually the exploratory dry hole cost).

By dividing the losses per failure into the total capital available we have the maximum number of successive failures we could endure before gambler's ruin. If we had found anything before reaching this point the question would be moot (on the assumption that the revenues from the discoveries would in turn provide capital to continue the search). If we had found nothing after drilling, say, n successive failures we would have to drop out of the play for lack of additional drilling capital.

The problem, then, is to assess the likelihood of this latter outcome occurring, and trying to devise alternative strategies which will make this likelihood of gambler's ruin as low as possible.

Conceptually the problem can be represented as a partial decision tree, as in Figure 11.1. In this figure the symbols $P(D_1)$, $P(F_1)$ represent

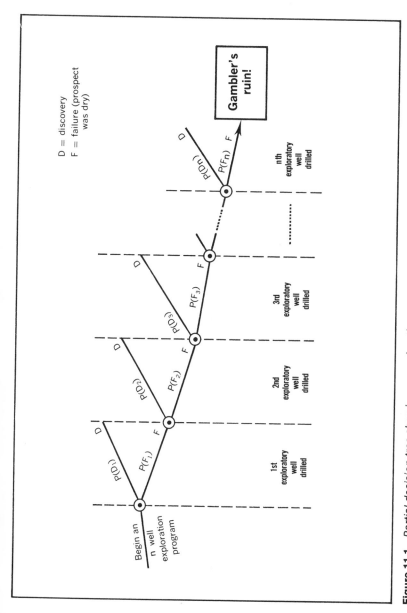

Figure 11.1 *Partial decision tree showing an exploration program in which n consecutive failures (dry holes) would result in gambler's ruin. A discovery prior to the nth well would make the issue of gambler's ruin moot because revenues from the discovery could be used to finance any additional exploration that may be warranted. Drilling of n consecutive failures would result in exhaustion of available drilling capital without having generated any revenues (a nice way to say we had experienced gambler's ruin!).*

the probabilities of discovery and failure (respectively) on the first well, $P(D_2)$, $P(F_2)$ the respective probabilities of discovery and failure on the second well drilled, etc. The tree is general in the sense that it is applicable for the situation where the wells are a series of dependent events (sampling without replacement) and also the situation where the wells can be represented as a series of independent events (sampling with replacement). The probability of a gambler's ruin run of n consecutive failures is given as:

$$\text{Probability of } n \text{ consecutive failures} = P(F_1) \times P(F_2) \times P(F_3) \times \ldots \times P(F_n) \quad (11.1)$$

The only other outcome which can occur is at least one discovery in the sequence of up to n wells — the situation when the issue of gambler's ruin becomes moot. Consequently we can say that Equation 11.1 gives the probability of gambler's ruin, and the complementary probability is the likelihood of *not* reaching a point of gambler's ruin. We should also note that Equation 11.1 consists of multiplying together a series of probability terms. Each of these probability terms will be a decimal fraction less than one. Whenever we multiply together two decimal fractions less than one the product will always be a decimal fraction smaller than either of the two probability terms.

The more terms in the product the lower will be the product. Which is to say the more times the game is played the less likely will be the chances of a run of consecutive failures. To minimize the chance of gambler's ruin we need to make n large in Equation 11.1. This would require either a very large amount of money or devising an investment strategy of taking a small piece of a large number of deals — the normal practice for independent oil operators with limited capital.

Let's illustrate all this with an example. Suppose we have $5,000,-000 available to explore for oil in a new exploration area having the following parameters:

1. Exploratory dry hole costs = $1,000,000 each
2. Number of prospects (anomalies): N = 20
   Fraction of prospects expected to contain enough oil to be classified as a significant commercial discovery: p = 0.25

From this data it follows that the maximum number of consecutive failures we could endure before reaching gambler's ruin is $5,000,000/

$1,000,000 = 5$. Also, the estimated number of possible discoveries is $N \times p = 20 \times 0.25 = 5$. The corresponding number of prospects likely to be dry is $N \times (1 - p) = 20 \times (1 - 0.25) = 15$. *What is the likelihood of gambler's ruin?*

All we need to do is solve Equation 11.1 using the following conditional probability terms:

Probability Term (Using Symbols of Figure 11.1)	Meaning of the Probability Term	Numerical Value for Example
$P(F_1)$	Probability 1st well is dry	15/20
$P(F_2)$	Probability 2nd well is dry, given 1st one was dry	14/19
$P(F_3)$	Probability 3rd well is dry, given first two were dry	13/18
$P(F_4)$	Probability 4th well is dry, given first three were dry	12/17
$P(F_5)$	Probability 5th well is dry, given first four were dry	11/16

Solving Equation 11.1 gives the probability of gambler's ruin as:

$$\left. \begin{array}{l} \text{Probability of} \\ n = 5 \text{ consecutive} \\ \text{failures} \end{array} \right\} = P(F_1) \times P(F_2) \times P(F_3) \times P(F_4) \times P(F_5)$$
$$= \frac{15}{20} \times \frac{14}{19} \times \frac{13}{18} \times \frac{12}{17} \times \frac{11}{16} \cong \underline{\underline{0.194}}$$

We would have about a 19% chance of exhausting our available drilling money without having made a discovery.

Now suppose we found a joint interest partner who agreed to share in the drilling on a 50/50 basis. This would mean our proportionate cost of any dry holes would be reduced to $500,000. We could therefore participate in the drilling of up to 10 consecutive failures before reaching gambler's ruin. *What's the likelihood of this occurring?* Well, this is pretty simple to calculate also. We merely need to determine the likelihood of 10 consecutive failures. The first five terms in the computation are exactly as given above. By following the same logic as before the succeeding terms are: $P(F_6) = 10/15$, $P(F_7) = 9/14$, $P(F_8) = 8/13$, $P(F_9) = 7/12$, and $P(F_{10}) = 6/11$. Substituting all these terms into Equation 11.1 gives:

Probability of
n = 10 consecutive $\Big\}$ = P(F$_1$) × P(F$_2$) × P(F$_3$) × . . . × P(F$_9$) × P(F$_{10}$)
failures

$$= \frac{15}{20} \times \frac{14}{19} \times \frac{13}{18} \times \frac{12}{17} \times \frac{11}{16} \times \frac{10}{15} \times \frac{9}{14} \times \frac{8}{13} \times \frac{7}{12}$$

$$\times \frac{6}{11} \cong \underline{0.0162}$$

By participating in twice as many wells we reduced our chances of gamblers ruin by 12 — from 0.194 down to 0.0162. This result, of course, confirms our earlier statement that one hedge against gambler's ruin is to participate in more ventures. If you have a capital constraint this means taking only a proportionate interest in many wells rather than full ownership in a few wells.

To show the effects of higher likelihoods of failure suppose our original hypothesis about the fraction of the prospects having oil was p = 0.10, rather than p = 0.25. This would imply that 18 of the 20 prospects would likely be dry. If we drill the wells with full ownership we could afford a maximum of five consecutive failures. Under these conditions the likelihood of gambler's ruin is:

Probability of
n = 5 consecutive
failures, given   = P(F$_1$) × P(F$_2$) × P(F$_3$) × P(F$_4$) × P(F$_5$)
N = 20
p = 0.1

$$= \frac{18}{20} \times \frac{17}{19} \times \frac{16}{18} \times \frac{15}{17} \times \frac{14}{16} \cong \underline{0.553}$$

When the chances of discovery are less, the probability of reaching gambler's ruin in n wells increases. For the case of p = 0.10 it increases from 0.194 to 0.553. In this case it is more likely that we lose everything than it is to make at least one discovery!

As a final comment to this example, if N was very large we mentioned back in Chapter 7 that the sequence of a series of wells could be thought of as being approximately a series of independent trials. Using this approximation the probabilities of failure on each successive term would be constant. That is, P(F$_2$) = P(F$_3$) = P(F$_4$) = . . . = P(F$_n$) = P(F$_1$).

For the first question the solution of Equation 11.1 would have been:

Probability of
n = 5 consecutive
failures, given
that the wells $= [P(F_1)]^5 = \left[\dfrac{15}{20}\right]^5 = \underline{\underline{0.237}}$
are a series of
independent trials

Generalizing at this point we can conclude that in a given basin we can minimize the likelihood of a gambler's ruin run of consecutive failures by participating in as many exploratory wells as possible. This requires increasingly larger amounts of money, or, if constrained by a capital limit, a partial or diluted ownership in a larger number of wells. All of which is to say if we have a capital constraint we lessen our chances of gambler's ruin by putting a few eggs in many baskets rather all of our eggs in a few baskets.

This, of course, is not a startling revelation by any means. Independent operators having very limited capital assets have been using this strategy all along (that is, those who are still in business!). But it's comforting to know that once in a while we can confirm with numbers our intuitive feelings about such matters. And where it is necessary to evaluate the magnitude of the chance of gambler's ruin, the computation methods we've just described should prove helpful.

II. PROFITABILITY CRITERIA BASED ON DISCOUNTING— A SECOND LOOK

In this section we speak to two issues relating to the concept of discounting and present value. The first is whether or not any discounted measure of value, such as rate of return or NPV profit, is a realistic indicator of value for projects having long extended cash flows (in the range of 20–40 years). The second issue is whether any form of discounted measures of value have meaning if there are only limited future reinvestment opportunities, as opposed to an unlimited supply of future reinvestment opportunities.

Although these are actually two separate questions it will turn out that the resolution of the first question will get us involved in the issues realted to the second question. And as we will see there are some aspects of these two questions which simply do not have an answer at this time. Consequently these questions represent open issues as related to the overall state-of-the-art of decision analysis in petroleum exploration. Even though the questions can't be resolved completely at the moment they are, nonetheless, very important concerns of many executive management in the industry.

Let's look at the question of long extended cash flow projects first. Anyone who has worked much with discounting is fully aware of the fact that discount factors tend to approach zero as time increases. In the year 20, for example, the 10% discount factor (for midyear cash flows) is 0.156, at a 20% discount rate the factor is 0.029, and for 30% it is only 0.006. As the discounting rate increases the discount factors approach zero in even less time. This has the effect of giving little or no present-value to income received beyond 20 or 25 years. On an NPV basis the equivalent effect is as if the well or field simply ceased to produce at about that time.

This fact has been well known by everyone but generally of no particular concern until the advent in recent years of exploration in remote and/or very expensive areas requiring long delays for building physical equipment (pipelines, offshore platforms, etc.). Examples include North Slope of Alaska, Northern Canada, North Sea area, etc. In these areas the first drop of oil may not even begin to flow until 8–10 years after discovery. And with an anticipated producing life of, say, 25 years we end up with cash flow revenue schedules spread out over 35 years of time. Revenue received during most of the producing life of these delayed schedules will have virtually no value today on an NPV basis.

The upshot of all this is a concern that, because of the characteristic of discount factors tending to zero after about year 20, perhaps a discounted measure of value such as rate of return or NPV is not a realistic measure of value to use for the long, extended cash flow projects we frequently encounter today. If it is not a valid measure of valuation what alternative methods should we consider using?

One school of thought follows roughly the following bit of logic. They reason that when the revenue does not begin until 10 or 12 years after discovery we make a mistake to discount these revenues all the way back to time zero. The result is that nearly all the revenue cash flows are multiplied by such low discount factors (due to the time delay) that the NPV sum underestimates the true value of the project.

The resolution of this, in their view, is to move the common point in time used for discounting from time zero to a point coinciding with when the revenues begin to be received. If this point is 10 years from now the analyst would find the total net discounted value of the cash flows (at a discount rate of i_0) as of year 10. All cash flows received after year 10

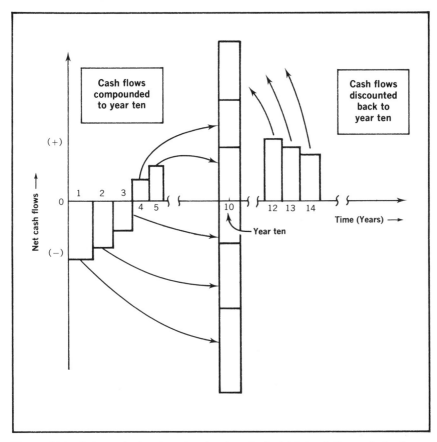

Figure 11.2 *Schematic drawing showing how to find total net discounted value as of year 10. All cash flows received or disbursed prior to year 10 are compounded up to year 10 at i_0 rate. All cash flows received or disbursed after year 10 are discounted back to year 10 at i_0 rate.*

would be discounted back to year 10. Revenue received in year 11 would be discounted back through just one year. For year 12 the revenue would be discounted back through 2 years, etc.

All cash flows received or disbursed before year 10 would be compounded, or appreciated, up to year 10, as in Figure 11.2. When all cash flows had been compounded or discounted to year 10 the analyst would sum all the discounted values (algebraically) to yield the total net discounted value as of year 10.

The result will, of course, be a much larger number than an NPV using the same discount rate. This is because the revenues are not as severely discounted. The proponents of this view say the much larger discounted value as of year 10 more accurately reflects the true value of the project. The numerical values for an offshore prospect having a long delay before the start of production might be as follows:

- Total net discounted value as of year $10 = \$44.1$ million ($i_0 = 16\%$)
- Total net discounted value as of time zero (NPV) = \$10.0 million ($i_0 = 16\%$)

Their claim is that the larger \$44.1 million is more representative value of the long extended cash flows of the project.

Is it really better? Have we solved our analysis problem for long extended cash flows by moving the time reference point from time zero to time equal to year 10 or 12? The answers to these questions is an emphatic *NO!* We have done absolutely nothing except massage our numbers so that the total discounted value appears larger numerically. In fact, if our time-value of money was 16% we should be indifferent between receiving \$10 million now or \$44.1 as of year 10.

The reason is that if we took the smaller amount now and invested it at 16% it would appreciate in value to exactly \$44.1 million ten years from now! Those who propose this scheme would not have had to even go through all the calculations described in Figure 11.2. Rather, all they would have had to do is compound the NPV at i_0 for 10 years and calculate the \$44.1 million immediately. Absolutely nothing has been changed. In using criteria based on discounting (rate of return, net discounted value at i_0, etc.) it makes absolutely no difference how we select the common point in time to use in the calculations.

We usually select time zero because it simplifies the computations, but we could use as our common time reference any point we wish — year 2, year 10, year 17.5, etc.

So this is one proposed resolution which is of no consequence whatever. Which puts us right back to "square one" as far as the original question is concerned. To make a long story short the answer to the question as to whether discounted criteria are meaningful measures of value is *yes* — *provided* we have an unlimited supply of future reinvestment opportunities yielding earnings rates equal to i_0, the discount rate.

As long as we have some place to reinvest the cash flows when they are received the later years beyond year 20 will be taken care of by the revenues received and reinvested prior to that time. Consequently if we have two projects of the following parameters we are justified in preferring project A over B. As long as we can reinvest the future revenues at i_0 we should only be concerned with maximizing NPV.

Project A:	Prospect having 3 million barrels reserves, a relatively short life (15 years)	NPV@ $i_0 = \underline{+\$2.7 \text{ million}}$
Project B:	Prospect having 100 million barrels reserves to be received over the next 45 years	NPV@ $i_0 = \underline{+\$1.8 \text{ million}}$

If it means accepting a low return project having a rapid depletion schedule (and hence a high NPV) over a project having a very long depletion schedule (and hence lower NPV) then fine. The revenues received quickly from Project A can be immediately reinvested and these continual reinvestments in succeeding years will result in years 20, 30, 40, etc. taking care of themselves in due course. It's all based on the hard cold facts of discounting formulae. Assigning very little value today to receiving money 20 or 30 years from now should cause us no concern. It will take care of itself by virtue of the assumption of immediate reinvestment.

But therein lies the catch! Do we have an unlimited supply of future reinvestment opportunities with regard to petroleum exploration investments? There are many managers in the business today who feel we have reached a point where we no longer have unlimited reinvestment assumptions. Rather, we only have a limited supply of places to invest future revenues in petroleum exploration. If this hypothesis is true then all bets are off regarding what we just said about discounted measures of value. In fact, if we only have limited reinvestment opportunities the whole concept of discounting falls down. Which leads us inextricably into the second question we posed at the beginning of this section.

The symptoms which suggest we've reached the point of having only limited future reinvestment opportunities are pretty obvious and well known:

■ Exploration in areas of extreme difficulty and high cost such as Arctic regions of North America, North Sea, etc. Would we as an industry be investing these huge sums of money into the Arctic area if we could find equivalent reserves in shallow, easy to drill areas such as central Kansas?

■ As you read the parameters of Project A and B did you say to yourself, "I'd prefer B because 100 million barrels of oil is worth more to me than 3 million barrels, even though NPV is less?" If you did, what does that suggest?

■ Would we be stressing energy conservation and development of alternative sources of energy today if we had an unlimited supply of new exploration areas awaiting the drill?

If the hypothesis of only limited future reinvestment opportunities is true it changes our whole perspective of value. It means, for example, that we may have to accept a long-life project having sub-optimal value (using present analysis criteria) just to insure a supply of oil 20 or 30 years from now for our refining and marketing operations. But if we begin investing in expensive long range exploration projects we may expose ourselves to short-term liquidity problems resulting from directing near-term revenues into expenditures needed to assure long-term revenues. In fact, there is a chain reaction of problems relating to the new perspective, almost as if knocking over one domino resulted in a whole chain of dominoes falling down.

If the hypothesis is true what alternative value system is available to evaluate the limited investment opportunities as they present themselves? None that I know of. This is clearly one of the current unanswered issues of decision analysis. It is an area where we all need to do some hard thinking in the days ahead. Until we can conclude something more definite about all I can do is explain one approach that has been tried, and suggest a scheme to help management better judge exploration investments until the time occurs when we develop a better valuation parameter.

One approach which has been suggested to account for the constraint of limited future reinvestment opportunities is based on the following logic. The first step is to make an assessment of projected income from wells and fields now on production, and an assessment of desired corporate income over the next 30 or 40 years. Such as assessment might appear as in Figure 11.3. The lower declining curve represents projected income from present producing operations assuming no further capital expenditures for exploration.

The upper curve is an expression of where we would like to be in terms of corporate earnings. This would be specified by executive management. The gap between these two curves represents the new

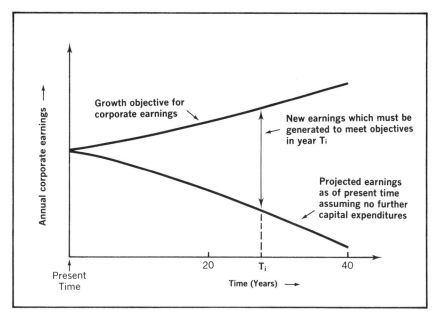

Figure 11.3 *Graph of projected earnings and corporate objectives over the next 40 years.*

earnings which must be generated by subsequent exploration. A graph such as Figure 11.3 is nothing new—it's just an inventory of where we are and where we want to be in terms of corporate earnings over the next 40 years.

The next step in the logic is to discard any form of discounting. All future cash flows would be treated on a whole-dollar basis. A dollar received in year T_i would be worth its full value toward filling the gap between the two curves in year T_i. We had previously used the discount rate as an expression of corporate objectives, but since corporate objectives are now considered by virtue of the upper curve of the graph we no longer need discounting to express corporate objectives.

Given the inventory of future earnings (Figure 11.3) and the premise that all future revenues would be treated on a whole-dollar basis, the specific details of the new criterion are as follows:

1. The initial screening of projects is on the basis of whether the project reaches payout. If it does it will contribute new dollars of income in the future to fill the gap between the curves. Thus,

all projects which are estimated to at least reach payout are candidates for further consideration. If a project does not reach payout it's rejected outright, (in the same way we reject such projects using present day criteria).

2. The ratio of total net profit to total investment (undiscounted) is computed and compared to some minimum value α. Setting $\alpha = 1.5$, for example, eliminates the situation of spending $100 million in a project which returns $100 million by year 10 (reaches payout) and then produces $1 dollar per year over the following 20 years. The ratio would have to exceed a previously determined minimum value of α to be acceptable.

3. Finally, during the period before payout we would have money invested in, but not yet recovered from the project. This means, for example, that if we were operating with borrowed capital we'd have a bank loan outstanding until we had reached payout. Bank loans require interest payments (as we all know), and to account for the cost of borrowed capital interest charges on money invested in, but not yet recovered from the project, would be added to the investment total as an additional debit.

These interest charges would be treated as additional out-of-pocket cash expenditures. These additional interest charges would have the effect of increasing total investment, reducing total net profit, and deferring the time required to reach payout.

4. As a result of all these considerations the accept/reject parameter is:

$$\left[\frac{\text{Total Net Profit}}{\text{Total Investment} + \text{interest charges}} \right] \geq \alpha$$

If the ratio exceeds α the project is accepted. If it is less than α it is rejected.

And that's all there is to it! A simple profit-to-investment ratio (adjusted for interest charges on money invested in but not yet recovered from the project) and a test as to whether the ratio exceeds α. How do you react to that? At first glance it seems so simple as to be somewhat frightening. Could you justify the gamble to explore for oil in some new remote area on the basis of a simple ratio exceeding α?

Well indeed there are a number of drawbacks to the approach. For example, who would determine the value of α? How much should it be? No clues here that I know of. Some have suggested its numerical value would depend on the degree of desperation of the management. The more desperate they perceived the situation to be (regarding the limited number of new exploration areas) the lower its value. We really have no help that I know of to determine whether α should be 20, 8.5, 1.6, or whatever.

Another probable result of such a decision strategy would be early acceptance of whatever limited exploration prospects come along which could help to fill the gap. Nowadays these projects are generally in remote areas requiring that huge amounts of capital be tied up for substantial lengths of time before the start of production.

Where would we obtain all the capital? It's certainly conceivable to envision the situation where we plow our near-term earnings back into these long-term projects. And one could easily envision a liquidity problem such that in an effort to secure long-term sources of crude we went bankrupt in two years! Indeed, there are now articles in financial literature describing techniques to evaluate investing in short-term, suboptimal projects just to generate additional cash flows to plow back into the financing of longer term projects (in lieu of borrowing the money). Handling the capital needs would require a very delicate balancing act indeed!

I am aware of one petroleum exploration firm who decided to try this series of logic and this new approach for selecting exploration prospects. They decided to start out in a modest fashion and construct the inventory graph such as Figure 11.3 only 5 years into the future. They extrapolated lease decline curves to establish the lower curve and their management specified a corporate objective to define the upper curve. But they never proceeded any further than this graph.

The reason was that when they had established the gap between the two curves which had to be filled by new exploration ventures they concluded there simply weren't enough viable exploration prospects left to fill the gap. As a result, they promptly decided to diversify and began investing in non-petroleum projects such as real estate, television stations, etc. The effect of the diversification, of course, was to fill the void of limited reinvestment opportunities by opening up their firm to non-exploration investments.

While the approach has certain intuitive appeal to speak to the issue of how to evaluate projects given the constraint of only limited reinvestment opportunities it certainly can't be considered the complete answer. And until such time that we can devise a realistic measure of value I think the analyst should concentrate on giving management an NPV *plus* the undiscounted profit-to-investment ratio we talked about at the beginning of Chapter 2. The reason is that while NPV may not assign much value to long extended cash flows the profit/investment ratio will.

In fact, if we had given the ratio for Prospects A and B in the earlier example it might have given management a much better clue as to the possible ultimate value of Prospect B. Even though the NPV values are about equal the ratios suggest that Prospect B will make a much greater amount of new money per dollar invested as compared to Prospect A. This should raise a red flag to the manager that the NPV of Prospect B may not be the whole story.

Prospect A	Prospect B
3 million barrels short life	100 million barrels long life
NPV @ i_0 = +$2.7 million	NPV @ i_0 = +$1.8 million
$\dfrac{\text{Profit}}{\text{Invest.}}$ = <u>1.62</u>	$\dfrac{\text{Profit}}{\text{Invest.}}$ = <u>14.3</u>

In summary, I think we can safely agree that we as an industry do not have a totally valid measure of worth to use if we have only limited future reinvestment opportunities. But one way to identify projects which might generate huge amounts of new capital (even though it may be well in the future) is to present to management an undiscounted profit/investment ratio in conjunction with the usual discounted parameters. It's a bit of irony, I guess, that we've come full circle back to a parameter which in Chapter 2 was generally discredited for not telling us anything about the time value of money. Whatever we said about it initially should now be modified in terms of its value as a means to identify the merits of projects having very long, extended cash flow schedules.

III. MAXIMIZING EXPECTED MONETARY VALUE – IS IT THE ONLY DIMENSION OF VALUE MANAGEMENT MUST CONSIDER?

In our discussions from Chapter 3 onward in this book we have repeatedly suggested that an expected monetary value (EMV, expected net present value profit, etc.) is the parameter the decision maker should use to compare and select investments having risk and uncertainty. The merits of expected monetary value rested on the fact that it included all pertinent dimensions of profitability as well as the degree of risk and uncertainty. Consequently, if we had a list of available investment projects

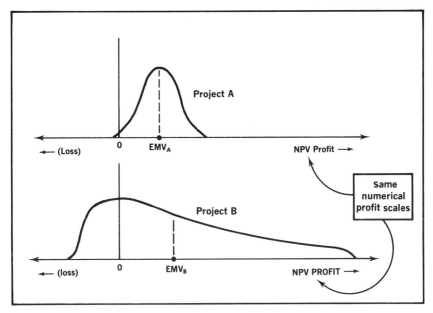

Figure 11.4 *Probability Distributions of two investment projects. Each distribution has the same numerical scale for the abscissa. $EMV_B > EMV_A$.*

but only limited capital we would presumably rank the projects by their respective expected values.

At the top of the list would be the project having the highest (positive) expected monetary value, and the bottom of the list would represent the project having the lowest (positive) expected monetary value. Our capital budgeting scheme might be to proceed down the list accepting each project having successively lower EMV's until we had exhausted our available investment capital.

But is it really quite that simple? For example, suppose we had to make a choice between Projects A and B of Figure 11.4. Project B has the higher EMV and, all other factors equal, would be preferred over Project A. But Project A has virtually no chance of losing money, whereas Project B has a substantial likelihood of resulting in a loss (the area to the left of NPV = 0). Isn't it conceivable that if we wanted to minimize our chances of losing money we might prefer Project A over B?

Isn't it conceivable that we might be willing to trade the higher EMV and the possibility of a large gain from Project B for the slightly lower

EMV of Project A knowing we would, by this trade, essentially eliminate the chance of a monetary loss? Certainly it is – particularly if the decision maker is conservative in his attitudes about taking risks and/or has a limited amount of money.

If this possibility exists the obvious consequence is that perhaps ranking projects by expected values alone isn't the whole story. It may, in fact, suggest that management may have a dual objective in selecting decisions under uncertainty: to maximize expected gain, and at the same time minimize exposure to loss.

One way to incorporate such a dual objective strategy is by use of preference curves of the type we mentioned in Chapter 5. Indeed one of the merits of preference theory was its ability to balance the relative desires for financial gains against the feelings of displeasure associated with financial losses. If we were using expected preference values (rather than trying to maximize expected *monetary* values) the above questions would have become moot. But we also mentioned in Chapter 5 that we didn't yet have a completely suitable way of defining preference curves, resulting in the preference theory being just that – a theory or concept, but not a practical reality. In the absence of a viable way to obtain and use the preference theory concepts what other alternative do we have for honoring a strategy of maximizing expected gains and minimizing exposure to losses?

One approach was proposed by Hertz in 1968 (Reference 11.3). He stated that most managements would like to have investment policies that both maximize financial result over the long run and minimize uncertainty, or risk. In his work he defined uncertainty as the variability, or spread, of a profit distribution about its mean value. He went on to propose what he called the 'efficiency' concept whereby various investments were quantified as to their expected values (EMV) and their variability about the mean (expressed as the distribution σ divided by its μ). The objective then was to select those projects which would maximize expected gain and minimize the variability about the mean.

The concept Mr. Hertz proposed has definite merit. However, his work included several assumptions which are not directly applicable to petroleum exploration. For one thing the profit distributions of his investment choices were assumed to be normally distributed. This rarely occurs for drilling prospects. Secondly, his concern with minimizing variability is probably not the main concern of the exploration decision maker. I think most explorationists are less concerned with variability about the mean than about whether the variability extends over to the

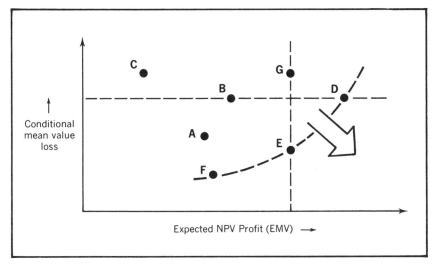

Figure 11.5 *Graph of EMV versus conditional mean loss for seven drilling prospects. Conditional mean value loss represents the weighted average loss given the prospect ended up losing money. EMV is defined in the usual manner. A strategy to maximize EMV and minimize conditional mean value losses suggests an investment strategy which attempts to move in the downward, diagonal direction of the arrow. The three prospects along the dashed line (Prospects F, E, and D) represent the optimal choices.*

loss portion of the random variable scale. Losses are the primary concern—their magnitudes and their relative likelihoods of occurrence.

Consequently, a slightly revised approach that might be more relevant to the exploration sector is to prepare for management a graph such as Figure 11.5. Each of the seven plotted points represents an exploratory drilling prospect. The abscissa of the graph is EMV computed in the usual manner (e.g. the mean value of NPV profit distribution derived from a simulation analysis).

The dimension on the ordinate is the weighted average loss from the prospect given that a loss is incurred. Conceptually, it represents the knife-edge balancing point of the portion of the profit distribution to the left of $NPV = 0$, as shown in Figure 11.6. To compute its numerical value from a simulation analysis merely requires that we program the computer to find the arithmetic average value of all the profits which were negative on the repetitive simulation passes. Knowing the conditional mean value loss and the usual EMV for each prospect we have the coordinate data to plot the point for each drilling prospect in Figure 11.5.

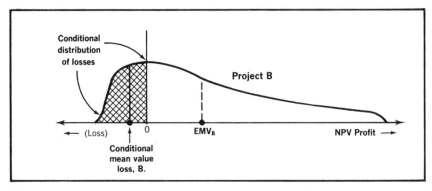

Figure 11.6 *Profit distribution of drilling prospect B that was originally given in Figure 11.4. The cross-hatched portion of the profit distribution represents the conditional distribution given a loss. The mean value of this conditional distribution is the parameter on the vertical axis of Figure 11.5.*

Once the data points have been plotted for each prospect the graph of Figure 11.5 can then be used by management to select those projects which will maximize EMV and at the same time minimize expected losses given the project results in losses. The optimal choices lie along the dashed curve and include projects F, E, and D.

The reason this dashed line represents the optimal choices is that the dual management strategy seeks to move horizontally to the right as far as possible (maximizing EMV) and to move downward as far as possible (minimizing expected losses). The net effect is a sort of vector in the diagonal, downward direction of the arrow. The dashed line represents the leading edge of projects as we proceed as far as possible in the diagonal direction.

That projects along the dashed curve are indeed optimal is easy to prove. Any project above and to the left of the dashed curve is suboptimal because it has a lower EMV and/or a higher expected value loss. For instance, for a given expected value such as EMV_E, Project G also has the same EMV. But it has a higher expected value loss than E so it is less desirable. For a given value of expected losses, say for example the value corresponding to Projects B and D, the choice of D is preferred over B because it has a higher positive EMV.

What is suggested here is that if we have a suite of available investment opportunities we can make a graph such as Figure 11.5 using the EMV of each project and its corresponding conditional mean value loss. Then if management is seeking to select projects from the suite he can choose those along the line which bounds the plotted points along the diagonal downward vector direction.

As such the graph helps management balance the dual objective of trying to maximize EMV and at the same time minimize the level of expected losses if the project results in a loss. The only new data needed is the conditional mean value loss for each prospect. As we mentioned earlier this number is readily computed by finding the arithmetic average of all the profit values (a value of profit is computed on each simulation pass) which are negative.

Why was conditional mean value loss chosen as the parameter management would seek to minimize in prospect analysis? The reason is two-fold: he is concerned with minimizing the magnitude of possible losses, *and* the probability of losses. Either parameter by itself is not sufficient. For example, we could have entered on the vertical scale the maximum possible loss of each prospect (the minimum value of the profit distribution). If our goal was to minimize the maximum possible loss then of two projects having the following maximum possible losses we would prefer Project 2, all other factors equal.

Project 1: Maximum loss: −$5 MM
Project 2: Maximum loss: −$4 MM

But suppose the likelihood of a $5 MM loss in Project 1 is 0.001, and the likelihood of a $4 MM loss from Project 2 is 0.75. Would Project 2 still look more favorable? Probably not. Thus, we can probably rule out a strategy of maximizing EMV and minimizing the maximum possible loss because it does not recognize whether the maximum possible loss is extremely unlikely or quite likely.

We could also have used the parameter "probability of loss" on the vertical ordinate. Then the strategy would be to maximize EMV and minimize the probability of loss. Sounds good — but this approach is also flawed. We could have two projects, 3 and 4 having the following data, and prefer 4 because it has the lower probability of loss.

Project 3: Probability of loss: 0.35 $\left\{ \begin{array}{l} \text{The left tails of the} \\ \text{profit distribution;} \\ \text{the probability of NPV} < 0 \end{array} \right.$
Project 4: Probability of loss: 0.30

But suppose the magnitudes of loss from Project 3 are, say, −$100 and the magnitudes of losses from Project 4 are in the range of −$10 MM. We may prefer a 35% chance of losing $100 to a 30% chance of losing $10 million dollars. So we see that probabilities by themselves are also not sufficient because the strategy of just minimizing probability of loss does not recognize the magnitudes of losses.

If we are trying to minimize our exposure to losses we need to consider both the magnitudes of potential losses as well as their corresponding conditional probabilities of occurrence. And this indeed is what expected values are — hence, the suggestion to use the conditional mean value losses as the vertical ordinate in a dual-strategy graph such as Figure 11.5. The numerical value of conditional mean value loss is based on the magnitudes of losses and their corresponding likelihoods of occurrence, given a loss.

This parameter eliminates the obvious flaws of using just maximum loss or just the area of the profit distribution to the left of NPV = 0. Strategies to minimize expected value losses have exactly the same connotation and interpretation from an expected value point of view as maximizing expected monetary value profit.

As a final comment, we should note that this approach does not tell the decision maker which of the projects along the dashed line is preferable. If he doesn't have enough capital to accept F, E, and D, he will have to choose from among these three by assessing the maximum expected value loss he will accept. Obviously D is better from an EMV standpoint, but it also has higher expected losses if the project fails. So he still is faced with a judgment about balancing EMV with the magnitudes of expected losses.

The value of the graph is to initially screen out F, E, and D as being preferable to all the other available projects. By preparing graphs such as Figure 11.5 the analyst can help the decision maker see the relative effects of EMV and expected value losses. Since the graph is so easy to plot (requiring only one additional computation for each prospect) it probably can be used on a fairly routine basis for selecting prospects with a limited capital constraint.

Suggestions for
Further Reading

THE first series of references are general in nature, dealing with the broad subject of decision analysis. References on specific topics are listed under the appropriate chapter headings.

(1.1) Hertz, David B., *New Power for Management,* McGraw-Hill Book Company, New York, 1969

A general overview of the rapidly expanding discipline of management science. Written for the executive to describe new decision-making techniques, computer systems, and corporate information systems.

(1.2) Springer, C. H., et. al., *Mathematics for Management Series,* Richard D. Irwin, Inc., Homewood, Illinois, 1965

A four volume, paperback series covering virtually all aspects of the new mathematical techniques to evaluate decision choices. Vol. I — Basic Mathematics, Vol. II — Advanced Methods and Models, Vol. III — Statistical Inference, and Vol. IV — Probabilistic Models. Many self-study problems are included to assist the reader in understanding the techniques described.

(1.3) Grayson, C. Jackson, Jr., *Decisions Under Uncertainty, Drilling Decisions by Oil and Gas Operators,* Harvard Business School, Division of Research, Boston, 1960

Chapters 9, 10, and 11 are particularly recommended.

(1.4) Schlaifer, Robert, *Probability and Statistics for Business Decisions,* McGraw-Hill Book Company, New York, 1959

An introduction to decision making under conditions of uncertainty.

(1.5) Schlaifer, Robert, *Analysis of Decision Under Uncertainty,* McGraw-Hill Book Company, New York, 1969

(1.6) Mao, James C. T., *Quantitative Analysis of Financial Decisions,* The Macmillan Company, New York, 1969

(1.7) Weaver, Warren, *Lady Luck,* Anchor Books Science Study Series S30, Doubleday and Company, Garden City, New York, 1963 (paperback)

An excellent introduction to the theory of probability.

(1.8) Dorn, William S. and Greenberg, Herbert J., *Mathematics and Computing: with FORTRAN Programming,* John Wiley and Sons, Inc. New York, 1967

An excellent introduction to linear programming, random numbers, probability theory, and FORTRAN Programming.

(1.9) Solomon, Ezra (editor), *The Management of Corporate Capital,* The Free Press of Glencoe, New York, 1959
    A collection of important articles on corporate finance and capital budgeting.
(1.10) Wright, M. G., *Discounted Cash Flow,* McGraw-Hill Book Company, New York, 1967
(1.11) Campbell, John M., *Oil Property Evaluation,* Prentice-Hall, Inc., Englewood Cliffs, N.J., 1959
(1.12) Porter, Stanley P., *Petroleum Accounting Practices,* McGraw-Hill Book Company, New York, 1965
    A useful reference on taxation and accounting procedures of the oil business.
(1.13) McCray, A. W., *Petroleum Evaluations and Economic Decisions,* Prentice-Hall, Inc., Englewood Cliffs, N.J., 1975.

Chapter 2 – Measures of Profitability

(2.1) Dean, Joel, "Measuring the Productivity of Capital," Harvard Business Review, January – February 1954.
(2.2) Solomon, *op. cit.*
(2.3) Jackson, E. G., "Here's a way to figure your real cost-of-capital," Oil and Gas Journal, January 9, 1967, p. 124–126.
(2.4) Reul, Raymond I., "Which Investment Appraisal Technique Should You Use," Chemical Engineering, April 22, 1968, p. 212–218.
(2.5) Kaitz, Melvin, "Percentage Gain on Investment – An Investment Decision Yardstick," Journal of Petroleum Technology, May, 1967, p. 679–687.
(2.6) Garies, D. F., "A Short-Cut Method for Determining Present Worth of Future Income and Rate of Return on Investments in Oil-Producing Properties," paper presented at 40th Annual Meeting of Society of Petroleum Engineers, Denver, October 1965 (SPE Preprint 1255, available from SPE office in Dallas on microfilm).
(2.7) *Oil and Gas Property Evaluation and Reserve Estimates,* Society of Petroleum Engineers of AIME, Petroleum Reprint Series No. 3, Part II, p. 152–174.
(2.8) Schoemaker, R. P., "A Graphical Short-cut for Rate of Return Determinations" and "Special Applications," a series appearing in July–September, 1963 issues of World Oil.
(2.9) Campbell, J. M., "Optimization of Capital Expenditures in Petroleum Investments," Journal of Petroleum Technology, July 1962, p. 708–714.
(2.10) Wansbrough, R. S., "An Approach to the Evaluation of Oil-Production Capital Investment Risks," Journal of Petroleum Technology, Sept. 1960, p. 25–30.
(2.11) Hardin, George C. and Mygdal, Karl, "Geologic Success and Economic

Failure," AAPG Bulletin, Vol. 52, No. 11 (November, 1968), p. 2079–2091.

(2.12) Wright, *op. cit.*

(2.13) Thuesen, H. G., *Engineering Economy,* Prentice-Hall, Inc., Englewood Cliffs, N.J., 1957 (Second Edition).

(2.14) Phillips, Charles E., "The Relationship Between Rate of Return, Payout and Ultimate Return in Oil and Gas Properties," Journal of Petroleum Technology, September 1958, p. 26–29.

(2.15) Northern, I. G., "Investment Decisions in Petroleum Exploration and Producing," Journal of Petroleum Technology, July 1964, p. 727–731.

(2.16) Swalm, Ralph O., "Capital Expenditure Analysis—A Bibliography," The Engineering Economist, Vol. 13—Number 2, Winter 1968, p. 105–129 (An excellent list of references on Capital Budgeting. Not all of the references are petroleum related, however.)

(2.17) Pegels, C. Carl, "A Comparison of Decision Criteria for Capital Investment Decisions," The Engineering Economist, Vol. 13, No. 4, Summer 1968, p. 211–220.

(2.18) Silbergh, Michael, and Brons, Folkert, "Profitability Analysis of Leveraged Transactions," Journal of Petroleum Technology, March 1973, p. 319–328.

(2.19) _____, "Profitability Analysis—Where Are We Now?," Journal of Petroleum Technology, January 1972, p. 90–100. (This article has an extensive list of references on the general subject of profitability analysis.)

(2.20) Megill, R. E., *An Introduction to Exploration Economics,* The Petroleum Publishing Company, Tulsa, 1971

(2.21) Henry, Arthur J., "Appraisal of Income Acceleration Projects," Journal of Petroleum Technology, April 1972, p. 393–398.

(2.22) Capen, E. C., R. V. Clapp, and W. W. Phelps, "Growth Rate-A Rate-of-Return Measure of Investment Efficiency," Journal of Petroleum Technology, May, 1976, p. 531–543.

(2.23) El Banbi, H. A. and El Mekkawy, A. S., "Simplified profit index aids economic evaluations," World Oil, January 1973, p. 79–82.

(2.24) Phillips, Charles E., "The Appreciation of Equity Concept and Its Relationship to Multiple Rates of Return," Journal of Petroleum Technology, February 1965, p. 159–163.

Chapter 3—The Expected Value Concept

(3.1) Laplace, Pierre Simon, Marquis de, *A Philosophical Essay on Probabilities,* translated from 6th French Edition by F. W. Truscott and F. L. Emory, Dover Publications, Inc., New York, 1951 (Laplace's essay was originally published in 1814.)

(3.2) Bross, Irwin D. J., *Design for Decision,* MacMillan Company, New York, 1953.

(3.3) Grayson, *op. cit.,* Chapters 9 and 11.

(3.4) Levinson, Horace C., *Chance, Luck, and Statistics,* Dover Publications, Inc., New York, 1963 (An enjoyable, nontechnical book about chance.)

(3.5) Schlaifer, *op. cit.*

(3.6) Townes, Harrison L., "Using Economics in Exploration Decisions," Transactions of the West Texas Geological Society's Symposium on Economics and the Petroleum Geologist held in Midland, Texas, October 20–21, 1966. Publication No. 66–53, pgs. 108–117.

(3.7) _____, "How to Use Economics in Exploration Decision-Making," World Oil, April 1966, pgs. 105–109.

(3.8) Raiffa, Howard, *Decision Analysis — Introductory Lectures on Choices Under Uncertainty,* Addison-Wesley, Reading, Mass., 1968 (paperback)

(3.9) Ore, Oystein, *Cardano, The Gambling Scholar,* Dover Publications, Inc., New York, 1953

(3.10) Brown, Rex V., "Do Managers Find Decision Theory Useful?", Harvard Business Review, May–June, 1970, p. 78–89.

(3.11) Newendorp, Paul D., "Expected Value — A Logic for Decision Making," paper presented at SPE 1971 Symposium on Petroleum Economics and Evaluation, Dallas, March 8–9, 1971, and published in the Symposium Proceedings.

Chapter 4 — Decision Tree Analysis

(4.1) Byrne, R., et. al., "Some New Approaches to Risk," U.S. Government Report No. AD-653,727, May 1967.

(4.2) Dorn and Greenberg, *op. cit.,* pp. 455–464.

(4.3) Grayson, *op. cit.,* Chapter 11

(4.4) Hammond, J. S., "Better Decisions with Preference Theory," Harvard Business Review, Vol. 45, No. 6, (November–December 1967), pp. 123–141.

(4.5) Magee, John F., "Decision Trees for Decision Making," Harvard Business Review, Vol. 42, No. 4 (July–August, 1964), pp. 126–138.

(4.6) _____, "How to Use Decision Trees in Capital Investment," Harvard Business Review, Vol. 42, No. 5 (September–October, 1964), pp. 79–96.

(4.7) Schlaifer, *op. cit.*

Chapter 5 — Preference Theory Concepts

(5.1) von Neumann, John, and Morgenstern, Oskar, *Theory of Games and Economic Behavior,* Princeton University Press, Princeton, New Jersey, 3rd Edition, 1953.

Chapter I includes their development of the mathematical theory of utility.

(5.2) Grayson, *op. cit.,* Chapter 10.

(5.3) Newendorp, Paul D., *Application of Utility Theory to Drilling Investment Decisions,* Doctoral Dissertation, University of Oklahoma, Norman, Oklahoma, 1967. (Copies available from University Microfilms, Inc., Ann Arbor, Michigan, Publication No. 67-15, 898.)

(5.4) Green, Paul E., "Risk Attitudes and Chemical Investment Decisions," Chemical Engineering Progress, Vol. 59, No. 1 (January, 1963), pp. 35–40.

Describes author's experiment to construct utility curves for various management levels in a large chemical company.

(5.5) Hammond, John S. III, "Better Decisions with Preference Theory," Harvard Business Review, Vol. 45, No. 6 (November–December 1967), pp. 123–141.

(5.6) Swalm, Ralph O., "Utility Theory — Insights into Risk Taking," Harvard Business Review, Vol. 44, No. 6 (November–December, 1966), pp. 123–136.

(5.7) Smith, Marvin B., "Parametric Utility Functions for Decisions Under Uncertainty," paper presented at 47th Annual Meeting of Society of Petroleum Engineers, San Antonio, Texas, Oct. 8–11, 1972 (SPE Preprint No. 3973)

(5.8) Newendorp, Paul D., "A Decision Criterion for Drilling Investments," paper presented at 43rd Annual Meeting of Society of Petroleum Engineers, Houston, Texas, Sept. 29–Oct. 2, 1968 (SPE Preprint No. 2219)

(5.9) Pratt, John W., "Risk Aversion in the Small and in the Large," Econometrica, (1964), pp. 122–136.

Chapter 6 — Basic Principles of Probability and Statistics

(6.1) Weaver, *op. cit.*

(6.2) Aitchison, J. and Brown, J. A. C., *The Lognormal Distribution,* Cambridge University Press, London, 1957.

(6.3) Kaufman, Gordon M., *Statistical Decision and Related Techniques in Oil and Gas Exploration,* Prentice-Hall, Inc., Englewood Cliffs, N.J., 1963.

(6.4) _____, "Statistical Analysis of the Size Distribution of Oil and Gas Fields," SPE Paper No. 1096 presented at the 1965 Symposium on

Petroleum Economics and Evaluation, Dallas, Texas, March 4–5, 1965. (Available on microfilm from SPE office in Dallas.)

(6.5) Krumbein, W. C. and Graybill, F. A., *An Introduction to Statistical Models in Geology*, McGraw-Hill Book Company, New York, 1965.

(6.6) Oil and Gas Journal – "Giant Fields of Venezuela," July 29, 1968, pp. 144–152. (Contains statistical data on field sizes.)

(6.7) McCrossan, R. G., "An Analysis of Size Frequency Distributions of Oil and Gas Reserves of Western Canada," Geological Survey of Canada, Report of Activities, Part B, Nov. 1967–March 1968, pp. 64–68.

(6.8) Trudgen, Patricia, and Hoffman, Frank, "Statistically Analyzing Core Data," Journal of Petroleum Technology, August 1967, pp. 497–503.

(6.9) Adler, Irving, *Probability and Statistics for Everyman*, Signet Science Library T2850, 1963. (A paperback which covers the basic aspects of probability and statistics.)

(6.10) Lahee, F. H., "How Many Fields Really Pay Off?" Oil and Gas Journal, Sept. 17, 1956, pp. 369–371.

(6.11) Stanley, L. T., *Practical Statistics for Petroleum Engineers*, The Petroleum Publishing Company, Tulsa, 1973

(6.12) Koch, George S., Jr., and Link, Richard F., *Statistical Analysis of Geological Data*, John Wiley & Sons, Inc., New York, 1970

(6.13) Burr, Irving W., *Engineering Statistics and Quality Control*, McGraw-Hill Book Company, Inc., New York, 1953

(6.14) Newendorp, Paul D., "Bayesian Analysis – A Method for Updating Risk Estimates," Journal of Petroleum Technology, February 1972, pp. 193–198.

(6.15) Grayson, C. Jackson, Jr., "Bayesian Analysis – A New Approach to Statistical Decision-Making," Journal of Petroleum Technology, June 1962, pp. 603–607.

(6.16) Davis, John C., *Statistics and data analysis in geology*, John Wiley & Sons, Inc., New York, 1973

Chapter 7 – Petroleum Exploration Risk Analysis Methods

(7.1) Hayward, J. T., "Probabilities and Wildcats tested Through Mathematical Manipulation," Oil and Gas Journal, Vol. 33, No. 26 (November 15, 1934), pp. 129–131.

(7.2) Mabra, D. Allen, Jr., "Here's a New Way to Evaluate Drilling Prospects," Oil and Gas Journal, Vol. 55, No. 5 (Feb. 4, 1957), pp. 186–188.

(7.3) Schwade, Irving T., "Geologic Quantification. Description-Numbers-Success Ratio," AAPG Bulletin, Vol. 51, No. 7 (July, 1967), pp. 1225–1239.

(7.4) Arps, J. J. and Roberts, T. G., "Economics of Drilling for Cretaceous Oil on East Flank of Denver-Julesburg Basin," Bulletin of AAPG, Vol. 42, No. 11, November, 1958, pp. 2549–2566.

(7.5) Campbell, W. M. and Schuh, F. J., "Risk Analysis: Over-all Chance of Success Related to Number of Ventures," SPE Paper No. 218 presented at 36th Annual Fall Meeting of Society of Petroleum Engineers, October 8–11, 1961, Dallas. The paper is reprinted in "Oil and Gas Property Evaluation and Reserve Estimates," Society of Petroleum Engineers Reprint Series No. 3 (Revised 1970 Edition), pp. 168–175.

(7.6) Walstrom, John E., "A Statistical Method for Evaluating Functions Containing Indeterminate Variables and Its Application to Recoverable Reserves Calculations and Water Saturation Determinations," Geological Sciences, Stanford University Publications, Vol. IX, No. 2, p. 823.

(7.7) Northern, I. G., "Risk, Probability and Decision-Making in Oil and Gas Operations," Journal of Canadian Petroleum Technology, Vol. 6, No. 4 (December, 1967), pp. 150–154.

(7.8) Oil and Gas Journal, "Wildcat success ratio gets worse," July 14, 1969, pp. 50–51.

(7.9) Drew, L. J., "Grid-Drilling Exploration and Its Application to the Search for Petroleum," Economic Geology, Vol. 62, 1967, pp. 698–710.

(7.10) _____, "Estimation of Petroleum Exploration Success and the Effects of Resource Base Exhaustion Via a Simulation Model," U.S. Geological Survey Bulletin 1328, Washington D.C., (1974)

(7.11) Dougherty, E. L., "Application of Optimization Methods to Oilfield Problems: Proved, Probable, Possible," SPE Preprint No. 3978, paper presented at 47th Annual Fall Meeting of the Society of Petroleum Engineers, San Antonio, Texas, Oct. 8–11, 1972 (The article contains an extensive list of references relating to use of quantitative analysis methods in the petroleum industry.)

(7.12) Smith, Marvin B., "Probability Estimates for Petroleum Drilling Decisions," Journal of Petroleum Technology, June 1974, pp. 687–695.

(7.13) Demirmen, Ferruh, "Probabilistic Study of Oil Occurrence Based on Geologic Structure in Stafford County, South-Central Kansas," Kansas Geological Survey publication, Lawrence, Kansas, 1973

(7.14) Prelat, Alfredo, "Statistical Estimation of Wildcat Well Outcome Probabilities by Visual Analysis of Structure Contour Maps of Stafford County, Kansas," Kansas Geological Survey publication, Lawrence, Kansas, 1974.

(7.15) Megill, Robert E., An Introduction To Risk Analysis, The Petroleum Publishing Co., Tulsa, 1977.

(7.16) Harbaugh, J. W., J. H. Doveton, and J. C. Davis, Probability Methods in Oil Exploration, Wiley-Interscience, New York, 1977.

(7.17) Gebelein, C. A., Conrad E. Pearson, and Michael Silbergh, "Assessing Political Risk to Foreign Oil Investment Ventures," SPE 6335 presented at 1977 Economics and Evaluation Symposium of the Society of Petroleum Engineers, Dallas, Texas, February 21–22, 1977, and published in the Symposium Proceedings, p. 55–64.

Chapter 8 – Risk Analysis Using Simulation Techniques

(8.1) Hertz, David B., "Risk Analysis in Capital Investment," Harvard Business Review, Vol. 42, No. 1 (January–February, 1964), pp. 95–106.

(8.2) Walstrom, J. E., Mueller, T. D., and McFarlane, R. C., "Evaluating Uncertainty in Engineering Calculations," Journal of Petroleum Technology, December, 1967, pp. 1595–1603.

(8.3) Newendorp, Paul D. and Root, Paul J., "Risk Analysis in Drilling Investment Decisions," Journal of Petroleum Technology, June, 1968, pp. 579–585.

(8.4) Smith, Marvin B., "Estimate Reserves by Using Computer Simulation Method," Oil and Gas Journal, March 11, 1968, pp. 81–84.

(8.5) Thorngren, J. T., "Probability Technique Improves Investment Analysis," Chemical Engineering, Vol. 74, No. 17, August 14, 1967, pp. 143–151.

(8.6) Hess, S. W. and Quigley, H. A., "Analysis of Risk in Investments Using Monte Carlo Techniques," Chemical Engineering Progress Symposium Series No. 42: Statistics and Numerical Methods in Chemical Engineering, Vol. 59, 1963, pp. 55–63 (published by American Institute of Chemical Engineers, New York)

(8.7) International Business Machines Corp., White Plains, N.Y., "Random Number Generation and Testing," Publication No. C20-8011-0, 1959.

(8.8) Mize, Joe H. and Cox, J. Grady, Essentials of Simulation, Prentice-Hall, Englewood Cliffs, New Jersey, 1968.

(8.9) Sprow, Frank B., "Evaluation of Research Expenditures Using Triangular Distribution Functions and Monte Carlo Methods," Industrial and Engineering Chemistry, Vol. 59, No. 7 (July, 1967), pp. 35–38.

(8.10) Klausner, Robert F., "The Evaluation of Risk in Marine Capital Investments," The Engineering Economist, Vol. 14, No. 4, Summer 1969, p. 183–214.
(Marine capital investments in this article include ships, barges, tugs, etc.)

(8.11) Breiman, Leo, "The Kinds of Randomness," Science and Technology, December 1968, pp. 34–44.

(8.12) McCarron, J. K., "Computer simulation may lower costs for offshore pipelines," Oil and Gas Journal, February 17, 1969, pp. 84–93.

(8.13) McCray, A. W., "Evaluation of Exploratory Drilling Ventures by Statistical Decision Methods," Journal of Petroleum Technology, September 1969, pp. 1199–1209.

(8.14) Walstrom, J. E., Brown, A. A., and Harvey, R. P., "An Analysis of Uncertainty in Directional Surveying," Journal of Petroleum Technology, April, 1969, pp. 515–523.

(8.15) Schlaifer, Robert, Computer Programs for Elementary Decision Analysis, Division of Research, Harvard University, Boston, 1971.

(8.16) Davidson, L. B., and D. O. Cooper, "A Simple Way of Developing a Probability Distribution of Present Value," Journal of Petroleum Technology, September 1976, p. 1069–1078.

(8.17) Capen, E. C., "The Difficulty of Assessing Uncertainty," Journal of Petroleum Technology, August 1976, p. 843–850.

(8.18) Anderson, M. L., "Application of Risk Analysis to Enhanced Recovery Pilot Testing Decisions," SPE 6352 presented at 1977 Economics and Evaluation Symposium of the Society of Petroleum Engineers, Dallas, Texas, February 21–22, 1977, and published in the Symposium Proceedings, p. 233–242.

(8.19) Kleijnen, Jack P. C., Statistical Techniques in Simulation, Marcel Dekker, Inc., New York, 1974.

Chapter 9 — Implementing Risk Analysis Methods

(9.1) Grayson, C. Jackson, Jr., "Management science and business practice," Harvard Business Review, July–August 1973, pp. 41–48.

(9.2) Brown, Rex V., "Do managers find decision theory useful?," Harvard Business Review, May–June 1970, pp. 78–89.

(9.3) Behrman, Robert G., Jr., "A Suggestion for Exploration Management," AAPG Bulletin, v. 57, no. 1 (January 1973), pp. 207–209.

(9.4) Carter, E. Eugene, "What are the risks in risk analysis?," Harvard Business Review, July–August 1972, pp. 72–82.

(9.5) Raiffa, Howard, Decision Analysis, Addison-Wesley, Reading, Mass., 1968.

Chapter 11 — Special Problems and Open Issues

(11.1) Arps, J. J., and J. L. Arps, "Prudent Risk-Taking," Journal of Petroleum Technology, July 1974, pp. 711–716.

(11.2) Arps, J. J., "The Profitability of Exploration Ventures," Economics of Petroleum Exploration, Development, and Property Valuation (Southwestern Legal Foundation), Prentice-Hall, Inc., Englewood Cliffs, N.J., 1961, pp. 153–173.

(11.3) Hertz, David B., "Investment policies that pay off," Harvard Business Review, January–February 1968, pp. 96–108.

Glossary
of Terms and
Abbreviations

FOR those who might be unfamiliar with some of the terminology used in this book the following glossary of terms should prove helpful. Following the glossary is a listing of most of the abbreviations used, as well as some of the mathematical symbols that have been given in the book.

Glossary of Exploration Terms

Acreage
The leased area for which exclusive drilling rights are held.

Anomaly
A deviation in the geologic structure or stratigraphy of a basin. Usually used in the sense of a seismic anomaly, an apparent structure observed from seismic records.

Backin option
An agreement between two or more companies owning working interest in a proposed well whereby one party chooses not to participate in the drilling of the well. The other parties then proceed to drill the well and recover a previously determined amount of revenue, at which time the carried party then "Backs in" and shares proportionately in future revenues. The clause permitting a working interest owner the right to do this is called a penalty clause, or backin option. The carried party shares in no costs, revenues, or risk until the well has recovered the previously determined amount of revenue (usually well costs plus an additional penalty). If the well is dry the carried party does not have to pay any of the dry hole costs.

Block of acreage
A leased area (or prospective area for leasing) usually in one contiguous piece, or block

Bonus
A cash payment made to the owner of the mineral rights (oil and gas) to acquire exclusive drilling rights

on a block of acreage or leases. The cash payments offered at competitive bid sales of leases, such as off-shore.

Buy acreage   To obtain leases, or the leasehold drilling rights

Cash flow   A movement of money into or out of the firm's treasury. Expenditures are usually termed as negative cash flows, and revenues (or receipts) are considered positive cash flows.

Concession   An agreement for exclusive drilling rights within a specified geographical area. A lease, or an operating agreement area.

Delineation well   A well drilled after discovery of a field to try to define, or delineate the productive limits of the newly discovered field.

Development well   A well drilled in a field after discovery to complete development of the field. Sometimes called exploitation, or field wells.

Discounted cash flow A cash flow which has been discounted at a given interest rate to find the cash flow's equivalent value at an earlier point in time. Usually the point of time is taken as time zero, or the time of analysis and decision making. As such the discounted cash flows are sometimes called present values of the cash flows.

Drop acreage   A decision to cancel, relinquish, or terminate a lease. To give up or surrender the drilling rights within a geographic area. Usually it means the acreage appears to have no further value for petroleum exploration.

Expendible well   A well drilled for information with no specific intention to produce the well if it encounters hydrocarbons. In offshore areas they are the wells drilled from a floating drillship after a discovery to assess the size of the field. The well may later be re-entered from a permanent platform or equipped with seabed production facilities if feasible.

Exploratory well   A well drilled to determine if a structure or prospect has oil or gas in commercial quantities. Sometimes called a wildcat.

Farmout   An assignment of ownership and drilling rights of a leased area to another party, usually in exchange for a proportion of any oil or gas found and produced, called an overriding royalty interest (ORI). The party who farms out does not stand any risk or costs of drilling the well, and the retained overriding royalty interest is normally free and clear of all operating expenses. The party to whom the drilling rights have been assigned is said to have "farmed in." Companies usually farm out their rights in a leased area if they feel the drilling prospects are poor (by their standards) and/or they prefer to invest their exploration capital elsewhere.

Gross reserves

The total recoverable reserves of a well or field without regard to proportionate ownership of parties drilling the wells and owning mineral rights.

Lease

An agreement authorizing exclusive rights for exploring for and producing of any hydrocarbons found. The agreement is between the company wishing to explore for oil and gas and the owner of the mineral rights. In return for assigning the exclusive drilling rights the mineral owner reserves or receives a portion of any oil or gas found. This portion is normally called a royalty. The terms lease and leasehold rights are used interchangeably.

Lease bonus

A cash payment or bonus paid to the owner of the mineral rights to induce the owner to grant the party exclusive drilling rights.

Net reserves

The portion of gross recoverable reserves owned by a company. If the company is the sole owner of the lease net reserves are simply gross reserves minus the royalty portion of reserves payable to the party owning the mineral rights. If the company owns a partial interest in the lease its net reserves are the proportionate portion of gross reserves minus royalty.

Operating area

A concession or lease granting exclusive drilling rights within a geographical area.

Overriding royalty interest (ORI)

The portion of any oil and gas found on a farmout which is retained and payable to the party originally owning the lease.

Penalty clause

See backin option

Play	An area of concentrated exploration activity and/or interest within a sedimentary basin.
Project	An investment opportunity, a drilling prospect
Prospect	An area under which is thought to exist a geologic trap having oil or gas deposits. A seismic anomaly, for example. The area being considered to locate and drill an exploratory well.
Recoverable reserves	The portion of oil and gas in place which can be produced to the surface.
Royalty	The portion of oil and gas produced that is paid to the owner of the mineral rights under the terms of the lease or operating agreement.
Spacing pattern	The density with which development wells can be drilled in areas regulated by government agencies. Primarily applicable in the U.S.
Township	A measurement of land in certain states in the U.S. It measures six miles by six miles and contains 36 square miles.
Tract	A block of acreage or area, as in offshore areas.
Wildcat	An exploratory well
Working interest (WI)	A share, or proportionate ownership of leasehold rights. Working interest can vary up to 100%, and an owner of a working interest shares in all costs, risks, and revenues in the same proportion as his working interest bears to the total working interests in a lease.

Abbreviations

BAF	Barrels per acre-foot. A measure of the amount of oil recovered per unit of volume of reservoir rock, where the unit of volume is expressed as acre-feet
Bbls.	Barrels
Bbls/acre-foot	Barrels recovered per acre-foot
BCF	Billion cubic feet, a measure of gas reserves, or volume
BOPD	Barrels oil per day
DCF	Discounted cash flow, as in DCF rate of return

DPR	Discounted profit-to-investment ratio. A measure of profitability. Sometimes called PVI, present value index
EMV	Expected monetary value
EOL	Expected opportunity loss
EV	Expected value
IP	Initial potential of a well or field, such as Bbls. per day
M	Meaning thousand. 100 M Bbls. = 100,000 Bbls.
MM	Meaning millions. $10MM Profit = $10,000,000 Profit
MCF	Thousand cubic feet—a measure of gas volume
MCF/acre-foot	Gas recovery per acre-foot
NPV	Net present value. The algebraic sum of all cash flows of a project when discounted to time zero at a given discount rate. Also called present worth, or present worth profit
ORI	Overriding royalty interest
PVI	Present value index, same as DPR
PW	Present worth, as in present worth profit. Same term as NPV
QED	That which was to be shown or demonstrated
ROI	Return-on-investment. A profit-to-investment ratio
WI	Working interest

Symbols

i_0	Discounting rate used to compute NPV (net present value) profit. Usually taken to represent the anticipated rate at which future cash revenues can be reinvested. Sometimes called the average opportunity rate.
μ	Greek letter mu. Usually used to denote the mean value of a probability distribution.
σ	Greek letter sigma. Usually used to denote the standard deviation about the mean of a probability distribution.
Σ	Symbol used to denote a summation
$\left(C_x^n\right)$	Symbol which stands for the "combination of n things taken x at a time," and is numerically equal to the number of mutually exclusive ways x outcomes can occur in n trials. It is computed as

$$\left(C_x^n\right) = \frac{n!}{x!(n-x)!}$$

!	Symbol for factorial. n! is defined as the product of all of the integers up to and including n; that is, $n! = 1 \times 2 \times 3 \times 4 \times \ldots \times n$. 0! is defined as 1.
$\therefore$	Symbol used in proofs or derivations to denote the word "therefore"
$<$	Symbol which means less than, as $x < 10$.
$>$	Symbol which means greater than, as $y > x$
$\leq$	Symbol which means less than or equal to
$\geq$	Symbol which means greater than or equal to
P(A)	The probability of event A occurring
P(A + B)	The probability of event A and/or event B occurring
P(AB)	The probability of events A and B occurring
P(A | B)	The conditional probability of event A occurring given that event B has occurred.

Conversion Table

Barrel—unless listed differently is an oil barrel—42 gallons
Gallon—unless listed differently means U.S. gallon

Multiply	By	To Obtain
Acres	4047	Square meters
Acres	0.001562	Square miles
Acres	0.4047	Hectare
Acre-feet	43,560	Cubic feet
Acre-feet	7,758	Barrels-oil
Barrels-oil	42	Gallons
Barrels-oil	5.61	Cubic feet
Barrels-oil	0.159	Cubic meters
Barrels per day (BOPD) (36°API oil)	0.134	Metric tons oil per day
Cubic feet	0.02832	Cubic meters
Cubic feet	7.4805	Gallons
Cubic feet	28.32	Liters
Cubic meters	35.31	Cubic feet
Cubic meters	264.2	Gallons
Cubic meters	6.2897	Barrels-oil
Feet	0.3048	Meters
Gallons	0.1337	Cubic feet
Gallons	0.003785	Cubic meters
Gallons (U.S.)	3.785	Litres
Gallons, Imperial	1.20095	Gallons (U.S.)
Gallons (U.S.)	0.83267	Gallons, Imperial
Hectare	2.471	Acres
Hectare	0.003861	Square miles
Kilograms	2.205	Pounds
Kilometers	0.6214	Miles
Meter	3.281	Feet
Metric tons	2205	Pounds
Metric tons	1000	Kilograms
Metric ton	7.454	Barrels of 36° API oil
Miles	1.609	Kilometers
Pounds	0.4536	Kilograms
Pounds	0.0004536	Metric tons

Pounds per square inch (psi)	0.06804	Atmospheres
Square meters	0.0002471	Acres
Square meters	3.861×10^{-7}	Square miles
Square miles	640	Acres
Square miles	2.59	Square kilometers
Square mile	259	Hectares
Square kilometer	100	Hectares
Square kilometers	0.3861	Square miles
Township*	36	Square miles
Township*	93.24	Square Kilometers

* Township is a land measurement used in many areas of the U.S. It is usually a square tract of land six miles on each side.

Index of
Appendix Sections

THE following pages contain various tables and examples of special computations applicable to petroleum exploration decision analysis. Each appendix section is identified by letter (A, B, C, etc.). At the beginning of each appendix section there is a brief explanation of how the tables of that section are used and interpreted.

APPENDIX A Discount Factors for Annual Compounding and Mid-Year Cash Flows

THE discount factors given in this series of tables have been computed on the basis of annual compounding and the assumption that the cash flows will be received or disbursed at the midpoint of each year. The table is easy to use — simply locate the column having the desired discounting rate and read the discount factors from that column for the appropriate year. For example, to find the 12% discount factor to use to discount revenue received in year 7 read under the "12%" column a value of 0.479. Thus a cash revenue of, say, $10,000 received at the midpoint of year 7 would have a present value (discounted at 12%) of $10,000 × 0.479 = $4,790.

The discount factors in this table were computed from the following relationship:

$$\text{MID-YEAR DISCOUNT FACTOR} = \left[\frac{1}{(1+i)^{n-0.5}}\right]$$

where i is the discount rate expressed as a decimal fraction and n is the year in which the cash flow is received or disbursed.

The factors in this table are frequently used to discount a series of annual cash flows projected for a well or field. In this context the assumption that the entire annual cash flow is received as a lump sum payment at the midpoint of the year is, of course, not an exact analog of the manner in which revenues would be received throughout the year. The assumption is justified on the basis of achieving greatly simplified discounting calculations. This assumption causes only minor differences (as compared to the more precise monthly receipts), particularly at the lower discount rates of about 20% or less.

APPENDIX A
DISCOUNT FACTORS FOR ANNUAL COMPOUNDING AND MID-YEAR CASH FLOW

DISCOUNT RATE (Expressed as percentage)

Year	3%	4%	5%	6%	7%	8%	9%
1	.985	.981	.976	.971	.967	.962	.958
2	.957	.943	.929	.916	.903	.891	.879
3	.929	.907	.885	.864	.844	.825	.806
4	.902	.872	.843	.816	.789	.764	.740
5	.875	.838	.803	.769	.738	.707	.679
6	.850	.806	.765	.726	.689	.655	.623
7	.825	.775	.728	.685	.644	.606	.571
8	.801	.745	.694	.646	.602	.561	.524
9	.778	.717	.661	.609	.563	.520	.481
10	.755	.689	.629	.575	.526	.481	.441
11	.733	.662	.599	.542	.491	.446	.405
12	.712	.637	.571	.512	.459	.413	.371
13	.691	.612	.543	.483	.429	.382	.341
14	.671	.589	.518	.455	.401	.354	.312
15	.651	.566	.493	.430	.375	.328	.287
16	.632	.544	.469	.405	.350	.303	.263
17	.614	.524	.447	.382	.327	.281	.241
18	.596	.503	.426	.361	.306	.260	.221
19	.579	.484	.406	.340	.286	.241	.203
20	.562	.465	.386	.321	.267	.223	.186
21	.546	.448	.368	.303	.250	.206	.171
22	.530	.430	.350	.286	.233	.191	.157
23	.514	.414	.334	.270	.218	.177	.144
24	.499	.398	.318	.254	.204	.164	.132
25	.485	.383	.303	.240	.191	.152	.121
26	.471	.368	.288	.226	.178	.141	.111
27	.457	.354	.274	.214	.166	.130	.102
28	.444	.340	.261	.201	.156	.120	.093
29	.431	.327	.249	.190	.145	.112	.086
30	.418	.314	.237	.179	.136	.103	.079
31	.406	.302	.226	.169	.127	.096	.072
32	.394	.291	.215	.160	.119	.089	.066
33	.383	.280	.205	.151	.111	.082	.061
34	.371	.269	.195	.142	.104	.076	.056
35	.361	.258	.186	.134	.097	.070	.051
36	.350	.248	.177	.126	.091	.065	.047
37	.340	.239	.169	.119	.085	.060	.043
38	.330	.230	.160	.112	.079	.056	.039
39	.320	.221	.153	.106	.074	.052	.036
40	.311	.212	.146	.100	.069	.048	.033
41	.302	.204	.139	.094	.065	.044	.030
42	.293	.196	.132	.089	.060	.041	.028
43	.285	.189	.126	.084	.056	.038	.026
44	.276	.182	.120	.079	.053	.035	.024
45	.268	.175	.114	.075	.049	.033	.022
46	.261	.168	.109	.071	.046	.030	.020
47	.253	.161	.103	.067	.043	.028	.018
48	.246	.155	.099	.063	.040	.026	.017
49	.238	.149	.094	.059	.038	.024	.015
50	.232	.144	.089	.056	.035	.022	.014

580

APPENDIX A
DISCOUNT FACTORS FOR ANNUAL COMPOUNDING AND MID-YEAR CASH FLOW

DISCOUNT RATE (Expressed as percentage)

Year	10%	11%	12%	13%	14%	15%	16%
1	.953	.949	.945	.941	.937	.933	.928
2	.867	.855	.844	.833	.822	.811	.800
3	.788	.770	.753	.737	.721	.705	.690
4	.716	.694	.673	.652	.632	.613	.595
5	.651	.625	.601	.577	.555	.533	.513
6	.592	.563	.536	.511	.486	.464	.442
7	.538	.507	.479	.452	.427	.403	.381
8	.489	.457	.427	.400	.374	.351	.329
9	.445	.412	.382	.354	.328	.305	.283
10	.404	.371	.341	.313	.288	.265	.244
11	.368	.334	.304	.277	.253	.231	.210
12	.334	.301	.272	.245	.222	.200	.181
13	.304	.271	.243	.217	.194	.174	.156
14	.276	.244	.217	.192	.171	.152	.135
15	.251	.220	.193	.170	.150	.132	.116
16	.228	.198	.173	.150	.131	.115	.100
17	.208	.179	.154	.133	.115	.100	.086
18	.189	.161	.138	.118	.101	.087	.074
19	.171	.145	.123	.104	.089	.075	.064
20	.156	.131	.110	.092	.078	.066	.055
21	.142	.118	.098	.082	.068	.057	.048
22	.129	.106	.087	.072	.060	.050	.041
23	.117	.096	.078	.064	.052	.043	.035
24	.106	.086	.070	.057	.046	.037	.031
25	.097	.078	.062	.050	.040	.033	.026
26	.088	.070	.056	.044	.035	.028	.023
27	.080	.063	.050	.039	.031	.025	.020
28	.073	.057	.044	.035	.027	.021	.017
29	.066	.051	.040	.031	.024	.019	.015
30	.060	.046	.035	.027	.021	.016	.013
31	.055	.041	.032	.024	.018	.014	.011
32	.050	.037	.028	.021	.016	.012	.010
33	.045	.034	.025	.019	.014	.011	.008
34	.041	.030	.022	.017	.012	.009	.007
35	.037	.027	.020	.015	.011	.008	.006
36	.034	.025	.018	.013	.010	.007	.005
37	.031	.022	.016	.012	.008	.006	.004
38	.028	.020	.014	.010	.007	.005	.004
39	.025	.018	.013	.009	.006	.005	.003
40	.023	.016	.011	.008	.006	.004	.003
41	.021	.015	.010	.007	.005	.903	.002
42	.019	.013	.009	.006	.004	.003	.002
43	.017	.012	.008	.006	.004	.003	.002
44	.016	.011	.007	.005	.003	.002	.002
45	.014	.010	.006	.004	.003	.002	.001
46	.013	.009	.006	.004	.003	.002	.001
47	.012	.008	.005	.003	.002	.002	.001
48	.011	.007	.005	.003	.002	.001	.001
49	.010	.006	.004	.003	.002	.001	.001
50	.009	.006	.004	.002	.002	.001	.001

DISCOUNT FACTORS FOR ANNUAL COMPOUNDING AND MID-YEAR CASH FLOW

DISCOUNT RATE (Expressed as percentage)

Year	17%	18%	19%	20%	25%	30%	35%
1	.925	.921	.917	.913	.894	.877	.861
2	.790	.780	.770	.761	.716	.675	.638
3	.675	.661	.647	.634	.572	.519	.472
4	.577	.560	.544	.528	.458	.399	.350
5	.493	.475	.457	.440	.366	.307	.259
6	.422	.402	.384	.367	.293	.236	.192
7	.360	.341	.323	.306	.234	.182	.142
8	.308	.289	.271	.255	.188	.140	.105
9	.263	.245	.228	.212	.150	.108	.078
10	.225	.208	.192	.177	.120	.083	.058
11	.192	.176	.161	.147	.096	.064	.043
12	.164	.149	.135	.123	.077	.049	.032
13	.141	.126	.114	.102	.061	.038	.023
14	.120	.107	.096	.085	.049	.029	.017
15	.103	.091	.080	.071	.039	.022	.013
16	.088	.077	.067	.059	.031	.017	.010
17	.075	.065	.057	.049	.025	.013	.007
18	.064	.055	.048	.041	.020	.010	.005
19	.055	.047	.040	.034	.016	.008	.004
20	.047	.040	.034	.029	.013	.006	.003
21	.040	.034	.028	.024	.010	.005	.002
22	.034	.028	.024	.020	.008	.004	.002
23	.029	.024	.020	.017	.007	.003	.001
24	.025	.020	.017	.014	.005	.002	.001
25	.021	.017	.014	.011	.004	.002	.001
26	.018	.015	.102	.010	.003	.001	
27	.016	.012	.010	.008	.003	.001	
28	.013	.011	.008	.007	.002	.001	
29	.011	.009	.007	.006	.002	.001	
30	.010	.008	.006	.005	.001		
31	.008	.006	.005	.004	.001		
32	.007	.005	.004	.003	.001		
33	.006	.005	.004	.003	.001		
34	.005	.004	.003	.002	.001		
35	.004	.003	.002	.002			
36	.004	.003	.002	.002			
37	.003	.002	.002	.001			
38	.003	.002	.001	.001			
39	.002	.002	.001	.001			
40	.002	.001	.001	.001			
41	.002	.001	.001	.001			
42	.001	.001	.001	.001			
43	.001	.001	.001				
44	.001	.001	.001				
45	.001	.001					
46	.001	.001					
47	.001						
48	.001						
49							
50							

APPENDIX A
DISCOUNT FACTORS FOR ANNUAL COMPOUNDING AND MID-YEAR CASH FLOW

DISCOUNT RATE (Expressed as percentage)

Year	40%	45%	50%	55%
1	.845	.830	.817	.803
2	.604	.573	.544	.518
3	.431	.395	.363	.334
4	.308	.272	.242	.216
5	.220	.188	.161	.139
6	.157	.130	.108	.090
7	.112	.089	.072	.058
8	.080	.062	.048	.037
9	.057	.043	.032	.024
10	.041	.029	.021	.016
11	.029	.020	.014	.010
12	.021	.014	.009	.006
13	.015	.010	.006	.004
14	.011	.007	.004	.003
15	.008	.005	.003	.002
16	.005	.003	.002	.001
17	.004	.002	.001	.001
18	.003	.002	.001	
19	.002	.001	.001	
20	.001	.001		
21	.001			
22	.001			
23	.001			
24				

APPENDIX B Discount Factors for Annual Compounding and Year-End Cash Flows

THE discount factors in this table have been computed on the basis of annual compounding and the assumption that the cash flows will be received or disbursed at the end of the year. To use this table locate the column having the desired discount rate and read the discount factor from that column for the appropriate year. For example, to find the 8% discount factor to use to find the present value of revenue received at the end of year 5 (that is, to be received exactly five years hence) locate the column of 8% discount factors. Under this heading read a discount factor of 0.681 corresponding to year 5. Thus, receipt of $10,000 at the end of year 5 has an equivalent present value of $10,000 × 0.681 = $6810.

The discount factors in this table were computed from the following relationship:

$$\text{YEAR-END DISCOUNT FACTOR} = \left[\frac{1}{(1+i)^n} \right]$$

where i is the discount rate expressed as a decimal fraction and n is the year in which the cash flow is received or disbursed.

DISCOUNT FACTORS FOR ANNUAL COMPOUNDING AND YEAR-END CASH FLOWS

DISCOUNT RATE (Expressed as percentage)

Year	3%	4%	5%	6%	7%	8%	9%	10%	11%	12%	13%
1	.971	.962	.952	.943	.935	.926	.917	.909	.901	.893	.885
2	.943	.925	.907	.890	.873	.857	.842	.826	.812	.797	.783
3	.915	.889	.864	.840	.816	.794	.772	.751	.731	.712	.693
4	.888	.855	.823	.792	.763	.735	.708	.683	.659	.636	.613
5	.863	.822	.784	.747	.713	.681	.650	.621	.593	.567	.543
6	.837	.790	.746	.705	.666	.630	.596	.564	.535	.507	.480
7	.813	.760	.711	.665	.623	.583	.547	.513	.482	.452	.425
8	.789	.731	.677	.627	.582	.540	.502	.467	.434	.404	.376
9	.766	.703	.645	.592	.544	.500	.460	.424	.391	.361	.333
10	.744	.676	.614	.558	.508	.463	.422	.386	.352	.322	.295
11	.722	.650	.585	.527	.475	.429	.388	.350	.317	.287	.261
12	.701	.625	.557	.497	.444	.397	.356	.319	.286	.257	.231
13	.681	.601	.530	.469	.415	.368	.326	.290	.258	.229	.204
14	.661	.577	.505	.442	.388	.340	.299	.263	.232	.205	.181
15	.642	.555	.481	.417	.362	.315	.275	.239	.209	.183	.160
16	.623	.534	.458	.394	.339	.292	.252	.218	.188	.163	.141
17	.605	.513	.436	.371	.317	.270	.231	.198	.170	.146	.125
18	.587	.494	.416	.350	.296	.250	.212	.180	.153	.130	.111
19	.570	.475	.396	.331	.277	.232	.194	.164	.138	.116	.098
20	.554	.456	.377	.312	.258	.215	.178	.149	.124	.104	.087
21	.538	.439	.359	.294	.242	.199	.164	.135	.112	.093	.077
22	.522	.422	.342	.278	.226	.184	.150	.123	.101	.083	.068
23	.507	.406	.326	.262	.211	.170	.138	.112	.091	.074	.060
24	.492	.390	.310	.247	.197	.158	.126	.102	.082	.066	.053
25	.478	.375	.295	.233	.184	.146	.116	.092	.074	.059	.047
26	.464	.361	.281	.220	.172	.135	.106	.084	.066	.053	.042
27	.450	.347	.268	.207	.161	.125	.098	.076	.060	.047	.037
28	.437	.333	.255	.196	.150	.116	.090	.069	.054	.042	.033
29	.424	.321	.243	.185	.141	.107	.082	.063	.048	.037	.029
30	.412	.308	.231	.174	.131	.099	.075	.057	.044	.033	.026
31	.400	.296	.220	.164	.123	.092	.069	.052	.039	.030	.023
32	.388	.285	.210	.155	.115	.085	.063	.047	.035	.027	.020
33	.377	.274	.200	.146	.107	.079	.058	.043	.032	.024	.018
34	.366	.264	.190	.138	.100	.073	.053	.039	.029	.021	.016
35	.355	.253	.181	.130	.094	.068	.049	.036	.026	.019	.014
36	.345	.244	.173	.123	.088	.063	.045	.032	.023	.017	.012
37	.335	.234	.164	.116	.082	.058	.041	.029	.021	.015	.011
38	.325	.225	.157	.109	.076	.054	.038	.027	.019	.013	.010
39	.316	.217	.149	.103	.071	.050	.035	.024	.017	.012	.009
40	.307	.208	.142	.097	.067	.046	.032	.022	.015	.011	.008
41	.298	.200	.135	.092	.062	.043	.029	.020	.014	.010	.007
42	.289	.193	.129	.087	.058	.039	.027	.018	.012	.009	.006
43	.281	.185	.123	.082	.055	.037	.025	.017	.011	.008	.005
44	.272	.178	.117	.077	.051	.034	.023	.015	.010	.007	.005
45	.264	.171	.111	.073	.048	.031	.021	.014	.009	.006	.004
46	.257	.165	.106	.069	.044	.029	.019	.012	.008	.005	.004
47	.249	.158	.101	.065	.042	.027	.017	011	.007	.005	.003
48	.242	.152	.096	.061	.039	.025	.016	.010	.007	.004	.003
49	.235	.146	.092	.058	.036	.023	.015	.009	.006	.004	.003
50	.228	.141	.087	.054	.034	.021	.013	.009	.005	.003	.002

DISCOUNT FACTORS FOR ANNUAL COMPOUNDING AND YEAR-END CASH FLOWS

DISCOUNT RATE (Expressed as percentage)

Year	14%	15%	16%	17%	18%	19%	20%	21%	22%	23%	24%	25%
1	.877	.870	.862	.855	.847	.840	.833	.826	.820	.813	.806	.800
2	.769	.756	.743	.731	.718	.706	.694	.683	.672	.661	.650	.640
3	.675	.658	.641	.624	.609	.593	.579	.564	.551	.537	.524	.512
4	.592	.572	.552	.534	.516	.499	.482	.467	.451	.437	.423	.410
5	.519	.497	.476	.456	.437	.419	.402	.386	.370	.355	.341	.328
6	.456	.432	.410	.390	.370	.352	.335	.319	.303	.289	.275	.262
7	.400	.376	.354	.333	.314	.296	.279	.263	.249	.235	.222	.210
8	.351	.327	.305	.285	.266	.249	.233	.218	.204	.191	.179	.168
9	.308	.284	.263	.243	.225	.209	.194	.180	.167	.155	.144	.134
10	.270	.247	.227	.208	.191	.176	.162	.149	.137	.126	.116	.107
11	.237	.215	.195	.178	.162	.148	.135	.123	.112	.103	.094	.086
12	.208	.187	.168	.152	.137	.124	.112	.102	.092	.083	.076	.069
13	.182	.163	.145	.130	.116	.104	.093	.084	.075	.068	.061	.055
14	.160	.141	.125	.111	.099	.088	.078	.069	.062	.055	.049	.044
15	.140	.123	.108	.095	.084	.074	.065	.057	.051	.045	.040	.035
16	.123	.107	.093	.081	.071	.062	.054	.047	.042	.036	.032	.028
17	.108	.093	.080	.069	.060	.052	.045	.039	.034	.030	.026	.023
18	.095	.081	.069	.059	.051	.044	.038	.032	.028	.024	.021	.018
19	.083	.070	.060	.051	.043	.037	.031	.027	.023	.020	.017	.014
20	.073	.061	.051	.043	.037	.031	.026	.022	.019	.016	.014	.012
21	.064	.053	.044	.037	.031	.026	.022	.018	.015	.013	.011	.009
22	.056	.046	.038	.032	.026	.022	.018	.015	.013	.011	.009	.007
23	.049	.040	.033	.027	.022	.018	.015	.012	.010	.009	.007	.006
24	.043	.035	.028	.023	.019	.015	.013	.010	.008	.007	.006	.005
25	.038	.030	.024	.020	.016	.013	.010	.009	.007	.006	.005	.004
26	.033	.026	.021	.017	.014	.011	.009	.007	.006	.005	.004	.003
27	.029	.023	.018	.014	.011	.009	.007	.006	.005	.004	.003	.002
28	.026	.020	.016	.012	.010	.008	.006	.005	.004	.003	.002	.002
29	.022	.017	.014	.011	.008	.006	.005	.004	.003	.002	.002	.002
30	.020	.015	.012	.009	.007	.005	.004	.003	.003	.002	.002	.001
31	.017	.013	.010	.008	.006	.005	.004	.003	.002	.002	.001	.001
32	.015	.011	.009	.007	.005	.004	.003	.002	.002	.001	.001	.001
33	.013	.010	.007	.006	.004	.003	.002	.002	.001	.001	.001	.001
34	.012	.009	.006	.005	.004	.003	.002	.002	.001	.001	.001	.001
35	.010	.008	.006	.004	.003	.002	.002	.001	.001	.001	.001	
36	.009	.007	.005	.004	.003	.002	.001	.001	.001	.001		
37	.008	.006	.004	.003	.002	.002	.001	.001	.001			
38	.007	.005	.004	.003	.002	.001	.001	.001	.001			
39	.006	.004	.003	.002	.002	.001	.001	.001				
40	.005	.004	.003	.002	.001	.001	.001					
41	.005	.003	.002	.002	.001	.001	.001					
42	.004	.003	.002	.001	.001	.001						
43	.004	.002	.002	.001	.001	.001						
44	.003	.002	.001	.001	.001							
45	.003	.002	.001	.001	.001							
46	.002	.002	.001	.001								
47	.002	.001	.001	.001								
48	.002	.001	.001	.001								
49	.002	.001	.001									
50	.001	.001	.001									

APPENDIX B
DISCOUNT FACTORS FOR ANNUAL COMPOUNDING AND YEAR-END CASH FLOWS

DISCOUNT RATE (Expressed as percentage)

Year	26%	27%	28%	29%	30%	35%	40%	45%	50%	55%	60%	
1	.794	.787	.781	.775	.769	.741	.714	.690	.667	.645	.625	
2	.630	.620	.610	.601	.592	.549	.510	.476	.444	.416	.391	
3	.500	.488	.477	.466	.455	.406	.364	.328	.296	.269	.244	
4	.397	.384	.373	.361	.350	.301	.260	.226	.198	.173	.153	
5	.315	.303	.291	.280	.269	.223	.186	.156	.132	.112	.095	
6	.250	.238	.227	.217	.207	.165	.133	.108	.088	.072	.060	
7	.198	.188	.178	.168	.159	.122	.095	.074	.059	.047	.037	
8	.157	.148	.139	.130	.123	.091	.068	.051	.039	.030	.023	
9	.125	.116	.108	.101	.094	.067	.048	.035	.026	.019	.015	
10	.099	.092	.085	.078	.073	.050	.035	.024	.017	.012	.009	
11	.079	.072	.066	.061	.056	.037	.025	.017	.012	.008	.006	
12	.062	.057	.052	.047	.043	.027	.018	.012	.008	.005	.004	
13	.050	.045	.040	.037	.033	.020	.013	.008	.005	.003	.002	
14	.039	.035	.032	.028	.025	.015	.009	.006	.003	.002	.001	
15	.031	.028	.025	.022	.020	.011	.006	.004	.002	.001	.001	
16	.025	.022	.019	.017	.015	.008	.005	.003	.002	.001	.001	
17	.020	.017	.015	.013	.012	.006	.003	.002	.001	.001		
18	.016	.014	.012	.010	.009	.005	.002	.001	.001			
19	.012	.011	.009	.008	.007	.003	.002	.001				
20	.010	.008	.007	.006	.005	.002	.001	.001				
21	.008	.007	.006	.005	.004	.002	.001					
22	.006	.005	.004	.004	.003	.001	.001					
23	.005	.004	.003	.003	.002	.001						
24	.004	.003	.003	.002	.002	.001						
25	.003	.003	.002	.002	.001	.001						
26	.002	.002	.002	.001	.001							
27	.002	.002	.001	.001	.001							
28	.002	.001	.001	.001	.001							
29	.001	.001	.001	.001								
30	.001	.001	.001									
31	.001	.001										
32	.001											
33												
34												
35												
36												
37												
38												
39												
40												
41												
42												
43												
44												
45												
46												
47												
48												
49												
50												

Discount and Compound Interest Factors Based on Continuous Compounding

THIS table can be used for most discounting and compounding calculations that are based on continuous compounding. The table is reproduced with permission from a *World Oil* series by R. P. Shoemaker appearing in the July–September, 1963 issues.

Recall from Chapter 2 that the basic compound interest equation was

$$C(1 + i)^n = S$$

where

$$i = \left(1 + \frac{j}{m}\right)^m - 1$$

The term $(1 + i)^n$ is called a compound interest factor. If the interest period is made increasingly small it can be shown that in the limiting case of compounding continuously (infinite number of interest periods per year) the following equivalence holds:

$$(1 + i)^n \rightarrow e^{jn}$$

where e is the base of the natural logarithms, j is the nominal annual interest rate expressed as a decimal fraction, and n is the length of time in years separating the values C and S.

Thus, for continuous compounding the basic time value of money relationships are:

$$C(e^{jn}) = S \qquad \text{Compound interest equation}$$

$$C = \frac{S}{e^{jn}} = S(e^{-jn}) \qquad \text{Present value equation}$$

Some companies have taken the position that since money is entering and leaving the treasury continuously, continuous compounding is therefore a more representative measure of their firm's time-value of money. Continuous compounding also has some computational advantages over other types of compounding schedules. The tables in this Appendix can be used for a variety of purposes, however only their use for discounting and compounding will be discussed here.

The table is arranged in a series of four columns identified as x, (A), (B), and (C). The term x represents the exponent of the compound interest and discount factor terms and will be equivalent to the product jn. Again, remember that n is the number of years separating the values C and S, and j is the nominal annual interest or discount rate, expressed as a decimal fraction.

Column (A) contains continuous compound interest factors. It is used to convert a cash amount C at time zero to its equivalent value n years in the future for the case where the investment earns interest at an annual (nominal) rate of "j percent" compounded continuously.

EXAMPLE: $1,000 is invested at a nominal rate of 5% compounded continuously. What will the investment be worth 10 years from now?

$$C = \$1,000, \; j = 0.05, \; n = 10 \text{ years}, \; S = ?$$

We use the compound interest equation to determine S:

$$C(e^{jn}) = S$$

To use the tables we must compute the product jn and set it equal to x.

$$jn = x = (0.05)(10) = 0.5$$

Entering the tables for a value of $x = 0.5$ read from Column (A) the term 1.6487. That factor is $e^{jn} = e^{0.5}$, the compound interest factor. Thus

$$(\$1,000)(1.6487) = S = \underline{\$1,648.70}$$

The investment of $1,000 would be worth $1,648.70 at the end of 10 years, compounded continuously at a nominal rate of 5%.

Column (B) is the reciprocal of Column (A) and contains discount factors. It is used to convert a future cash revenue, S, to its equivalent present worth, C, using continuous discounting at a nominal rate of "j percent" per year.

EXAMPLE: $10,000 is received 8 years from now. What is its equivalent present worth discounted continuously at a nominal rate of 15%?

$$S = \$10,000, \; n = 8 \text{ years}, \; j = 0.15, \; C = ?$$

For this calculation we solve the present value equation:

$$C = S(e^{-jn})$$

Again we must convert the product jn to x.

$$jn = x = (0.15)(8) = 1.20$$

Entering the tables for a value of $x = 1.20$ read from Column (B) the term 0.301194, the discount factor. Thus:

$$C = (\$10,000)(0.301194) = \underline{\$3,011.94}$$

When using Columns (A) and (B) the value of n must be in units of years, but it does not have to be an integer number. For instance if we wish to find the present value of a cash flow received exactly $7\frac{1}{4}$ years from now the value of n would be 7.25. It should, of course, be obvious that a specific value of n is required to use Columns (A) and (B), and as such, the factors are applicable to only single value, lump sum cash flows. Columns (A) and (B) cannot be used if the cash flows are received uniformly over a period of time. (See Appendix D for discount factors for uniform annual cash flows.)

Column (C) can be used to find the overall composite discount factor which, when multiplied by the total, accumulated cash flows received at a constant rate over a period of n years, gives the present worth of the total money received.

EXAMPLE: $2,000 per year is received continuously over a period of 6 years. What is the present worth, discounted continuously at a nominal rate of 12%?

Total accumulated revenue: $2,000/year $\times$ 6 years = $12,000

Set the product jn equal to x to use the tables. Here j equals the discount rate, 0.12, and *n is the total length of time* in which the uniform cash flows were received, 6 years.

$$jn = (0.12)(6) = 0.72 = x$$

Entering the tables for a value of $x = 0.72$ read from Column (C) the value 0.712844. The present value of the total $12,000, discounted continuously at 12% is

$$(\$12,000)(0.712844) = \underline{\$8,554.41}$$

Additional Notes about the Exponential Tables of Appendix C

a. The tables can also be used to evaluate logarithms of numbers – a necessary part of trying to compute the value of a number raised to a power without a slide rule. (Example: evaluating $(1 + 0.12)^{7.5}$, the compound interest factor for a $7\frac{1}{2}$ year investment compounded at 12% per year.) The procedures for doing this are given in Appendix I.

b. The tables of Appendix C can also be used to compute total oil production from a well on exponential decline, as well as the total present value of revenues from the production. This is a very useful evaluation shortcut that eliminates the need to break the decline production into a series of annual cash flows. See Appendix I.

c. The discount factors are given here to six decimal places. This accuracy usually exceeds the accuracy of the original cash flow projections. Therefore, it is usually sufficient to just use three decimal places for discount factors obtained in Appendix C.

d. As a matter of interest, the *effective* annual interest rate, i, corresponding to continuous compounding is:

$$i = e^j - 1$$

This compares to Equation 2.3 for computing the effective interest rate for any other type of compounding schedule.

Exponential Functions

x	(A) e^x	(B) e^{-x}	(C) $\dfrac{1-e^{-x}}{x}$
0.00	1.0000	1.000000	1.000000
0.01	1.0101	0.990050	0.995016
0.02	1.0202	0.980199	0.990066
0.03	1.0305	0.970446	0.985149
0.04	1.0408	0.960789	0.980264
0.05	1.0513	0.951229	0.975311
0.06	1.0618	0.941765	0.970591
0.07	1.0725	0.932394	0.965802
0.08	1.0833	0.923116	0.961040
0.09	1.0942	0.913931	0.956320
0.10	1.1052	0.904837	0.951626
0.11	1.1163	0.895834	0.946962
0.12	1.1275	0.886921	0.942330
0.13	1.1388	0.878095	0.937727
0.14	1.1503	0.869358	0.933155
0.15	1.1618	0.860708	0.928614
0.16	1.1735	0.852144	0.924101
0.17	1.1853	0.843665	0.919619
0.18	1.1972	0.835270	0.915166
0.19	1.2092	0.826959	0.910741
0.20	1.2214	0.818731	0.906346
0.21	1.2337	0.810584	0.901980
0.22	1.2461	0.802519	0.897642
0.23	1.2586	0.794534	0.893332
0.24	1.2712	0.786628	0.889051
0.25	1.2840	0.778801	0.884797
0.26	1.2969	0.771052	0.880571
0.27	1.3100	0.763380	0.876372
0.28	1.3231	0.755784	0.872201
0.29	1.3364	0.748264	0.868057
0.30	1.3499	0.740818	0.863939
0.31	1.3634	0.733447	0.859849
0.32	1.3771	0.726149	0.855784
0.33	1.3910	0.718924	0.851746
0.34	1.4049	0.711771	0.847734
0.35	1.4191	0.704688	0.843748
0.36	1.4333	0.697676	0.839788
0.37	1.4477	0.690735	0.835852
0.38	1.4623	0.683862	0.831942
0.39	1.4770	0.677057	0.828059
0.40	1.4918	0.670320	0.824200
0.41	1.5068	0.663650	0.820365
0.42	1.5220	0.657047	0.816555
0.43	1.5373	0.650509	0.812768
0.44	1.5527	0.644037	0.809008
0.45	1.5683	0.637628	0.805271
0.46	1.5841	0.631284	0.801557

x	(A) e^x	(B) e^{-x}	(C) $\dfrac{1-e^{-x}}{x}$
1.00	2.7183	0.367880	0.632121
1.01	2.7456	0.364219	0.629986
1.02	2.7732	0.360595	0.628866
1.03	2.8011	0.357007	0.624265
1.04	2.8292	0.353455	0.621678
1.05	2.8577	0.349938	0.619107
1.06	2.8864	0.346456	0.616551
1.07	2.9154	0.343009	0.614011
1.08	2.9447	0.339596	0.611486
1.09	2.9743	0.336217	0.608976
1.10	3.0042	0.332871	0.606481
1.11	3.0344	0.329559	0.604001
1.12	3.0649	0.326280	0.601536
1.13	3.0957	0.323033	0.599086
1.14	3.1268	0.319819	0.596650
1.15	3.1582	0.316637	0.594229
1.16	3.1899	0.313486	0.591822
1.17	3.2220	0.310367	0.589430
1.18	3.2544	0.307279	0.587052
1.19	3.2871	0.304221	0.584688
1.20	3.3201	0.301194	0.582338
1.21	3.3535	0.298197	0.580002
1.22	3.3872	0.295230	0.577680
1.23	3.4212	0.292293	0.575372
1.24	3.4556	0.289384	0.573077
1.25	3.4903	0.286505	0.570796
1.26	3.5254	0.283654	0.568529
1.27	3.5609	0.280832	0.566274
1.28	3.5966	0.278038	0.564034
1.29	3.6328	0.275271	0.561806
1.30	3.6693	0.272532	0.559591
1.31	3.7062	0.269820	0.557389
1.32	3.7434	0.267136	0.555201
1.33	3.7810	0.264477	0.553025
1.34	3.8190	0.261846	0.550862
1.35	3.8574	0.259240	0.548711
1.36	3.8962	0.256661	0.546573
1.37	3.9353	0.254107	0.544448
1.38	3.9749	0.251579	0.542335
1.39	4.0149	0.249076	0.540234
1.40	4.0552	0.246597	0.538145
1.41	4.0960	0.244144	0.536069
1.42	4.1371	0.241714	0.533952
1.43	4.1787	0.239309	0.531952
1.44	4.2207	0.236928	0.529902
1.45	4.2631	0.234571	0.527883
1.46	4.3060	0.232236	0.525866

x	(A) e^x	(B) e^{-x}	(C) $\dfrac{1-e^{-x}}{x}$
2.00	7.3890	0.135335	0.432332
2.01	7.4456	0.133989	0.430852
2.02	7.5383	0.132656	0.429379
2.03	7.6141	0.131336	0.427914
2.04	7.6906	0.130029	0.426457
2.05	7.7679	0.128735	0.425007
2.06	7.8460	0.127454	0.423566
2.07	7.9248	0.126186	0.422133
2.08	8.0045	0.124930	0.420707
2.09	8.0849	0.123687	0.419289
2.10	8.1662	0.122457	0.417878
2.11	8.2482	0.121238	0.416475
2.12	8.3311	0.120032	0.415060
2.13	8.4149	0.118837	0.413691
2.14	8.4994	0.117655	0.412311
2.15	8.5849	0.116484	0.410938
2.16	8.6711	0.115325	0.409572
2.17	8.7583	0.114178	0.408213
2.18	8.8463	0.113042	0.406862
2.19	8.9352	0.111917	0.405518
2.20	9.0250	0.110803	0.404180
2.21	9.1157	0.109701	0.402850
2.22	9.2073	0.108609	0.401527
2.23	9.2999	0.107528	0.400212
2.24	9.3933	0.106459	0.398903
2.25	9.4877	0.105399	0.397600
2.26	9.5831	0.104351	0.396305
2.27	9.6794	0.103312	0.395017
2.28	9.7767	0.102284	0.393735
2.29	9.8749	0.101267	0.392460
2.30	9.9742	0.100259	0.391192
2.31	10.0744	0.099261	0.389930
2.32	10.1757	0.098274	0.388675
2.33	10.2779	0.097296	0.387427
2.34	10.3812	0.096328	0.386185
2.35	10.4856	0.095369	0.384949
2.36	10.5909	0.094420	0.383720
2.37	10.6974	0.093481	0.382498
2.38	10.8049	0.092551	0.381281
2.39	10.9135	0.091630	0.380071
2.40	11.0232	0.090718	0.378868
2.41	11.1340	0.089815	0.377670
2.42	11.2459	0.088922	0.376479
2.43	11.3589	0.088037	0.375293
2.44	11.4725	0.087161	0.374114
2.45	11.5683	0.086294	0.372941
2.46	11.7048	0.085435	0.371774

Table of e^x, e^{-x}, and $(1-e^{-x})/x$

x	e^x	e^{-x}	$(1-e^{-x})/x$
0.50	1.6487	0.606531	0.786939
0.51	1.6653	0.600496	0.783342
0.52	1.6820	0.594521	0.779767
0.53	1.6989	0.588605	0.776217
0.54	1.7160	0.582748	0.772688
0.55	1.7333	0.576950	0.769182
0.56	1.7507	0.571209	0.765698
0.57	1.7683	0.565525	0.762237
0.58	1.7860	0.559898	0.758797
0.59	1.8040	0.554327	0.755378
0.60	1.8221	0.548812	0.751980
0.61	1.8404	0.543351	0.748605
0.62	1.8589	0.537944	0.745252
0.63	1.8776	0.532592	0.741918
0.64	1.8965	0.527292	0.738606
0.65	1.9155	0.522046	0.735314
0.66	1.9348	0.516851	0.732044
0.67	1.9542	0.511709	0.728793
0.68	1.9739	0.506617	0.725563
0.69	1.9937	0.501576	0.722354
0.70	2.0138	0.496585	0.719164
0.71	2.0340	0.491644	0.715994
0.72	2.0544	0.486752	0.712844
0.73	2.0751	0.481909	0.709714
0.74	2.0959	0.477114	0.706603
0.75	2.1170	0.472367	0.703511
0.76	2.1383	0.467666	0.700439
0.77	2.1598	0.463013	0.697386
0.78	2.1815	0.458406	0.694351
0.79	2.2034	0.453845	0.691336
0.80	2.2255	0.449329	0.688339
0.81	2.2479	0.444858	0.685360
0.82	2.2705	0.440432	0.682400
0.83	2.2933	0.436049	0.679457
0.84	2.3164	0.431711	0.676535
0.85	2.3396	0.427415	0.673630
0.86	2.3632	0.423162	0.670742
0.87	2.3869	0.418952	0.667872
0.88	2.4109	0.414783	0.665020
0.89	2.4351	0.410656	0.662185
0.90	2.4596	0.406570	0.659367
0.91	2.4843	0.402524	0.656567
0.92	2.5093	0.398519	0.653784
0.93	2.5345	0.394554	0.651018
0.94	2.5600	0.390628	0.648268
0.95	2.5857	0.386741	0.645536
0.96	2.6117	0.382893	0.642820
0.97	2.6379	0.379083	0.640121
0.98	2.6645	0.375311	0.637438
0.99	2.6912	0.371577	0.634771
1.00	2.7183	0.367880	0.632121

x	e^x	e^{-x}	$(1-e^{-x})/x$
1.50	4.4817	0.223130	0.517913
1.51	4.5267	0.220910	0.515954
1.52	4.5722	0.218712	0.514005
1.53	4.6182	0.216536	0.512065
1.54	4.6646	0.214381	0.510142
1.55	4.7115	0.212248	0.508227
1.56	4.7588	0.210136	0.506323
1.57	4.8066	0.208045	0.504430
1.58	4.8550	0.205975	0.502647
1.59	4.9037	0.203926	0.500676
1.60	4.9530	0.201897	0.498815
1.61	5.0028	0.199888	0.496964
1.62	5.0531	0.197899	0.495124
1.63	5.1039	0.195930	0.493295
1.64	5.1552	0.193980	0.491476
1.65	5.2070	0.192050	0.489667
1.66	5.2593	0.190139	0.487868
1.67	5.3122	0.188247	0.486080
1.68	5.3656	0.186374	0.484301
1.69	5.4195	0.184520	0.482533
1.70	5.4739	0.182684	0.480774
1.71	5.5290	0.180866	0.479026
1.72	5.5845	0.179066	0.477287
1.73	5.6407	0.177285	0.475558
1.74	5.6973	0.175521	0.473839
1.75	5.7546	0.173774	0.472129
1.76	5.8124	0.172045	0.470429
1.77	5.8709	0.170333	0.468737
1.78	5.9299	0.168638	0.467057
1.79	5.9895	0.166960	0.465385
1.80	6.0496	0.165299	0.463723
1.81	6.1104	0.163654	0.462070
1.82	6.1719	0.162026	0.460426
1.83	6.2339	0.160414	0.458791
1.84	6.2965	0.158818	0.457156
1.85	6.3598	0.157237	0.455546
1.86	6.4237	0.155673	0.453940
1.87	6.4883	0.154124	0.452340
1.88	6.5535	0.152590	0.450750
1.89	6.6194	0.151072	0.449168
1.90	6.6859	0.149569	0.447596
1.91	6.7531	0.148081	0.446031
1.92	6.8210	0.146607	0.444475
1.93	6.8895	0.145148	0.442929
1.94	6.9588	0.143704	0.441390
1.95	7.0287	0.142274	0.439860
1.96	7.0993	0.140859	0.438338
1.97	7.1707	0.139457	0.436824
1.98	7.2427	0.138069	0.435319
1.99	7.3155	0.136696	0.433822
2.00	7.3890	0.135335	0.432332

x	e^x	e^{-x}	$(1-e^{-x})/x$
2.50	12.1825	0.082085	0.367166
2.51	12.3049	0.081268	0.366029
2.52	12.4286	0.080460	0.364897
2.53	12.5535	0.079659	0.363771
2.54	12.6797	0.078866	0.362651
2.55	12.8071	0.078082	0.361537
2.56	12.9358	0.077305	0.360428
2.57	13.0658	0.076536	0.359325
2.58	13.1971	0.075774	0.358227
2.59	13.3298	0.075020	0.357135
2.60	13.4637	0.074274	0.356049
2.61	13.5990	0.073535	0.354966
2.62	13.7357	0.072803	0.353892
2.63	13.8738	0.072079	0.352824
2.64	14.0132	0.071361	0.351757
2.65	14.1540	0.070651	0.350698
2.66	14.2963	0.069948	0.349644
2.67	14.4400	0.069252	0.348595
2.68	14.5851	0.068563	0.347551
2.69	14.7317	0.067881	0.346513
2.70	14.8797	0.067206	0.345479
2.71	15.0293	0.066537	0.344451
2.72	15.1803	0.065875	0.343428
2.73	15.3329	0.065219	0.342411
2.74	15.4870	0.064570	0.341398
2.75	15.6426	0.063926	0.340390
2.76	15.7998	0.063292	0.339387
2.77	15.9586	0.062662	0.338389
2.78	16.1190	0.062039	0.337396
2.79	16.2810	0.061421	0.336408
2.80	16.4446	0.060810	0.335425
2.81	16.6099	0.060205	0.334447
2.82	16.7768	0.059606	0.333473
2.83	16.9455	0.059013	0.332504
2.84	17.1158	0.058426	0.331540
2.85	17.2878	0.057844	0.330581
2.86	17.4615	0.057269	0.329626
2.87	17.6370	0.056699	0.328676
2.88	17.8143	0.056135	0.327731
2.89	17.9933	0.055576	0.326790
2.90	18.1741	0.055023	0.325854
2.91	18.3566	0.054476	0.324922
2.92	18.5413	0.053934	0.323995
2.93	18.7276	0.053397	0.323073
2.94	18.9158	0.052866	0.322155
2.95	19.1060	0.052340	0.321241
2.96	19.2960	0.051819	0.320331
2.97	19.4919	0.051303	0.319426
2.98	19.6878	0.050793	0.318526
2.99	19.8857	0.050287	0.317630
3.00	20.0855	0.049787	0.316738

x	(A) e^x	(B) e^{-x}	(C) $\dfrac{1-e^{-x}}{x}$
3.00	20.0855	0.049787	0.316738
3.01	20.2874	0.049292	0.315850
3.02	20.4913	0.048801	0.314966
3.03	20.6972	0.048316	0.314087
3.04	20.9052	0.047835	0.313212
3.05	21.1153	0.047359	0.312341
3.06	21.3276	0.046888	0.311475
3.07	21.5419	0.046421	0.310612
3.08	21.7584	0.045959	0.309753
3.09	21.9771	0.045502	0.308899
3.10	22.1979	0.045049	0.308049
3.11	22.4210	0.044601	0.307202
3.12	22.6464	0.044157	0.306360
3.13	22.8740	0.043718	0.305521
3.14	23.1039	0.043283	0.304687
3.15	23.3361	0.042852	0.303856
3.16	23.5706	0.042426	0.303030
3.17	23.8075	0.042004	0.302207
3.18	24.0467	0.041586	0.301388
3.19	24.2884	0.041172	0.300573
3.20	24.5325	0.040762	0.299762
3.21	24.7791	0.040357	0.298954
3.22	25.0281	0.039955	0.298151
3.23	25.2796	0.039558	0.297351
3.24	25.5337	0.039164	0.296554
3.25	25.7903	0.038774	0.295762
3.26	26.0495	0.038388	0.294973
3.27	26.3113	0.038005	0.294188
3.28	26.5758	0.037628	0.293406
3.29	26.8429	0.037254	0.292628
3.30	27.1126	0.036883	0.291854
3.31	27.3851	0.036516	0.291083
3.32	27.6603	0.036153	0.290315
3.33	27.9383	0.035793	0.289552
3.34	28.2191	0.035437	0.288791
3.35	28.5027	0.035084	0.288035
3.36	28.7892	0.034735	0.287281
3.37	29.0785	0.034390	0.286531
3.38	29.3708	0.034047	0.285785
3.39	29.6659	0.033709	0.285042
3.40	29.9641	0.033373	0.284302
3.41	30.2652	0.033041	0.283566
3.42	30.5694	0.032712	0.282833
3.43	30.8766	0.032387	0.282103
3.44	31.1869	0.032065	0.281377
3.45	31.5004	0.031746	0.280653
3.46	31.8170	0.031430	0.279934

x	(A) e^x	(B) e^{-x}	(C) $\dfrac{1-e^{-x}}{x}$
4.00	54.5981	0.018316	0.245621
4.01	55.1468	0.018133	0.244855
4.02	55.7011	0.017953	0.244290
4.03	56.2609	0.017778	0.243728
4.04	56.8263	0.017597	0.243169
4.05	57.3974	0.017422	0.242612
4.06	57.9743	0.017249	0.242104
4.07	58.5569	0.017077	0.241504
4.08	59.1455	0.016907	0.240955
4.09	59.7399	0.016739	0.240406
4.10	60.3403	0.016573	0.239860
4.11	60.9467	0.016408	0.239317
4.12	61.5592	0.016245	0.238776
4.13	62.1779	0.016083	0.238237
4.14	62.8028	0.015923	0.237702
4.15	63.4340	0.015764	0.237165
4.16	64.0715	0.015605	0.236633
4.17	64.7155	0.015452	0.236103
4.18	65.3659	0.015299	0.235574
4.19	66.0228	0.015146	0.235049
4.20	66.6864	0.014996	0.234525
4.21	67.3566	0.014846	0.234003
4.22	68.0335	0.014698	0.233484
4.23	68.7173	0.014552	0.232966
4.24	69.4079	0.014408	0.232451
4.25	70.1055	0.014264	0.231938
4.26	70.8100	0.014122	0.231427
4.27	71.5217	0.013982	0.230918
4.28	72.2405	0.013843	0.230411
4.29	72.9665	0.013705	0.229906
4.30	73.6999	0.013569	0.229403
4.31	74.4406	0.013434	0.228902
4.32	75.1887	0.013300	0.228403
4.33	75.9444	0.013168	0.227906
4.34	76.7076	0.013037	0.227411
4.35	77.4786	0.012907	0.226916
4.36	78.2572	0.012778	0.226427
4.37	79.0477	0.012651	0.225940
4.38	79.8381	0.012525	0.225451
4.39	80.6405	0.012401	0.224966
4.40	81.4510	0.012277	0.224492
4.41	82.2696	0.012155	0.224001
4.42	83.0964	0.012034	0.223522
4.43	83.9315	0.011914	0.223044
4.44	84.7751	0.011796	0.222566
4.45	85.6271	0.011679	0.222095
4.46	86.4677	0.011562	0.221623

x	(A) e^x	(B) e^{-x}	(C) $\dfrac{1-e^{-x}}{x}$
5.00	148.4132	0.006738	0.198652
5.01	149.9050	0.006671	0.198269
5.02	151.4116	0.006605	0.197887
5.03	152.9333	0.006539	0.197507
5.04	154.4703	0.006474	0.197126
5.05	156.0228	0.006409	0.196751
5.06	157.5905	0.006346	0.196374
5.07	159.1747	0.006282	0.195999
5.08	160.7744	0.006220	0.195626
5.09	162.3902	0.006158	0.195254
5.10	164.0223	0.006097	0.194863
5.11	165.6707	0.006036	0.194513
5.12	167.3358	0.005976	0.194145
5.13	169.0175	0.005917	0.193778
5.14	170.7162	0.005858	0.193413
5.15	172.4319	0.005799	0.193049
5.16	174.1649	0.005742	0.192686
5.17	175.9153	0.005685	0.192324
5.18	177.6833	0.005628	0.191964
5.19	179.4691	0.005572	0.191605
5.20	181.2728	0.005517	0.191247
5.21	183.0946	0.005462	0.190890
5.22	184.9347	0.005407	0.190535
5.23	186.7934	0.005354	0.190181
5.24	188.6707	0.005300	0.189828
5.25	190.5669	0.005247	0.189477
5.26	192.4621	0.005195	0.189126
5.27	194.4166	0.005144	0.188777
5.28	196.3705	0.005092	0.188429
5.29	198.3441	0.005042	0.188083
5.30	200.3375	0.004992	0.187737
5.31	202.3509	0.004942	0.187393
5.32	204.3846	0.004893	0.187050
5.33	206.4387	0.004844	0.186709
5.34	208.5134	0.004796	0.186368
5.35	210.6091	0.004748	0.186028
5.36	212.7257	0.004701	0.185690
5.37	214.8637	0.004654	0.185353
5.38	217.0231	0.004608	0.185017
5.39	219.2042	0.004562	0.184682
5.40	221.4073	0.004517	0.184349
5.41	223.6325	0.004472	0.184016
5.42	225.8802	0.004427	0.183685
5.43	228.1502	0.004383	0.183355
5.44	230.4431	0.004339	0.183026
5.45	232.7591	0.004296	0.182698

3.49	32.7859	0.030501	0.277793					5.49	242.2582	0.004128	0.181397
3.50	33.1154	0.030197	0.277066	4.50	90.0171	0.011109	0.219754	5.50	244.6924	0.004087	0.181075
3.51	33.4483	0.029897	0.276363	4.51	90.9218	0.010998	0.219291	5.51	247.1516	0.004047	0.180754
3.52	33.7844	0.029599	0.275662	4.52	91.8356	0.010889	0.218830	5.52	249.6355	0.004006	0.180434
3.53	34.1240	0.029305	0.274984	4.53	92.7585	0.010781	0.218371	5.53	252.1444	0.003966	0.180115
3.54	34.4669	0.029013	0.274290	4.54	93.6908	0.010673	0.217913	5.54	254.6785	0.003927	0.179797
3.55	34.8133	0.028725	0.273599	4.55	94.6324	0.010567	0.217458	5.55	257.2381	0.003887	0.179480
3.56	35.1632	0.028439	0.272910	4.56	95.5835	0.010462	0.217004	5.56	259.8234	0.003849	0.179164
3.57	35.5166	0.028156	0.272225	4.57	96.5441	0.010356	0.216552	5.57	262.4347	0.003810	0.178849
3.58	35.8735	0.027876	0.271543	4.58	97.5144	0.010255	0.216102	5.58	265.0722	0.003773	0.178535
3.59	36.2341	0.027598	0.270864	4.59	98.4944	0.010153	0.215653	5.59	267.7362	0.003735	0.178223
3.60	36.5982	0.027324	0.270188	4.60	99.4843	0.010052	0.215206	5.60	270.4270	0.003698	0.177911
3.61	36.9660	0.027052	0.269515	4.61	100.4841	0.009952	0.214761	5.61	273.1449	0.003661	0.177600
3.62	37.3376	0.026783	0.268845	4.62	101.4940	0.009853	0.214318	5.62	275.8901	0.003625	0.177291
3.63	37.7128	0.026516	0.268177	4.63	102.5141	0.009755	0.213876	5.63	278.6628	0.003589	0.176982
3.64	38.0918	0.026252	0.267513	4.64	103.5444	0.009658	0.213436	5.64	281.4634	0.003553	0.176675
3.65	38.4747	0.025991	0.266852	4.65	104.5850	0.009562	0.212997	5.65	284.2922	0.003518	0.176368
3.66	38.8613	0.025733	0.266193	4.66	105.6361	0.009466	0.212561	5.66	287.1494	0.003483	0.176063
3.67	39.2519	0.025476	0.265538	4.67	106.6978	0.009372	0.212126	5.67	290.0353	0.003448	0.175759
3.68	39.6464	0.025223	0.264885	4.68	107.7701	0.009279	0.211692	5.68	292.9502	0.003414	0.175455
3.69	40.0448	0.024972	0.264235	4.69	108.8532	0.009187	0.211261	5.69	295.8944	0.003380	0.175153
3.70	40.4473	0.024724	0.263588	4.70	109.9472	0.009095	0.210831	5.70	298.8682	0.003346	0.174851
3.71	40.8538	0.024478	0.262944	4.71	111.0522	0.009005	0.210402	5.71	301.8719	0.003313	0.174551
3.72	41.2644	0.024234	0.262303	4.72	112.1683	0.008915	0.209976	5.72	304.9058	0.003280	0.174252
3.73	41.6791	0.023993	0.261664	4.73	113.2956	0.008826	0.209550	5.73	307.9702	0.003247	0.173953
3.74	42.0980	0.023754	0.261028	4.74	114.4343	0.008739	0.209127	5.74	311.0654	0.003215	0.173656
3.75	42.5211	0.023518	0.260395	4.75	115.5844	0.008652	0.208705	5.75	314.1916	0.003183	0.173359
3.76	42.9484	0.023284	0.259765	4.76	116.7460	0.008566	0.208284	5.76	317.3493	0.003151	0.173064
3.77	43.3801	0.023052	0.259137	4.77	117.9193	0.008480	0.207866	5.77	320.5388	0.003120	0.172769
3.78	43.8160	0.022823	0.258513	4.78	119.1044	0.008396	0.207448	5.78	323.7602	0.003089	0.172476
3.79	44.2564	0.022596	0.257890	4.79	120.3015	0.008312	0.207033	5.79	327.0141	0.003058	0.172183
3.80	44.7012	0.022371	0.257271	4.80	121.5105	0.008230	0.206619	5.80	330.3007	0.003028	0.171892
3.81	45.1504	0.022148	0.256654	4.81	122.7317	0.008148	0.206206	5.81	333.6203	0.002997	0.171601
3.82	45.6042	0.021928	0.256040	4.82	123.9652	0.008067	0.205795	5.82	336.9732	0.002968	0.171311
3.83	46.0625	0.021710	0.255428	4.83	125.2111	0.007987	0.205386	5.83	340.3599	0.002938	0.171023
3.84	46.5255	0.021494	0.254819	4.84	126.4695	0.007907	0.204978	5.84	343.7806	0.002909	0.170735
3.85	46.9930	0.021280	0.254213	4.85	127.7405	0.007828	0.204571	5.85	347.2356	0.002880	0.170448
3.86	47.4653	0.021068	0.253609	4.86	129.0243	0.007750	0.204167	5.86	350.7254	0.002851	0.170162
3.87	47.9424	0.020858	0.253008	4.87	130.3211	0.007673	0.203763	5.87	354.2503	0.002823	0.169877
3.88	48.4242	0.020651	0.252410	4.88	131.6308	0.007597	0.203361	5.88	357.8106	0.002795	0.169593
3.89	48.9109	0.020445	0.251814	4.89	132.9537	0.007521	0.202961	5.89	361.4067	0.002767	0.169309
3.90	49.4024	0.020242	0.251220	4.90	134.2900	0.007447	0.202562	5.90	365.0389	0.002739	0.169027
3.91	49.8989	0.020041	0.250629	4.91	135.6396	0.007372	0.202164	5.91	368.7076	0.002712	0.168746
3.92	50.4004	0.019841	0.250041	4.92	137.0028	0.007299	0.201768	5.92	372.4132	0.002685	0.168465
3.93	50.9070	0.019644	0.249455	4.93	138.3797	0.007226	0.201374	5.93	376.1560	0.002658	0.168186
3.94	51.4186	0.019449	0.248871	4.94	139.7705	0.007155	0.200981	5.94	379.9365	0.002632	0.167907
3.95	51.9353	0.019255	0.248290	4.95	141.1752	0.007083	0.200589	5.95	383.7549	0.002606	0.167629
3.96	52.4573	0.019063	0.247711	4.96	142.5940	0.007013	0.200199	5.96	387.6117	0.002580	0.167352
3.97	52.9845	0.018873	0.247135	4.97	144.0271	0.006943	0.199810	5.97	391.5073	0.002554	0.167076
3.98	53.5170	0.018686	0.246561	4.98	145.4746	0.006874	0.199423	5.98	395.4420	0.002529	0.166801
3.99	54.0549	0.018500	0.245990	4.99	146.9367	0.006806	0.199037	5.99	399.4163	0.002504	0.166527
4.00	54.5981	0.018316	0.245421	5.00	148.4134	0.006738	0.198652	6.00	403.4305	0.002479	0.166253

Exponential Functions

x	(A) e^x	(B) e^{-x}	(C) $\dfrac{1-e^{-x}}{x}$
6.00	403.4305	0.002479	0.166253
6.01	407.4851	0.002454	0.165981
6.02	411.5804	0.002430	0.165709
6.03	415.7169	0.002405	0.165438
6.04	419.8949	0.002382	0.165166
6.05	424.1150	0.002358	0.164899
6.06	428.3774	0.002334	0.164631
6.07	432.6827	0.002311	0.164364
6.08	437.0313	0.002288	0.164097
6.09	441.4235	0.002265	0.163831
6.10	445.8599	0.002243	0.163567
6.11	450.3409	0.002221	0.163303
6.12	454.8669	0.002198	0.163039
6.13	459.4385	0.002177	0.162777
6.14	464.0559	0.002155	0.162515
6.15	468.7198	0.002133	0.162255
6.16	473.4305	0.002112	0.161995
6.17	478.1886	0.002091	0.161735
6.18	482.9945	0.002070	0.161477
6.19	487.8487	0.002050	0.161220
6.20	492.7517	0.002029	0.160963
6.21	497.7039	0.002009	0.160707
6.22	502.7060	0.001989	0.160452
6.23	507.7583	0.001969	0.160197
6.24	512.8614	0.001950	0.159944
6.25	518.0157	0.001930	0.159691
6.26	523.2219	0.001911	0.159439
6.27	528.4804	0.001892	0.159188
6.28	533.7917	0.001873	0.158937
6.29	539.1564	0.001855	0.158687
6.30	544.5751	0.001836	0.158439
6.31	550.0462	0.001818	0.158190
6.32	555.5763	0.001800	0.157943
6.33	561.1600	0.001782	0.157696
6.34	566.7997	0.001764	0.157450
6.35	572.4962	0.001747	0.157205
6.36	578.2499	0.001729	0.156961
6.37	584.0614	0.001712	0.156717
6.38	589.9314	0.001695	0.156474
6.39	595.8603	0.001678	0.156232
6.40	601.8488	0.001662	0.155990
6.41	607.8975	0.001645	0.155749
6.42	614.0070	0.001629	0.155509
6.43	620.1779	0.001612	0.155270
6.44	626.4109	0.001596	0.155031
6.45	632.7064	0.001581	0.154794

x	(A) e^x	(B) e^{-x}	(C) $\dfrac{1-e^{-x}}{x}$
7.00	1096.6606	0.000912	0.142727
7.01	1107.6620	0.000903	0.142524
7.02	1118.7642	0.000894	0.142323
7.03	1130.0382	0.000885	0.142121
7.04	1141.3953	0.000876	0.141921
7.05	1152.6865	0.000867	0.141721
7.06	1164.4530	0.000859	0.141521
7.07	1176.1559	0.000850	0.141322
7.08	1187.9765	0.000842	0.141124
7.09	1199.9158	0.000833	0.140926
7.10	1211.9732	0.000825	0.140729
7.11	1224.1557	0.000817	0.140532
7.12	1236.4687	0.000809	0.140336
7.13	1248.8853	0.000801	0.140140
7.14	1261.4368	0.000793	0.139944
7.15	1274.1144	0.000785	0.139750
7.16	1286.9195	0.000777	0.139556
7.17	1299.8332	0.000769	0.139362
7.18	1312.9189	0.000762	0.139170
7.19	1326.1120	0.000754	0.138977
7.20	1339.4396	0.000747	0.138785
7.21	1352.9012	0.000739	0.138594
7.22	1366.4980	0.000732	0.138403
7.23	1380.2316	0.000725	0.138212
7.24	1394.1031	0.000717	0.138022
7.25	1408.1141	0.000710	0.137833
7.26	1422.2658	0.000703	0.137644
7.27	1436.5598	0.000696	0.137456
7.28	1450.9975	0.000689	0.137268
7.29	1465.5802	0.000682	0.137080
7.30	1480.3095	0.000676	0.136894
7.31	1495.1869	0.000669	0.136707
7.32	1510.2137	0.000662	0.136521
7.33	1525.3916	0.000656	0.136336
7.34	1540.7221	0.000649	0.136151
7.35	1556.2066	0.000643	0.135967
7.36	1571.8467	0.000636	0.135782
7.37	1587.6440	0.000630	0.135600
7.38	1603.8001	0.000624	0.135417
7.39	1619.7165	0.000617	0.135235
7.40	1635.9849	0.000611	0.135052
7.41	1652.4369	0.000605	0.134871
7.42	1669.0042	0.000599	0.134690
7.43	1685.8183	0.000593	0.134510
7.44	1702.7610	0.000587	0.134330
7.45	1719.8741	0.000581	0.134150

x	(A) e^x	(B) e^{-x}	(C) $\dfrac{1-e^{-x}}{x}$
8.00	2980.9764	0.000335	0.124958
8.01	3010.9361	0.000332	0.124802
8.02	3041.1968	0.000329	0.124647
8.03	3071.7617	0.000326	0.124492
8.04	3102.6337	0.000322	0.124338
8.05	3133.8161	0.000319	0.124184
8.06	3165.3118	0.000316	0.124030
8.07	3197.1241	0.000313	0.123877
8.08	3229.2560	0.000310	0.123724
8.09	3261.7110	0.000307	0.123571
8.10	3294.4921	0.000304	0.123419
8.11	3327.6026	0.000301	0.123267
8.12	3361.0460	0.000298	0.123116
8.13	3394.8254	0.000295	0.122965
8.14	3428.9444	0.000292	0.122814
8.15	3463.4062	0.000289	0.122664
8.16	3498.2144	0.000286	0.122514
8.17	3533.3724	0.000283	0.122364
8.18	3568.8836	0.000280	0.122215
8.19	3604.7520	0.000277	0.122066
8.20	3640.9808	0.000275	0.121918
8.21	3677.5737	0.000272	0.121769
8.22	3714.5343	0.000269	0.121622
8.23	3751.8665	0.000267	0.121474
8.24	3789.5738	0.000264	0.121327
8.25	3827.6600	0.000261	0.121180
8.26	3866.1291	0.000259	0.121034
8.27	3904.9847	0.000256	0.120888
8.28	3944.2309	0.000254	0.120742
8.29	3983.8715	0.000251	0.120597
8.30	4023.9105	0.000249	0.120452
8.31	4064.3519	0.000246	0.120307
8.32	4105.1998	0.000244	0.120163
8.33	4146.4582	0.000241	0.120019
8.34	4188.1312	0.000239	0.119875
8.35	4230.2231	0.000236	0.119732
8.36	4272.7380	0.000234	0.119589
8.37	4315.6802	0.000232	0.119446
8.38	4359.0540	0.000229	0.119304
8.39	4402.8637	0.000227	0.119162
8.40	4447.1137	0.000225	0.119021
8.41	4491.8085	0.000223	0.118879
8.42	4536.9524	0.000220	0.118739
8.43	4582.5500	0.000218	0.118598
8.44	4628.6059	0.000216	0.118458
8.45	4675.1247	0.000214	0.118318

Numerical table (four-figure argument with three function columns per argument range). Arguments run from 6.49 to 9.00.

x = 8.49 – 9.00

x			
8.49	0.117761	0.000206	4865.9223
8.50	0.117623	0.000203	4914.7992
8.51	0.117485	0.000201	4964.1943
8.52	0.117347	0.000199	5014.0059
8.53	0.117210	0.000197	5064.4768
8.54	0.117073	0.000195	5115.3762
8.55	0.116936	0.000194	5166.7892
8.56	0.116800	0.000192	5218.7169
8.57	0.116664	0.000190	5271.1664
8.58	0.116526	0.000188	5324.1431
8.59	0.116395	0.000186	5377.6522
8.60	0.116259	0.000184	5431.6992
8.61	0.116123	0.000182	5486.2899
8.62	0.115986	0.000180	5541.4279
8.63	0.115854	0.000179	5597.1208
8.64	0.115720	0.000177	5653.3725
8.65	0.115587	0.000175	5710.1914
8.66	0.115453	0.000173	5767.5804
8.67	0.115320	0.000172	5825.5462
8.68	0.115188	0.000170	5884.0946
8.69	0.115055	0.000168	5943.2314
8.70	0.114923	0.000167	6002.9625
8.71	0.114792	0.000165	6063.2939
8.72	0.114660	0.000163	6124.2313
8.73	0.114529	0.000162	6185.7819
8.74	0.114398	0.000160	6247.9508
8.75	0.114267	0.000158	6310.7444
8.76	0.114137	0.000157	6374.1692
8.77	0.114007	0.000155	6438.2313
8.78	0.113878	0.000154	6502.9373
8.79	0.113748	0.000152	6568.2937
8.80	0.113619	0.000151	6634.3069
8.81	0.113490	0.000149	6700.9835
8.82	0.113361	0.000148	6768.3303
8.83	0.113234	0.000146	6836.3538
8.84	0.113106	0.000145	6905.0611
8.85	0.112978	0.000143	6974.4589
8.86	0.112851	0.000142	7044.5541
8.87	0.112724	0.000141	7115.3538
8.88	0.112597	0.000139	7186.8651
8.89	0.112470	0.000138	7259.0952
8.90	0.112344	0.000136	7332.0510
8.91	0.112218	0.000135	7405.7402
8.92	0.112092	0.000134	7480.1698
8.93	0.111967	0.000132	7555.3477
8.94	0.111842	0.000131	7631.2811
8.95	0.111717	0.000130	7707.9775
8.96	0.111593	0.000128	7785.4449
8.97	0.111468	0.000127	7863.6907
8.98	0.111344	0.000126	7942.7230
8.99	0.111221	0.000125	8022.5497
9.00	0.111097	0.000123	8103.1785

x = 7.49 – 8.00

x			
7.49	0.133437	0.000559	1790.0634
7.50	0.133259	0.000553	1808.0547
7.51	0.133081	0.000548	1826.2259
7.52	0.132907	0.000542	1844.5798
7.53	0.132731	0.000537	1863.1181
7.54	0.132555	0.000531	1881.8427
7.55	0.132381	0.000526	1900.7555
7.56	0.132206	0.000521	1919.8584
7.57	0.132032	0.000516	1939.1533
7.58	0.131858	0.000511	1958.6421
7.59	0.131686	0.000505	1978.3268
7.60	0.131513	0.000500	1998.2093
7.61	0.131341	0.000495	2018.2916
7.62	0.131169	0.000491	2038.5757
7.63	0.130998	0.000486	2059.0637
7.64	0.130827	0.000481	2079.7576
7.65	0.130657	0.000476	2100.6595
7.66	0.130487	0.000471	2121.7715
7.67	0.130317	0.000467	2143.0956
7.68	0.130148	0.000462	2164.6341
7.69	0.129979	0.000457	2186.3890
7.70	0.129811	0.000453	2208.3625
7.71	0.129643	0.000448	2230.5569
7.72	0.129476	0.000444	2252.9974
7.73	0.129309	0.000439	2275.6171
7.74	0.129143	0.000435	2298.4674
7.75	0.128977	0.000431	2321.5676
7.76	0.128811	0.000426	2344.9199
7.77	0.128646	0.000422	2368.4867
7.78	0.128481	0.000418	2392.2904
7.79	0.128316	0.000414	2416.3303
7.80	0.128152	0.000410	2440.6176
7.81	0.127989	0.000406	2465.1464
7.82	0.127826	0.000402	2489.9215
7.83	0.127663	0.000398	2514.9956
7.84	0.127501	0.000394	2540.2212
7.85	0.127339	0.000390	2565.7509
7.86	0.127177	0.000386	2591.5370
7.87	0.127016	0.000382	2617.5624
7.88	0.126855	0.000378	2643.8895
7.89	0.126694	0.000374	2670.4610
7.90	0.126535	0.000371	2697.2996
7.91	0.126376	0.000367	2724.4079
7.92	0.126217	0.000363	2751.7886
7.93	0.126058	0.000360	2779.4445
7.94	0.125900	0.000356	2807.3784
7.95	0.125742	0.000353	2835.5930
7.96	0.125584	0.000349	2864.0911
7.97	0.125427	0.000346	2892.8757
7.98	0.125270	0.000342	2921.9495
7.99	0.125114	0.000339	2951.3156
8.00	0.124998	0.000335	2980.9764

x = 6.49 – 7.00

x			
6.49	0.153849	0.001519	658.5278
6.50	0.153615	0.001503	665.1445
6.51	0.153381	0.001488	671.8293
6.52	0.153148	0.001474	678.5814
6.53	0.152916	0.001459	685.4012
6.54	0.152684	0.001444	692.2897
6.55	0.152453	0.001430	699.2473
6.56	0.152223	0.001416	706.2749
6.57	0.151994	0.001402	713.3731
6.58	0.151765	0.001388	720.5427
6.59	0.151538	0.001374	727.7843
6.60	0.151309	0.001360	735.0987
6.61	0.151082	0.001347	742.4866
6.62	0.150856	0.001333	749.9488
6.63	0.150630	0.001320	757.4859
6.64	0.150405	0.001307	765.0988
6.65	0.150181	0.001294	772.7882
6.66	0.149956	0.001281	780.5549
6.67	0.149735	0.001268	788.3997
6.68	0.149512	0.001256	796.3232
6.69	0.149291	0.001243	804.3265
6.70	0.149070	0.001231	812.4101
6.71	0.148850	0.001219	820.5750
6.72	0.148630	0.001207	828.8220
6.73	0.148411	0.001195	837.1518
6.74	0.148192	0.001183	845.5654
6.75	0.147975	0.001171	854.0635
6.76	0.147757	0.001159	862.6470
6.77	0.147541	0.001148	871.3168
6.78	0.147325	0.001136	880.0737
6.79	0.147110	0.001125	888.9187
6.80	0.146895	0.001114	897.8525
6.81	0.146681	0.001103	906.8761
6.82	0.146467	0.001092	915.9904
6.83	0.146254	0.001081	925.1963
6.84	0.146042	0.001070	934.4947
6.85	0.145831	0.001059	943.8866
6.86	0.145620	0.001049	953.3729
6.87	0.145409	0.001038	962.9545
6.88	0.145199	0.001028	972.6324
6.89	0.144989	0.001018	982.4075
6.90	0.144781	0.001008	992.2810
6.91	0.144573	0.000998	1002.2536
6.92	0.144366	0.000988	1012.3265
6.93	0.144159	0.000978	1022.5006
6.94	0.143953	0.000968	1032.7769
6.95	0.143747	0.000959	1043.1565
6.96	0.143542	0.000949	1053.6405
6.97	0.143337	0.000940	1064.2298
6.98	0.143133	0.000930	1074.9255
6.99	0.142930	0.000921	1085.7288
7.00	0.142727	0.000912	1096.6406

APPENDIX D Discount Factors Based on Continuous Compounding and Uniform (Continuous) Annual Cash Flows

THE tables in this appendix are based on the same continuous compounding principle as the tables in Appendix C. However, this appendix contains discount factors for annual cash flows which are assumed to be received uniformly (continuously) throughout the year. Its use is straightforward—simply find the column headed by the particular discount rate being considered, then read down the column for the discount factor applicable to the specified year of receipt.

EXAMPLE: What is the present worth of $10,000 received uniformly in year 12, discounted continuously at a nominal annual rate of 10%?

Under the column labeled "10%," read a discount factor of 0.317 for income received uniformly in year 12. Therefore, the present worth of the $10,000 is

$$(\$10,000)(0.317) = \$3,170.$$

Note the distinction of uniform cash receipts. If an annual cash flow projection is made, the next step is to apply some form of discounting. To do this an assumption must be made as to when the money is actually received in any given year. Alternatives in making this assumption include mid-year, year-end, quarterly, monthly, etc. The tables of this appendix are based on receipt of the cash flow continuously and at a constant, uniform rate throughout the year.

The discount factors given in this table were computed from the following relationship:

$$\text{Discount Factor for Uniform Annual Cash Flows and Continuous Discounting} = \frac{e^j - 1}{(j)(e^{nj})}$$

where

j = nominal annual discounting rate, decimal fraction
n = year corresponding to period of uniform annual cash flow
e = base of natural logarithm, 2.7183

The discount factor used in the above example, 0.317, was computed from this relationship with $j = 0.10$ and $n = 12$ years.

598

DISCOUNT FACTORS BASED ON CONTINUOUS COMPOUNDING AND UNIFORM (CONTINUOUS) ANNUAL CASH FLOWS

DISCOUNT RATE (Expressed as percentage)

Year	3%	4%	5%	6%	7%	8%	9%	10%	11%
1	.985	.980	.975	.971	.966	.961	.956	.952	.947
2	.956	.942	.928	.914	.901	.887	.874	.861	.848
3	.928	.905	.883	.861	.840	.819	.799	.779	.760
4	.900	.869	.840	.811	.783	.756	.730	.705	.681
5	.874	.835	.799	.764	.730	.698	.667	.638	.610
6	.848	.803	.760	.719	.681	.644	.610	.577	.546
7	.823	.771	.723	.677	.635	.595	.557	.522	.489
8	.799	.741	.687	.638	.592	.549	.509	.473	.439
9	.775	.712	.654	.601	.552	.507	.466	.428	.393
10	.752	.684	.622	.566	.514	.468	.425	.387	.352
11	.730	.657	.592	.533	.480	.432	.389	.350	.315
12	.708	.631	.563	.502	.447	.399	.355	.317	.282
13	.687	.607	.535	.472	.417	.368	.325	.287	.253
14	.667	.583	.509	.445	.389	.340	.297	.259	.227
15	.647	.560	.484	.419	.363	.314	.271	.235	.203
16	.628	.538	.461	.395	.338	.290	.248	.212	.182
17	.610	.517	.438	.372	.315	.267	.227	.192	.163
18	.592	.497	.417	.350	.294	.247	.207	.174	.146
19	.574	.477	.397	.330	.274	.228	.189	.157	.131
20	.557	.458	.377	.310	.255	.210	.173	.142	.117
21	.541	.441	.359	.292	.238	.194	.158	.129	.105
22	.525	.423	.341	.275	.222	.179	.145	.117	.094
23	.509	.407	.325	.259	.207	.165	.132	.105	.084
24	.494	.391	.209	.244	.193	.153	.121	.095	.075
25	.480	.375	.294	.230	.180	.141	.110	.086	.068
26	.465	.361	.280	.217	.168	.130	.101	.078	.061
27	.452	.347	.266	.204	.157	.120	.092	.071	.054
28	.438	.333	.253	.192	.146	.111	.084	.064	.049
29	.425	.320	.241	.181	.136	.102	.077	.058	.044
30	.413	.307	.229	.170	.127	.094	.070	.052	.039
31	.401	.295	.218	.160	.118	.087	.064	.047	.035
32	.389	.284	.207	.151	.110	.081	.059	.043	.031
33	.377	.273	.197	.142	.103	.074	.054	.039	.028
34	.366	.262	.187	.134	.096	.069	.049	.035	.025
35	.355	.252	.178	.126	.089	.063	.045	.032	.023
36	.345	.242	.170	.119	.083	.058	.041	.029	.020
37	.335	.232	.161	.112	.078	.054	.038	.026	.018
38	.325	.223	.153	.105	.073	.050	.034	.024	.016
39	.315	.214	.146	.099	.068	.046	.031	.021	.015
40	.306	.206	.139	.094	.063	.042	.029	.019	.013
41	.297	.198	.132	.088	.059	.039	.026	.017	.012
42	.288	.190	.126	.083	.055	.036	.024	.016	.010
43	.279	.183	.119	.078	.051	.033	.022	.014	.009
44	.271	.176	.114	.074	.048	.031	.020	.013	.008
45	.263	.169	.108	.069	.044	.028	.018	.012	.008
46	.255	.162	.103	.065	.041	.026	.017	.011	.007
47	.248	.156	.098	.061	.039	.024	.015	.010	.006
48	.241	.150	.093	.058	.036	.022	.014	.009	.005
49	.233	.144	.089	.055	.034	.021	.013	.008	.005
50	.227	.138	.084	.051	.031	.019	.012	.007	.004

DISCOUNT FACTORS BASED ON CONTINUOUS COMPOUNDING AND UNIFORM (CONTINUOUS) ANNUAL CASH FLOWS

DISCOUNT RATE (Expressed as percentage)

Year	12%	13%	14%	15%	16%	17%	18%	19%	20%	21%
1	.942	.938	.933	.929	.924	.920	.915	.911	.906	.902
2	.836	.823	.811	.799	.788	.776	.764	.753	.742	.731
3	.741	.723	.705	.688	.671	.655	.639	.623	.608	.593
4	.657	.635	.613	.592	.572	.552	.533	.515	.497	.480
5	.583	.558	.533	.510	.487	.466	.446	.426	.407	.389
6	.517	.490	.463	.439	.415	.393	.372	.352	.333	.316
7	.459	.430	.403	.378	.354	.332	.311	.291	.273	.256
8	.407	.378	.350	.325	.302	.280	.260	.241	.224	.207
9	.361	.331	.305	.280	.257	.236	.217	.199	.183	.168
10	.320	.291	.265	.241	.219	.199	.181	.165	.150	.136
11	.284	.256	.230	.207	.187	.168	.151	.136	.123	.111
12	.252	.224	.200	.178	.159	.142	.126	.113	.100	.090
13	.223	.197	.174	.154	.136	.120	.106	.093	.082	.073
14	.198	.173	.151	.132	.115	.101	.088	.077	.067	.059
15	.176	.152	.131	.114	.098	.085	.074	.064	.055	.048
16	.156	.133	.114	.098	.084	.072	.062	.053	.045	.039
17	.138	.117	.099	.084	.071	.061	.051	.044	.037	.031
18	.123	.103	.086	.073	.061	.051	.043	.036	.030	.025
19	.109	.090	.075	.062	.052	.043	.036	.030	.025	.021
20	.096	.079	.065	.054	.044	.036	.030	.025	.020	.017
21	.086	.070	.057	.046	.038	.031	.025	.020	.017	.014
22	.076	.061	.049	.040	.032	.026	.021	.017	.014	.011
23	.067	.054	.043	.034	.027	.022	.017	.014	.011	.009
24	.060	.047	.037	.030	.023	.018	.015	.012	.009	.007
25	.053	.041	.032	.025	.020	.016	.012	.010	.008	.006
26	.047	.036	.028	.022	.017	.013	.010	.008	.006	.005
27	.042	.032	.025	.019	.014	.011	.009	.007	.005	.004
28	.037	.028	.021	.016	.012	.009	.007	.005	.004	.003
29	.033	.025	.019	.014	.011	.008	.006	.005	.003	.003
30	.029	.022	.016	.012	.009	.007	.005	.004	.003	.002
31	.026	.019	.014	.010	.008	.006	.004	.003	.002	.002
32	.023	.017	.012	.009	.007	.005	.004	.003	.002	.001
33	.020	.015	.011	.008	.006	.004	.003	.002	.002	.001
34	.018	.013	.009	.007	.005	.003	.002	.002	.001	.001
35	.016	.011	.008	.006	.004	.003	.002	.001	.001	.001
36	.014	.010	.007	.005	.003	.002	.002	.001	.001	.001
37	.013	.009	.006	.004	.003	.002	.001	.001	.001	.001
38	.011	.008	.005	.004	.003	.002	.001	.001	.001	
39	.010	.007	.005	.003	.002	.001	.001	.001	.001	
40	.009	.006	.004	.003	.002	.001	.001	.001		
41	.008	.005	.004	.002	.002	.001	.001	.001		
42	.007	.005	.003	.002	.001	.001	.001			
43	.006	.004	.003	.002	.001	.001	.001			
44	.005	.004	.002	.002	.001	.001				
45	.005	.003	.002	.001	.001	.001				
46	.004	.003	.002	.001	.001					
47	.004	.002	.002	.001	.001					
48	.003	.002	.001	.001	.001					
49	.003	.002	.001	.001						
50	.003	.002	.001	.001						

DISCOUNT FACTORS BASED ON CONTINUOUS COMPOUNDING AND UNIFORM (CONTINUOUS) ANNUAL CASH FLOWS

DISCOUNT RATE (Expressed as percentage)

Year	22%	23%	24%	25%	26%	27%	28%	29%	30%	35%
1	.898	.893	.889	.885	.881	.876	.872	.868	.864	.844
2	.720	.710	.699	.689	.679	.669	.659	.650	.640	.595
3	.578	.564	.550	.537	.524	.511	.498	.486	.474	.419
4	.464	.448	.433	.418	.404	.390	.377	.364	.351	.295
5	.372	.356	.340	.326	.311	.298	.285	.272	.260	.208
6	.299	.283	.268	.254	.240	.227	.215	.204	.193	.147
7	.240	.225	.211	.197	.185	.173	.163	.152	.143	.103
8	.192	.179	.166	.154	.143	.132	.123	.114	.106	.073
9	.154	.142	.130	.120	.110	.101	.093	.085	.078	.051
10	.124	.113	.103	.093	.085	.077	.070	.064	.058	.036
11	.100	.090	.081	.073	.065	.059	.053	.048	.043	.026
12	.080	.071	.063	.057	.050	.045	.040	.036	.032	.018
13	.064	.057	.050	.044	.039	.034	.030	.027	.024	.013
14	.051	.045	.039	.034	.030	.026	.023	.020	.018	.009
15	.041	.036	.031	.027	.023	.020	.017	.015	.013	.006
16	.033	.028	.024	.021	.018	.015	.013	.011	.010	.004
17	.027	.023	.019	.016	.014	.012	.010	.008	.007	.003
18	.021	.018	.015	.013	.011	.009	.008	.006	.005	.002
19	.017	.014	.012	.010	.008	.007	.006	.005	.004	.002
20	.014	.011	.009	.008	.006	.005	.004	.004	.003	.001
21	.011	.009	.007	.006	.005	.004	.003	.003	.002	.001
22	.009	.007	.006	.005	.004	.003	.002	.002	.002	.001
23	.007	.006	.005	.004	.003	.002	.002	.002	.001	
24	.006	.005	.004	.003	.002	.002	.001	.001	.001	
25	.005	.004	.003	.002	.002	.001	.001	.001	.001	
26	.004	.003	.002	.002	.001	.001	.001	.001	.001	
27	.003	.002	.002	.001	.001	.001	.001	.001		
28	.002	.002	.001	.001	.001	.001	.001			
29	.002	.001	.001	.001	.001	.001				
30	.002	.001	.001	.001	.001					
31	.001	.001	.001	.001						
32	.001	.001	.001							
33	.001	.001								
34	.001	.001								
35	.001									
36										
37										
38										
39										
40										
41										
42										
43										
44										
45										
46										
47										
48										
49										
50										

APPENDIX D

DISCOUNT FACTORS BASED ON CONTINUOUS COMPOUNDING AND
UNIFORM (CONTINUOUS) ANNUAL CASH FLOWS

DISCOUNT RATE (Expressed as percentage)

Year	40%	45%	50%	55%	60%	
1	.824	.805	.787	.769	.752	
2	.553	.514	.477	.444	.413	
3	.370	.327	.290	.256	.227	
4	.248	.209	.176	.148	.124	
5	.166	.133	.107	.085	.068	
6	.112	.085	.065	.049	.037	
7	.075	.054	.039	.028	.021	
8	.050	.035	.024	.016	.011	
9	.034	.022	.014	.009	.006	
10	.023	.014	.009	.005	.003	
11	.015	.009	.005	.003	.002	
12	.010	.006	.003	.002	.001	
13	.007	.004	.002	.001	.001	
14	.005	.002	.001	.001		
15	.003	.002	.001			
16	.002	.001				
17	.001	.001				
18	.001					
19	.001					
20						
21						
22						
23						
24						

APPENDIX E Expected Opportunity Loss Concept

IN Chapter 3 we briefly mentioned that there are certain instances in which it is meaningful to express the conditional values received in an expected value computation as opportunity losses, rather than the usual monetary profits. One such instance relates to the determination of how much money can be spent to obtain additional information to reduce the uncertainty of a given set of decision choices. Or more precisely, how much could we afford to spend to obtain perfect information. Questions of this type can be resolved by using the expected opportunity loss concept.

The idea is widely used in quality control and sequential sampling theories but has rather limited application in the petroleum exploration context. One possible use is illustrated in the following numerical example. This example should serve to demonstrate the use and interpretation of expected opportunity loss as well as to show that expected opportunity loss and expected monetary value are complementary methods of analysis. And finally, this example will give us the chance to define a new term—the cost of uncertainty.

Suppose we are considering a drilling prospect having the decision alternatives, possible outcomes, and monetary profits and losses given in Table E-1.

Table E-1

Table Showing Possible Decision Alternatives, Outcomes, and Monetary Profits and Losses

| Possible Outcome | Probability Outcome Will Occur | Decision Alternative | | | | |
		Drill (100% WI)	Drill (50% WI)	Farmout	Backin with 50% WI	Pass up Deal Entirely
Dry Hole	0.60	−$ 60,000	−$ 30,000	0	0	0
50 M Bbls.	0.10	−$ 20,000	−$ 10,000	+$ 5,000	0	0
100 M Bbls.	0.15	+$ 40,000	+$ 20,000	+$ 10,000	+$ 20,000	0
400 M Bbls.	0.10	+$400,000	+$200,000	+$ 60,000	+$200,000	0
800 M Bbls.	0.05	+$800,000	+$400,000	+$120,000	+$400,000	0
	1.00					

603

If it was simply a question of trying to determine which of the five alternatives to select we would make the usual EMV computation, the results of which are summarized below:

Decision Alternative	Expected Monetary Value Profit
1. Drill with full 100% working interest (WI) ownership	+$48,000 Preferred choice
2. Drill, but with only a 50% working interest ownership in the well	+$24,000
3. Farm out leasehold rights and retain an override	+$14,000
4. Exercise the penalty clause and subsequent backin option with 50% ownership	+$43,000
5. Pass up the deal entirely, i.e. do nothing	0

But now suppose that after reviewing these options the decision maker asks whether we could afford to run a detailed seismic survey on the prospect before deciding which alternative to select. A valid question, of course, but to resolve the answer we must recast the problem into the context of opportunity loss.

First we must define a few terms and restate some previous definitions. Recall from our earlier discussions on Bayesian Analysis that the term "perfect information" meant that we knew the true state of nature. That is, if we had perfect information we would know with certainty whether this drilling prospect had 50 M barrels reserves, or 100 M barrels, etc. — or was a dry hole. There would be a 1.0 in Column 2 of Table E-1 opposite the true state of nature and zeros in the column for the remaining four possible outcomes.

Now consider the situation where we decided to drill with full 100% WI ownership and found reserves of only 50 M barrels. Our actual monetary profit would be −$20,000. *If* we had, in fact, known beforehand that this was the state of nature our best strategy would have been to farm out, resulting in a profit of +$5,000. The difference between what we achieved under conditions of uncertainty and the optimal amount we could have received if we had had perfect information is called the opportunity loss.

DEFINITION: Opportunity loss is the difference between an actual profit or loss and the profit or loss which would have resulted if the decision maker had had perfect information at the time he made the decision.

Opportunity losses are perfectly valid value criteria to use for the conditional values received in an expected value computation. Having specified the values received as opportunity losses it is then meaningful to compute an expected opportunity loss for each decision alternative. The interpretation of these expected opportunity loss (EOL) values will provide the insight to respond to the decision maker's question about whether he can afford to purchase additional seismic information.

The first step is to identify the optimal strategies for each state of nature if, in fact, the decision maker had perfect information. In addition, we must denote the maximum gain that would be received from each of these optimal strategies. Table E-2 is simply a repeat of the profits shown in Table E-1. The circled entries represent the maximum profit that could be achieved for each possible outcome, or state of nature. For example, if we had perfect information and knew the prospect to be a dry hole our best strategy would be to farm out, take a backin option, or do nothing—each resulting in zero profit (the circled numbers in the row opposite Dry Hole). If we knew with certainty the well had 800 M barrels our best strategy would be to drill with 100% WI—resulting in an $800,000 profit.

Table E-2
Table Showing Profits and Losses for Example Drilling Prospect

Possible Outcome	Decision Alternative				
	Drill (100% WI)	Drill (50% WI)	Farmout	Backin	Pass up Deal
Dry Hole	−$ 60,000	−$ 30,000	0	0	0
50 M Bbls.	−$ 20,000	−$ 10,000	+$ 5,000	0	0
100 M Bbls.	+$ 40,000	+$ 20,000	+$ 10,000	+$ 20,000	0
400 M Bbls.	+$400,000	+$200,000	+$ 60,000	+$200,000	0
800 M Bbls.	+$800,000	+$400,000	+$120,000	+$400,000	0

Next we construct an opportunity loss table as shown in Table E-3. Each entry in this table is computed by subtracting the corresponding entry in Table E-2 from the circled, or maximum, entry *in the same row* of Table E-2. For example, if the true state of nature was 50 M barrels and we had drilled with 100% WI the difference, or opportunity loss, between the $20,000 loss and the maximum we could have received of +$5,000 from a farmout is +$5,000 − (−$20,000) = $25,000. Table E-3 is constructed a row at a time. For each row the corresponding entries of Table E-2 are subtracted from the maximum (circled) entries of that row and the differences entered in Table E-3 as opportunity losses. Getting from the profits of Table E-2 to the opportunity losses of Table E-3 is a

Table E-3

Table Showing Conditional Opportunity Losses, As Computed from Table E-2

Possible Outcome	Decision Alternative				
	Drill (100% WI)	Drill (50% WI)	Farmout	Backin	Pass up Deal
Dry Hole	$60,000	$ 30,000	0	0	0
50 M Bbls.	$25,000	$ 15,000	0	$ 5,000	$ 5,000
100 M Bbls.	0	$ 20,000	$ 30,000	$ 20,000	$ 40,000
400 M Bbls.	0	$200,000	$340,000	$200,000	$400,000
800 M Bbls.	0	$400,000	$680,000	$400,000	$800,000

crucial step—so be sure you can verify how all the entries of Table E-3 were obtained.

Finally we make an expected value computation by computing the expected opportunity loss for each outcome and then finding the sum of these products for each alternative. The calculation is identical to an EMV computation except in this instance we are multiplying probabilities by conditional opportunity losses, rather than monetary value profits. These EOL computations are shown in Table E-4 and summarized below:

Decision Alternative	Expected Opportunity Loss (EOL)
Drill (100% WI)	$38,500
Drill (50% WI)	$62,500
Farmout	$72,500
Backin option	$43,500
Pass up deal	$86,500

These expected opportunity loss values for each decision alternative represent the expected amount that profit is below the expectation we would have received if we had had perfect information. If we had perfect information our expected opportunity loss would be zero. The greater the value of EOL the greater is the difference between this optimum and that which we can expect given the uncertainty. From a project selection standpoint we would, therefore, want to choose the project which *minimizes* the expected opportunity losses.

The EOL values above are also defined as the "cost of uncertainty" for each alternative. As such the lowest EOL represents the maximum amount we could afford to spend to secure perfect information. So for our particular problem

TABLE E-4

EXPECTED OPPORTUNITY LOSS COMPUTATIONS. CONDITIONAL OPPORTUNITY LOSSES ARE FROM Table E-3. THE EXPECTED OPPORTUNITY LOSSES ARE COMPUTED BY MULTIPLYING THE PROBABILITIES BY THE CONDITIONAL OPPORTUNITY LOSSES.

		DECISION ALTERNATIVE									
		Drill (100%WI)		Drill (50%WI)		Farm Out		Backin		Pass Up Deal	
Possible Outcome	Prob. Outcome Occur	Conditional Opportunity Loss	Expected Opportunity Loss	Conditional Opportunity Loss	Expected Opportunity Loss	Conditional Opportunity Loss	Expected Opportunity Loss	Conditional Opportunity Loss	Expected Opportunity Loss	Conditional Opportunity Loss	Expected Opportunity Loss
Dry Hole	0.60	$60,000	$36,000	$30,000	$18,000	0	0	0	0	0	0
50M Bbls.	0.10	$25,000	$2,500	$15,000	$1,500	0	0	$5,000	$500	$5,000	$500
100M Bbls.	0.15	0	0	$20,000	$3,000	$30,000	$4,500	$20,000	$3,000	$40,000	$6,000
400M Bbls.	0.10	0	0	$200,000	$20,000	$340,000	$34,000	$200,000	$20,000	$400,000	$40,000
800M Bbls.	0.05	0	0	$400,000	$20,000	$680,000	$34,000	$400,000	$20,000	$800,000	$40,000
	1.00		$38,500		$62,500		$72,500		$43,500		$86,500
			(EOL)		(EOL)		(EOL)		(EOL)		(EOL)

607

we can interpret these findings to mean that our decision maker could afford to spend as much as $38,500 for a seismic survey in an effort to reduce uncertainty.

Now of course we all realize that in this context of oil prospecting there is no such thing as perfect information. Even if we ran the seismic we would have no guarantee that the results would tell us which of the five possible outcomes is the true state of nature. Yet seismic can be helpful to us. If we obtained seismic we might, for example, determine that the prospect isn't as large as originally thought and rule out 800 M barrels as a possible outcome. Or it may show no structure (trapping mechanism) with the conclusion that we'd probably get a dry hole if we drilled. The point here is that if we feel the seismic may provide additional information to reduce uncertainty (even though being short of perfect information) the maximum we could pay for the survey would be $38,500. The cost of uncertainty thus represents the maximum amount we could pay to obtain perfect information.

GENERAL COMMENTS ABOUT EXPECTED OPPORTUNITY LOSS

1. Expected opportunity loss is the complement of EMV. Maximizing EMV will lead to the same decision choice as minimizing EOL. In our example this should be obvious by comparing EMV with EOL values. The Drill (100%) option has the highest EMV and the lowest EOL. The strategy to pass up the deal has the lowest EMV and the highest EOL. Opportunity loss computations are not required if the decision maker is only concerned about which alternative to select. He can use EMV directly for this purpose. But if he chose to make his decision on the basis of opportunity loss he should select the alternative having the lowest EOL.

Decision Alternative	Expected Monetary Value Profit (EMV)	EMV Ranking	Expected Opportunity Loss (EOL)	EOL Ranking
Drill (100% WI)	+$48,000	(1)	$38,500	(1)
Drill (50% WI)	+$24,000	(3)	$62,500	(3)
Farmout	+$14,000	(4)	$72,500	(4)
Backin Option	+$43,000	(2)	$43,500	(2)
Pass up Deal	0	(5)	$86,000	(5)

The reason I mention this option is that there are some types of situations where it may be easier for the analyst and decision maker to think in terms of opportunity losses directly, without first thinking in terms of profits. Examples might include:

a. If we don't drill this well we might miss out on the discovery of a ____ million barrel field.

b. What will we lose if we don't get into a new exploratory play in _____ Basin?

c. What potential losses would be incurred by drainage from offset wells if we don't drill the well?

In instances such as these the analyst may choose to express the conditional values received as opportunity losses directly. And in these cases the decision maker should accept the alternative which minimizes expected opportunity loss.

2. The entries in the conditional opportunity loss table (Table E-3) will always have a positive sign. This is a consequence of the manner in which the opportunity loss values were computed. There is no such thing as a negative opportunity loss.

3. The minimum EOL value is called the "cost of uncertainty" and represents the upper limit one can afford to spend to get perfect (or better) information. It does not mean we should spend this much. By all means if we can obtain the information at much less cost we should do so. It only represents an upper bound to the amount that could be spent for perfect information. The specific numerical value of EOL, or cost of uncertainty, relates to the analyst's specific view of uncertainty. That is, the EOL of $38,500 relates to his present estimates of the probabilities of occurrence in Column 2 of Table E-4. If the probabilities in Column 2 had different values the EOL values would all be different than those computed in Table E-4. The probabilities of Column 2 represent his perception about the states of nature and their relative likelihoods of occurrence, and the EOL values calculated apply only to these specific perceptions of uncertainty.

4. Finally we would point out there is, in fact, an algebraic relationship between EMV and EOL. Namely, $EMV_i + EOL_i = K$, where i is an index for each decision alternative and K is a constant. For instance, if we add the EMV and EOL for "drill with 100% WI" we get $48,000 + $38,500 = $86,500. If we add the corresponding values for each of the other four

Table E-5

Outcome (State of Nature)	Prob. of Occur.	Profit from Optimal Choice, Given Knowledge of the State of Nature	Expected Profit
Dry Hole	0.60	0 (Farmout, Backin, or Pass Up)	0
50 M Bbls.	0.10	$5,000 (Farmout)	0.10 × $ 5,000 = $ 500
100 M Bbls.	0.15	$40,000 (Drill, 100% WI)	0.15 × $ 40,000 = $ 6,000
400 M Bbls.	0.10	$400,000 (Drill, 100% WI)	0.10 × $400,000 = $40,000
800 M Bbls.	0.05	$800,000 (Drill, 100% WI)	0.05 × $800,000 = $40,000
	1.00		$86,500

alternatives the sum will be the same, $86,500. This suggests that the sum might have a meaning. And indeed it does. The sum of the EMV and EOL is the expected profit the decision maker could achieve if he could choose his decision alternative *after* learning the true state of nature. This is shown in Table E-5.

The difference between this sum and the EMV he would expect by making his choice *before* having perfect information is the EOL. That is:

$$(\$86,500) \quad - \quad (\$48,000) \quad = \quad (\$38,500)$$

| His expectation if he decided after receiving perfect information | His expectation if he decides before he has perfect infor- mation (EMV) | Value of obtaining perfect information (EOL) |

This computation (finding the sum of EMV and EOL) is not usually made in the course of decision making under uncertainty, but it may be useful to cast additional insight on what the expected opportunity loss concept is and how it is used to determine an upper limiting value that can be spent to secure perfect information.

Cumulative Binomial Probability Tables

THE tables in this appendix section can be used to solve the binomial probability equation given as equation 6.11 in Chapter 6.

$$\text{Binomial Probability of x successes in n trials} = \left(C_x^n \right) (p)^x (1-p)^{n-x} \qquad (6.11)$$

Where

x = Number of successes ($0 \leq x \leq n$)
n = Number of trials
p = Probability of success on any given trial ($0 \leq p \leq 1.0$)

The tables here are for $n = 2$–20, 25, 50, and 100 trials. For a complete set of cumulative binomial probabilities for all values of n up to 150 refer to Ordnance Corps Pamphlet ORDP 20–1, *Tables of the Cumulative Binomial Probabilities,* September, 1952, published by the Office of Technical Service, Dept. of Commerce, Washington, D.C. Most engineering and mathematics libraries will have this publication (or other publications listing complete binomial probabilities).

These tables give the cumulative probability of *x or more* successes in n trials for specified values of p. To find the probability of *exactly* x successes use the following rule:

$$\text{Binomial Probability of exactly x successes in n trials} = \left(\begin{array}{c} \text{Probability of} \\ \text{x or more} \\ \text{successes} \end{array} \right) - \left(\begin{array}{c} \text{Probability of} \\ [x+1] \text{ or more} \\ \text{successes} \end{array} \right)$$

This relation will hold for all values of x, including $x = 0$. For values of p greater than 0.50 redefine the problem such that successes are defined as failures.

Examples

1. What is the probability of one success in $n = 5$ trials, where $p = 0.15$?

   From the table for $n = 5$ trials, $x = 1$, and $p = 0.15$ read the probability of 1 or more successes as 0.5563. The probability of $x + 1$ (that is $1 + 1 = 2$) successes or more is read from the table as 0.1648. (It's the entry just below the first number we read from the table.) Using the given rule, the probability of *exactly* 1 success is the difference between these values: $(0.5563 - 0.1648) = 0.3915$. This is exactly the same value we computed in the numerical example relating to binomial probabilities in Chapter 6. Referring to Figure 6.28, the 0.5563 is the sum of all of the probability spikes of 1 or more successes – the right portion of the distribution. The second number, 0.1648, is the sum of all the spikes of 2 or more. When we subtract the latter number from the former the remainder is simply the probability spike above x = 1 success, and this corresponds to the probability of exactly 1 success in the five trials.

2. Consider a process of 50 Bernoulli trials having a value of p = 0.23. What is the probability of nine or more successes?

From the table for n = 50, x = 9, and p = 0.23 read *0.8437*. This is the probability of nine or more successes in 50 trials.

What is the probability of exactly nine successes in the 50 trials?

The cumulative probability of 9 or more successes was just determined to be 0.8437. Entering the table again for n = 50, x = 9 + 1 = 10, and p = 0.23 read the cumulative probability of 10 or more successes as 0.7436. The probability of exactly nine successes is the difference: $0.8437 - 0.7436 = 0.1001$. (Note again that the probability of exactly x successes in these tables is the difference between the cumulative probability of at least x successes and at least x + 1 successes.)

If the probability of a specific outcome is greater than 0.5 the tables cannot be used directly. However, if the problem is rephrased in terms of complementary probabilities the tables can still be used.

3. Consider a 25 well development well drilling program in which the probability of success is estimated to be p = 0.7. Assuming Bernoulli trials, what is the probability of 19 successful completions?

Since the tables only extend to values of p = 0.5, we rephrase the question as "What is the binomial probability of 6 'successes' (dry holes, actually) in 25 trials where the probability of 'success' is $1.0 - 0.7 = 0.3$?" From the tables for n = 25, x = 6, and p = 0.3, read the probability of 6 or more as 0.8065 and the probability of 7 or more as 0.6593. Hence the probability of exactly 6 'successes' is $0.8065 - 0.6593 = 0.1472$. Getting 6 dry holes in 25 wells is exactly the same as getting 19 producers in 25 wells, so the answer to our original question is 0.1472.

CUMULATIVE BINOMINAL PROBABILITY TABLES

Entries in this table represent the cumulative probability of x or more successes in n trials for specified values of p.

n = 2

x^p	.01	.02	.03	.04	.05	.06	.07	.08	.09	.10
1	.0199	.0396	.0591	.0784	.0975	.1164	.1351	.1536	.1719	.1900
2	.0001	.0004	.0009	.0016	.0025	.0036	.0049	.0064	.0081	.0100

x^p	.11	.12	.13	.14	.15	.16	.17	.18	.19	.20
1	.2079	.2256	.2431	.2604	.2775	.2944	.3111	.3276	.3439	.3600
2	.0121	.0144	.0169	.0196	.0225	.0256	.0289	.0324	.0361	.0400

x^p	.21	.22	.23	.24	.25	.26	.27	.28	.29	.30
1	.3759	.3916	.4071	.4224	.4375	.4524	.4671	.4816	.4959	.5100
2	.0441	.0484	.0529	.0576	.0625	.0676	.0729	.0784	.0841	.0900

x^p	.31	.32	.33	.34	.35	.36	.37	.38	.39	.40
1	.5239	.5376	.5511	.5644	.5775	.5904	.6031	.6156	.6279	.6400
2	.0961	.1024	.1089	.1156	.1225	.1296	.1369	.1444	.1521	.1600

x^p	.41	.42	.43	.44	.45	.46	.47	.48	.49	.50
1	.6519	.6636	.6751	.6864	.6975	.7084	.7191	.7296	.7399	.7500
2	.1681	.1764	.1849	.1936	.2025	.2116	.2209	.2304	.2401	.2500

n = 3

x^p	.01	.02	.03	.04	.05	.06	.07	.08	.09	.10
1	.0297	.0588	.0873	.1153	.1426	.1694	.1956	.2213	.2464	.2710
2	.0003	.0012	.0026	.0047	.0073	.0104	.0140	.0182	.0228	.0280
3				.0001	.0001	.0002	.0003	.0005	.0007	.0010

x^p	.11	.12	.13	.14	.15	.16	.17	.18	.19	.20
1	.2950	.3185	.3415	.3639	.3859	.4073	.4282	.4486	.4686	.4880
2	.0336	.0397	.0463	.0533	.0608	.0686	.0769	.0855	.0946	.1040
3	.0013	.0017	.0022	.0027	.0034	.0041	.0049	.0058	.0069	.0080

x^p	.21	.22	.23	.24	.25	.26	.27	.28	.29	.30
1	.5070	.5254	.5435	.5610	.5781	.5948	.6110	.6268	.6421	.6570
2	.1138	.1239	.1344	.1452	.1563	.1676	.1793	.1913	.2035	.2160
3	.0093	.0106	.0122	.0138	.0156	.0176	.0197	.0220	.0244	.0270

x^p	.31	.32	.33	.34	.35	.36	.37	.38	.39	.40
1	.6715	.6856	.6992	.7125	.7254	.7379	.7500	.7617	.7730	.7840
2	.2287	.2417	.2548	.2682	.2818	.2955	.3094	.3235	.3377	.3520
3	.0298	.0328	.0359	.0393	.0429	.0467	.0507	.0549	.0593	.0640

x^p	.41	.42	.43	.44	.45	.46	.47	.48	.49	.50
1	.7946	.8049	.8148	.8244	.8336	.8425	.8511	.8594	.8673	.8750
2	.3665	.3810	.3957	.4104	.4253	.4401	.4551	.4700	.4850	.5000
3	.0689	.0741	.0795	.0852	.0911	.0973	.1038	.1106	.1176	.1250

n = 4

x^P	.01	.02	.03	.04	.05	.06	.07	.08	.09	.10
1	.0394	.0776	.1147	.1507	.1855	.2193	.2519	.2836	.3143	.3439
2	.0006	.0023	.0052	.0091	.0140	.0199	.0267	.0344	.0430	.0523
3			.0001	.0002	.0005	.0008	.0013	.0019	.0027	.0037
4									.0001	.0001

x^P	.11	.12	.13	.14	.15	.16	.17	.18	.19	.20
1	.3726	.4003	.4271	.4530	.4780	.5021	.5254	.5479	.5695	.5904
2	.0624	.0732	.0847	.0968	.1095	.1228	.1366	.1509	.1656	.1808
3	.0049	.0063	.0079	.0098	.0120	.0144	.0171	.0202	.0235	.0272
4	.0001	.0002	.0003	.0004	.0005	.0007	.0008	.0010	.0013	.0016

x^P	.21	.22	.23	.24	.25	.26	.27	.28	.29	.30
1	.6105	.6298	.6485	.6664	.6836	.7001	.7160	.7313	.7459	.7599
2	.1963	.2122	.2285	.2450	.2617	.2787	.2959	.3132	.3307	.3483
3	.0312	.0356	.0403	.0453	.0508	.0566	.0628	.0694	.0763	.0837
4	.0019	.0023	.0028	.0033	.0039	.0046	.0053	.0061	.0071	.0081

x^P	.31	.32	.33	.34	.35	.36	.37	.38	.39	.40
1	.7733	.7862	.7985	.8103	.8215	.8322	.8425	.8522	.8615	.8704
2	.3660	.3837	.4015	.4193	.4370	.4547	.4724	.4900	.5075	.5248
3	.0915	.0996	.1082	.1171	.1265	.1362	.1464	.1569	.1679	.1792
4	.0092	.0105	.0119	.0134	.0150	.0168	.0187	.0209	.0231	.0256

x^P	.41	.42	.43	.44	.45	.46	.47	.48	.49	.50
1	.8788	.8868	.8944	.9017	.9085	.9150	.9211	.9269	.9323	.9375
2	.5420	.5590	.5759	.5926	.6090	.6252	.6412	.6569	.6724	.6875
3	.1909	.2030	.2155	.2283	.2415	.2550	.2689	.2831	.2977	.3125
4	.0283	.0311	.0342	.0375	.0410	.0448	.0488	.0531	.0576	.0625

n = 5

x^P	.01	.02	.03	.04	.05	.06	.07	.08	.09	.10
1	.0490	.0961	.1413	.1846	.2262	.2661	.3043	.3409	.3760	.4095
2	.0010	.0038	.0085	.0148	.0226	.0319	.0425	.0544	.0674	.0815
3		.0001	.0003	.0006	.0012	.0020	.0031	.0045	.0063	.0086
4						.0001	.0001	.0002	.0003	.0005

x^P	.11	.12	.13	.14	.15	.16	.17	.18	.19	.20
1	.4416	.4723	.5016	.5296	.5563	.5818	.6061	.6293	.6513	.6723
2	.0965	.1125	.1292	.1467	.1648	.1835	.2027	.2224	.2424	.2627
3	.0112	.0143	.0179	.0220	.0266	.0318	.0375	.0437	.0505	.0579
4	.0007	.0009	.0013	.0017	.0022	.0029	.0036	.0045	.0055	.0067
5				.0001	.0001	.0001	.0001	.0002	.0002	.0003

x^P	.21	.22	.23	.24	.25	.26	.27	.28	.29	.30
1	.6923	.7113	.7293	.7464	.7627	.7781	.7927	.8065	.8196	.8319
2	.2833	.3041	.3251	.3461	.3672	.3883	.4093	.4303	.4511	.4718
3	.0659	.0744	.0836	.0933	.1035	.1143	.1257	.1376	.1501	.1631
4	.0081	.0097	.0114	.0134	.0156	.0181	.0208	.0238	.0272	.0308
5	.0004	.0005	.0006	.0008	.0010	.0012	.0014	.0017	.0021	.0024

x\P	.31	.32	.33	.34	.35	.36	.37	.38	.39	.40
1	.8436	.8546	.8650	.8748	.8840	.8926	.9008	.9084	.9155	.9222
2	.4923	.5125	.5325	.5522	.5716	.5906	.6093	.6276	.6455	.6630
3	.1766	.1905	.2050	.2199	.2352	.2509	.2670	.2835	.3003	.3174
4	.0347	.0390	.0436	.0486	.0540	.0598	.0660	.0726	.0796	.0870
5	.0029	.0034	.0039	.0045	.0053	.0060	.0069	.0079	.0090	.0102

x\P	.41	.42	.43	.44	.45	.46	.47	.48	.49	.50
1	.9285	.9344	.9398	.9449	.9497	.9541	.9582	.9620	.9655	.9688
2	.6801	.6967	.7129	.7286	.7438	.7585	.7728	.7865	.7998	.8125
3	.3349	.3525	.3705	.3886	.4069	.4253	.4439	.4625	.4813	.5000
4	.0949	.1033	.1121	.1214	.1312	.1415	.1522	.1635	.1753	.1875
5	.0116	.0131	.0147	.0165	.0185	.0206	.0229	.0255	.0282	.0313

n = 6

x\P	01	.02	.03	.04	.05	.06	.07	.08	.09	.10
1	.0585	.1142	.1670	.2172	.2649	.3101	.3530	.3936	.4321	.4686
2	.0015	.0057	.0125	.0216	.0328	.0459	.0608	.0773	.0952	.1143
3		.0002	.0005	.0012	.0022	.0038	.0058	.0085	.0118	.0159
4					.0001	.0002	.0003	.0005	.0008	.0013
5										.0001

x\P	.11	.12	.13	.14	.15	.16	.17	.18	.19	.20
1	.5030	.5356	.5664	.5954	.6229	.6487	.6731	.6960	.7176	.7379
2	.1345	.1556	.1776	.2003	.2235	.2472	.2713	.2956	.3201	.3446
3	.0206	.0261	.0324	.0395	.0473	.0560	.0655	.0759	.0870	.0989
4	.0018	.0025	.0034	.0045	.0059	.0075	.0094	.0116	.0141	.0170
5	.0001	.0001	.0002	.0003	.0004	.0005	.0007	.0010	.0013	.0016
6										.0001

x\P	.21	.22	.23	.24	.25	.26	.27	.28	.29	.30
1	.7569	.7748	.7916	.8073	.8220	.8358	.8487	.8607	.8719	.8824
2	.3692	.3937	.4180	.4422	.4661	.4896	.5128	.5356	.5580	.5798
3	.1115	.1250	.1391	.1539	.1694	.1856	.2023	.2196	.2374	.2557
4	.0202	.0239	.0280	.0326	.0376	.0431	.0492	.0557	.0628	.0705
5	.0020	.0025	.0031	.0038	.0046	.0056	.0067	.0079	.0093	.0109
6	.0001	.0001	.0001	.0002	.0002	.0003	.0004	.0005	.0006	.0007

x\P	.31	.32	.33	.34	.35	.36	.37	.38	.39	.40
1	.8921	.9011	.9095	.9173	.9246	.9313	.9375	.9432	.9485	.9533
2	.6012	.6220	.6422	.6619	.6809	.6994	.7172	.7343	.7508	.7667
3	.2744	.2936	.3130	.3328	.3529	.3732	.3937	.4143	.4350	.4557
4	.0787	.0875	.0969	.1069	.1174	.1286	.1404	.1527	.1657	.1792
5	.0127	.0148	.0170	.0195	.0223	.0254	.0288	.0325	.0365	.0410
6	.0009	.0011	.0013	.0015	.0018	.0022	.0026	.0030	.0035	.0041

x\P	.41	.42	.43	.44	.45	.46	.47	.48	.49	.50
1	.9578	.9619	.9657	.9692	.9723	.9752	.9778	.9802	.9824	.9844
2	.7819	.7965	.8105	.8238	.8364	.8485	.8599	.8707	.8810	.8906
3	.4764	.4971	.5177	.5382	.5585	.5786	.5985	.6180	.6373	.6563
4	.1933	.2080	.2232	.2390	.2553	.2721	.2893	.3070	.3252	.3438
5	.0458	.0510	.0566	.0627	.0692	.0762	.0837	.0917	.1003	.1094
6	.0048	.0055	.0063	.0073	.0083	.0095	.0108	.0122	.0138	.0156

n = 7

$x\backslash P$	.01	.02	.03	.04	.05	.06	.07	.08	.09	.10
1	.0679	.1319	.1920	.2486	.3017	.3515	.3983	.4422	.4832	.5217
2	.0020	.0079	.0171	.0294	.0444	.0618	.0813	.1026	.1255	.1497
3		.0003	.0009	.0020	.0038	.0063	.0097	.0140	.0193	.0257
4				.0001	.0002	.0004	.0007	.0012	.0018	.0027
5								.0001	.0001	.0002

$x\backslash P$	.11	.12	.13	.14	.15	.16	.17	.18	.19	.20
1	.5577	.5913	.6227	.6521	.6794	.7049	.7286	.7507	.7712	.7903
2	.1750	.2012	.2281	.2556	.2834	.3115	.3396	.3677	.3956	.4233
3	.0331	.0416	.0513	.0620	.0738	.0866	.1005	.1154	.1313	.1480
4	.0039	.0054	.0072	.0094	.0121	.0153	.0189	.0231	.0279	.0333
5	.0003	.0004	.0006	.0009	.0012	.0017	.0022	.0029	.0037	.0047
6					.0001	.0001	.0001	.0002	.0003	.0004

$x\backslash P$	.21	.22	.23	.24	.25	.26	.27	.28	.29	.30
1	.8080	.8243	.8395	.8535	.8665	.8785	.8895	.8997	.9090	.9176
2	.4506	.4775	.5040	.5298	.5551	.5796	.6035	.6266	.6490	.6706
3	.1657	.1841	.2033	.2231	.2436	.2646	.2861	.3081	.3304	.3529
4	.0394	.0461	.0536	.0617	.0706	.0802	.0905	.1016	.1134	.1260
5	.0058	.0072	.0088	.0107	.0129	.0153	.0181	.0213	.0248	.0288
6	.0005	.0006	.0008	.0011	.0013	.0017	.0021	.0026	.0031	.0038
7						.0001	.0001	.0001	.0002	.0002

$x\backslash P$	.31	.32	.33	.34	.35	.36	.37	.38	.39	.40
1	.9255	.9328	.9394	.9454	.9510	.9560	.9606	.9648	.9686	.9720
2	.6914	.7113	.7304	.7487	.7662	.7828	.7987	.8137	.8279	.8414
3	.3757	.3987	.4217	.4447	.4677	.4906	.5134	.5359	.5581	.5801
4	.1394	.1534	.1682	.1837	.1998	.2167	.2341	.2521	.2707	.2898
5	.0332	.0380	.0434	.0492	.0556	.0625	.0701	.0782	.0869	.0963
6	.0046	.0055	.0065	.0077	.0090	.0105	.0123	.0142	.0164	.0188
7	.0003	.0003	.0004	.0005	.0006	.0008	.0009	.0011	.0014	.0016

$x\backslash P$	.41	.42	.43	.44	.45	.46	.47	.48	.49	.50
1	.9751	.9779	.9805	.9827	.9848	.9866	.9883	.9897	.9910	.9922
2	.8541	.8660	.8772	.8877	.8976	.9068	.9153	.9233	.9307	.9375
3	.6017	.6229	.6436	.6638	.6836	.7027	.7213	.7393	.7567	.7734
4	.3094	.3294	.3498	.3706	.3917	.4131	.4346	.4563	.4781	.5000
5	.1063	.1169	.1282	.1402	.1529	.1663	.1803	.1951	.2105	.2266
6	.0216	.0246	.0279	.0316	.0357	.0402	.0451	.0504	.0562	.0625
7	.0019	.0023	.0027	.0032	.0037	.0044	.0051	.0059	.0068	.0078

n = 8

$x\backslash P$	01	.02	.03	.04	.05	.06	.07	.08	.09	.10
1	.0773	.1492	.2163	.2786	.3366	.3904	.4404	.4868	.5297	.5695
2	.0027	.0103	.0223	.0381	.0572	.0792	.1035	.1298	.1577	.1869
3	.0001	.0004	.0013	.0031	.0058	.0096	.0147	.0211	.0289	.0381
4			.0001	.0002	.0004	.0007	.0013	.0022	.0034	.0050
5							.0001	.0001	.0003	.0004

x\P	11	.12	.13	.14	.15	.16	.17	.18	.19	.20
1	.6063	.6404	.6718	.7008	.7275	.7521	.7748	.7956	.8147	.8322
2	.2171	.2480	.2794	.3111	.3428	.3744	.4057	.4366	.4670	.4967
3	.0487	.0608	.0743	.0891	.1052	.1226	.1412	.1608	.1815	.2031
4	.0071	.0097	.0129	.0168	.0214	.0267	.0328	.0397	.0476	.0563
5	.0007	.0010	.0015	.0021	.0029	.0038	.0050	.0065	.0083	.0104
6		.0001	.0001	.0002	.0002	.0003	.0005	.0007	.0009	.0012
7									.0001	.0001

x\P	.21	.22	.23	.24	.25	.26	.27	.28	.29	.30
1	.8483	.8630	.8764	.8887	.8999	.9101	.9194	.9278	.9354	.9424
2	.5257	.5538	.5811	.6075	.6329	.6573	.6807	.7031	.7244	.7447
3	.2255	.2486	.2724	.2967	.3215	.3465	.3718	.3973	.4228	.4482
4	.0659	.0765	.0880	.1004	.1138	.1281	.1433	.1594	.1763	.1941
5	.0129	.0158	.0191	.0230	.0273	.0322	.0377	.0438	.0505	.0580
6	.0016	.0021	.0027	.0034	.0042	.0052	.0064	.0078	.0094	.0113
7	.0001	.0002	.0002	.0003	.0004	.0005	.0006	.0008	.0010	.0013
8									.0001	.0001

x\P	.31	.32	.33	.34	.35	.36	.37	.38	.39	.40
1	.9486	.9543	.9594	.9640	.9681	.9719	.9752	.9782	.9808	.9832
2	.7640	.7822	.7994	.8156	.8309	.8452	.8586	.8711	.8828	.8936
3	.4736	.4987	.5236	.5481	.5722	.5958	.6189	.6415	.6634	.6846
4	.2126	.2319	.2519	.2724	.2936	.3153	.3374	.3599	.3828	.4059
5	.0661	.0750	.0846	.0949	.1061	.1180	.1307	.1443	.1586	.1737
6	.0134	.0159	.0187	.0218	.0253	.0293	.0336	.0385	.0439	.0498
7	.0016	.0020	.0024	.0030	.0036	.0043	.0051	.0061	.0072	.0085
8	.0001	.0001	.0001	.0002	.0002	.0003	.0004	.0004	.0005	.0007

x\P	.41	.42	.43	.44	.45	.46	.47	.48	.49	.50
1	.9853	.9872	.9889	.9903	.9916	.9928	.9938	.9947	.9954	.9961
2	.9037	.9130	.9216	.9295	.9368	.9435	.9496	.9552	.9602	.9648
3	.7052	.7250	.7440	.7624	.7799	.7966	.8125	.8276	.8419	.8555
4	.4292	.4527	.4762	.4996	.5230	.5463	.5694	.5922	.6146	.6367
5	.1895	.2062	.2235	.2416	.2604	.2798	.2999	.3205	.3416	.3633
6	.0563	.0634	.0711	.0794	.0885	.0982	.1086	.1198	.1318	.1445
7	.0100	.0117	.0136	.0157	.0181	.0208	.0239	.0272	.0310	.0352
8	.0008	.0010	.0012	.0014	.0017	.0020	.0024	.0028	.0033	.0039

n = 9

x\P	01	.02	.03	.04	.05	.06	.07	.08	.09	.10
1	.0865	.1663	.2398	.3075	.3698	.4270	.4796	.5278	.5721	.6126
2	.0034	.0131	.0282	.0478	.0712	.0978	.1271	.1583	.1912	.2252
3	.0001	.0006	.0020	.0045	.0084	.0138	.0209	.0298	.0405	.0530
4			.0001	.0003	.0006	.0013	.0023	.0037	.0057	.0083
5						.0001	.0002	.0003	.0005	.0009
6										.0001

x^P	.11	.12	.13	.14	.15	.16	.17	.18	.19	.20
1	.6496	.6835	.7145	.7427	.7684	.7918	.8131	.8324	.8499	.8658
2	.2599	.2951	.3304	.3657	.4005	.4348	.4685	.5012	.5330	.5638
3	.0672	.0833	.1009	.1202	.1409	.1629	.1861	.2105	.2357	.2618
4	.0117	.0158	.0209	.0269	.0339	.0420	.0512	.0615	.0730	.0856
5	.0014	.0021	.0030	.0041	.0056	.0075	.0098	.0125	.0158	.0196
6	.0001	.0002	.0003	.0004	.0006	.0009	.0013	.0017	.0023	.0031
7						.0001	.0001	.0002	.0002	.0003

x^P	.21	.22	.23	.24	.25	.26	.27	.28	.29	.30
1	.8801	.8931	.9048	.9154	.9249	.9335	.9411	.9480	.9542	.9596
2	.5934	.6218	.6491	.6750	.6997	.7230	.7452	.7660	.7856	.8040
3	.2885	.3158	.3434	.3713	.3993	.4273	.4552	.4829	.5102	.5372
4	.0994	.1144	.1304	.1475	.1657	.1849	.2050	.2260	.2478	.2703
5	.0240	.0291	.0350	.0416	.0489	.0571	.0662	.0762	.0870	.0988
6	.0040	.0051	.0065	.0081	.0100	.0122	.0149	.0179	.0213	.0253
7	.0004	.0006	.0008	.0010	.0013	.0017	.0022	.0028	.0035	.0043
8			.0001	.0001	.0001	.0001	.0002	.0003	.0003	.0004

x^P	.31	.32	.33	.34	.35	.36	.37	.38	.39	.40
1	.9645	.9689	.9728	.9762	.9793	.9820	.9844	.9865	.9883	.9899
2	.8212	.8372	.8522	.8661	.8789	.8908	.9017	.9118	.9210	.9295
3	.5636	.5894	.6146	.6390	.6627	.6856	.7076	.7287	.7489	.7682
4	.2935	.3173	.3415	.3662	.3911	.4163	.4416	.4669	.4922	.5174
5	.1115	.1252	.1398	.1553	.1717	.1890	.2072	.2262	.2460	.2666
6	.0298	.0348	.0404	.0467	.0536	.0612	.0696	.0787	.0886	.0994
7	.0053	.0064	.0078	.0094	.0112	.0133	.0157	.0184	.0215	.0250
8	.0006	.0007	.0009	.0011	.0014	.0017	.0021	.0026	.0031	.0038
9				.0001	.0001	.0001	.0001	.0002	.0002	.0003

x^P	.41	.42	.43	.44	.45	.46	.47	.48	.49	.50
1	.9913	.9926	.9936	.9946	.9954	.9961	.9967	.9972	.9977	.9980
2	.9372	.9442	.9505	.9563	.9615	.9662	.9704	.9741	.9775	.9805
3	.7866	.8039	.8204	.8359	.8505	.8642	.8769	.8889	.8999	.9102
4	.5424	.5670	.5913	.6152	.6386	.6614	.6836	.7052	.7260	.7461
5	.2878	.3097	.3322	.3551	.3786	.4024	.4265	.4509	.4754	.5000
6	.1109	.1233	.1366	.1508	.1658	.1817	.1985	.2161	.2346	.2539
7	.0290	.0334	.0383	.0437	.0498	.0564	.0637	.0717	.0804	.0898
8	.0046	.0055	.0065	.0077	.0091	.0107	.0125	.0145	.0169	.0195
9	.0003	.0004	.0005	.0006	.0008	.0009	.0011	.0014	.0016	.0020

$$n = 10$$

x^P	01	.02	.03	.04	.05	.06	.07	.08	.09	.10
1	.0956	.1829	.2626	.3352	.4013	.4614	.5160	.5656	.6106	.6513
2	.0043	.0162	.0345	.0582	.0861	.1176	.1517	.1879	.2254	.2639
3	.0001	.0009	.0028	.0062	.0115	.0188	.0283	.0401	.0540	.0702
4			.0001	.0004	.0010	.0020	.0036	.0058	.0088	.0128
5					.0001	.0002	.0003	.0006	.0010	.0016
6									.0001	.0001

x\P	.11	.12	.13	.14	.15	.16	.17	.18	.19	.20
1	.6882	.7215	.7516	.7787	.8031	.8251	.8448	.8626	.8784	.8926
2	.3028	.3417	.3804	.4184	.4557	.4920	.5270	.5608	.5932	.6242
3	.0884	.1087	.1308	.1545	.1798	.2064	.2341	.2628	.2922	.3222
4	.0178	.0239	.0313	.0400	.0500	.0614	.0741	.0883	.1039	.1209
5	.0025	.0037	.0053	.0073	.0099	.0130	.0168	.0213	.0266	.0328
6	.0003	.0004	.0006	.0010	.0014	.0020	.0027	.0037	.0049	.0064
7		.0001	.0001	.0001	.0002	.0003	.0004	.0006	.0009
8									.0001	.0001

x\P	.21	.22	.23	.24	.25	.26	.27	.28	.29	.30
1	.9053	.9166	.9267	.9357	.9437	.9508	.9570	.9626	.9674	.9718
2	.6536	.6815	.7079	.7327	.7560	.7778	.7981	.8170	.8345	.8507
3	.3526	.3831	.4137	.4442	.4744	.5042	.5335	.5622	.5901	.6172
4	.1391	.1587	.1794	.2012	.2241	.2479	.2726	.2979	.3239	.3504
5	.0399	.0479	.0569	.0670	.0781	.0904	.1037	.1181	.1337	.1503
6	.0082	.0104	.0130	.0161	.0197	.0239	.0287	.0342	.0404	.0473
7	.0012	.0016	.0021	.0027	.0035	.0045	.0056	.0070	.0087	.0106
8	.0001	.0002	.0002	.0003	.0004	.0006	.0007	.0010	.0012	.0016
9							.0001	.0001	.0001	.0001

x\P	.31	.32	.33	.34	.35	.36	.37	.38	.39	.40
1	.9755	.9789	.9818	.9843	.9865	.9885	.9902	.9916	.9929	.9940
2	.8656	.8794	.8920	.9035	.9140	.9236	.9323	.9402	.9473	.9536
3	.6434	.6687	.6930	.7162	.7384	.7595	.7794	.7983	.8160	.8327
4	.3772	.4044	.4316	.4589	.4862	.5132	.5400	.5664	.5923	.6177
5	.1679	.1867	.2064	.2270	.2485	.2708	.2939	.3177	.3420	.3669
6	.0551	.0637	.0732	.0836	.0949	.1072	.1205	.1348	.1500	.1662
7	.0129	.0155	.0185	.0220	.0260	.0305	.0356	.0413	.0477	.0548
8	.0020	.0025	.0032	.0039	.0048	.0059	.0071	.0086	.0103	.0123
9	.0002	.0003	.0003	.0004	.0005	.0007	.0009	.0011	.0014	.0017
10								.0001	.0001	.0001

x\P	.41	.42	.43	.44	.45	.46	.47	.48	.49	.50
1	.9949	.9957	.9964	.9970	.9975	.9979	.9983	.9986	.9988	.9990
2	.9594	.9645	.9691	.9731	.9767	.9799	.9827	.9852	.9874	.9893
3	.8483	.8628	.8764	.8889	.9004	.9111	.9209	.9298	.9379	.9453
4	.6425	.6665	.6898	.7123	.7340	.7547	.7745	.7933	.8112	.8281
5	.3922	.4178	.4436	.4696	.4956	.5216	.5474	.5730	.5982	.6230
6	.1834	.2016	.2207	.2407	.2616	.2832	.3057	.3288	.3526	.3770
7	.0626	.0712	.0806	.0908	.1020	.1141	.1271	.1410	.1560	.1719
8	.0146	.0172	.0202	.0236	.0274	.0317	.0366	.0420	.0480	.0547
9	.0021	.0025	.0031	.0037	.0045	.0054	.0065	.0077	.0091	.0107
10	.0001	.0002	.0002	.0003	.0003	.0004	.0005	.0006	.0008	.0010

n = 11

x\P	.01	.02	.03	.04	.05	.06	.07	.08	.09	.10
1	.1047	.1993	.2847	.3618	.4312	.4937	.5499	.6004	.6456	.6862
2	.0052	.0195	.0413	.0692	.1019	.1382	.1772	.2181	.2601	.3026
3	.0002	.0012	.0037	.0083	.0152	.0248	.0370	.0519	.0695	.0896
4			.0002	.0007	.0016	.0030	.0053	.0085	.0129	.0185
5					.0001	.0003	.0005	.0010	.0017	.0028
6								.0001	.0002	.0003

$x\backslash P$	.11	.12	.13	.14	.15	.16	.17	.18	.19	.20
1	.7225	.7549	.7839	.8097	.8327	.8531	.8712	.8873	.9015	.9141
2	.3452	.3873	.4286	.4689	.5078	.5453	.5811	.6151	.6474	.6779
3	.1120	.1366	.1632	.1915	.2212	.2521	.2839	.3164	.3494	.3826
4	.0256	.0341	.0442	.0560	.0694	.0846	.1013	.1197	.1397	.1611
5	.0042	.0061	.0087	.0119	.0159	.0207	.0266	.0334	.0413	.0504
6	.0005	.0008	.0012	.0018	.0027	.0037	.0051	.0068	.0090	.0117
7		.0001	.0001	.0002	.0003	.0005	.0007	.0010	.0014	.0020
8							.0001	.0001	.0002	.0002

$x\backslash P$	.21	.22	.23	.24	.25	.26	.27	.28	.29	.30
1	.9252	.9350	.9436	.9511	.9578	.9636	.9686	.9730	.9769	.9802
2	.7065	.7333	.7582	.7814	.8029	.8227	.8410	.8577	.8730	.8870
3	.4158	.4488	.4814	.5134	.5448	.5753	.6049	.6335	.6610	.6873
4	.1840	.2081	.2333	.2596	.2867	.3146	.3430	.3719	.4011	.4304
5	.0607	.0723	.0851	.0992	.1146	.1313	.1493	.1685	.1888	.2103
6	.0148	.0186	.0231	.0283	.0343	.0412	.0490	.0577	.0674	.0782
7	.0027	.0035	.0046	.0059	.0076	.0095	.0119	.0146	.0179	.0216
8	.0003	.0005	.0007	.0009	.0012	.0016	.0021	.0027	.0034	.0043
9			.0001	.0001	.0001	.0002	.0002	.0003	.0004	.0006

$x\backslash P$	.31	.32	.33	.34	.35	.36	.37	.38	.39	.40
1	.9831	.9856	.9878	.9896	.9912	.9926	.9938	.9948	.9956	.9964
2	.8997	.9112	.9216	.9310	.9394	.9470	.9537	.9597	.9650	.9698
3	.7123	.7361	.7587	.7799	.7999	.8186	.8360	.8522	.8672	.8811
4	.4598	.4890	.5179	.5464	.5744	.6019	.6286	.6545	.6796	.7037
5	.2328	.2563	.2807	.3059	.3317	.3581	.3850	.4122	.4397	.4672
6	.0901	.1031	.1171	.1324	.1487	.1661	.1847	.2043	.2249	.2465
7	.0260	.0309	.0366	.0430	.0501	.0581	.0670	.0768	.0876	.0994
8	.0054	.0067	.0082	.0101	.0122	.0148	.0177	.0210	.0249	.0293
9	.0008	.0010	.0013	.0016	.0020	.0026	.0032	.0039	.0048	.0059
10	.0001	.0001	.0001	.0002	.0002	.0003	.0004	.0005	.0006	.0007

$x\backslash P$	.41	.42	.43	.44	.45	.46	.47	.48	.49	.50
1	.9970	.9975	.9979	.9983	.9986	.9989	.9991	.9992	.9994	.9995
2	.9739	.9776	.9808	.9836	.9861	.9882	.9900	.9916	.9930	.9941
3	.8938	.9055	.9162	.9260	.9348	.9428	.9499	.9564	.9622	.9673
4	.7269	.7490	.7700	.7900	.8089	.8266	.8433	.8588	.8733	.8867
5	.4948	.5223	.5495	.5764	.6029	.6288	.6541	.6787	.7026	.7256
6	.2690	.2924	.3166	.3414	.3669	.3929	.4193	.4460	.4729	.5000
7	.1121	.1260	.1408	.1568	.1738	.1919	.2110	.2312	.2523	.2744
8	.0343	.0399	.0461	.0532	.0610	.0696	.0791	.0895	.1009	.1133
9	.0072	.0087	.0104	.0125	.0148	.0175	.0206	.0241	.0282	.0327
10	.0009	.0012	.0014	.0018	.0022	.0027	.0033	.0040	.0049	.0059
11	.0001	.0001	.0001	.0001	.0002	.0002	.0002	.0003	.0004	.0005

n = 12

x\p	.01	.02	.03	.04	.05	.06	.07	.08	.09	.10
1	.1136	.2153	.3062	.3873	.4596	.5241	.5814	.6323	.6775	.7176
2	.0062	.0231	.0486	.0809	.1184	.1595	.2033	.2487	.2948	.3410
3	.0002	.0015	.0048	.0107	.0196	.0316	.0468	.0652	.0866	.1109
4		.0001	.0003	.0010	.0022	.0043	.0075	.0120	.0180	.0256
5				.0001	.0002	.0004	.0009	.0016	.0027	.0043
6							.0001	.0002	.0003	.0005
7										.0001

x\p	.11	.12	.13	.14	.15	.16	.17	.18	.19	.20
1	.7530	.7843	.8120	.8363	.8578	.8766	.8931	.9076	.9202	.9313
2	.3867	.4314	.4748	.5166	.5565	.5945	.6304	.6641	.6957	.7251
3	.1377	.1667	.1977	.2303	.2642	.2990	.3344	.3702	.4060	.4417
4	.0351	.0464	.0597	.0750	.0922	.1114	.1324	.1552	.1795	.2054
5	.0065	.0095	.0133	.0181	.0239	.0310	.0393	.0489	.0600	.0726
6	.0009	.0014	.0022	.0033	.0046	.0065	.0088	.0116	.0151	.0194
7	.0001	.0002	.0003	.0004	.0007	.0010	.0015	.0021	.0029	.0039
8					.0001	.0001	.0002	.0003	.0004	.0006
9										.0001

x\p	.21	.22	.23	.24	.25	.26	.27	.28	.29	.30
1	.9409	.9493	.9566	.9629	.9683	.9730	.9771	.9806	.9836	.9862
2	.7524	.7776	.8009	.8222	.8416	.8594	.8755	.8900	.9032	.9150
3	.4768	.5114	.5450	.5778	.6093	.6397	.6687	.6963	.7225	.7472
4	.2326	.2610	.2904	.3205	.3512	.3824	.4137	.4452	.4765	.5075
5	.0866	.1021	.1192	.1377	.1576	.1790	.2016	.2254	.2504	.2763
6	.0245	.0304	.0374	.0453	.0544	.0646	.0760	.0887	.1026	.1178
7	.0052	.0068	.0089	.0113	.0143	.0178	.0219	.0267	.0322	.0386
8	.0008	.0011	.0016	.0021	.0028	.0036	.0047	.0060	.0076	.0095
9	.0001	.0001	.0002	.0003	.0004	.0005	.0007	.0010	.0013	.0017
10						.0001	.0001	.0001	.0002	.0002

x\p	.31	.32	.33	.34	.35	.36	.37	.38	.39	.40
1	.9884	.9902	.9918	.9932	.9943	.9953	.9961	.9968	.9973	.9978
2	.9256	.9350	.9435	.9509	.9576	.9634	.9685	.9730	.9770	.9804
3	.7704	.7922	.8124	.8313	.8487	.8648	.8795	.8931	.9054	.9166
4	.5381	.5681	.5973	.6258	.6533	.6799	.7053	.7296	.7528	.7747
5	.3032	.3308	.3590	.3876	.4167	.4459	.4751	.5043	.5332	.5618
6	.1343	.1521	.1711	.1913	.2127	.2352	.2588	.2833	.3087	.3348
7	.0458	.0540	.0632	.0734	.0846	.0970	.1106	.1253	.1411	.1582
8	.0118	.0144	.0176	.0213	.0255	.0304	.0359	.0422	.0493	.0573
9	.0022	.0028	.0036	.0045	.0056	.0070	.0086	.0104	.0127	.0153
10	.0003	.0004	.0005	.0007	.0008	.0011	.0014	.0018	.0022	.0028
11				.0001	.0001	.0001	.0001	.0002	.0002	.0003

x\p	.41	.42	.43	.44	.45	.46	.47	.48	.49	.50
1	.9982	.9986	.9988	.9990	.9992	.9994	.9995	.9996	.9997	.9998
2	.9834	.9860	.9882	.9901	.9917	.9931	.9943	.9953	.9961	.9968
3	.9267	.9358	.9440	.9513	.9579	.9637	.9688	.9733	.9773	.9807
4	.7953	.8147	.8329	.8498	.8655	.8801	.8934	.9057	.9168	.9270
5	.5899	.6175	.6443	.6704	.6956	.7198	.7430	.7652	.7862	.8062
6	.3616	.3889	.4167	.4448	.4731	.5014	.5297	.5577	.5855	.6128
7	.1765	.1959	.2164	.2380	.2607	.2843	.3089	.3343	.3604	.3872
8	.0662	.0760	.0869	.0988	.1117	.1258	.1411	.1575	.1751	.1938
9	.0183	.0218	.0258	.0304	.0356	.0415	.0481	.0555	.0638	.0730
10	.0035	.0043	.0053	.0065	.0079	.0095	.0114	.0137	.0163	.0193
11	.0004	.0005	.0007	.0009	.0011	.0014	.0017	.0021	.0026	.0032
12				.0001	.0001	.0001	.0001	.0001	.0002	.0002

n = 13

x\P	.01	.02	.03	.04	.05	.06	.07	.08	.09	.10
1	.1225	.2310	.3270	.4118	.4867	.5526	.6107	.6617	.7065	.7458
2	.0072	.0270	.0564	.0932	.1354	.1814	.2298	.2794	.3293	.3787
3	.0003	.0020	.0062	.0135	.0245	.0392	.0578	.0799	.1054	.1339
4		.0001	.0005	.0014	.0031	.0060	.0103	.0163	.0242	.0342
5				.0001	.0003	.0007	.0013	.0024	.0041	.0065
6						.0001	.0001	.0003	.0005	.0009
7									.0001	.0001

x\P	.11	.12	.13	.14	.15	.16	.17	.18	.19	.20
1	.7802	.8102	.8364	.8592	.8791	.8963	.9113	.9242	.9354	.9450
2	.4270	.4738	.5186	.5614	.6017	.6396	.6751	.7080	.7384	.7664
3	.1651	.1985	.2337	.2704	.3080	.3463	.3848	.4231	.4611	.4983
4	.0464	.0609	.0776	.0967	.1180	.1414	.1667	.1939	.2226	.2527
5	.0097	.0139	.0193	.0260	.0342	.0438	.0551	.0681	.0827	.0991
6	.0015	.0024	.0036	.0053	.0075	.0104	.0139	.0183	.0237	.0300
7	.0002	.0003	.0005	.0008	.0013	.0019	.0027	.0038	.0052	.0070
8			.0001	.0001	.0002	.0003	.0004	.0006	.0009	.0012
9								.0001	.0001	.0002

x\P	.21	.22	.23	.24	.25	.26	.27	.28	.29	.30
1	.9533	.9604	.9666	.9718	.9762	.9800	.9833	.9860	.9883	.9903
2	.7920	.8154	.8367	.8559	.8733	.8889	.9029	.9154	.9265	.9363
3	.5347	.5699	.6039	.6364	.6674	.6968	.7245	.7505	.7749	.7975
4	.2839	.3161	.3489	.3822	.4157	.4493	.4826	.5155	.5478	.5794
5	.1173	.1371	.1585	.1816	.2060	.2319	.2589	.2870	.3160	.3457
6	.0375	.0462	.0562	.0675	.0802	.0944	.1099	.1270	.1455	.1654
7	.0093	.0120	.0154	.0195	.0243	.0299	.0365	.0440	.0527	.0624
8	.0017	.0024	.0032	.0043	.0056	.0073	.0093	.0118	.0147	.0182
9	.0002	.0004	.0005	.0007	.0010	.0013	.0018	.0024	.0031	.0040
10			.0001	.0001	.0001	.0002	.0003	.0004	.0005	.0007
11									.0001	.0001

x\P	.31	.32	.33	.34	.35	.36	.37	.38	.39	.40
1	.9920	.9934	.9945	.9955	.9963	.9970	.9975	.9980	.9984	.9987
2	.9450	.9527	.9594	.9653	.9704	.9749	.9787	.9821	.9849	.9874
3	.8185	.8379	.8557	.8720	.8868	.9003	.9125	.9235	.9333	.9421
4	.6101	.6398	.6683	.6957	.7217	.7464	.7698	.7917	.8123	.8314
5	.3760	.4067	.4376	.4686	.4995	.5301	.5603	.5899	.6188	.6470
6	.1867	.2093	.2331	.2581	.2841	.3111	.3388	.3673	.3962	.4256
7	.0733	.0854	.0988	.1135	.1295	.1468	.1654	.1853	.2065	.2288
8	.0223	.0271	.0326	.0390	.0462	.0544	.0635	.0738	.0851	.0977
9	.0052	.0065	.0082	.0102	.0126	.0154	.0187	.0225	.0270	.0321
10	.0009	.0012	.0015	.0020	.0025	.0032	.0040	.0051	.0063	.0078
11	.0001	.0001	.0002	.0003	.0003	.0005	.0006	.0008	.0010	.0013
12							.0001	.0001	.0001	.0001

x\P	.41	.42	.43	.44	.45	.46	.47	.48	.49	.50
1	.9990	.9992	.9993	.9995	.9996	.9997	.9997	.9998	.9998	.9999
2	.9895	.9912	.9928	.9940	.9951	.9960	.9967	.9974	.9979	.9983
3	.9499	.9569	.9630	.9684	.9731	.9772	.9808	.9838	.9865	.9888
4	.8492	.8656	.8807	.8945	.9071	.9185	.9288	.9381	.9464	.9539
5	.6742	.7003	.7254	.7493	.7721	.7935	.8137	.8326	.8502	.8666
6	.4552	.4849	.5146	.5441	.5732	.6019	.6299	.6573	.6838	.7095
7	.2524	.2770	.3025	.3290	.3563	.3842	.4127	.4415	.4707	.5000
8	.1114	.1264	.1426	.1600	.1788	.1988	.2200	.2424	.2659	.2905
9	.0379	.0446	.0520	.0605	.0698	.0803	.0918	.1045	.1183	.1334
10	.0096	.0117	.0141	.0170	.0203	.0242	.0287	.0338	.0396	.0461
11	.0017	.0021	.0027	.0033	.0041	.0051	.0063	.0077	.0093	.0112
12	.0002	.0002	.0003	.0004	.0005	.0007	.0009	.0011	.0014	.0017
13							.0001	.0001	.0001	.0001

n = 14

x\P	.01	.02	.03	.04	.05	.06	.07	.08	.09	.10
1	.1313	.2464	.3472	.4353	.5123	.5795	.6380	.6888	.7330	.7712
2	.0084	.0310	.0645	.1059	.1530	.2037	.2564	.3100	.3632	.4154
3	.0003	.0025	.0077	.0167	.0301	.0478	.0698	.0958	.1255	.1584
4		.0001	.0006	.0019	.0042	.0080	.0136	.0214	.0315	.0441
5				.0002	.0004	.0010	.0020	.0035	.0059	.0092
6						.0001	.0002	.0004	.0008	.0015
7									.0001	.0002

x\P	.11	.12	.13	.14	.15	.16	.17	.18	.19	.20
1	.8044	.8330	.8577	.8789	.8972	.9129	.9264	.9379	.9477	.9560
2	.4658	.5141	.5599	.6031	.6433	.6807	.7152	.7469	.7758	.8021
3	.1939	.2315	.2708	.3111	.3521	.3932	.4341	.4744	.5138	.5519
4	.0594	.0774	.0979	.1210	.1465	.1742	.2038	.2351	.2679	.3018
5	.0137	.0196	.0269	.0359	.0467	.0594	.0741	.0907	.1093	.1298
6	.0024	.0038	.0057	.0082	.0115	.0157	.0209	.0273	.0349	.0439
7	.0003	.0006	.0009	.0015	.0022	.0032	.0046	.0064	.0087	.0116
8		.0001	.0001	.0002	.0003	.0005	.0008	.0012	.0017	.0024
9						.0001	.0001	.0002	.0003	.0004

x\P	.21	.22	.23	.24	.25	.26	.27	.28	.29	.30
1	.9631	.9691	.9742	.9786	.9822	.9852	.9878	.9899	.9917	.9932
2	.8259	.8473	.8665	.8837	.8990	.9126	.9246	.9352	.9444	.9525
3	.5887	.6239	.6574	.6891	.7189	.7467	.7727	.7967	.8188	.8392
4	.3366	.3719	.4076	.4432	.4787	.5136	.5479	.5813	.6137	.6448
5	.1523	.1765	.2023	.2297	.2585	.2884	.3193	.3509	.3832	.4158
6	.0543	.0662	.0797	.0949	.1117	.1301	.1502	.1718	.1949	.2195
7	.0152	.0196	.0248	.0310	.0383	.0467	.0563	.0673	.0796	.0933
8	.0033	.0045	.0060	.0079	.0103	.0132	.0167	.0208	.0257	.0315
9	.0006	.0008	.0011	.0016	.0022	.0029	.0038	.0050	.0065	.0083
10	.0001	.0001	.0002	.0002	.0003	.0005	.0007	.0009	.0012	.0017
11						.0001	.0001	.0001	.0002	.0002

x\P	.31	.32	.33	.34	.35	.36	.37	.38	.39	.40
1	.9945	.9955	.9963	.9970	.9976	.9981	.9984	.9988	.9990	.9992
2	.9596	.9657	.9710	.9756	.9795	.9828	.9857	.9881	.9902	.9919
3	.8577	.8746	.8899	.9037	.9161	.9271	.9370	.9457	.9534	.9602
4	.6747	.7032	.7301	.7556	.7795	.8018	.8226	.8418	.8595	.8757
5	.4486	.4813	.5138	.5458	.5773	.6080	.6378	.6666	.6943	.7207
6	.2454	.2724	.3006	.3297	.3595	.3899	.4208	.4519	.4831	.5141
7	.1084	.1250	.1431	.1626	.1836	.2059	.2296	.2545	.2805	.3075
8	.0381	.0458	.0545	.0643	.0753	.0876	.1012	.1162	.1325	.1501
9	.0105	.0131	.0163	.0200	.0243	.0294	.0353	.0420	.0497	.0583
10	.0022	.0029	.0037	.0048	.0060	.0076	.0095	.0117	.0144	.0175
11	.0003	.0005	.0006	.0008	.0011	.0014	.0019	.0024	.0031	.0039
12		.0001	.0001	.0001	.0001	.0002	.0003	.0003	.0005	.0006
13										.0001

x\P	.41	.42	.43	.44	.45	.46	.47	.48	.49	.50
1	.9994	.9995	.9996	.9997	.9998	.9998	.9999	.9999	.9999	.9999
2	.9934	.9946	.9956	.9964	.9971	.9977	.9981	.9985	.9988	.9991
3	.9661	.9713	.9758	.9797	.9830	.9858	.9883	.9903	.9921	.9935
4	.8905	.9039	.9161	.9270	.9368	.9455	.9532	.9601	.9661	.9713
5	.7459	.7697	.7922	.8132	.8328	.8510	.8678	.8833	.8974	.9102
6	.5450	.5754	.6052	.6344	.6627	.6900	.7163	.7415	.7654	.7880
7	.3355	.3643	.3937	.4236	.4539	.4843	.5148	.5451	.5751	.6047
8	.1692	.1896	.2113	.2344	.2586	.2840	.3105	.3380	.3663	.3953
9	.0680	.0789	.0910	.1043	.1189	.1348	.1520	.1707	.1906	.2120
10	.0212	.0255	.0304	.0361	.0426	.0500	.0583	.0677	.0782	.0898
11	.0049	.0061	.0076	.0093	.0114	.0139	.0168	.0202	.0241	.0287
12	.0008	.0010	.0013	.0017	.0022	.0027	.0034	.0042	.0053	.0065
13	.0001	.0001	.0001	.0002	.0003	.0003	.0004	.0006	.0007	.0009
14										.0001

n = 15

x^P	.01	.02	.03	.04	.05	.06	.07	.08	.09	.10
1	.1399	.2614	.3667	.4579	.5367	.6047	.6633	.7137	.7570	.7941
2	.0096	.0353	.0730	.1191	.1710	.2262	.2832	.3403	.3965	.4510
3	.0004	.0030	.0094	.0203	.0362	.0571	.0829	.1130	.1469	.1841
4		.0002	.0008	.0024	.0055	.0104	.0175	.0273	.0399	.0556
5			.0001	.0002	.0006	.0014	.0028	.0050	.0082	.0127
6					.0001	.0001	.0003	.0007	.0013	.0022
7								.0001	.0002	.0003

x^P	.11	.12	.13	.14	.15	.16	.17	.18	.19	.20
1	.8259	.8530	.8762	.8959	.9126	.9269	.9389	.9490	.9576	.9648
2	.5031	.5524	.5987	.6417	.6814	.7179	.7511	.7813	.8085	.8329
3	.2238	.2654	.3084	.3520	.3958	.4392	.4819	.5234	.5635	.6020
4	.0742	.0959	.1204	.1476	.1773	.2092	.2429	.2782	.3146	.3518
5	.0187	.0265	.0361	.0478	.0617	.0778	.0961	.1167	.1394	.1642
6	.0037	.0057	.0084	.0121	.0168	.0227	.0300	.0387	.0490	.0611
7	.0006	.0010	.0015	.0024	.0036	.0052	.0074	.0102	.0137	.0181
8	.0001	.0001	.0002	.0004	.0006	.0010	.0014	.0021	.0030	.0042
9					.0001	.0001	.0002	.0003	.0005	.0008
10									.0001	.0001

x^P	.21	.22	.23	.24	.25	.26	.27	.28	.29	.30
1	.9709	.9759	.9802	.9837	.9866	.9891	.9911	.9928	.9941	.9953
2	.8547	.8741	.8913	.9065	.9198	.9315	.9417	.9505	.9581	.9647
3	.6385	.6731	.7055	.7358	.7639	.7899	.8137	.8355	.8553	.8732
4	.3895	.4274	.4650	.5022	.5387	.5742	.6086	.6416	.6732	.7031
5	.1910	.2195	.2495	.2810	.3135	.3469	.3810	.4154	.4500	.4845
6	.0748	.0905	.1079	.1272	.1484	.1713	.1958	.2220	.2495	.2704
7	.0234	.0298	.0374	.0463	.0566	.0684	.0817	.0965	.1130	.1311
8	.0058	.0078	.0104	.0135	.0173	.0219	.0274	.0338	.0413	.0500
9	.0011	.0016	.0023	.0031	.0042	.0056	.0073	.0094	.0121	.0152
10	.0002	.0003	.0004	.0006	.0008	.0011	.0015	.0021	.0028	.0037
11			.0001	.0001	.0001	.0002	.0002	.0003	.0005	.0007
12									.0001	.0001

x^P	.31	.32	.33	.34	.35	.36	.37	.38	.39	.40
1	.9962	.9969	.9975	.9980	.9984	.9988	.9990	.9992	.9994	.9995
2	.9704	.9752	.9794	.9829	.9858	.9883	.9904	.9922	.9936	.9948
3	.8893	.9038	.9167	.9281	.9383	.9472	.9550	.9618	.9678	.9729
4	.7314	.7580	.7829	.8060	.8273	.8469	.8649	.8813	.8961	.9095
5	.5187	.5523	.5852	.6171	.6481	.6778	.7062	.7332	.7587	.7827
6	.3084	.3393	.3709	.4032	.4357	.4684	.5011	.5335	.5654	.5968
7	.1509	.1722	.1951	.2194	.2452	.2722	.3003	.3295	.3595	.3902
8	.0599	.0711	.0837	.0977	.1132	.1302	.1487	.1687	.1902	.2131
9	.0190	.0236	.0289	.0351	.0422	.0504	.0597	.0702	.0820	.0950
10	.0048	.0062	.0079	.0099	.0124	.0154	.0190	.0232	.0281	.0338
11	.0009	.0012	.0016	.0022	.0028	.0037	.0047	.0059	..0075	.0093
12	.0001	.0002	.0003	.0004	.0005	.0006	.0009	.0011	.0015	.0019
13					.0001	.0001	.0001	.0002	.0002	.0003

$x\backslash^P$	.41	.42	.43	.44	.45	.46	.47	.48	.49	.50
1	.9996	.9997	.9998	.9998	.9999	.9999	.9999	.9999	1.0000	1.0000
2	.9958	.9966	.9973	.9979	.9983	.9987	.9990	.9992	.9994	.9995
3	.9773	.9811	.9843	.9870	.9893	.9913	.9929	.9943	.9954	.9963
4	.9215	.9322	.9417	.9502	.9576	.9641	.9697	.9746	.9788	.9824
5	.8052	.8261	.8454	.8633	.8796	.8945	.9080	.9201	.9310	.9408
6	.6274	.6570	.6856	.7131	.7392	.7641	.7875	.8095	.8301	.8491
7	.4214	.4530	.4847	.5164	.5478	.5789	.6095	.6394	.6684	.6964
8	.2374	.2630	.2898	.3176	.3465	.3762	.4065	.4374	.4686	.5000
9	.1095	.1254	.1427	.1615	.1818	.2034	.2265	.2510	.2767	.3036
10	.0404	.0479	.0565	.0661	.0769	.0890	.1024	.1171	.1333	.1509
11	.0116	.0143	.0174	.0211	.0255	.0305	.0363	.0430	.0506	.0592
12	.0025	.0032	.0040	.0051	.0063	.0079	.0097	.0119	.0145	.0176
13	.0004	.0005	.0007	.0009	.0011	.0014	.0018	.0023	.0029	.0037
14			.0001	.0001	.0001	.0002	.0002	.0003	.0004	.0005

n = 16

$x\backslash^P$	.01	.02	.03	.04	.05	.06	.07	.08	.09	.10
1	.1485	.2762	.3857	.4796	.5599	.6284	.6869	.7366	.7789	.8147
2	.0109	.0399	.0818	.1327	.1892	.2489	.3098	.3701	.4289	.4853
3	.0005	.0037	.0113	.0242	.0429	.0673	.0969	.1311	.1694	.2108
4		.0002	.0011	.0032	.0070	.0132	.0221	.0342	.0496	.0684
5			.0001	.0003	.0009	.0019	.0038	.0068	.0111	.0170
6					.0001	.0002	.0005	.0010	.0019	.0033
7							.0001	.0001	.0003	.0005
8										.0001

$x\backslash^P$	.11	.12	.13	.14	.15	.16	.17	.18	.19	.20
1	.8450	.8707	.8923	.9105	.9257	.9386	.9493	.9582	.9657	.9719
2	.5386	.5885	.6347	.6773	.7161	.7513	.7830	.8115	.8368	.8593
3	.2545	.2999	.3461	.3926	.4386	.4838	.5277	.5698	.6101	.6482
4	.0907	.1162	.1448	.1763	.2101	.2460	.2836	.3223	.3619	.4019
5	.0248	.0348	.0471	.0618	.0791	.0988	.1211	.1458	.1727	.2018
6	.0053	.0082	.0120	.0171	.0235	.0315	.0412	.0527	.0662	.0817
7	.0009	.0015	.0024	.0038	.0056	.0080	.0112	.0153	.0204	.0267
8	.0001	.0002	.0004	.0007	.0011	.0016	.0024	.0036	.0051	.0070
9				.0001	.0001	.0002	.0003	.0004	.0007	.0010	.0015
10								.0001	.0001	.0002	.0002

$x\backslash^P$	.21	.22	.23	.24	.25	.26	.27	.28	.29	.30
1	.9770	.9812	.9847	.9876	.9900	.9919	.9935	.9948	.9958	.9967
2	.8791	.8965	.9117	.9250	.9365	.9465	.9550	.9623	.9686	.9739
3	.6839	.7173	.7483	.7768	.8029	.8267	.8482	.8677	.8851	.9006
4	.4418	.4814	.5203	.5583	.5950	.6303	.6640	.6959	.7260	.7541
5	.2327	.2652	.2991	.3341	.3698	.4060	.4425	.4788	.5147	.5501
6	.0992	.1188	.1405	.1641	.1897	.2169	.2458	.2761	.3077	.3402
7	.0342	.0432	.0536	.0657	.0796	.0951	.1125	.1317	.1526	.1753
8	.0095	.0127	.0166	.0214	.0271	.0340	.0420	.0514	.0621	.0744
9	.0021	.0030	.0041	.0056	.0075	.0098	.0127	.0163	.0206	.0257
10	.0004	.0006	.0008	.0012	.0016	.0023	.0031	.0041	.0055	.0071
11	.0001	.0001	.0001	.0002	.0003	.0004	.0006	.0008	.0011	.0016
12						.0001	.0001	.0001	.0002	.0003

x^P	.31	.32	.33	.34	.35	.36	.37	.38	.39	.40
1	.9974	.9979	.9984	.9987	.9990	.9992	.9994	.9995	.9996	.9997
2	.9784	.9822	.9854	.9880	.9902	.9921	.9936	.9948	.9959	.9967
3	.9144	.9266	.9374	.9467	.9549	.9620	.9681	.9734	.9778	.9817
4	.7804	.8047	.8270	.8475	.8661	.8830	.8982	.9119	.9241	.9349
5	.5846	.6181	.6504	.6813	.7108	.7387	.7649	.7895	.8123	.8334
6	.3736	.4074	.4416	.4759	.5100	.5438	.5770	.6094	.6408	.6712
7	.1997	.2257	.2531	.2819	.3119	.3428	.3746	.4070	.4398	.4728
8	.0881	.1035	.1205	.1391	.1594	.1813	.2048	.2298	.2562	.2839
9	.0317	.0388	.0470	.0564	.0671	.0791	.0926	.1076	.1242	.1423
10	.0092	.0117	.0148	.0185	.0229	.0280	.0341	.0411	.0491	.0583
11	.0021	.0028	.0037	.0048	.0062	.0079	.0100	.0125	.0155	.0191
12	.0004	.0005	.0007	.0010	.0013	.0017	.0023	.0030	.0038	.0049
13		.0001	.0001	.0001	.0002	.0003	.0004	.0005	.0007	.0009
14								.0001	.0001	.0001

x^P	.41	.42	.43	.44	.45	.46	.47	.48	.49	.50
1	.9998	.9998	.9999	.9999	.9999	.9999	1.0000	1.0000	1.0000	1.0000
2	.9974	.9979	.9984	.9987	.9990	.9992	.9994	.9995	.9997	.9997
3	.9849	.9876	.9899	.9918	.9934	.9947	.9958	.9966	.9973	.9979
4	.9444	.9527	.9600	.9664	.9719	.9766	.9806	.9840	.9869	.9894
5	.8529	.8707	.8869	.9015	.9147	.9265	.9370	.9463	.9544	.9616
6	.7003	.7280	.7543	.7792	.8024	.8241	.8441	.8626	.8795	.8949
7	.5058	.5387	.5711	.6029	.6340	.6641	.6932	.7210	.7476	.7728
8	.3128	.3428	.3736	.4051	.4371	.4694	.5019	.5343	.5665	.5982
9	.1619	.1832	.2060	.2302	.2559	.2829	.3111	.3405	.3707	.4018
10	.0687	.0805	.0936	.1081	.1241	.1416	.1607	.1814	.2036	.2272
11	.0234	.0284	.0342	.0409	.0486	.0574	.0674	.0786	.0911	.1051
12	.0062	.0078	.0098	.0121	.0149	.0183	.0222	.0268	.0322	.0384
13	.0012	.0016	.0021	.0027	.0035	.0044	.0055	.0069	.0086	.0106
14	.0002	.0002	.0003	.0004	.0006	.0007	.0010	.0013	.0016	.0021
15					.0001	.0001	.0001	.0001	.0002	.0003

n = 17

x^P	.01	.02	.03	.04	.05	.06	.07	.08	.09	.10
1	.1571	.2907	.4042	.5004	.5819	.6507	.7088	.7577	.7988	.8332
2	.0123	.0446	.0909	.1465	.2078	.2717	.3362	.3995	.4604	.5182
3	.0006	.0044	.0134	.0286	.0503	.0782	.1118	.1503	.1927	.2382
4		.0003	.0014	.0040	.0088	.0164	.0273	.0419	.0603	.0826
5			.0001	.0004	.0012	.0026	.0051	.0089	.0145	.0221
6					.0001	.0003	.0007	.0015	.0027	.0047
7							.0001	.0002	.0004	.0008
8										.0001

x^P	.11	.12	.13	.14	.15	.16	.17	.18	.19	.20
1	.8621	.8862	.9063	.9230	.9369	.9484	.9579	.9657	.9722	.9775
2	.5723	.6223	.6682	.7099	.7475	.7813	.8113	.8379	.8613	.8818
3	.2858	.3345	.3836	.4324	.4802	.5266	.5711	.6133	.6532	.6904
4	.1087	.1383	.1710	.2065	.2444	.2841	.3251	.3669	.4091	.4511
5	.0321	.0446	.0598	.0778	.0987	.1224	.1487	.1775	.2087	.2418
6	.0075	.0114	.0166	.0234	.0319	.0423	.0548	.0695	.0864	.1057
7	.0014	.0023	.0037	.0056	.0083	.0118	.0163	.0220	.0291	.0377
8	.0002	.0004	.0007	.0011	.0017	.0027	.0039	.0057	.0080	.0109
9		.0001	.0001	.0002	.0003	.0005	.0008	.0012	.0018	.0026
10						.0001	.0001	.0002	.0003	.0005
11										.0001

x\P	.21	.22	.23	.24	.25	.26	.27	.28	.29	.30
1	.9818	.9854	.9882	.9906	.9925	.9940	.9953	.9962	.9970	.9977
2	.8996	.9152	.9285	.9400	.9499	.9583	.9654	.9714	.9765	.9807
3	.7249	.7567	.7859	.8123	.8363	.8578	.8771	.8942	.9093	.9226
4	.4927	.5333	.5728	.6107	.6470	.6814	.7137	.7440	.7721	.7981
5	.2766	.3128	.3500	.3879	.4261	.4643	.5023	.5396	.5760	.6113
6	.1273	.1510	.1770	.2049	.2347	.2661	.2989	.3329	.3677	.4032
7	.0479	.0598	.0736	.0894	.1071	.1268	.1485	.1721	.1976	.2248
8	.0147	.0194	.0251	.0320	.0402	.0499	.0611	.0739	.0884	.1046
9	.0037	.0051	.0070	.0094	.0124	.0161	.0206	.0261	.0326	.0403
10	.0007	.0011	.0016	.0022	.0031	.0042	.0057	.0075	.0098	.0127
11	.0001	.0002	.0003	.0004	.0006	.0009	.0013	.0018	.0024	.0032
12				.0001	.0001	.0002	.0002	.0003	.0005	.0007
13									.0001	.0001

x\P	.31	.32	.33	.34	.35	.36	.37	.38	.39	.40
1	.9982	.9986	.9989	.9991	.9993	.9995	.9996	.9997	.9998	.9998
2	.9843	.9872	.9896	.9917	.9933	.9946	.9957	.9966	.9973	.9979
3	.9343	.9444	.9532	.9608	.9673	.9728	.9775	.9815	.9849	.9877
4	.8219	.8437	.8634	.8812	.8972	.9115	.9241	.9353	.9450	.9536
5	.6453	.6778	.7087	.7378	.7652	.7906	.8142	.8360	.8559	.8740
6	.4390	.4749	.5105	.5458	.5803	.6139	.6465	.6778	.7077	.7361
7	.2536	.2838	.3153	.3479	.3812	.4152	.4495	.4839	.5182	.5522
8	.1227	.1426	.1642	.1877	.2128	.2395	.2676	.2971	.3278	.3595
9	.0492	.0595	.0712	.0845	.0994	.1159	.1341	.1541	.1757	.1989
10	.0162	.0204	.0254	.0314	.0383	.0464	.0557	.0664	.0784	.0919
11	.0043	.0057	.0074	.0095	.0120	.0151	.0189	.0234	.0286	.0348
12	.0009	.0013	.0017	.0023	.0030	.0040	.0051	.0066	.0084	.0106
13	.0002	.0002	.0003	.0004	.0006	.0008	.0011	.0015	.0019	.0025
14				.0001	.0001	.0001	.0002	.0002	.0003	.0005
15										.0001

x\P	.41	.42	.43	.44	.45	.46	.47	.48	.49	.50
1	.9999	.9999	.9999	.9999	1.0000	1.0000	1.0000	1.0000	1.0000	1.0000
2	.9984	.9987	.9990	.9992	.9994	.9996	.9997	.9998	.9998	.9999
3	.9900	.9920	.9935	.9948	.9959	.9968	.9975	.9980	.9985	.9988
4	.9610	.9674	.9729	.9776	.9816	.9849	.9877	.9901	.9920	.9936
5	.8904	.9051	.9183	.9301	.9404	.9495	.9575	.9644	.9704	.9755
6	.7628	.7879	.8113	.8330	.8529	.8712	.8878	.9028	.9162	.9283
7	.5856	.6182	.6499	.6805	.7098	.7377	.7641	.7890	.8122	.8338
8	.3920	.4250	.4585	.4921	.5257	.5590	.5918	.6239	.6552	.6855
9	.2238	.2502	.2780	.3072	.3374	.3687	.4008	.4335	.4667	.5000
10	.1070	.1236	.1419	.1618	.1834	.2066	.2314	.2577	.2855	.3145
11	.0420	.0503	.0597	.0705	.0826	.0962	.1112	.1279	.1462	.1662
12	.0133	.0165	.0203	.0248	.0301	.0363	.0434	.0517	.0611	.0717
13	.0033	.0042	.0054	.0069	.0086	.0108	.0134	.0165	.0202	.0245
14	.0006	.0008	.0011	.0014	.0019	.0024	.0031	.0040	.0050	.0064
15	.0001	.0001	.0002	.0002	.0003	.0004	.0005	.0007	.0009	.0012
16							.0001	.0001	.0001	.0001

$$n = 18$$

x\P	.01	.02	.03	.04	.05	.06	.07	.08	.09	.10
1	.1655	.3049	.4220	.5204	.6028	.6717	.7292	.7771	.8169	.8499
2	.0138	.0495	.1003	.1607	.2265	.2945	.3622	.4281	.4909	.5497
3	.0007	.0052	.0157	.0333	.0581	.0898	.1275	.1702	.2168	.2662
4		.0004	.0018	.0050	.0109	.0201	.0333	.0506	.0723	.0982
5			.0002	.0006	.0015	.0034	.0067	.0116	.0186	.0282
6				.0001	.0002	.0005	.0010	.0021	.0038	.0064
7							.0001	.0003	.0006	.0012
8									.0001	.0002

x\P	.11	.12	.13	.14	.15	.16	.17	.18	.19	.20
1	.8773	.8998	.9185	.9338	.9464	.9566	.9651	.9719	.9775	.9820
2	.6042	.6540	.6992	.7398	.7759	.8080	.8362	.8609	.8824	.9009
3	.3173	.3690	.4206	.4713	.5203	.5673	.6119	.6538	.6927	.7287
4	.1282	.1618	.1986	.2382	.2798	.3229	.3669	.4112	.4554	.4990
5	.0405	.0558	.0743	.0959	.1206	.1482	.1787	.2116	.2467	.2836
6	.0102	.0154	.0222	.0310	.0419	.0551	.0708	.0889	.1097	.1329
7	.0021	.0034	.0054	.0081	.0118	.0167	.0229	.0306	.0400	.0513
8	.0003	.0006	.0011	.0017	.0027	.0041	.0060	.0086	.0120	.0163
9		.0001	.0002	.0003	.0005	.0008	.0013	.0020	.0029	.0043
10					.0001	.0001	.0002	.0004	.0006	.0009
11								.0001	.0001	.0002

x\P	.21	.22	.23	.24	.25	.26	.27	.28	.29	.30
1	.9856	.9886	.9909	.9928	.9944	.9956	.9965	.9973	.9979	.9984
2	.9169	.9306	.9423	.9522	.9605	.9676	.9735	.9784	.9824	.9858
3	.7616	.7916	.8187	.8430	.8647	.8839	.9009	.9158	.9288	.9400
4	.5414	.5825	.6218	.6591	.6943	.7272	.7578	.7860	.8119	.8354
5	.3220	.3613	.4012	.4414	.4813	.5208	.5594	.5968	.6329	.6673
6	.1586	.1866	.2168	.2488	.2825	.3176	.3538	.3907	.4281	.4656
7	.0645	.0799	.0974	.1171	.1390	.1630	.1891	.2171	.2469	.2783
8	.0217	.0283	.0363	.0458	.0569	.0699	.0847	.1014	.1200	.1409
9	.0060	.0083	.0112	.0148	.0193	.0249	.0316	.0395	.0488	.0596
10	.0014	.0020	.0028	.0039	.0054	.0073	.0097	.0127	.0164	.0210
11	.0003	.0004	.0006	.0009	.0012	.0018	.0025	.0034	.0046	.0061
12		.0001	.0001	.0002	.0002	.0003	.0005	.0007	.0010	.0014
13						.0001	.0001	.0001	.0002	.0003

x\P	.31	.32	.33	.34	.35	.36	.37	.38	.39	.40
1	.9987	.9990	.9993	.9994	.9996	.9997	.9998	.9998	.9999	.9999
2	.9886	.9908	.9927	.9942	.9954	.9964	.9972	.9978	.9983	.9987
3	.9498	.9581	.9652	.9713	.9764	.9807	.9843	.9873	.9897	.9918
4	.8568	.8759	.8931	.9083	.9217	.9335	.9439	.9528	.9606	.9672
5	.7001	.7309	.7598	.7866	.8114	.8341	.8549	.8737	.8907	.9058
6	.5029	.5398	.5759	.6111	.6450	.6776	.7086	.7379	.7655	.7912
7	.3111	.3450	.3797	.4151	.4509	.4867	.5224	.5576	.5921	.6257
8	.1633	.1878	.2141	.2421	.2717	.3027	.3349	.3681	.4021	.4366
9	.0720	.0861	.1019	.1196	.1391	.1604	.1835	.2084	.2350	.2632
10	.0264	.0329	.0405	.0494	.0597	.0714	.0847	.0997	.1163	.1347
11	.0080	.0104	.0133	.0169	.0212	.0264	.0325	.0397	.0480	.0576
12	.0020	.0027	.0036	.0047	.0062	.0080	.0102	.0130	.0163	.0203
13	.0004	.0005	.0008	.0011	.0014	.0019	.0026	.0034	.0044	.0058
14	.0001	.0001	.0001	.0002	.0003	.0004	.0005	.0007	.0010	.0013
15						.0001	.0001	.0001	.0002	.0002

x\P	.41	.42	.43	.44	.45	.46	.47	.48	.49	.50
1	.9999	.9999	1.0000	1.0000	1.0000	1.0000	1.0000	1.0000	1.0000	1.0000
2	.9990	.9992	.9994	.9996	.9997	.9998	.9998	.9999	.9999	.9999
3	.9934	.9948	.9959	.9968	.9975	.9981	.9985	.9989	.9991	.9993
4	.9729	.9777	.9818	.9852	.9880	.9904	.9923	.9939	.9952	.9962
5	.9193	.9313	.9418	.9510	.9589	.9658	.9717	.9767	.9810	.9846
6	.8151	.8372	.8573	.8757	.8923	.9072	.9205	.9324	.9428	.9519
7	.6582	.6895	.7193	.7476	.7742	.7991	.8222	.8436	.8632	.8811
8	.4713	.5062	.5408	.5750	.6085	.6412	.6728	.7032	.7322	.7597
9	.2928	.3236	.3556	.3885	.4222	.4562	.4906	.5249	.5591	.5927
10	.1549	.1768	.2004	.2258	.2527	.2812	.3110	.3421	.3742	.4073
11	.0686	.0811	.0951	.1107	.1280	.1470	.1677	.1902	.2144	.2403
12	.0250	.0307	.0372	.0449	.0537	.0638	.0753	.0883	.1028	.1189
13	.0074	.0094	.0118	.0147	.0183	.0225	.0275	.0334	.0402	.0481
14	.0017	.0022	.0029	.0038	.0049	.0063	.0079	.0100	.0125	.0154
15	.0003	.0004	.0006	.0007	.0010	.0013	.0017	.0023	.0029	.0038
16		.0001	.0001	.0001	.0001	.0002	.0003	.0004	.0005	.0007
17									.0001	.0001

n = 19

x\P	.01	.02	.03	.04	.05	.06	.07	.08	.09	.10
1	.1738	.3188	.4394	.5396	.6226	.6914	.7481	.7949	.8334	.8649
2	.0153	.0546	.1100	.1751	.2453	.3171	.3879	.4560	.5202	.5797
3	.0009	.0061	.0183	.0384	.0665	.1021	.1439	.1908	.2415	.2946
4		.0005	.0022	.0061	.0132	.0243	.0398	.0602	.0853	.1150
5			.0002	.0007	.0020	.0044	.0085	.0147	.0235	.0352
6				.0001	.0002	.0006	.0014	.0029	.0051	.0086
7						.0001	.0002	.0004	.0009	.0017
8								.0001	.0001	.0003

x\P	.11	.12	.13	.14	.15	.16	.17	.18	.19	.20
1	.8908	.9119	.9291	.9431	.9544	.9636	.9710	.9770	.9818	.9856
2	.6342	.6835	.7277	.7669	.8015	.8318	.8581	.8809	.9004	.9171
3	.3488	.4032	.4568	.5089	.5587	.6059	.6500	.6910	.7287	.7631
4	.1490	.1867	.2275	.2708	.3159	.3620	.4085	.4549	.5005	.5449
5	.0502	.0685	.0904	.1158	.1444	.1762	.2107	.2476	.2864	.3267
6	.0135	.0202	.0290	.0401	.0537	.0700	.0891	.1110	.1357	.1631
7	.0030	.0048	.0076	.0113	.0163	.0228	.0310	.0411	.0532	.0676
8	.0005	.0009	.0016	.0026	.0041	.0061	.0089	.0126	.0173	.0233
9	.0001	.0002	.0003	.0005	.0008	.0014	.0021	.0032	.0047	.0067
10				.0001	.0001	.0002	.0004	.0007	.0010	.0016
11							.0001	.0001	.0002	.0003

x\P	.21	.22	.23	.24	.25	.26	.27	.28	.29	.30
1	.9887	.9911	.9930	.9946	.9958	.9967	.9975	.9981	.9985	.9989
2	.9313	.9434	.9535	.9619	.9690	.9749	.9797	.9837	.9869	.9896
3	.7942	.8222	.8471	.8692	.8887	.9057	.9205	.9333	.9443	.9538
4	.5877	.6285	.6671	.7032	.7369	.7680	.7965	.8224	.8458	.8668
5	.3681	.4100	.4520	.4936	.5346	.5744	.6129	.6498	.6848	.7178
6	.1929	.2251	.2592	.2950	.3322	.3705	.4093	.4484	.4875	.5261
7	.0843	.1034	.1248	.1487	.1749	.2032	.2336	.2657	.2995	.3345
8	.0307	.0396	.0503	.0629	.0775	.0941	.1129	.1338	.1568	.1820
9	.0093	.0127	.0169	.0222	.0287	.0366	.0459	.0568	.0694	.0839
10	.0023	.0034	.0047	.0066	.0089	.0119	.0156	.0202	.0258	.0326
11	.0005	.0007	.0011	.0016	.0023	.0032	.0044	.0060	.0080	.0105
12	.0001	.0001	.0002	.0003	.0005	.0007	.0010	.0015	.0021	.0028
13				.0001	.0001	.0001	.0002	.0003	.0004	.0006
14									.0001	.0001

x\P	.31	.32	.33	.34	.35	.36	.37	.38	.39	.40
1	.9991	.9993	.9995	.9996	.9997	.9998	.9998	.9999	.9999	.9999
2	.9917	.9935	.9949	.9960	.9969	.9976	.9981	.9986	.9989	.9992
3	.9618	.9686	.9743	.9791	.9830	.9863	.9890	.9913	.9931	.9945
4	.8856	.9022	.9169	.9297	.9409	.9505	.9588	.9659	.9719	.9770
5	.7486	.7773	.8037	.8280	.8500	.8699	.8878	.9038	.9179	.9304
6	.5641	.6010	.6366	.6707	.7032	.7339	.7627	.7895	.8143	.8371
7	.3705	.4073	.4445	.4818	.5188	.5554	.5913	.6261	.6597	.6919
8	.2091	.2381	.2688	.3010	.3344	.3690	.4043	.4401	.4762	.5122
9	.1003	.1186	.1389	.1612	.1855	.2116	.2395	.2691	.3002	.3325
10	.0405	.0499	.0608	.0733	.0875	.1035	.1213	.1410	.1626	.1861
11	.0137	.0176	.0223	.0280	.0347	.0426	.0518	.0625	.0747	.0885
12	.0038	.0051	.0068	.0089	.0114	.0146	.0185	.0231	.0287	.0352
13	.0009	.0012	.0017	.0023	.0031	.0041	.0054	.0070	.0091	.0116
14	.0002	.0002	.0003	.0005	.0007	.0009	.0013	.0017	.0023	.0031
15			.0001	.0001	.0001	.0002	.0002	.0003	.0005	.0006
16									.0001	.0001

x\P	.41	.42	.43	.44	.45	.46	.47	.48	.49	.50
1	1.0000	1.0000	1.0000	1.0000	1.0000	1.0000	1.0000	1.0000	1.0000	1.0000
2	.9994	.9995	.9996	.9997	.9998	.9999	.9999	.9999	.9999	1.0000
3	.9957	.9967	.9974	.9980	.9985	.9988	.9991	.9993	.9995	.9996
4	.9813	.9849	.9878	.9903	.9923	.9939	.9952	.9963	.9971	.9978
5	.9413	.9508	.9590	.9660	.9720	.9771	.9814	.9850	.9879	.9904
6	.8579	.8767	.8937	.9088	.9223	.9342	.9446	.9537	.9615	.9682
7	.7226	.7515	.7787	.8039	.8273	.8488	.8684	.8862	.9022	.9165
8	.5480	.5832	.6176	.6509	.6831	.7138	.7430	.7706	.7964	.8204
9	.3660	.4003	.4353	.4706	.5060	.5413	.5762	.6105	.6439	.6762
10	.2114	.2385	.2672	.2974	.3290	.3617	.3954	.4299	.4648	.5000
11	.1040	.1213	.1404	.1613	.1841	.2087	.2351	.2631	.2928	.3238
12	.0429	.0518	.0621	.0738	.0871	.1021	.1187	.1372	.1575	.1796
13	.0146	.0183	.0227	.0280	.0342	.0415	.0500	.0597	.0709	.0835
14	.0040	.0052	.0067	.0086	.0109	.0137	.0171	.0212	.0261	.0318
15	.0009	.0012	.0016	.0021	.0028	.0036	.0046	.0060	.0076	.0096
16	.0001	.0002	.0003	.0004	.0005	.0007	.0010	.0013	.0017	.0022
17				.0001	.0001	.0001	.0001	.0002	.0003	.0004

$$n = 20$$

x\P	.01	.02	.03	.04	.05	.06	.07	.08	.09	.10
1	.1821	.3324	.4562	.5580	.6415	.7099	.7658	.8113	.8484	.8784
2	.0169	.0599	.1198	.1897	.2642	.3395	.4131	.4831	.5484	.6083
3	.0010	.0071	.0210	.0439	.0755	.1150	.1610	.2121	.2666	.3231
4		.0006	.0027	.0074	.0159	.0290	.0471	.0706	.0993	.1330
5			.0003	.0010	.0026	.0056	.0107	.0183	.0290	.0432
6				.0001	.0003	.0009	.0019	.0038	.0068	.0113
7						.0001	.0003	.0006	.0013	.0024
8								.0001	.0002	.0004
9										.0001

x\P	.11	.12	.13	.14	.15	.16	.17	.18	.19	.20
1	.9028	.9224	.9383	.9510	.9612	.9694	.9759	.9811	.9852	.9885
2	.6624	.7109	.7539	.7916	.8244	.8529	.8773	.8982	.9159	.9308
3	.3802	.4369	.4920	.5450	.5951	.6420	.6854	.7252	.7614	.7939
4	.1710	.2127	.2573	.3041	.3523	.4010	.4496	.4974	.5439	.5886
5	.0610	.0827	.1083	.1375	.1702	.2059	.2443	.2849	.3271	.3704
6	.0175	.0260	.0370	.0507	.0673	.0870	.1098	.1356	.1643	.1958
7	.0041	.0067	.0103	.0153	.0219	.0304	.0409	.0537	.0689	.0867
8	.0008	.0014	.0024	.0038	.0059	.0088	.0127	.0177	.0241	.0321
9	.0001	.0002	.0005	.0008	.0013	.0021	.0033	.0049	.0071	.0100
10			.0001	.0001	.0002	.0004	.0007	.0011	.0017	.0026
11						.0001	.0001	.0002	.0004	.0006
12									.0001	.0001

x\P	.21	.22	.23	.24	.25	.26	.27	.28	.29	.30
1	.9910	.9931	.9946	.9959	.9968	.9976	.9982	.9986	.9989	.9992
2	.9434	.9539	.9626	.9698	.9757	.9805	.9845	.9877	.9903	.9924
3	.8230	.8488	.8716	.8915	.9087	.9237	.9365	.9474	.9567	.9645
4	.6310	.6711	.7085	.7431	.7748	.8038	.8300	.8534	.8744	.8929
5	.4142	.4580	.5014	.5439	.5852	.6248	.6625	.6981	.7315	.7625
6	.2297	.2657	.3035	.3427	.3828	.4235	.4643	.5048	.5447	.5836
7	.1071	.1301	.1557	.1838	.2142	.2467	.2810	.3169	.3540	.3920
8	.0419	.0536	.0675	.0835	.1018	.1225	.1455	.1707	.1982	.2277
9	.0138	.0186	.0246	.0320	.0409	.0515	.0640	.0784	.0948	.1133
10	.0038	.0054	.0075	.0103	.0139	.0183	.0238	.0305	.0385	.0480
11	.0009	.0013	.0019	.0028	.0039	.0055	.0074	.0100	.0132	.0171
12	.0002	.0003	.0004	.0006	.0009	.0014	.0019	.0027	.0038	.0051
13			.0001	.0001	.0002	.0003	.0004	.0006	.0009	.0013
14							.0001	.0001	.0002	.0003

x\P	.31	.32	.33	.34	.35	.36	.37	.38	.39	.40
1	.9994	.9996	.9997	.9998	.9998	.9999	.9999	.9999	.9999	1.0000
2	.9940	.9953	.9964	.9972	.9979	.9984	.9988	.9991	.9993	.9995
3	.9711	.9765	.9811	.9848	.9879	.9904	.9924	.9940	.9953	.9964
4	.9092	.9235	.9358	.9465	.9556	.9634	.9700	.9755	.9802	.9840
5	.7911	.8173	.8411	.8626	.8818	.8989	.9141	.9274	.9390	.9490
6	.6213	.6574	.6917	.7242	.7546	.7829	.8090	.8329	.8547	.8744
7	.4305	.4693	.5079	.5460	.5834	.6197	.6547	.6882	.7200	.7500
8	.2591	.2922	.3268	.3624	.3990	.4361	.4735	.5108	.5478	.5841
9	.1340	.1568	.1818	.2087	.2376	.2683	.3005	.3341	.3688	.4044
10	.0591	.0719	.0866	.1032	.1218	.1424	.1650	.1897	.2163	.2447
11	.0220	.0279	.0350	.0434	.0532	.0645	.0775	.0923	.1090	.1275
12	.0069	.0091	.0119	.0154	.0196	.0247	.0308	.0381	.0466	.0565
13	.0018	.0025	.0034	.0045	.0060	.0079	.0102	.0132	.0167	.0210
14	.0004	.0006	.0008	.0011	.0015	.0021	.0028	.0037	.0049	.0065
15	.0001	.0001	.0001	.0002	.0003	.0004	.0006	.0009	.0012	.0016
16						.0001	.0001	.0002	.0002	.0003

x\P	.41	.42	.43	.44	.45	.46	.47	.48	.49	.50
1	1.0000	1.0000	1.0000	1.0000	1.0000	1.0000	1.0000	1.0000	1.0000	1.0000
2	.9996	.9997	.9998	.9998	.9999	.9999	.9999	1.0000	1.0000	1.0000
3	.9972	.9979	.9984	.9988	.9991	.9993	.9995	.9996	.9997	.9998
4	.9872	.9898	.9920	.9937	.9951	.9962	.9971	.9977	.9983	.9987
5	.9577	.9651	.9714	.9767	.9811	.9848	.9879	.9904	.9924	.9941
6	.8921	.9078	.9217	.9340	.9447	.9539	.9619	.9687	.9745	.9793
7	.7780	.8041	.8281	.8501	.8701	.8881	.9042	.9186	.9312	.9423
8	.6196	.6539	.6868	.7183	.7480	.7759	.8020	.8261	.8482	.8684
9	.4406	.4771	.5136	.5499	.5857	.6207	.6546	.6873	.7186	.7483
10	.2748	.3064	.3394	.3736	.4086	.4443	.4804	.5166	.5525	.5881
11	.1480	.1705	.1949	.2212	.2493	.2791	.3104	.3432	.3771	.4119
12	.0679	.0810	.0958	.1123	.1308	.1511	.1734	.1977	.2238	.2517
13	.0262	.0324	.0397	.0482	.0580	.0694	.0823	.0969	.1133	.1316
14	.0084	.0107	.0136	.0172	.0214	.0265	.0326	.0397	.0480	.0577
15	.0022	.0029	.0038	.0050	.0064	.0083	.0105	.0133	.0166	.0207
16	.0004	.0006	.0008	.0011	.0015	.0020	.0027	.0035	.0046	.0059
17	.0001	.0001	.0001	.0002	.0003	.0004	.0005	.0007	.0010	.0013
18						.0001	.0001	.0001	.0001	.0002

n = 25

$x \backslash P$	.01	.02	.03	.04	.05	.06	.07	.08	.09	.10
1	.2222	.3965	.5330	.6396	.7226	.7871	.8370	.8756	.9054	.9282
2	.0258	.0886	.1720	.2642	.3576	.4473	.5304	.6053	.6714	.7288
3	.0020	.0132	.0380	.0765	.1271	.1871	.2534	.3232	.3937	.4629
4	.0001	.0014	.0062	.0165	.0341	.0598	.0936	.1351	.1831	.2364
5		.0001	.0008	.0028	.0072	.0150	.0274	.0451	.0686	.0980
6			.0001	.0004	.0012	.0031	.0065	.0123	.0210	.0334
7					.0002	.0005	.0013	.0028	.0054	.0095
8						.0001	.0002	.0005	.0011	.0023
9								.0001	.0002	.0005
10										.0001

$x \backslash P$	.11	.12	.13	.14	.15	.16	.17	.18	.19	.20
1	.9457	.9591	.9692	.9770	.9828	.9872	.9905	.9930	.9948	.9962
2	.7779	.8195	.8543	.8832	.9069	.9263	.9420	.9546	.9646	.9726
3	.5291	.5912	.6483	.7000	.7463	.7870	.8226	.8533	.8796	.9018
4	.2934	.3525	.4123	.4714	.5289	.5837	.6352	.6829	.7266	.7660
5	.1331	.1734	.2183	.2668	.3179	.3707	.4241	.4772	.5292	.5793
6	.0499	.0709	.0965	.1268	.1615	.2002	.2425	.2875	.3347	.3833
7	.0156	.0243	.0359	.0509	.0695	.0920	.1185	.1488	.1827	.2200
8	.0041	.0070	.0113	.0173	.0255	.0361	.0495	.0661	.0859	.1091
9	.0009	.0017	.0030	.0050	.0080	.0121	.0178	.0252	.0348	.0468
10	.0002	.0004	.0007	.0013	.0021	.0035	.0055	.0083	.0122	.0173
11		.0001	.0001	.0003	.0005	.0009	.0015	.0024	.0037	.0056
12					.0001	.0002	.0003	.0006	.0010	.0015
13							.0001	.0001	.0002	.0004
14										.0001

$x \backslash P$	.21	.22	.23	.24	.25	.26	.27	.28	.29	.30
1	.9972	.9980	.9985	.9990	.9992	.9995	.9996	.9997	.9998	.9999
2	.9789	.9838	.9877	.9907	.9930	.9947	.9961	.9971	.9979	.9984
3	.9204	.9360	.9488	.9593	.9679	.9748	.9804	.9848	.9883	.9910
4	.8013	.8324	.8597	.8834	.9038	.9211	.9358	.9481	.9583	.9668
5	.6270	.6718	.7134	.7516	.7863	.8174	.8452	.8696	.8910	.9095
6	.4325	.4816	.5299	.5767	.6217	.6644	.7044	.7415	.7755	.8065
7	.2601	.3027	.3471	.3927	.4389	.4851	.5308	.5753	.6183	.6593
8	.1358	.1658	.1989	.2349	.2735	.3142	.3565	.3999	.4440	.4882
9	.0614	.0788	.0993	.1228	.1494	.1790	.2115	.2465	.2838	.3231
10	.0240	.0325	.0431	.0560	.0713	.0893	.1101	.1338	.1602	.1894
11	.0082	.0117	.0163	.0222	.0297	.0389	.0502	.0636	.0795	.0978
12	.0024	.0036	.0053	.0076	.0107	.0148	.0199	.0264	.0345	.0442
13	.0006	.0010	.0015	.0023	.0034	.0049	.0069	.0096	.0130	.0175
14	.0001	.0002	.0004	.0006	.0009	.0014	.0021	.0030	.0043	.0060
15			.0001	.0001	.0002	.0003	.0005	.0008	.0012	.0018
16						.0001	.0001	.0002	.0003	.0005
17									.0001	.0001

x\P	.31	.32	.33	.34	.35	.36	.37	.38	.39	.40
1	.9999	.9999	1.0000	1.0000	1.0000	1.0000	1.0000	1.0000	1.0000	1.0000
2	.9989	.9992	.9994	.9996	.9997	.9998	.9998	.9999	.9999	.9999
3	.9932	.9949	.9961	.9971	.9979	.9984	.9989	.9992	.9994	.9996
4	.9737	.9793	.9838	.9874	.9903	.9926	.9944	.9958	.9968	.9976
5	.9254	.9390	.9504	.9600	.9680	.9745	.9799	.9842	.9877	.9905
6	.8344	.8593	.8813	.9006	.9174	.9318	.9441	.9546	.9633	.9706
7	.6981	.7343	.7679	.7987	.8266	.8517	.8742	.8940	.9114	.9264
8	.5319	.5747	.6163	.6561	.6939	.7295	.7626	.7932	.8211	.8464
9	.3639	.4057	.4482	.4908	.5332	.5748	.6152	.6542	.6914	.7265
10	.2213	.2555	.2919	.3300	.3697	.4104	.4517	.4933	.5347	.5754
11	.1188	.1424	.1686	.1975	.2288	.2624	.2981	.3355	.3743	.4142
12	.0560	.0698	.0859	.1044	.1254	.1490	.1751	.2036	.2346	.2677
13	.0230	.0299	.0383	.0485	.0604	.0745	.0907	.1093	.1303	.1538
14	.0083	.0112	.0149	.0196	.0255	.0326	.0412	.0515	.0637	.0778
15	.0026	.0036	.0050	.0069	.0093	.0124	.0163	.0212	.0271	.0344
16	.0007	.0010	.0015	.0021	.0029	.0041	.0056	.0075	.0100	.0132
17	.0002	.0002	.0004	.0005	.0008	.0011	.0016	.0023	.0032	.0043
18			.0001	.0001	.0002	.0003	.0004	.0006	.0008	.0012
19						.0001	.0001	.0001	.0002	.0003
20										.0001

x\P	.41	.42	.43	.44	.45	.46	.47	.48	.49	.50
1	1.0000	1.0000	1.0000	1.0000	1.0000	1.0000	1.0000	1.0000	1.0000	1.0000
2	1.0000	1.0000	1.0000	1.0000	1.0000	1.0000	1.0000	1.0000	1.0000	1.0000
3	.9997	.9998	.9998	.9999	.9999	1.0000	1.0000	1.0000	1.0000	1.0000
4	.9983	.9987	.9991	.9993	.9995	.9997	.9998	.9998	.9999	.9999
5	.9927	.9945	.9958	.9969	.9977	.9983	.9988	.9991	.9994	.9995
6	.9767	.9816	.9856	.9888	.9914	.9934	.9950	.9963	.9972	.9980
7	.9394	.9505	.9599	.9677	.9742	.9796	.9840	.9876	.9904	.9927
8	.8692	.8894	.9071	.9227	.9361	.9477	.9575	.9658	.9727	.9784
9	.7593	.7897	.8177	.8431	.8660	.8865	.9046	.9205	.9343	.9461
10	.6151	.6535	.6902	.7250	.7576	.7880	.8160	.8415	.8646	.8852
11	.4548	.4956	.5363	.5765	.6157	.6538	.6902	.7249	.7574	.7878
12	.3029	.3397	.3780	.4174	.4574	.4978	.5382	.5780	.6171	.6550
13	.1797	.2080	.2387	.2715	.3063	.3429	.3808	.4199	.4598	.5000
14	.0941	.1127	.1336	.1569	.1827	.2109	.2413	.2740	.3086	.3450
15	.0431	.0535	.0656	.0797	.0960	.1145	.1353	.1585	.1841	.2122
16	.0171	.0220	.0280	.0353	.0440	.0543	.0663	.0803	.0964	.1148
17	.0058	.0078	.0103	.0134	.0174	.0222	.0281	.0352	.0438	.0539
18	.0017	.0023	.0032	.0044	.0058	.0077	.0102	.0132	.0170	.0216
19	.0004	.0006	.0008	.0012	.0016	.0023	.0031	.0041	.0055	.0073
20	.0001	.0001	.0002	.0003	.0004	.0005	.0008	.0011	.0015	.0020
21					.0001	.0001	.0002	.0002	.0003	.0005
22									.0001	.0001

n = 50

x^P	.01	.02	.03	.04	.05	.06	.07	.08	.09	.10
1	.3950	.6358	.7819	.8701	.9231	.9547	.9734	.9845	.9910	.9948
2	.0894	.2642	.5995	.5995	.7206	.8100	.8735	.9173	.9468	.9662
3	.0138	.0784	.1892	.3233	.4595	.5838	.6892	.7740	.8395	.8883
4	.0016	.0178	.0628	.1391	.2396	.3527	.4673	.5747	.6697	.7497
5	.0001	.0032	.0168	.0490	.1036	.1794	.2710	.3710	.4723	.5688
6		.0005	.0037	.0144	.0378	.0776	.1350	.2081	.2928	.3839
7		.0001	.0007	.0036	.0118	.0289	.0583	.1019	.1596	.2298
8			.0001	.0008	.0032	.0094	.0220	.0438	.0768	.1221
9				.0001	.0008	.0027	.0073	.0167	.0328	.0579
10					.0002	.0007	.0022	.0056	.0125	.0245
11						.0002	.0006	.0017	.0043	.0094
12							.0001	.0005	.0013	.0032
13								.0001	.0004	.0010
14									.0001	.0003
15										.0001

x^P	.11	.12	.13	.14	.15	.16	.17	.18	.19	.20
1	.9971	.9983	.9991	.9995	.9997	.9998	.9999	1.0000	1.0000	1.0000
2	.9788	.9869	.9920	.9951	.9971	.9983	.9990	.9994	.9997	.9998
3	.9237	.9487	.9661	.9779	.9858	.9910	.9944	.9965	.9979	.9987
4	.8146	.8655	.9042	.9330	.9540	.9688	.9792	.9863	.9912	.9943
5	.6562	.7320	.7956	.8472	.8879	.9192	.9428	.9601	.9726	.9815
6	.4760	.5647	.6463	.7186	.7806	.8323	.8741	.9071	.9327	.9520
7	.3091	.3935	.4789	.5616	.6387	.7081	.7686	.8199	.8624	.8966
8	.1793	.2467	.3217	.4010	.4812	.5594	.6328	.6996	.7587	.8096
9	.0932	.1392	.1955	.2605	.3319	.4071	.4832	.5576	.6280	.6927
10	.0435	.0708	.1074	.1537	.2089	.2718	.3403	.4122	.4849	.5563
11	.0183	.0325	.0535	.0824	.1199	.1661	.2203	.2813	.3473	.4164
12	.0069	.0135	.0242	.0402	.0628	.0929	.1309	.1768	.2300	.2893
13	.0024	.0051	.0100	.0179	.0301	.0475	.0714	.1022	.1405	.1861
14	.0008	.0018	.0037	.0073	.0132	.0223	.0357	.0544	.0791	.1106
15	.0002	.0006	.0013	.0027	.0053	.0096	.0164	.0266	.0411	.0607
16	.0001	.0002	.0004	.0009	.0019	.0038	.0070	.0120	.0197	.0308
17			.0001	.0003	.0007	.0014	.0027	.0050	.0087	.0144
18				.0001	.0002	.0005	.0010	.0019	.0036	.0063
19					.0001	.0001	.0003	.0007	.0013	.0025
20							.0001	.0002	.0005	.0009
21								.0001	.0002	.0003
22										.0001

x^p	.21	.22	.23	.24	.25	.26	.27	.28	.29	.30
1	1.0000	1.0000	1.0000	1.0000	1.0000	1.0000	1.0000	1.0000	1.0000	1.0000
2	.9999	.9999	1.0000	1.0000	1.0000	1.0000	1.0000	1.0000	1.0000	1.0000
3	.9992	.9995	.9997	.9998	.9999	1.0000	1.0000	1.0000	1.0000	1.0000
4	.9964	.9978	.9986	.9992	.9995	.9997	.9998	.9999	.9999	1.0000
5	.9877	.9919	.9948	.9967	.9979	.9987	.9992	.9995	.9997	.9998
6	.9663	.9767	.9841	.9893	.9930	.9954	.9970	.9981	.9988	.9993
7	.9236	.9445	.9603	.9720	.9806	.9868	.9911	.9941	.9961	.9975
8	.8523	.8874	.9156	.9377	.9547	.9676	.9772	.9842	.9892	.9927
9	.7505	.8009	.8437	.8794	.9084	.9316	.9497	.9635	.9740	.9817
10	.6241	.6870	.7436	.7934	.8363	.8724	.9021	.9260	.9450	.9598
11	.4864	.5552	.6210	.6822	.7378	.7871	.8299	.8663	.8965	.9211
12	.3533	.4201	.4878	.5544	.6184	.6782	.7329	.7817	.8244	.8610
13	.2383	.2963	.3585	.4233	.4890	.5539	.6163	.6749	.7287	.7771
14	.1490	.1942	.2456	.3023	.3630	.4261	.4901	.5534	.6145	.6721
15	.0862	.1181	.1565	.2013	.2519	.3075	.3669	.4286	.4912	.5532
16	.0462	.0665	.0926	.1247	.1631	.2075	.2575	.3121	.3703	.4308
17	.0229	.0347	.0508	.0718	.0983	.1306	.1689	.2130	.2623	.3161
18	.0105	.0168	.0259	.0384	.0551	.0766	.1034	.1359	.1741	.2178
19	.0045	.0075	.0122	.0191	.0287	.0418	.0590	.0809	.1080	.1406
20	.0018	.0031	.0054	.0088	.0139	.0212	.0314	.0449	.0626	.0848
21	.0006	.0012	.0022	.0038	.0063	.0100	.0155	.0232	.0338	.0478
22	.0002	.0004	.0008	.0015	.0026	.0044	.0071	.0112	.0170	.0251
23	.0001	.0001	.0003	.0006	.0010	.0018	.0031	.0050	.0080	.0123
24			.0001	.0002	.0004	.0007	.0012	.0021	.0035	.0056
25				.0001	.0001	.0002	.0004	.0008	.0014	.0024
26						.0001	.0002	.0003	.0005	.0009
27								.0001	.0002	.0003
28									.0001	.0001

x^p	.31	.32	.33	.34	.35	.36	.37	.38	.39	.40
1	1.0000	1.0000	1.0000	1.0000	1.0000	1.0000	1.0000	1.0000	1.0000	1.0000
2	1.0000	1.0000	1.0000	1.0000	1.0000	1.0000	1.0000	1.0000	1.0000	1.0000
3	1.0000	1.0000	1.0000	1.0000	1.0000	1.0000	1.0000	1.0000	1.0000	1.0000
4	1.0000	1.0000	1.0000	1.0000	1.0000	1.0000	1.0000	1.0000	1.0000	1.0000
5	.9999	.9999	1.0000	1.0000	1.0000	1.0000	1.0000	1.0000	1.0000	1.0000
6	.9996	.9997	.9998	.9999	.9999	1.0000	1.0000	1.0000	1.0000	1.0000
7	.9984	.9990	.9994	.9996	.9998	.9999	.9999	1.0000	1.0000	1.0000
8	.9952	.9969	.9980	.9987	.9992	.9995	.9997	.9998	.9999	.9999
9	.9874	.9914	.9942	.9962	.9975	.9984	.9990	.9994	.9996	.9998
10	.9710	.9794	.9856	.9901	.9933	.9955	.9971	.9981	.9988	.9992
11	.9409	.9563	.9683	.9773	.9840	.9889	.9924	.9949	.9966	.9978
12	.8916	.9168	.9371	.9533	.9658	.9753	.9825	.9878	.9916	.9943
13	.8197	.8564	.8873	.9130	.9339	.9505	.9635	.9736	.9811	.9867
14	.7253	.7732	.8157	.8524	.8837	.9097	.9310	.9481	.9616	.9720
15	.6131	.6698	.7223	.7699	.8122	.8491	.8805	.9069	.9286	.9460
16	.4922	.5530	.6120	.6679	.7199	.7672	.8094	.8462	.8779	.9045
17	.3734	.4328	.4931	.5530	.6111	.6664	.7179	.7649	.8070	.8439
18	.2666	.3197	.3760	.4346	.4940	.5531	.6105	.6653	.7164	.7631
19	.1786	.2220	.2703	.3227	.3784	.4362	.4949	.5533	.6101	.6644
20	.1121	.1447	.1826	.2257	.2736	.3255	.3805	.4376	.4957	.5535
21	.0657	.0882	.1156	.1482	.1861	.2289	.2764	.3278	.3824	.4390
22	.0360	.0503	.0685	.0912	.1187	.1513	.1890	.2317	.2788	.3299
23	.0184	.0267	.0379	.0525	.0710	.0938	.1214	.1540	.1916	.2340
24	.0087	.0133	.0196	.0282	.0396	.0544	.0730	.0960	.1236	.1562
25	.0039	.0061	.0094	.0141	.0207	.0295	.0411	.0560	.0748	.0978
26	.0016	.0026	.0042	.0066	.0100	.0149	.0216	.0305	.0423	.0573
27	.0006	.0011	.0018	.0029	.0045	.0070	.0106	.0155	.0223	.0314
28	.0002	.0004	.0007	.0012	.0019	.0031	.0048	.0074	.0110	.0160
29	.0001	.0001	.0002	.0004	.0007	.0012	.0020	.0032	.0050	.0076
30			.0001	.0002	.0003	.0005	.0008	.0013	.0021	.0034
31					.0001	.0002	.0003	.0005	.0008	.0014
32						.0001	.0001	.0002	.0003	.0005
33								.0001	.0001	.0002
34										.0001

n = 50

x\P	.41	.42	.43	.44	.45	.46	.47	.48	.49	.50
1	1.0000	1.0000	1.0000	1.0000	1.0000	1.0000	1.0000	1.0000	1.0000	1.0000
2	1.0000	1.0000	1.0000	1.0000	1.0000	1.0000	1.0000	1.0000	1.0000	1.0000
3	1.0000	1.0000	1.0000	1.0000	1.0000	1.0000	1.0000	1.0000	1.0000	1.0000
4	1.0000	1.0000	1.0000	1.0000	1.0000	1.0000	1.0000	1.0000	1.0000	1.0000
5	1.0000	1.0000	1.0000	1.0000	1.0000	1.0000	1.0000	1.0000	1.0000	1.0000
6	1.0000	1.0000	1.0000	1.0000	1.0000	1.0000	1.0000	1.0000	1.0000	1.0000
7	1.0000	1.0000	1.0000	1.0000	1.0000	1.0000	1.0000	1.0000	1.0000	1.0000
8	1.0000	1.0000	1.0000	1.0000	1.0000	1.0000	1.0000	1.0000	1.0000	1.0000
9	.9999	.9999	1.0000	1.0000	1.0000	1.0000	1.0000	1.0000	1.0000	1.0000
10	.9995	.9997	.9998	.9999	.9999	1.0000	1.0000	1.0000	1.0000	1.0000
11	.9986	.9991	.9994	.9997	.9998	.9999	.9999	1.0000	1.0000	1.0000
12	.9962	.9975	.9984	.9990	.9994	.9996	.9998	.9999	.9999	1.0000
13	.9908	.9938	.9958	.9973	.9982	.9989	.9993	.9996	.9997	.9998
14	.9799	.9858	.9902	.9933	.9955	.9970	.9981	.9988	.9992	.9995
15	.9599	.9707	.9789	.9851	.9896	.9929	.9952	.9968	.9980	.9987
16	.9265	.9443	.9585	.9696	.9780	.9844	.9892	.9926	.9950	.9967
17	.8757	.9025	.9248	.9429	.9573	.9687	.9774	.9839	.9888	.9923
18	.8051	.8421	.8740	.9010	.9235	.9418	.9565	.9680	.9769	.9836
19	.7152	.7617	.8037	.8406	.8727	.8998	.9225	.9410	.9559	.9675
20	.6099	.6638	.7143	.7608	.8026	.8396	.8718	.8991	.9219	.9405
21	.4965	.5539	.6099	.6635	.7138	.7602	.8020	.8391	.8713	.8987
22	.3840	.4402	.4973	.5543	.6100	.6634	.7137	.7599	.8018	.8389
23	.2809	.3316	.3854	.4412	.4981	.5548	.6104	.6636	.7138	.7601
24	.1936	.2359	.2826	.3331	.3866	.4422	.4989	.5554	.6109	.6641
25	.1255	.1580	.1953	.2375	.2840	.3343	.3876	.4431	.4996	.5561
26	.0762	.0992	.1269	.1593	.1966	.2386	.2850	.3352	.3885	.4439
27	.0432	.0584	.0772	.1003	.1279	.1603	.1975	.2395	.2858	.3359
28	.0229	.0320	.0439	.0591	.0780	.1010	.1286	.1609	.1981	.2399
29	.0113	.0164	.0233	.0325	.0444	.0595	.0784	.1013	.1289	.1611
30	.0052	.0078	.0115	.0166	.0235	.0327	.0446	.0596	.0784	.1013
31	.0022	.0034	.0053	.0079	.0116	.0167	.0236	.0327	.0445	.0595
32	.0009	.0014	.0022	.0035	.0053	.0079	.0116	.0166	.0234	.0325
33	.0003	.0005	.0009	.0014	.0022	.0035	.0053	.0078	.0114	.0164
34	.0001	.0002	.0003	.0005	.0009	.0014	.0022	.0034	.0052	.0077
35		.0001	.0001	.0002	.0003	.0005	.0008	.0014	.0021	.0033
36				.0001	.0001	.0002	.0003	.0005	.0008	.0013
37						.0001	.0001	.0002	.0003	.0005
38								.0001	.0001	.0002

n = 100

x\P	.01	.02	.03	.04	.05	.06	.07	.08	.09	.10
1	.6340	.8674	.9524	.9831	.9941	.9979	.9993	.9998	.9999	1.0000
2	.2642	.5967	.8054	.9128	.9629	.9848	.9940	.9977	.9991	.9997
3	.0794	.3233	.5802	.7679	.8817	.9434	.9742	.9887	.9952	.9981
4	.0184	.1410	.3528	.5705	.7422	.8570	.9256	.9633	.9827	.9922
5	.0034	.0508	.1821	.3711	.5640	.7232	.8368	.9097	.9526	.9763
6	.0005	.0155	.0808	.2116	.3840	.5593	.7086	.8201	.8955	.9424
7	.0001	.0041	.0312	.1064	.2340	.3936	.5557	.6968	.8060	.8828
8		.0009	.0106	.0475	.1280	.2517	.4012	.5529	.6872	.7939
9		.0002	.0032	.0190	.0631	.1463	.2660	.4074	.5506	.6791
10			.0009	.0068	.0282	.0775	.1620	.2780	.4125	.5487
11		.0002		.0022	.0115	.0376	.0908	.1757	.2882	.4168
12			.0007	.0043	.0168	.0469	.1028	.1876	.2970
13			.0002	.0015	.0069	.0224	.0559	.1138	.1982
14				.0005	.0026	.0099	.0282	.0645	.1239
15				.0001	.0009	.0041	.0133	.0341	.0726
16						.0003	.0016	.0058	.0169	.0399
17						.0001	.0006	.0024	.0078	.0206
18							.0002	.0009	.0034	.0100
19							.0001	.0003	.0014	.0046
20								.0001	.0005	.0020
21									.0002	.0008
22									.0001	.0003
23										.0001

x\P	.11	.12	.13	.14	.15	.16	.17	.18	.19	.20
1	1.0000	1.0000	1.0000	1.0000	1.0000	1.0000	1.0000	1.0000	1.0000	1.0000
2	.9999	1.0000	1.0000	1.0000	1.0000	1.0000	1.0000	1.0000	1.0000	1.0000
3	.9992	.9997	.9999	1.0000	1.0000	1.0000	1.0000	1.0000	1.0000	1.0000
4	.9966	.9985	.9994	.9998	.9999	1.0000	1.0000	1.0000	1.0000	1.0000
5	.9886	.9947	.9977	.9990	.9996	.9998	.9999	1.0000	1.0000	1.0000
6	.9698	.9848	.9926	.9966	.9984	.9993	.9997	.9999	1.0000	1.0000
7	.9328	.9633	.9808	.9903	.9953	.9978	.9990	.9996	.9998	.9999
8	.8715	.9239	.9569	.9766	.9878	.9939	.9970	.9986	.9994	.9997
9	.7835	.8614	.9155	.9508	.9725	.9853	.9924	.9962	.9982	.9991
10	.6722	.7743	.8523	.9078	.9449	.9684	.9826	.9908	.9953	.9977
11	.5471	.6663	.7663	.8440	.9006	.9393	.9644	.9800	.9891	.9943
12	.4206	.5458	.6611	.7591	.8365	.8939	.9340	.9605	.9773	.9874
13	.3046	.4239	.5446	.6566	.7527	.8297	.8876	.9289	.9567	.9747
14	.2076	.3114	.4268	.5436	.6526	.7469	.8234	.8819	.9241	.9531
15	.1330	.2160	.3173	.4294	.5428	.6490	.7417	.8177	.8765	.9196
16	.0802	.1414	.2236	.3227	.4317	.5420	.6458	.7370	.8125	.8715
17	.0456	.0874	.1492	.2305	.3275	.4338	.5414	.6429	.7327	.8077
18	.0244	.0511	.0942	.1563	.2367	.3319	.4357	.5408	.6403	.7288
19	.0123	.0282	.0564	.1006	.1628	.2424	.3359	.4374	.5403	.6379
20	.0059	.0147	.0319	.0614	.1065	.1689	.2477	.3395	.4391	.5398
21	.0026	.0073	.0172	.0356	.0663	.1121	.1745	.2525	.3429	.4405
22	.0011	.0034	.0088	.0196	.0393	.0710	.1174	.1797	.2570	.3460
23	.0005	.0015	.0042	.0103	.0221	.0428	.0754	.1223	.1846	.2611
24	.0002	.0006	.0020	.0051	.0119	.0246	.0462	.0796	.1270	.1891
25	.0001	.0003	.0009	.0024	.0061	.0135	.0271	.0496	.0837	.1314
26		.0001	.0004	.0011	.0030	.0071	.0151	.0295	.0528	.0875
27			.0001	.0005	.0014	.0035	.0081	.0168	.0318	.0558
28			.0001	.0002	.0006	.0017	.0041	.0091	.0184	.0342
29				.0001	.0003	.0008	.0020	.0048	.0102	.0200
30					.0001	.0003	.0009	.0024	.0054	.0112
31						.0001	.0004	.0011	.0027	.0061
32						.0001	.0002	.0005	.0013	.0031
33							.0001	.0002	.0006	.0016
34								.0001	.0003	.0007
35									.0001	.0003
36										.0001
37										.0001

$$n = 100$$

$x\backslash P$	.21	.22	.23	.24	.25	.26	.27	.28	.29	.30
1	1.0000	1.0000	1.0000	1.0000	1.0000	1.0000	1.0000	1.0000	1.0000	1.0000
2	1.0000	1.0000	1.0000	1.0000	1.0000	1.0000	1.0000	1.0000	1.0000	1.0000
3	1.0000	1.0000	1.0000	1.0000	1.0000	1.0000	1.0000	1.0000	1.0000	1.0000
4	1.0000	1.0000	1.0000	1.0000	1.0000	1.0000	1.0000	1.0000	1.0000	1.0000
5	1.0000	1.0000	1.0000	1.0000	1.0000	1.0000	1.0000	1.0000	1.0000	1.0000
6	1.0000	1.0000	1.0000	1.0000	1.0000	1.0000	1.0000	1.0000	1.0000	1.0000
7	1.0000	1.0000	1.0000	1.0000	1.0000	1.0000	1.0000	1.0000	1.0000	1.0000
8	.9999	1.0000	1.0000	1.0000	1.0000	1.0000	1.0000	1.0000	1.0000	1.0000
9	.9996	.9998	.9999	1.0000	1.0000	1.0000	1.0000	1.0000	1.0000	1.0000
10	.9989	.9995	.9998	.9999	1.0000	1.0000	1.0000	1.0000	1.0000	1.0000
11	.9971	.9986	.9993	.9997	.9999	.9999	1.0000	1.0000	1.0000	1.0000
12	.9933	.9965	.9983	.9992	.9996	.9998	.9999	1.0000	1.0000	1.0000
13	.9857	.9922	.9959	.9979	.9990	.9995	.9998	.9999	1.0000	1.0000
14	.9721	.9840	.9911	.9953	.9975	.9988	.9994	.9997	.9999	.9999
15	.9496	.9695	.9823	.9900	.9946	.9972	.9986	.9993	.9997	.9998
16	.9153	.9462	.9671	.9806	.9889	.9939	.9967	.9983	.9992	.9996
17	.8668	.9112	.9430	.9647	.9789	.9878	.9932	.9963	.9981	.9990
18	.8032	.8625	.9074	.9399	.9624	.9773	.9867	.9925	.9959	.9978
19	.7252	.7991	.8585	.9038	.9370	.9601	.9757	.9856	.9918	.9955
20	.6358	.7220	.7953	.8547	.9005	.9342	.9580	.9741	.9846	.9911
21	.5394	.6338	.7189	.7918	.8512	.8973	.9316	.9560	.9726	.9835
22	.4419	.5391	.6320	.7162	.7886	.8479	.8943	.9291	.9540	.9712
23	.3488	.4432	.5388	.6304	.7136	.7856	.8448	.8915	.9267	.9521
24	.2649	.3514	.4444	.5386	.6289	.7113	.7828	.8420	.8889	.9245
25	.1933	.2684	.3539	.4455	.5383	.6276	.7091	.7802	.8393	.8864
26	.1355	.1972	.2717	.3561	.4465	.5381	.6263	.7071	.7778	.8369
27	.0911	.1393	.2009	.2748	.3583	.4475	.5380	.6252	.7053	.7756
28	.0588	.0945	.1429	.2043	.2776	.3602	.4484	.5378	.6242	.7036
29	.0364	.0616	.0978	.1463	.2075	.2803	.3621	.4493	.5377	.6232
30	.0216	.0386	.0643	.1009	.1495	.2105	.2828	.3638	.4501	.5377
31	.0123	.0232	.0406	.0669	.1038	.1526	.2134	.2851	.3654	.4509
32	.0067	.0134	.0247	.0427	.0693	.1065	.1554	.2160	.2873	.3669
33	.0035	.0074	.0144	.0262	.0446	.0717	.1091	.1580	.2184	.2893
34	.0018	.0039	.0081	.0154	.0276	.0465	.0739	.1116	.1605	.2207
35	.0009	.0020	.0044	.0087	.0164	.0290	.0482	.0760	.1139	.1629
36	.0004	.0010	.0023	.0048	.0094	.0174	.0303	.0499	.0780	.1161
37	.0002	.0005	.0011	.0025	.0052	.0101	.0183	.0316	.0515	.0799
38	.0001	.0002	.0005	.0013	.0027	.0056	.0107	.0193	.0328	.0530
39		.0001	.0002	.0006	.0014	.0030	.0060	.0113	.0201	.0340
40			.0001	.0003	.0007	.0015	.0032	.0064	.0119	.0210
41				.0001	.0003	.0008	.0017	.0035	.0068	.0125
42				.0001	.0001	.0004	.0008	.0018	.0037	.0072
43					.0001	.0002	.0004	.0009	.0020	.0040
44						.0001	.0002	.0005	.0010	.0021
45							.0001	.0002	.0005	.0011
46								.0001	.0002	.0005
47									.0001	.0003
48										.0001
49										.0001

n = 100

x\P	.31	.32	.33	.34	.35	.36	.37	.38	.39	.40
1	1.0000	1.0000	1.0000	1.0000	1.0000	1.0000	1.0000	1.0000	1.0000	1.0000
2	1.0000	1.0000	1.0000	1.0000	1.0000	1.0000	1.0000	1.0000	1.0000	1.0000
3	1.0000	1.0000	1.0000	1.0000	1.0000	1.0000	1.0000	1.0000	1.0000	1.0000
4	1.0000	1.0000	1.0000	1.0000	1.0000	1.0000	1.0000	1.0000	1.0000	1.0000
5	1.0000	1.0000	1.0000	1.0000	1.0000	1.0000	1.0000	1.0000	1.0000	1.0000
6	1.0000	1.0000	1.0000	1.0000	1.0000	1.0000	1.0000	1.0000	1.0000	1.0000
7	1.0000	1.0000	1.0000	1.0000	1.0000	1.0000	1.0000	1.0000	1.0000	1.0000
8	1.0000	1.0000	1.0000	1.0000	1.0000	1.0000	1.0000	1.0000	1.0000	1.0000
9	1.0000	1.0000	1.0000	1.0000	1.0000	1.0000	1.0000	1.0000	1.0000	1.0000
10	1.0000	1.0000	1.0000	1.0000	1.0000	1.0000	1.0000	1.0000	1.0000	1.0000
11	1.0000	1.0000	1.0000	1.0000	1.0000	1.0000	1.0000	1.0000	1.0000	1.0000
12	1.0000	1.0000	1.0000	1.0000	1.0000	1.0000	1.0000	1.0000	1.0000	1.0000
13	1.0000	1.0000	1.0000	1.0000	1.0000	1.0000	1.0000	1.0000	1.0000	1.0000
14	1.0000	1.0000	1.0000	1.0000	1.0000	1.0000	1.0000	1.0000	1.0000	1.0000
15	.9999	1.0000	1.0000	1.0000	1.0000	1.0000	1.0000	1.0000	1.0000	1.0000
16	.9998	.9999	1.0000	1.0000	1.0000	1.0000	1.0000	1.0000	1.0000	1.0000
17	.9995	.9998	.9999	1.0000	1.0000	1.0000	1.0000	1.0000	1.0000	1.0000
18	.9989	.9995	.9997	.9999	.9999	1.0000	1.0000	1.0000	1.0000	1.0000
19	.9976	.9988	.9994	.9997	.9999	.9999	1.0000	1.0000	1.0000	1.0000
20	.9950	.9973	.9986	.9993	.9997	.9998	.9999	1.0000	1.0000	1.0000
21	.9904	.9946	.9971	.9985	.9992	.9996	.9998	.9999	1.0000	1.0000
22	.9825	.9898	.9942	.9968	.9983	.9991	.9996	.9998	.9999	1.0000
23	.9698	.9816	.9891	.9938	.9966	.9982	.9991	.9995	.9998	.9999
24	.9504	.9685	.9806	.9885	.9934	.9963	.9980	.9990	.9995	.9997
25	.9224	.9487	.9672	.9797	.9879	.9930	.9961	.9979	.9989	.9994
26	.8841	.9204	.9471	.9660	.9789	.9873	.9926	.9958	.9977	.9988
27	.8346	.8820	.9185	.9456	.9649	.9780	.9867	.9922	.9956	.9976
28	.7736	.8325	.8800	.9168	.9442	.9638	.9773	.9862	.9919	.9954
29	.7021	.7717	.8305	.8781	.9152	.9429	.9628	.9765	.9857	.9916
30	.6224	.7007	.7699	.8287	.8764	.9137	.9417	.9618	.9759	.9852
31	.5376	.6216	.6994	.7684	.8270	.8748	.9123	.9405	.9610	.9752
32	.4516	.5376	.6209	.6982	.7669	.8254	.8733	.9110	.9395	.9602
33	.3683	.4523	.5375	.6203	.6971	.7656	.8240	.8720	.9098	.9385
34	.2912	.3696	.4530	.5375	.6197	.6961	.7643	.8227	.8708	.9087
35	.2229	.2929	.3708	.4536	.5376	.6192	.6953	.7632	.8216	.8697
36	.1650	.2249	.2946	.3720	.4542	.5376	.6188	.6945	.7623	.8205
37	.1181	.1671	.2268	.2961	.3731	.4547	.5377	.6184	.6938	.7614
38	.0816	.1200	.1690	.2285	.2976	.3741	.4553	.5377	.6181	.6932
39	.0545	.0833	.1218	.1708	.2301	.2989	.3750	.4558	.5378	.6178
40	.0351	.0558	.0849	.1235	.1724	.2316	.3001	.3759	.4562	.5379
41	.0218	.0361	.0571	.0863	.1250	.1739	.2330	.3012	.3767	.4567
42	.0131	.0226	.0371	.0583	.0877	.1265	.1753	.2343	.3023	.3775
43	.0075	.0136	.0233	.0380	.0594	.0889	.1278	.1766	.2355	.3033
44	.0042	.0079	.0141	.0240	.0389	.0605	.0901	.1290	.1778	.2365
45	.0023	.0044	.0082	.0146	.0246	.0397	.0614	.0911	.1301	.1789
46	.0012	.0024	.0046	.0085	.0150	.0252	.0405	.0623	.0921	.1311
47	.0006	.0012	.0025	.0048	.0088	.0154	.0257	.0411	.0631	.0930
48	.0003	.0006	.0013	.0026	.0050	.0091	.0158	.0262	.0417	.0638
49	.0001	.0003	.0007	.0014	.0027	.0052	.0094	.0162	.0267	.0423
50	.0001	.0001	.0003	.0007	.0015	.0029	.0054	.0096	.0165	.0271
51		.0001	.0002	.0003	.0007	.0015	.0030	.0055	.0098	.0168
52		.0001	.0001	.0002	.0004	.0008	.0016	.0030	.0056	.0100
53			.0001	.0002	.0004	.0004	.0008	.0016	.0031	.0058
54				.0001	.0001	.0002	.0004	.0008	.0017	.0032
55						.0001	.0002	.0004	.0009	.0017
56							.0001	.0002	.0004	.0009
57								.0001	.0002	.0004
58									.0001	.0002
59										.0001

n = 100

x\P	.41	.42	.43	.44	.45	.46	.47	.48	.49	.50
1	1.0000	1.0000	1.0000	1.0000	1.0000	1.0000	1.0000	1.0000	1.0000	1.0000
2	1.0000	1.0000	1.0000	1.0000	1.0000	1.0000	1.0000	1.0000	1.0000	1.0000
3	1.0000	1.0000	1.0000	1.0000	1.0000	1.0000	1.0000	1.0000	1.0000	1.0000
4	1.0000	1.0000	1.0000	1.0000	1.0000	1.0000	1.0000	1.0000	1.0000	1.0000
5	1.0000	1.0000	1.0000	1.0000	1.0000	1.0000	1.0000	1.0000	1.0000	1.0000
6	1.0000	1.0000	1.0000	1.0000	1.0000	1.0000	1.0000	1.0000	1.0000	1.0000
7	1.0000	1.0000	1.0000	1.0000	1.0000	1.0000	1.0000	1.0000	1.0000	1.0000
8	1.0000	1.0000	1.0000	1.0000	1.0000	1.0000	1.0000	1.0000	1.0000	1.0000
9	1.0000	1.0000	1.0000	1.0000	1.0000	1.0000	1.0000	1.0000	1.0000	1.0000
10	1.0000	1.0000	1.0000	1.0000	1.0000	1.0000	1.0000	1.0000	1.0000	1.0000
11	1.0000	1.0000	1.0000	1.0000	1.0000	1.0000	1.0000	1.0000	1.0000	1.0000
12	1.0000	1.0000	1.0000	1.0000	1.0000	1.0000	1.0000	1.0000	1.0000	1.0000
13	1.0000	1.0000	1.0000	1.0000	1.0000	1.0000	1.0000	1.0000	1.0000	1.0000
14	1.0000	1.0000	1.0000	1.0000	1.0000	1.0000	1.0000	1.0000	1.0000	1.0000
15	1.0000	1.0000	1.0000	1.0000	1.0000	1.0000	1.0000	1.0000	1.0000	1.0000
16	1.0000	1.0000	1.0000	1.0000	1.0000	1.0000	1.0000	1.0000	1.0000	1.0000
17	1.0000	1.0000	1.0000	1.0000	1.0000	1.0000	1.0000	1.0000	1.0000	1.0000
18	1.0000	1.0000	1.0000	1.0000	1.0000	1.0000	1.0000	1.0000	1.0000	1.0000
19	1.0000	1.0000	1.0000	1.0000	1.0000	1.0000	1.0000	1.0000	1.0000	1.0000
20	1.0000	1.0000	1.0000	1.0000	1.0000	1.0000	1.0000	1.0000	1.0000	1.0000
21	1.0000	1.0000	1.0000	1.0000	1.0000	1.0000	1.0000	1.0000	1.0000	1.0000
22	1.0000	1.0000	1.0000	1.0000	1.0000	1.0000	1.0000	1.0000	1.0000	1.0000
23	1.0000	1.0000	1.0000	1.0000	1.0000	1.0000	1.0000	1.0000	1.0000	1.0000
24	.9999	.9999	1.0000	1.0000	1.0000	1.0000	1.0000	1.0000	1.0000	1.0000
25	.9997	.9999	.9999	1.0000	1.0000	1.0000	1.0000	1.0000	1.0000	1.0000
26	.9994	.9997	.9999	.9999	1.0000	1.0000	1.0000	1.0000	1.0000	1.0000
27	.9987	.9994	.9997	.9998	.9999	1.0000	1.0000	1.0000	1.0000	1.0000
28	.9975	.9987	.9993	.9997	.9998	.9999	1.0000	1.0000	1.0000	1.0000
29	.9952	.9974	.9986	.9993	.9996	.9998	.9999	1.0000	1.0000	1.0000
30	.9913	.9950	.9972	.9985	.9992	.9996	.9998	.9999	1.0000	1.0000
31	.9848	.9910	.9948	.9971	.9985	.9992	.9996	.9998	.9999	1.0000
32	.9746	.9844	.9907	.9947	.9970	.9984	.9992	.9996	.9998	.9999
33	.9594	.9741	.9840	.9905	.9945	.9969	.9984	.9991	.9996	.9998
34	.9376	.9587	.9736	.9837	.9902	.9944	.9969	.9983	.9991	.9996
35	.9078	.9368	.9581	.9732	.9834	.9900	.9942	.9968	.9983	.9991
36	.8687	.9069	.9361	.9576	.9728	.9831	.9899	.9941	.9967	.9982
37	.8196	.8678	.9061	.9355	.9571	.9724	.9829	.9897	.9941	.9967
38	.7606	.8188	.8670	.9054	.9349	.9567	.9721	.9827	.9896	.9940
39	.6927	.7599	.8181	.8663	.9049	.9345	.9563	.9719	.9825	.9895
40	.6176	.6922	.7594	.8174	.8657	.9044	.9341	.9561	.9717	.9824
41	.5380	.6174	.6919	.7589	.8169	.8653	.9040	.9338	.9558	.9716
42	.4571	.5382	.6173	.6916	.7585	.8165	.8649	.9037	.9335	.9557
43	.3782	.4576	.5383	.6173	.6913	.7582	.8162	.8646	.9035	.9334
44	.3041	.3788	.4580	.5385	.6172	.6912	.7580	.8160	.8645	.9033
45	.2375	.3049	.3794	.4583	.5387	.6173	.6911	.7579	.8159	.8644
46	.1799	.2384	.3057	.3799	.4587	.5389	.6173	.6911	.7579	.8159
47	.1320	.1807	.2391	.3063	.3804	.4590	.5391	.6174	.6912	.7579
48	.0938	.1328	.1815	.2398	.3069	.3809	.4593	.5393	.6176	.6914
49	.0644	.0944	.1335	.1822	.2404	.3074	.3813	.4596	.5395	.6178
50	.0428	.0650	.0950	.1341	.1827	.2409	.3078	.3816	.4599	.5398
51	.0275	.0432	.0655	.0955	.1346	.1832	.2413	.3082	.3819	.4602
52	.0170	.0278	.0436	.0659	.0960	.1350	.1836	.2417	.3084	.3822
53	.0102	.0172	.0280	.0439	.0662	.0963	.1353	.1838	.2419	.3006
54	.0059	.0103	.0174	.0282	.0441	.0664	.0965	.1355	.1840	.2421
55	.0033	.0059	.0104	.0175	.0284	.0443	.0666	.0967	.1356	.1841
56	.0017	.0033	.0060	.0105	.0176	.0285	.0444	.0667	.0967	.1356
57	.0009	.0018	.0034	.0061	.0106	.0177	.0286	.0444	.0667	.0967
58	.0004	.0009	.0018	.0034	.0061	.0106	.0177	.0286	.0444	.0666
59	.0002	.0005	.0009	.0018	.0034	.0061	.0106	.0177	.0285	.0443
60	.0001	.0002	.0005	.0009	.0018	.0034	.0061	.0106	.0177	.0284
61		.0001	.0002	.0005	.0009	.0018	.0034	.0061	.0106	.0176
62			.0001	.0002	.0005	.0009	.0018	.0034	.0061	.0105
63				.0001	.0002	.0005	.0009	.0018	.0034	.0060
64					.0001	.0002	.0005	.0009	.0018	.0033
65						.0001	.0002	.0005	.0009	.0018
66							.0001	.0002	.0004	.0009
67								.0001	.0002	.0004
68									.0001	.0002
69										.0001

APPENDIX G Random Number Tables

THIS appendix section contains five pages of random number tables taken with permission from *Handbook of Mathematical Functions, National Bureau of Standards Applied Mathematics Series — 55, June, 1964.* Another source containing an extensive list of random numbers is the book *A Million Random Digits with 100,000 Normal Deviates,* The Rand Corporation, Free Press of Glencoe, New York, 1955.

The use of random number tables is very simple. The only things which are needed to use the table are: a) a random entry, or starting point, and b) a consistent pattern or scheme to read the numbers after entering the table. If you have not memorized these five pages of tables a way to secure a random starting point which will probably be satisfactory is to turn to a page and more or less let your pencil come down on the page blindly. Given this starting point you can then begin to read sequences of 2-digit random numbers in any direction you choose. Up, down, across, the first 2 digits of the 5-digit numbers, the third and fourth digits, the first and fifth digits, or whatever pattern you wish. The only proviso is that the starting point must be randomly selected and the pattern or scheme for reading the random numbers from within the 5-digit entries must be followed in a consistent manner.

A more precise way to establish a random starting point in the tables is to use the following sequence of steps.

A. Select by any method one of the five pages of tables.
B. Without direction, bring a pencil point down on the page so as to hit a random digit.
C. Read the digit and those to the right of the digit in the same row. Let the digit itself specify the page to enter (0,1 equals first page, 2,3 equals second page, . . . 8,9 equals fifth page). Then look at the next four consecutive digits to the right of the initial digit. Let the first two digit combination designate the row to start and second two digit number the column. (If the two digit numbers are in the range 50–99, the upper left column and row represent 50 instead of 00.)

Example: The pencil is placed on the digit 2 and the next four digits to the right are 3853. The digit 2 specifies that we enter on the second page of tables. The row and column in which to start are row 38 and column 4.

Column Index

Row index	00 or 50	01 or 51	02 or 52	03 or 53	04 or 54	05 or 55	06 or 56	07 or 57	08 or 58	09 or 59	10 or 60	11 or 61	12 or 62	13 or 63	14 or 64	
00 or 50	2	6	6	8	7	7	4	2	2	3	4	3	5	4	6	. . .
01 or 51	6	0	6	7	5											
02 or 52	4	5	4	1	8											
03 or 53	6	9	8	7	2											
		.														
		.														
		.														
		.														

Random Number Table

D. Having established the page, row, and column, two digit random numbers are obtained by reading the starting number and the next number to the right. Subsequent two digit numbers are read from the same column below the previous numbers. For example, in Step C we determined that the starting point is the second page of the random number table, row 38, column 4. The initial two digit number is, therefore, 65. Successive two digit numbers are 06, 60, 84, 28, 27, 88, 52, 75, 55, 88, 73, 35, 74 (top two digits in columns 6 and 7), 75, etc.

E. Appropriate changes can be made in Step C for different problems. For example, if a larger random number table is used a different numbering scheme to determine the page to enter must be used. Steps (A), (B), and (C) merely provide a means of initially entering the table in a random manner.

This series of steps is designed to eliminate any bias in selecting the initial starting point in the tables. The numerical value located by dropping a pencil point in a (presumably) random position is then used to establish the actual starting point. Once a random starting point has been established any pattern can be used to read the sequence of random numbers so long as the pattern is followed consistently. The pattern or scheme should be stated or defined before selecting the random starting point.

2500 FIVE DIGIT RANDOM NUMBERS

55034	81217	90564	81943	11241	84512	12288	89862	00760	76159
25521	99536	43233	48786	49221	06960	31564	21458	88199	06312
85421	72744	97242	66383	00132	05661	96442	37388	57671	27916
61219	48390	47344	30413	39392	91365	56203	79204	05330	31196
20230	03147	58854	11650	28415	12821	58931	30508	65989	26675
95776	83206	56144	55953	89787	64426	08448	45707	80364	60262
07603	17344	01148	83300	96955	65027	31713	89013	79557	49755
00645	17459	78742	39005	36027	98807	72666	54484	68262	38827
62950	83162	61504	31557	80590	47893	72360	72720	08396	33674
79350	10276	81933	26347	08068	67816	06659	87917	74166	85519
48339	69834	59047	82175	92010	58446	69591	56205	95700	86211
05842	08439	79836	50957	32059	32910	15842	13918	41365	80115
25855	02209	07307	59942	71389	76159	11263	38787	61541	22606
25272	16152	82323	70718	98081	38631	91956	49909	76253	33970
73003	29058	17605	49298	47675	90445	68919	05676	23823	84892
81310	94430	22663	06584	38142	00146	17496	51115	61458	65790
10024	44713	59832	80721	63711	67882	25100	45345	55743	67618
84671	52806	89124	37691	20897	82339	22627	06142	05773	03547
29296	58162	21858	33732	94056	88806	54603	00384	66340	69232
51771	94074	70630	41286	90583	87680	13961	55627	23670	35109
42166	56251	60770	51672	36031	77273	85218	14812	90758	23677
78355	67041	22492	51522	31164	30450	27600	44428	96380	26772
09552	51347	33864	89018	73418	81538	77399	30448	97740	18158
15771	63127	34847	05660	06156	48970	55699	61818	91763	20821
13231	99058	93754	36730	44286	44326	15729	37500	47269	13333
50583	03570	38472	73236	67613	72780	78174	18718	99092	64114
99485	57330	10634	74905	90671	19643	69903	60950	17968	37217
54676	39524	73785	48864	69835	62798	65205	69187	05572	74741
99343	71549	10248	76036	31702	76868	88909	69574	27642	00336
35492	40231	34868	55356	12847	68093	52643	32732	67016	46784
98170	25384	03841	23920	47954	10359	70114	11177	63298	99903
02670	86155	56860	02592	01646	42200	79950	37764	82341	71952
36934	42879	81637	79952	07066	41625	96804	92388	88860	68580
56851	12778	24309	73660	84264	24668	16686	02239	66022	64133
05464	28892	14271	23778	88599	17081	33884	88783	39015	57118
15025	20237	63386	71122	06620	07415	94982	32324	79427	70387
95610	08030	81469	91066	88857	56583	01224	28097	19726	71465
09026	40378	05731	55128	74298	49196	31669	42605	30368	96424
81431	99955	52462	67667	97322	69808	21240	65921	12629	92896
21431	59335	58627	94822	65484	09641	41018	85100	16110	32077
95832	76145	11636	80284	17787	97934	12822	73890	66009	27521
99813	44631	43746	99790	86823	12114	31706	05024	28156	04202
77210	31148	50543	11603	50934	02498	09184	95875	85840	71954
13268	02609	79833	66058	80277	08533	28676	37532	70535	82356
44285	71735	26620	54691	14909	52132	81110	74548	78853	31996
70526	45953	79637	57374	05053	31965	33376	13232	85666	86615
88386	11222	25080	71462	09818	46001	19065	68981	18310	74178
83161	73994	17209	79441	64091	49790	11936	44864	86978	34538
50214	71721	33851	45144	05696	29935	12823	01594	08453	52825
97689	29341	67747	80643	13620	23943	49396	83686	37302	95350

These tables of random numbers are reprinted by permission from *Handbook of Mathematical Functions*, National Bureau of Standards Applied Mathematics Series-55, June, 1964.

DECISION ANALYSIS FOR PETROLEUM EXPLORATION

2500 FIVE DIGIT RANDOM NUMBERS

26687	74223	43546	45699	94469	82125	37370	23966	68926	37664
60675	75169	24510	15100	02011	14375	65187	10630	64421	66745
45418	98635	83123	98558	09953	60255	42071	40930	97992	93085
69872	48026	89755	28470	44130	59979	91063	28766	85962	77173
03765	86366	99539	44183	23886	89977	11964	51581	18033	56239
84686	57636	32326	19867	71345	42002	96997	84379	27991	21459
91512	49670	32556	85189	28023	88151	62896	95498	29423	38138
10737	49307	18307	22246	22461	10003	93157	66984	44919	30467
54870	19676	58367	20905	38324	00026	98440	37427	22896	37637
48967	49579	65369	74305	62085	39297	10309	23173	74212	32272
91430	79112	03685	05411	23027	54735	91550	06250	18705	18909
92564	29567	47476	62804	73428	04535	86395	12162	59647	97726
41734	12199	77441	92415	63542	42115	84972	12454	33133	48467
25251	78110	54178	78241	09226	87529	35376	90690	54178	08561
91657	11563	66036	28523	83705	09956	76610	88116	78351	50877
00149	84745	63222	50533	50159	60433	04822	49577	89049	16162
53250	73200	84066	59620	61009	38542	05758	06178	80193	26466
25587	17481	56716	49749	70733	32733	60365	14108	52573	39391
01176	12182	06882	27562	75456	54261	38564	89054	96911	88906
83531	15544	40834	20296	88576	47815	96540	79462	78666	25353
19902	98866	32805	61091	91587	30340	84909	64047	67750	87638
96516	78705	25556	35181	29064	49005	29843	68949	50506	45862
99417	56171	19848	24352	51844	03791	72127	57958	08366	43190
77699	57853	93213	27342	28906	31052	65815	21637	49385	75406
32245	83794	99528	05150	27246	48263	62156	62469	97048	16511
12874	72753	66469	13782	64330	00056	73324	03920	13193	19466
63899	41910	45484	55461	66518	82486	74694	07865	09724	76490
16255	43271	26540	41298	35095	32170	70625	66407	01050	44225
75553	30207	41814	74985	40223	91223	64238	73012	83100	92041
41772	18441	34685	13892	38843	69007	10362	84125	08814	66785
09270	01245	81765	06809	10561	10080	17482	05471	82273	06902
85058	17815	71551	36356	97519	54144	51132	83169	27373	68609
80222	87572	62758	14858	36350	23304	70453	21065	63812	29860
83901	88028	56743	25598	79349	47880	77912	52020	84305	02897
36303	57833	77622	02238	53285	77316	40106	38456	92214	54278
91543	63886	60539	96334	20804	72692	08944	02870	74892	22598
14415	33816	78231	87674	96473	44451	25098	29296	50679	07798
82465	07781	09938	66874	72128	99685	84329	14530	08410	45953
27306	39843	05634	96368	72022	01278	92830	40094	31776	41822
91960	82766	02331	08797	33858	21847	17391	53755	58079	48498
59284	96108	91610	07483	37943	96832	15444	12091	36690	58317
10428	96003	71223	21352	78685	55964	35510	94805	23422	04492
65527	41039	79574	05105	59588	02115	33446	56780	18402	36279
59688	43078	93275	31978	08768	84805	50661	18523	83235	50602
44452	10188	43565	46531	93023	07618	12910	60934	53403	18401
87275	82013	59804	78595	60553	14038	12096	95472	42736	08573
94155	93110	49964	27753	85090	77677	69303	66323	77811	22791
26488	76394	91282	03419	68758	89575	66469	97835	66681	03171
37073	34547	88296	68638	12976	50896	10023	27220	05785	77538
83835	89575	55956	93957	30361	47679	83001	35056	07103	63072

2500 FIVE DIGIT RANDOM NUMBERS

12367	23891	31506	90721	18710	89140	58595	99425	22840	08267
38890	30239	34237	22578	74420	22734	26930	40604	10782	80128
80788	55410	39770	93317	18270	21141	52085	78093	85638	81140
02395	77585	08854	23562	33544	45796	10976	44721	24781	09690
73720	70184	69112	71887	80140	72876	38984	23409	63957	44751
61383	17222	55234	18963	39006	93504	18273	49815	52802	69675
39161	44282	14975	97498	25973	33605	60141	30030	77677	49294
80907	74484	39884	19885	37311	04209	49675	39596	01052	43999
09052	65670	63660	34035	06578	87837	28125	48883	50482	55735
33425	24226	32043	60082	20418	85047	53570	32554	64099	52326
72651	69474	73648	71530	55454	19576	15552	20577	12124	50038
04142	32092	83586	61825	35482	32736	63403	91499	37196	02762
85226	14193	52213	60746	24414	57858	31884	51266	82293	73553
54888	03579	91674	59502	08619	33790	29011	85193	62262	28684
33258	51516	82032	45233	39351	33229	59464	65545	76809	16982
75973	15957	32405	82081	02214	57143	33526	47194	94526	73253
90638	75314	35381	34451	49246	11465	25102	71489	89883	99708
65061	15498	93348	33566	19427	66826	03044	97361	08159	47485
64420	07427	82233	97812	39572	07766	65844	29980	15533	90114
27175	17389	76963	75117	45580	99904	47160	55364	25666	25405
32215	30094	87276	56896	15625	32594	80663	08082	19422	80717
54209	58043	72350	89828	02706	16815	89985	37380	44032	59366
59286	66964	84843	71549	67553	33867	83011	66213	69372	23903
83872	58167	01221	95558	22196	65905	38785	01355	47489	28170
83310	57080	03366	80017	39601	40698	56434	64055	02495	50880
64545	29500	13351	78647	92628	19354	60479	57338	52133	07114
39269	00076	55489	01524	76568	22571	20328	84623	30188	43904
29763	05675	28193	65514	11954	78599	63902	21346	19219	90286
06310	02998	01463	27738	90288	17697	64511	39552	34694	03211
97541	47607	57655	59102	21851	44446	07976	54295	84671	78755
82968	85717	11619	97721	53513	53781	98941	38401	70939	11319
76878	34727	12524	90642	16921	13669	17420	84483	68309	85241
87394	78884	87237	92086	95633	66841	22906	64989	86952	54700
74040	12731	59616	33697	12592	44891	67982	72972	89795	10587
47896	41413	66431	70046	50793	45920	96564	67958	56369	44725
87778	71697	64148	54363	92114	34037	59061	62051	62049	33526
96977	63143	72219	80040	11990	47698	95621	72990	29047	85893
43820	13285	77811	81697	29937	70750	02029	32377	00556	86687
57203	83960	40096	39234	65953	59911	91411	55573	88427	45573
49065	72171	80939	06017	90323	63687	07932	99587	49014	26452
94250	84270	95798	13477	80139	26335	55169	73417	40766	45170
68148	81382	82383	18674	40453	92828	30042	37412	43423	45138
12208	97809	33619	28868	41646	16734	88860	32636	41985	84615
88317	89705	26119	12416	19438	65665	60989	59766	11418	18250
56728	80359	29613	63052	15251	44684	64681	42354	51029	77680
07138	12320	01073	19304	87042	58920	28454	81069	93978	66659
21188	64554	55618	36088	24331	84390	16022	12200	77559	75661
02154	12250	88738	43917	03655	21099	60805	63246	26842	35816
90953	85238	32771	07305	36181	47420	19681	33184	41386	03249
80103	91308	12858	41293	00325	15013	19579	91132	12720	92603

2500 FIVE DIGIT RANDOM NUMBERS

92630	78240	19267	95457	53497	23894	37708	79862	76471	66418
79445	78735	71549	44843	26104	67318	00701	34986	66751	99723
59654	71966	27386	50004	05358	94031	29281	18544	52429	06080
31524	49587	76612	39789	13537	48086	59483	60680	84675	53014
06348	76938	90379	51392	55887	71015	09209	79157	24440	30244
28703	51709	94456	48396	73780	06436	86641	69239	57662	80181
68108	89266	94730	95761	75023	48464	65544	96583	18911	16391
99938	90704	93621	66330	33393	95261	95349	51769	91616	33238
91543	73196	34449	63513	83834	99411	58826	40456	69268	48562
42103	02781	73920	56297	72678	12249	25270	36678	21313	75767
17138	27584	25296	28387	51350	61664	37893	05363	44143	42677
28297	14280	54524	21618	95320	38174	60579	08089	94999	78460
09331	56712	51333	06289	75345	08811	82711	57392	25252	30333
31295	04204	93712	51287	05754	79396	87399	51773	33075	97061
36146	15560	27592	42089	99281	59640	15221	96079	09961	05371
29553	18432	13630	05529	02791	81017	49027	79031	50912	09399
23501	22642	63081	08191	89420	67800	55137	54707	32945	64522
57888	85846	67967	07835	11314	01545	48535	17142	08552	67457
55336	71264	88472	04334	63919	36394	11196	92470	70543	29776
10087	10072	55980	64688	68239	20461	89381	93809	00796	95945
34101	81277	66090	88872	37818	72142	67140	50785	21380	16703
53362	44940	60430	22834	14130	96593	23298	56203	92671	15925
82975	66158	84731	19436	55790	69229	28661	13675	99318	76873
54827	84673	22898	08094	14326	87038	42892	21127	30712	48489
25464	59098	27436	89421	80754	89924	19097	67737	80368	08795
67609	60214	41475	84950	40133	02546	09570	45682	50165	15609
44921	70924	61295	51137	47596	86735	35561	76649	18217	63446
33170	30972	98130	95828	49786	13301	36081	80761	33985	68621
84687	85445	06208	17654	51333	02878	35010	67578	61574	20749
71886	56450	36567	09395	96951	35507	17555	35212	69106	01679
00475	02224	74722	14721	40215	21351	08596	45625	83981	63748
25993	38881	68361	59560	41274	69742	40703	37993	03435	18873
92882	53178	99195	93803	56985	53089	15305	50522	55900	43026
25138	26810	07093	15677	60688	04410	24505	37890	67186	62829
84631	71882	12991	83028	82484	90339	91950	74579	03539	90122
34003	92326	12793	61453	48121	74271	28363	66561	75220	35908
53775	45749	05734	86169	42762	70175	97310	73894	88606	19994
59316	97885	72807	54966	60859	11932	35265	71601	55577	67715
20479	66557	50705	26999	09854	52591	14063	30214	19890	19292
86180	84931	25455	26044	02227	52015	21820	50599	51671	65411
21451	68001	72710	40261	61281	13172	63819	48970	51732	54113
98062	68375	80089	24135	72355	95428	11808	29740	81644	86610
01788	64429	14430	94575	75153	94576	61393	96192	03227	32258
62465	04841	43272	68702	01274	05437	22953	18946	99053	41690
94324	31089	84159	92933	99989	89500	91586	02802	69471	68274
05797	43984	21575	09908	70221	19791	51578	36432	33494	79888
10395	14289	52185	09721	25789	38562	54794	04897	59012	89251
35177	56986	25549	59730	64718	52630	31100	62384	49483	11409
25633	89619	75882	98256	02126	72099	57183	55887	09320	73463
16464	48280	94254	45777	45150	68865	11382	11782	22695	41988

2500 FIVE DIGIT RANDOM NUMBERS

53479	81115	98036	12217	59526	40238	40577	39351	43211	69255
97344	70328	58116	91964	26240	44643	83287	97391	92823	77578
66023	38277	74523	71118	84892	13956	98899	92315	65783	59640
99776	75723	03172	43112	83086	81982	14538	26162	24899	20551
30176	48979	92153	38416	42436	26636	83903	44722	69210	69117
81874	83339	14988	99937	13213	30177	47967	93793	86693	98854
19839	90630	71863	95053	55532	60908	84108	55342	48479	63799
09337	33435	53869	52769	18801	25820	96198	66518	78314	97013
31151	58295	40823	41330	21093	93882	49192	44876	47185	81425
67619	52515	03037	81699	17106	64982	60834	85319	47814	08075
61946	48790	11602	83043	22257	11832	04344	95541	20366	55937
04811	64892	96346	79065	26999	43967	63485	93572	80753	96582
05763	39601	56140	25513	86151	78657	02184	29715	04334	15678
73260	56877	40794	13948	96289	90185	47111	66807	61849	44686
54909	09976	76580	02645	35795	44537	64428	35441	28318	99001
42583	36335	60068	04044	29678	16342	48592	25547	63177	75225
27266	27403	97520	23334	36453	33699	23672	45884	41515	04756
49843	11442	66682	36055	32002	78600	36924	59962	68191	62580
29316	40460	27076	69232	51423	58515	49920	03901	26597	33068
30463	27856	67798	16837	74273	05793	02900	63498	00782	35097
28708	84088	65535	44258	33869	82530	98399	26387	02836	36838
13183	50652	94872	28257	78547	55286	33591	61965	51723	14211
60796	76639	30157	40295	99476	28334	15368	42481	60312	42770
13486	46918	64683	07411	77842	01908	47796	65796	44230	77230
34914	94502	39374	34185	57500	22514	04060	94511	44612	10485
28105	04814	85170	86490	35695	03483	57315	63174	71902	71182
59231	45028	01173	08848	81925	71494	95401	34049	04851	65914
87437	82758	71093	36033	53582	25986	46005	42840	81683	21459
29046	01301	55343	65732	78714	43644	46248	53205	94868	48711
62035	71886	94506	15263	61435	10369	42054	68257	14385	79436
38856	80048	59973	73368	52876	47673	41020	82295	26430	87377
40666	43328	87379	86418	95841	25590	54137	94182	42308	07361
40588	90087	37379	08667	37256	20317	53316	50982	32900	32097
78237	86556	50276	20431	00243	02303	71029	49932	23245	00862
98247	67474	71455	69540	01169	03320	67017	92543	97977	52728
69977	78558	65430	32627	28312	61815	14598	79728	55699	91348
39843	23074	40814	03713	21891	96353	96806	24595	26203	26009
62880	87277	99895	99965	34374	42556	11679	99605	98011	48867
56138	64927	29454	52967	86624	62422	30163	76181	95317	39264
90804	56026	48994	64569	67465	60180	12972	03848	62582	93855
09665	44672	74762	33357	67301	80546	97659	11348	78771	45011
34756	50403	76634	12767	32220	34545	18100	53513	14521	72120
12157	73327	74196	26668	78087	53636	52304	00007	05708	63538
69384	07734	94451	76428	16121	09300	67417	68587	87932	38840
93358	64565	43766	45041	44930	69970	16964	08277	67752	60292
38879	35544	99563	85404	04913	62547	78406	01017	86187	22072
58314	60298	72394	69668	12474	93059	02053	29807	63645	12792
83568	10227	99471	74729	22075	10233	21575	20325	21317	57124
28067	91152	40568	33705	64510	07067	64374	26336	79652	31140
05730	75557	93161	80921	55873	54103	34801	83157	04534	81368

APPENDIX H Combining Distributions Analytically

IN Section I) of Chapter 8 we mention that one approach to determine the desired profit distribution is to combine all of the random variable distributions into the profit equation and solve analytically. We further mentioned that in general this is not a viable alternative, due to the mathematical complexities involved. There are, however, a few special situations in which we can operate with distributions in an analytic sense rather than having to use simulation. In this appendix section we will list some of these instances.

I. If x_1, x_2, x_3, . . . , x_n are n continuous random variables (each defined by a continuous probability distribution), and the equation of interest is:

$$y = x_1 \pm x_2 \pm x_3 \pm \ldots \pm x_n$$

Then

$$\mu_y = \mu_{x_1} \pm \mu_{x_2} \pm \mu_{x_3} \pm \ldots \pm \mu_{x_n}$$ $\boxed{\text{For independent and dependent random variables}}$

This relationship states that the mean of a sum (or difference) of n random variables is equal to the sum (or difference) of the means. This relationship applies when x_1, x_2, x_3, . . . , x_n are independent or dependent random variables.

If: $y = x_1 + x_2 + x_3 + \ldots + x_n$

then

$$\sigma_y^2 = \sigma_{x_1}^2 + \sigma_{x_2}^2 + \sigma_{x_3}^2 + \cdots + \sigma_{x_n}^2$$ For independent variables only

If: $y = x_1 - x_2$

then $\sigma_y^2 = \sigma_{x_1}^2 + \sigma_{x_2}^2$ For Independent variables only

The variance of a sum of n independent random variables is equal to the sum of the variances of the random variables. The variance of the difference between two independent random variables is equal to the *sum* of the variances of each variable. These relationships hold *only* if x_1, x_2, x_3, . . . , x_n are independent random variables. The relationships do not hold if two or more of the random variables are dependent. The corresponding relationships for dependent variables include additional terms to account for covariance.

II. If $x_1, x_2, x_3, \ldots, x_n$ are n independent random variables and the equation of interest is:

$$y = (x_1)(x_2)(x_3) \ldots (x_n)$$

Then

$$\mu_y = (\mu_{x_1})(\mu_{x_2})(\mu_{x_3}) \ldots (\mu_{x_n}) \quad \boxed{\text{For independent variables only}}$$

The mean of a product of n independent random variables is equal to the product of the means. This is true if, and only if, $x_1, x_2, x_3, \ldots, x_n$ are independent random variables. For dependent random variables the corresponding relationship includes covariance terms. The relationship for the variance of a product of random variables is rather complicated and involves covariance terms (or approximations) for both the case of independent and dependent variables.

III. If x_1 and x_2 are random variables and the equation of interest is:

$$y = \frac{x_1}{x_2}$$

Then

$$\mu_y \neq \frac{\mu_{x_1}}{\mu_{x_2}}$$

In general, no operations involving division of random variables can be simplified to the types of relationships given in I.) and II.), even if the random variables x_1 and x_2 are independent. The analytic operations involving division of random variables involve complicated terms dealing with covariance, etc. And some of these complicated relationships are only approximations at that! So if division of random variables occurs in the equation of interest about the only way to obtain the corresponding distribution of y (or its parameters μ_y and σ_y) is by simulation.

IV. *Expected NPV of a Series of Annual Cash Flows:*
If $CF_0, CF_1, CF_2, CF_3, \ldots, CF_n$ are distributions of cash flows occurring at time zero and at the ends of years 1, 2, 3, . . . , n respectively, where each CF distribution has a mean μ and standard deviation σ; and where NPV is defined as:

$$NPV = CF_0 + \frac{CF_1}{(1 + i_0)^1} + \frac{CF_2}{(1 + i_0)^2} + \frac{CF_3}{(1 + i_0)^3} + \ldots + \frac{CF_n}{(1 + i_0)^n}$$

Then

$$\mu_{NPV} = \mu_{CF_0} + \frac{\mu_{CF_1}}{(1 + i_0)^1} + \frac{\mu_{CF_2}}{(1 + i_0)^2} + \frac{\mu_{CF_3}}{(1 + i_0)^3} + \ldots + \frac{\mu_{CFn}}{(1 + i_0)^n}$$

This is to say if we want to compute an expected NPV we merely substitute the mean values of each of the annual cash flow distributions into the usual discounting equation. This relationship is valid no matter whether the annual cash flows are independent or dependent. i_0 is defined as the discounting rate, expressed as a decimal fraction. The corresponding relationship for the variance of the NPV distribution is a bit more complex. Those interested in finding the NPV distribution variance analytically (rather than by simulation) should consult pages 272–280 of Reference 1.6.

Miscellaneous Calculations

THIS appendix contains examples of a number of petroleum related calculations, including analysis of exponential decline curves using tables of Appendix C, calculation of composite discount factors, and raising numbers to a power using logarithms and the tables of Appendix C. For those interested in further study of decline curve analysis I recommend the article "Methods for Calculating Profitabilities" by Brons and McGarry which appears on p. 152–162 of Reference 2.7. They discuss exponential, hyperbolic, and harmonic decline curves in detail and have included numerous tables to simplify the calculations. We will only briefly discuss the exponential decline curve analysis here.

EXPONENTIAL DECLINE CURVE ANALYSIS

The production decline characteristics of many oil wells and fields follow what are called exponential declines. If oil production rate (on a logarithmic scale) is plotted as a function of time the decline performance will be a straight line. The slope of the decline curve is called the (true) exponential decline rate. it can be computed by the relation

$$a = \frac{\ln\left(\frac{q_i}{q_f}\right)}{t} \tag{1}$$

where a = exponential decline rate per year, decimal fraction. It is a constant which applies over the entire decline period.
q_i = production rate at beginning of any time period during the decline, Bbls. per year.
q_f = production rate at end of the time period, Bbls. per year.
t = no. of years between q_i and q_f.

Knowing this annual exponential decline rate we can compute the total production on decline with the following equation

$$\Delta N_D = \frac{1}{a}[q_i - q_f] \tag{2}$$

where ΔN_D is the total decline production, barrels, that is produced between the times corresponding to q_i and q_f .

651

Note that the units of q_i and q_f must be in barrels per year in equation (2). If the producing rates are expressed as barrels per day an additional multiplying conversion factor to convert to an annual basis is required.

EXAMPLE: An exponential decline curve has been extrapolated over a 16 year period to an economic limit of 5 BOPD. The producing rate at the start of the decline was 200 BOPD. How much oil is produced during the 16 year period of declining production?

$$t = 16 \text{ years}, \; q_i = 200 \frac{Bbl}{Day} \times 365 \frac{Days}{yr} = 73,000 \text{ Bbls/year}$$

$$q_f = 5 \frac{Bbl}{Day} \times 365 \frac{Days}{yr} = 1825 \text{ Bbls/year}$$

From Equation (1), the exponential decline rate is computed as

$$a = \frac{\ln\left(\frac{73,000}{1,825}\right)}{16 \text{ years}} = \frac{\ln (40)}{16} = 0.2306$$

The total decline production is computed as

$$\Delta N_D = \frac{1}{a} [q_i - q_f] = \frac{1}{0.2306} (73,000 - 1,825)$$

$$\Delta N_D = 308,000 \text{ Bbls.}$$

The exponential decline rate, a, computed by equation (1) is the only mathematically correct decline rate. There is, however, another definition which has been incorrectly called the exponential decline rate. To distinguish the two I would prefer to call the incorrect definition the "loss ratio." It is defined as

$$\text{"loss ratio"} = \frac{Q_1 - Q_2}{Q_1} \tag{3}$$

where Q_1 is the production (or revenue) rate at beginning of any one year interval, Q_2 is production (or revenue) rate at end of the same one year, and "loss ratio" is expressed as decimal fraction.

The true exponential decline rate can be computed from the "loss ratio" using equation (4):

$$a = -\ln (1 - \text{loss ratio}) \tag{4}$$

Or, the following table can be used to relate "loss ratio" to a.

If the exponential decline rate and the total length of decline are known the tables in Appendix C can be used to compute total oil produced on decline in lieu of solving equation (2). The procedure is to set the product $(a)(t) = x$, where a is the exponential decline rate and t is the total years on decline. Using this

Relation Between Exponential Decline Rate, a, and "Loss Ratio" (Both Terms Expressed As Decimal Fractions)

Loss Ratio (Defined by Equation 3)	a Exponential Decline Rate (Defined by Equation 1)
0	0
0.01	0.0101
0.02	0.0202
0.03	0.0305
0.04	0.0408
0.05	0.0513
0.06	0.0619
0.07	0.0726
0.08	0.0834
0.09	0.0941
0.10	0.1054
0.11	0.1165
0.12	0.1278
0.13	0.1393
0.14	0.1508
0.15	0.1625
0.16	0.1744
0.17	0.1863
0.18	0.1985
0.19	0.2107
0.20	0.2231
0.21	0.2357
0.22	0.2485
0.23	0.2614
0.24	0.2744
0.25	0.2877
0.26	0.3011
0.27	0.3147
0.28	0.3285
0.29	0.3425
0.30	0.3567
0.40	0.5108
0.50	0.6932

value of x enter Appendix C and read the corresponding entry under Column C. We will simply call this the "Column C Factor." Finally, total oil produced on decline is computed

Total oil produced in t years of decline = $(q_i)(t)(\text{"Column C Factor"})$

where q_i is the initial producing rate, barrels per year, as defined previously. We can try this procedure using the same example given previously.

$$q_i = 73,000 \text{ Bbls/year}$$
$$t = 16 \text{ years}$$
$$a = 0.2306$$
$$(a)(t) = (0.2306)\ (16) = 3.69$$

From Appendix C, the "Column C Factor" for a value of $x = 3.69$ is 0.264235. Thus the total oil produced in t years of decline is given as

$$\left(73,000\ \frac{\text{Bbls}}{\text{year}}\right)(16 \text{ years})\ (0.264235) = \underline{\underline{308,600}} \text{ Bbls.}$$

This is the same answer that we obtained using Equation (2).

The above procedure for computing oil produced on exponential decline using Appendix C is also applicable to the computation of total undiscounted revenue received. The only difference would be that q_i must be expressed in (\$/year) rather than (bbls/year). The rate, in dollars or barrels, represents the instantaneous rate as of the start of the decline period of t years. Note that if a well was producing 200 BOPD on the first day of the decline the instantaneous rate, converted to an annual rate, is 200 BOPD $\times$ 365 days/year = 73,000 Bbls/year. Actual production during the first year of decline will be less than 73,000 Bbls/year. The important point to remember is that q_i and q_f in equations (1) and (2) are instantaneous rates, even though they have units of barrels per year.

COMPUTING NPV OF DECLINE PRODUCTION

In any type of discounted cash flow analysis it will be necessary to find the net present value of income received during a period of exponential decline if a well is generating decline income. One way to make such a calculation would be to plot the decline curve on semilogarithmic paper (1n rate versus time) and estimate annual production rates for each year of decline. This series of cash flows would then be discounted using the appropriate discount factors and the sum of the discounted cash flows added to determine total NPV. The tables of Appendix C can be used to compute the NPV of total decline revenues by just a few simple steps of arithmetic. The procedure is as follows:

Define $k = a + j$ a = annual exponential decline rate
and $x = kt$ j = discount rate
                         t = years of decline production.

Using the above value of x, enter Appendix C tables and read the corresponding value under Column C. We will call this the "Column C discount Factor" for lack of a better name.

Finally, the NPV of the total decline production, as of the start of the decline period, discounted at j per cent per year is:

Net Present
Value of decline = $(q_i)(t)$("Column C discount factor")
Production

(here q_i is in units of \$/year)

EXAMPLE: Again referring to the previous example in which

$$q_i = 73{,}000 \text{ bbls/year}$$
$$q_f = 1{,}825 \text{ bbls/year}$$
$$t = 16 \text{ years}$$
$$a = 0.2306.$$

Compute the NPV as of the start of the decline, discounted at 10% per year, compounded continuously. Consider that the oil will be worth \$2.00/Bbl after local taxes, royalty, and operating costs.

$$k = a + j = 0.2306 + 0.1000 = 0.3306$$
$$x = kt = (0.3306)\,(16 \text{ years}) \cong 5.29$$

From Appendix C, read "Column C discount factor" corresponding to $x = 5.29$ of 0.188083.

$$q_i \text{ in \$/year} = \left(\frac{\$2.00}{\text{Bbl}}\right)\left(73{,}000\ \frac{\text{Bbls}}{\text{year}}\right) = \frac{\$146{,}000}{\text{year}}$$

Finally:

Net Present Value
of decline Prod.   = $(q_i)(t)$("Column C discount factor")

$$NPV = \left(\frac{\$146{,}000}{\text{year}}\right)(16 \text{ years})\,(0.188083)$$

$$NPV = \underline{\$439{,}000}$$

For those interested in mathematical equations, the total undiscounted value of income received during an exponential decline can be converted to NPV as of the start of the decline by multiplying by the following composite exponential decline discount factor:

$$\text{composite exp. decline d.f.} = \frac{a}{a+j}\left[\frac{1 - e^{-t(a+j)}}{1 - e^{-ta}}\right] \qquad (5)$$

EXAMPLE: Using the previous examples the total oil produced during the decline was 308,000 bbls. At \$2.00/bbl the total undiscounted value of the decline production would be 308,000 Bbls × \$2.00/bbl = \$616,000. For values of a = 0.2306, j = 0.10, and t = 16 years, the composite exponential decline discount factor is computed as

$$\text{composite exp. decline d.f.} = \frac{0.2306}{0.2306 + 0.10}\left[\frac{1 - e^{-16(.2306 + .10)}}{1 - e^{-16(.2306)}}\right]$$

$$= 0.712$$

Therefore, NPV, discounted at 10% is

$$(\$616{,}000)\,(0.712) = \underline{\$439{,}000}$$

(Note that this is the same answer obtained using the tables of Appendix C)

The exponential terms in the above discount factor equation can be read from Appendix C, Column B if you don't have a slide rule or a calculator handy. For example $e^{-16(.2306 + .10)}$ is $e^{-16(.3306)}$ or $e^{-5.29}$. Using Appendix C, read from Column B that $e^{-5.29} = 0.005042$. If the discounting of decline revenues is being done on a computer, the above calculation can be programmed as a single statement.

As a matter of interest, the equation for the composite discount factor which converts total income received at a constant rate to NPV, discounted continuously at j per cent per year, is

$$\begin{array}{l}\text{composite d.f. for}\\ \text{income received at}\\ \text{constant rate for}\\ \text{n years, discounted}\\ \text{at j discount rate}\end{array} = \frac{1 - e^{-jn}}{jn}$$

This is the equation which has been solved to obtain the composite discount factors of Column C in Appendix C. Its use was explained in part 3 of the description of how to use the tables of Appendix C.

Computing Composite NPV of Oil on a "Per-Barrel" Basis

It is sometimes desirable to compute a composite NPV factor in units of $/bbl. Such a factor, when multiplied by total barrels in a field, converts reserves to NPV in one step. To compute such a factor a representative production schedule must be assumed.

To illustrate how such a factor can be computed we will use the production schedule given in Figure (1). All of the necessary special computations have been described and/or computed previously in this Appendix section.

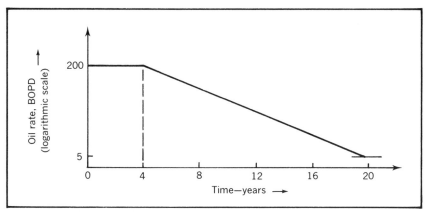

Figure 1 *A typical oil well production schedule.*

EXAMPLE: Assume that well production from a field can be represented by the production schedule of Figure (1). As shown by the graph the well produces at a constant rate of 200 BOPD for 4 years, then declines exponentially over the next 16 years to an economic limit of 5 BOPD. Assume that our firm's time value of money is 10% per year, and oil from this field is worth $2.00/bbl. net after local taxes, royalty, and operating expenses. Compute a composite NPV of a barrel of oil using continuous compounding.

Solution steps:

   a. Compute total oil produced
   b. Compute total undiscounted revenue
   c. Compute NPV @ 10% of total revenue
   d. Compute a composite discount factor which converts undiscounted revenue to NPV
   e. Find NPV per barrel by multiplying composite discount factor by the net value of a barrel oil.

a. *Compute total oil produced:*
   Total oil produced on decline was computed by several means earlier in this Appendix as 308,000 Bbls.
   Total oil produced at constant rate:

$$200 \, \frac{B}{D} \times 365 \, \frac{D}{yr} \times 4 \text{ yrs.} = 292,000 \text{ Bbls.}$$

Therefore, the total oil production over the entire 20 year production schedule is:

$$292,000 + 308,000 = 600,000 \text{ Bbls}$$

b. *Compute total undiscounted revenue:*

This is simply 600,000 Bbls $\times \dfrac{\$2.00}{Bbs} = \$1,200,000$.

c. *Compute NPV @ 10% of total revenue:*
NPV of decline production, *as of start of decline,* was computed previously as $439,000. But the start of decline is 4 years after the start of production, so we must discount the $439,000 back to time zero as a single lump-sum. Recall that we can use Column B of Appendix C to discount single payments by setting $x = jn$

$$x = jn = (0.10)(4 \text{ yrs.}) = 0.40$$

From Column B, for $x = 0.40$ read 0.6703. Thus the NPV of decline production, *as of time zero* is

$$(\$439,000)(e^{-(0.1)(4)}) = (\$439,000)(0.6703) = \underline{\$294,000}$$

The NPV of constant rate income is computed using Column C of Appendix C.

$x = jn$ where n is total period of constant rate production
$x = (0.10)(4 \text{ yrs.}) = 0.40$.

From Column C, for $x = 0.40$ read 0.8242. Thus NPV of constant rate income, as of time zero is

$$NPV = \left(292,000 \text{ Bbls} \times \frac{\$2.00}{Bbl}\right)(0.8242) = \underline{\$481,000}$$

Thus the combined NPV @ 10%, as of time zero is

$$\$294,000 + \$481,000 = \underline{\$775,000}$$

d. *Compute composite discount factor:*

$$\text{Total undiscounted revenue} = \$1,200,000$$
$$\text{Equivalent NPV @ } 10\% = \$775,000$$

Therefore each dollar, on a total composite basis, is worth

$$\frac{\$775,000}{\$1,200,000} = 0.645, \text{ or } \$0.645$$

The factor 0.645 is a composite discount factor for the given production schedule.

e. *Find NPV on a "per barrel basis:*
Each barrel was worth $2.00. From the above step we computed that each dollar received had a composite NPV @ 10% of $0.645. Thus the "NPV per barrel" is computed as

$$\frac{\$2.00}{Bbl} \times 0.645 = \underline{\frac{\$1.29}{Bbl}}$$

The interested reader can test his skill at the above series of computations with the following problem:

"A typical production schedule per well in the Plymouth field extends over 30 years. The well produces at a constant allowable rate of 200 BOPD for 11.5 years, then declines over the next 18.5 years to an economic limit of 5 BOPD. Assuming $j = 0.10$, continuous compounding, and net value of oil of \$2.00/Bbl, compute total oil produced and the NPV per barrel." (The answers are 1,200,000 Bbls and \$0.964 per barrel)

RAISING NUMBERS TO A POWER

Suppose we were making a discounted cash flow analysis and using a discount rate of 31% per year. The discount factor for year 10 would be $1/(1 + i)^n = 1/(1 + .31)^{10}$. In the absence of discount factor tables, how could we compute the factor without doing all the arithmetic to multiply 1.31 by itself ten times?

We can use the tables of Appendix C to raise numbers to a power by using logarithms. Rather than explain the mathematical reasons we will show how to raise numbers to a power in "cookbook" fashion:

We will call the number to $(1 + .31)^{10}$ y. Therefore we can write

$$y = (1.31)^{10}$$

Now we take the natural logarithm of both sides of the equation:

$$\ln y = \ln [(1.31)^{10}]$$

(Note: we could just as well have taken logarithms to the base 10, $\log_{10}$, but we do not have tables for logarithms to base 10 in this reference.)

Recalling from our high school algebra (?) the logarithm of a^b is simply b times the logarithm of a. Thus

$$\ln y = (10) \ln (1.31)$$

We can read the logarithm of 1.31 by entering the tables of Appendix C in Column A and reading the value of x corresponding to 1.31. From Appendix C we read $x = 0.27$. The $\ln (1.31)$ is 0.27. Thus we have

$$\ln y = (10)(0.27) = 2.7$$

From this we read "the logarithm of the answer is 2.7." Now we must find the number whose logarithm is 2.7. To do this we take the inverse logarithm of both sides

$$y = e^{2.7}$$

So we enter the tables of Appendix C again for $x = 2.7$ and read $e^{2.7}$ in Column A as 14.8797. That's our answer!

$$y = 14.8797 = (1.31)^{10}$$

In summary:

1. Given $y = (a)^b$ a can be any number greater than or less than 1.0.
   b can be any number, and does not have to be an integer value.
2. Take the logarithm of both sides:

$$\ln y = \ln [(a)^b] = b \ln(a)$$

3. Enter Column A or B of Appendix C to the value of a and read the corresponding value of x. Multiply this value of x by b.
4. Enter Appendix C with the value of x just computed and read y in Column A or B. If you entered Column A in step 3 then y will be in Column A. If you entered Column B (a was less than 1.0) then y will be in Column B.

Another Example

The probability of flipping 8 consecutive heads is $(0.5)^8$. What is this probability?

1. $y = (0.5)^8$ thus a = 0.5
                       b = 8.0
2. $\ln y = 8 \ln (0.5)$
3. From Appendix C, reading down Column B until reaching a = 0.5 read a corresponding value of x = 0.692 (interpolating)

$$(0.692)(8) = 5.536$$

4. Entering Appendix C for x = 5.536, read y = 0.00393 from Column B. Thus

$$(0.5)^8 \cong \underline{0.00393}$$

For practice, try these: $(1.20)^{7.5}$, $(0.85)^9$, $(0.81)^{0.5}$.

SPIDER DIAGRAMS (SENSITIVITY GRAPHS)

A convenient and useful way to present the results of having made a sensitivity analysis are graphs which are called spider diagrams. The name comes from the fact that they sometimes appear to resemble the web of a spider. Suppose we have the relationship

$$\text{Profit} = f(A, B, C, D, E, F)$$

in which the dependent variable Profit is a function of six variables, A, B, C, D, E, and F. Suppose further we computed a base case value of profit using specific base case values of each variable. This base case profit, labeled Profit Base Case, is plotted on the ordinate of the spider diagram of Figure 2 versus the value

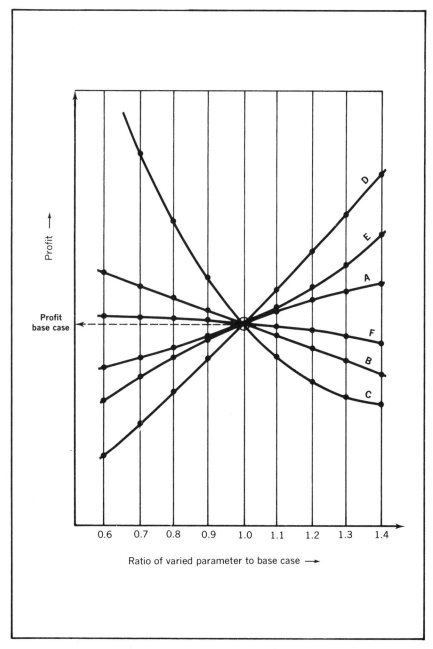

Figure 2. *Spider (sensitivity) diagram for the equation, Profit = f(A, B, C, D, E, F) Each curve represents the effects on profit of various values of the variable expressed as ratios of its base case value.*

1.0 on the abscissa. Next, the value of A is reduced to 0.9 of its base case value and the equation of profit again solved (holding all other variables at their base case value). This computed value of profit is plotted versus the ratio of 0.9 for parameter A.

The process is then repeated using a value of A equal to 0.8 of its base case value, then 0.7 of its base case value, then 0.6 of its base case value, then 1.1 of its base case value, etc. For each of these calculations the values of B, C, D, E, and F are held at their original base case values. The resulting profit values are plotted versus the corresponding ratios of A to its base case to give the curve in Figure 2 labeled A. This curve shows the effect on profit of different values of A, all other factors being held constant at their base case values.

The curves for variables B, C, D, E, and F are obtained in exactly the same manner and plotted on the same graph, as in Figure 2.

The interpretation of such a spider diagram is as follows. As the curves for the different variables become steeper (more vertical) we can conclude that given changes of the variable (from its base case value) will result in large changes of the dependent variable (profit in this case). As the curves become flatter (more horizontal) the implication is that given changes in the value of the variable cause very little change in the dependent variable.

From Figure 2 we could conclude that changes in variables C, D, and E will cause the greatest change in profit. Conversely, variables F and B cause very little change in profit, even if their numerical values should increase or decrease by as much as 40% from their base case values. So if we were concerned about the variability of profit if the variables deviate in value from their base case we would be the most concerned about variables C, D, and E, and the least concerned about variables F and B.

Graphical displays such as Figure 2 are useful ways to communicate the relative sensitivities of the different variables on the corresponding value of a dependent variable such as profit.

Index